General biology

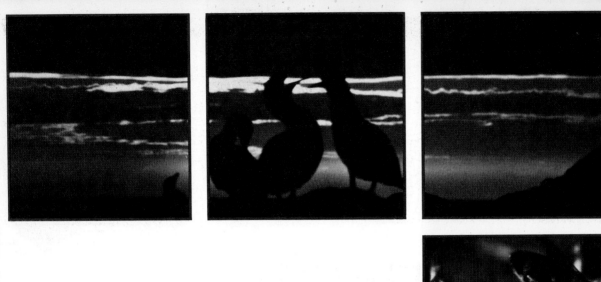

General biology

GEORGE B. NOLAND

Professor of Biology,
University of Dayton,
Dayton, Ohio

TENTH EDITION

with **894** illustrations

THE C. V. MOSBY COMPANY

ST. LOUIS • TORONTO • LONDON   1979

Cover photographs courtesy Dr. Cleveland P. Hickman, Jr.

TENTH EDITION

Previous editions copyrighted 1939, 1940, 1946, 1952, 1958, 1962, 1966, 1970, 1975

Printed in the United States of America

The C. V. Mosby Company
11830 Westline Industrial Drive, St. Louis, Missouri 63141

Library of Congress Cataloging in Publication Data

Noland, George B
    General biology.

    Includes bibliographies and index.
    1.   Biology.   I.   Title.
QH308.2.N64   1979   574   78-27065
ISBN 0-8016-3673-6

CB/VH/VH   9   8   7   6   5   4   3   2   1   02/A/240

PREFACE

The choice of a textbook for a beginning biology class is influenced by many factors. At the top of the list are the background and interests of the individual instructor. This is probably followed by the background and future plans of the student taking the course. I am heartened by the comment of a young secretary, Maureen Becker, who exclaimed after typing part of the manuscript, "Red-haired men, spinach cells, and cats! I can understand this text!" Maybe we can learn something from the feelings of an English major.

The scope of biology is so vast that all of it cannot be covered adequately in a single text or a beginning course; there must be an emphasis on one point of view throughout a text with the elimination of, or at least the lack of, emphasis on other areas. The emphasis in this text is on the organism. I believe that living things should not be forgotten amid the principles that explain how they exist. Physiology of the cell takes place inside a living organism that exists in nature as well as in the laboratory. The gene results in the expression of a phenotype in vivo as well as in vitro.

The opinion of a friend of mine, "What do I care what it is? The first thing I do with it is to grind it up anyhow," may be reasonable for a biochemist, but it is only part of the story for a biologist. In short, the living organism must be considered along with the processes and principles that have enabled it to survive. I assume that this view is shared by the professors who have adopted this text for their classes.

For this edition, although the general plan of the previous edition has been retained, the primary effort was directed toward making the text more readable. For this reason Chapter 2 on the early history of biology and Chapter 4 on basic chemistry have been moved to the Appendix. The five kingdom system of classification that was introduced in the last edition is now used throughout the text. This required a few changes in the arrangement and sequence of organisms.

The text now consists of seven parts. Part one, Introductory Biology, introduces science, biology, and scientific procedure. The chapters then lead into the characteristics of life in general, introduction to cell biology, and principles of reproduction and development in various organisms. Nutrients and basic materials are then considered. Finally, the whole-organism orientation of the text is reestablished by a chapter on diversity and the kinds of living organisms.

Part two, Microbiology, now includes viruses, monera, simple algae, protozoa, and fungi. Part three, Plant Biology, and Part four, Animal Biology, then continue the structural, functional, and evolutionary relationships among the range of living forms. Generally the phylum-by-phylum approach has been retained.

Part five, The Biology of Man, includes chapters on the human form, human metabolism, human control mechanisms, and human reproduction. This is followed by Part six, Heredity and Evolution, which includes the principles of classical and molecular genetics and evolution. Finally, Part seven, Organism and Environment, covers the interrelationships among organisms and their environment: ecology and animal behavior.

The final material now includes a glossary and two appendices. The glossary includes basic terms, prefixes, suffixes, and combining forms used in biology. Appendix A contains the basic chemistry that was formerly included in Chapter 4. Appendix B contains material on the early history and development of biology that was formerly Chapter 2.

As always, the patience and understanding of my wife and children were especially helpful to me during the many months devoted to this revision.

George B. Noland

CONTENTS

PART TWO

MICROBIOLOGY

PART THREE
PLANT BIOLOGY

PART FOUR

ANIMAL BIOLOGY

PART FIVE

THE BIOLOGY OF MAN

PART SIX

HEREDITY AND EVOLUTION

PART SEVEN

ORGANISM AND ENVIRONMENT

PART ONE

Introductory biology

The science of biology
Characteristics of life
Cells
The cell cycle and tissue formation
Patterns of reproduction
Growth and development
Nutrients and basic materials
The kinds of living things

1 The science of biology

Before beginning the formal study of biology, it might be well to consider who is reading this text. Two groups of students can be readily recognized. The first and perhaps most obvious is composed of those who realize a need to study biology as a preparation for their life's work. These are biology majors and medical technology, predental, and premedical students. They also include increasing numbers of students of psychology, sociology, and engineering. Biology also serves as a foundation for professions such as pharmacy, nursing, agriculture, forestry, education, entomology, horticulture, and landscape gardening. Preparation for such professions involves the interrelationship of all the sciences: chemistry, biology, physics, geology, geography, psychology, paleontology, and many others.

However, most students in the introductory classes are not biology majors. Moreover, many are taking a biology course only to fulfill a science requirement. And many choose biology simply to avoid chemistry or physics, or some other science. These students' interests are not in science, but in the world of ideas and service. Very likely, the best expression of their attitudes is that they do not openly object either to biology or to science in general.

This attitude is not new or even unusual. Some years ago the English author C. P. Snow, called attention to it in his book, *Two Cultures and the Scientific Revolution*. He thought that there was a widening gap between those whose beliefs and attitudes are shaped by facts and experiments and those persons of a more humanistic and subjective inclination. C. P. Snow, and many others, saw a growing danger that these two groups would no longer be able to communicate with each other.

WHY STUDY BIOLOGY?

At the risk of seeming overly dramatic, the answer to this question may well be "in order to survive!" Many people believe that the field of biology may be able to bridge the gap discussed in Snow's book.

If we consider some of the problems confronting people in the United States as well as in many other countries, we will find that they are con-

Shellfish-associated gastroenteritis—New Haven, Connecticut*

An outbreak of gastroenteritis occurred on November 16, 1968, following a shellfish sanitation association meeting in New Haven; 17 persons became ill after a cocktail party where mixed drinks, raw oysters and hard clams, potato chips, cheese snacks, and a hot sauce were served. The illness was characterized by nausea, vomiting, fever, and diarrhea with 16 of the 17 persons having diarrhea. Of 12 families surveyed with index cases, five reported secondary cases with these same symptoms. One other case occurred in the man who delivered the clams to the party and who consumed a dozen of them on November 13. He became ill 32 hours later on November 14.

Food histories from 23 persons at the party showed that 19 ate clams and 20 ate oysters. The one person who ate only clams became ill while the two persons who ate only oysters did not become ill. Of all food items, only the clams were found to be significantly associated with illness.

The clams were part of 11½ bushels of cherrystones and little necks harvested by a shellfish dealer on November 12 from a bed 3 miles southwest of Norwalk, Connecticut. One bushel was used at the party and the other 10½ were sent to a retail market in Yonkers, New York. On November 22, investigation showed that the clam bed was contaminated and, as a result, was closed to further harvesting. From December 2 through 6, samples taken twice a day from the bed also revealed abnormal contamination. In late October, waters near the bed had been of acceptable quality. The Yonkers health department was notified, but no increase in gastroenteritis cases was noted in Yonkers in November.

The probable source for the contamination was a sewage treatment plant located in eastern Norwalk that discharges the plant effluent 4.4 miles upstream from the bed. On November 11, the plant had no electricity for 4½ hours. Consequently, because the storage capacity of the system is 2 hours, there was a major overflow of the storage system for at least 2½ hours.* During the power failure, attempts were made to chlorinate the major overflow points with hypochlorite powder.

Because of the possibility of hepatitis developing following the ingestion of contaminated raw shellfish, most of the persons at the party received immune serum globulin and were observed for 2 to 6 weeks. Frozen stool specimens for virus isolation and rectal swabs for bacterial culture were also solicited, and additional clams were harvested from the contaminated beds for virus studies.

---

*From Morbidity and mortality weekly report, United States Public Health Service, Feb. 8, 1969.

*The possibility of a similar problem exists whenever there is a power failure anywhere in the country. Since this report, raw sewage has been dumped in San Francisco, Miami, Cleveland, and other cities. (See also p. 147.)

cerned with various aspects of biology. Newspaper and television reports ephasize that air and water pollution, inadequate nutrition, expanding population, pesticide poisoning, and destruction of natural resources are now part of our daily lives. (See box above and p. 147.)

That we are concerned with issues such as pollution and pesticides such as DDT and Kepone is a good example of biology in our lives. The extra cost of new cars because of emission controls is another example. When we talk about food shortages, miner's lung, cancer and carcinogens, water and air pollution, as well as minerals and nutrition, and the possible side effects of medicines, we are acknowledging our absolute need to consider the basic biologic factors influencing the quality of life.

Legal and moral aspects of test tube babies, abortion and heart, kidney, and lung transplants continue to plague the average person as well as the Congress, the courts, and the churches. We are at the point now where individual legislators are actually forced to vote on the question of when a life begins and when it ends.

A long and involved court case slowed, and possibly halted completely, the construction of a TVA dam on the Middle Tennessee River. It was discovered that a small fish, called a snail darter, occurred in a 12-mile stretch of the river and nowhere else. The fish was thought to be an endangered species and construction was halted. Arguments on this case are still going on.

Recent laboratory experiments have demonstrated that it is possible to take the DNA of one species and combine it with that of another. The combination enables the second species to produce something that it was unable to produce with its own genetic makeup. This recombinant DNA research may have tremendous benefits in agriculture and medicine, but along with the potential benefits there is the fear that indiscriminant research will lead to new types of disease-producing organisms. Some of the problems that surround this issue include questions of commercial ownership and whether research in this area should be federally licensed and controlled for safety reasons.

It should be obvious by now that many of the apparently biologically based problems are also moral, ethical, economic, and political problems. There is a good possibility that these same problems, or others similar to them, are being studied in many different departments of one's school. Table 1-1 shows some examples of biologic research.

The failure to appreciate and understand the true nature of science has caused much misunderstanding and criticism of the value of its methods. Scientists can use only the tools that are available to them—they can investigate the chemical and physical processes inherent in living things and attempt to explain life in terms of such investigations. The most powerful tool used by all scientists is the scientific method. There may be variations in the steps to be followed when using the scientific method, but the following steps are representative.

SCIENTIFIC METHOD

A course in biology or any other science should give the student an idea of the aim and nature of science, the methods employed, and its value and limitations. *Science* attempts to observe and describe facts and to relate them to each other; its conclusions are always subject to revision in the light of newly discovered facts.

There are numerous popular concepts concerning what science either tries to do or can do. Some people think that science can do anything and can solve all problems. Others are fearful or apprehensive of science.

Scientific credibility suffers when individual scientists differ on the *uses* of scientific data concerning environmental quality, food, population, causes of cancer, arrest and cure of cancer, care of the aged, health care, energy availability and use, and other concerns. This is especially apparent when we are given conflicting reports from agencies such as the Food and Drug Adminis-

Table 1-1 Biology-related problems in other fields

Department	Example of biologically oriented research
History	Effect of epidemics on ancient civilizations
Political science	Regulation of water and air pollution
Psychology	Biochemical control of mental illness
Sociology	Population control and society
Philosophy	Biomedical ethics
Theology	Morality of organ transplants; abortion, genetic ''engineering''
Anthropology	Blood groups and genes in native groups
Engineering	Artificial organs, bioengineering
Physics	Physical properties of complex macromolecules
Chemistry	Biochemistry of pesticides

tration and various manufacturers. Often, the apparent conflict is caused by the release of incomplete data by a news source; in many cases a ruling must be made even if all of the data are not yet available.

Steps in the scientific method

1. RECOGNITION OF THE PROBLEM

Previously unnoticed problems or conditions occur constantly in our work and in our daily living. These may be simple or extremely complex, requiring sophisticated techniques and experiments for their solution. The recognition of a specific problem may be stimulated (1) by mere general curiosity, (2) by an actual need for the solution of the problem, or (3) by reading or thinking about a similar problem.

Indeed, several famous discoveries are reputed to have been stimulated by such things as being hit by a falling apple, by daydreaming while watching bubbles rise in a glass, and by a nightmare after a late evening. Most, however, are the result of hard work. The study that follows this section was based on an old legend about red-haired people.

In the solution of any problem a main objective should be to ascertain the truth. In part, at least, this depends on an attitude in which problems are approached with an open, unprejudiced mind and with as much objectivity and detachment as possible.

2. ACCURATE PRELIMINARY OBSERVATION

This includes studying available information to determine what is already known about the condition being investigated. It also includes preliminary investigations and gathering information from other sources. This step is naturally not exhaustive in its scope but merely lays the foundation for the next step.

3. FORMULATION OF A HYPOTHESIS

A hypothesis might be considered a guess, speculation, or assumption that is a tentative ex-

planation of the problem. It is sometimes called a working hypothesis because we work from it. Sometimes only one hypothesis for a given problem can be suggested, whereas for another many hypotheses are evident. Each is considered in turn, those that are not proved are eliminated, and possibly new ones are substituted as progress is made.

4. TESTING THE HYPOTHESIS

Another important phase is the decision on methods of investigation. In some problems devising the proper methods of investigation may require broad practical training, imagination, special techniques, or even elaborate equipment or apparatus. In all cases the data and information must be sufficiently extensive to reduce the chance effects of unusual differences or variations.

A hypothesis may be tested (1) through additional observations and investigations, (2) through controlled scientific experiments, if such can be performed in the particular problem, or (3) by a combination of these.

It is essential to utilize a control group of organisms under conditions identical with the experimental group, except that the one condition being examined is not applied to the control group. All factors, except the one we are attempting to discover, are duplicated carefully in the controls. Only one variable should be permitted between the experimental and control groups at any one time.

5. EVALUATION OF THE COLLECTED DATA

As relevant data and information are collected, they must be precisely recorded. All measurements, records, interpretations of data, or "case histories" must be accurate and sufficiently comprehensive to be reliable. The investigator must be honest and faithful, his or her observations must be accurate, and records must contain all data relevant to the problem. Sometimes graphs, tables, and summaries are valuable in this step of the technique.

6. DRAWING CONCLUSIONS

After the information and data are recorded in such ways as to give accurate and meaningful revelations, they must be interpreted correctly. This means checking the hypothesis. Great care must be taken not to draw conclusions that are broader than the collected facts will support.

7. REPEATABILITY AND REPORTING THE RESULTS

We have all heard the lament of the professor, "publish or perish." The "publish" part of this statement is an absolute insistence that the results of experiments be made available to the scientific community. This is done by presenting papers and discussions before groups of workers in the same field. Here, the results may be questioned, observations may be made on the techniques used, and interpretations may be made. The report may then be published in a scholarly journal, such as *Science, Genetics, American Journal of Botany, Experimental Cell Research,* or any of several thousand more.

Other workers may attempt to repeat the experiment in their laboratories. If repeatable in these different settings, the results have been verified. The data become part of biologic literature and are available to others interested in the same topics. The investigation discussed in the following section was published in an English journal, *The Practitioner.*

Much of the material in Chapter 32, Genes and Gene Action, is basically a treatise on the use of the scientific method.

An example of the scientific method

The following comments refer to similarly numbered sections of the article, "Blood coagulation and platelet function in red-haired men," on the next pages. Refer to it before continuing with this section.

1 The study of science proceeds as an outcome of a natural curiosity and a human need to know. In the red-haired study the question to be answered was "Do red-haired subjects bleed more easily than non–red-haired subjects?"

2 The first procedure was to answer still another question, "What does 'red-haired' mean?" In determining this, a study was made of the methods used by other workers. As can be seen, several methods for distinguishing red hair had been in use. Both the method and a reference to the procedure are given. One procedure, apparently the most accurate, was chosen. A second group of subjects, none of whom had red hair, was selected as a control. This second group served as a standard of comparison for the results of testing the red-haired group. The question was now rephrased as, "Does the blood of the experimental group (red hair) clot more slowly than the blood of the control group (non–red hair)?"

3 The details of the procedure were then given. The researchers were saying, "This is how we ran the test to determine if our subjects were really red-haired. This is what we mean when we say 'red-haired.'"

Next, by referring to the library, they determined the kinds of tests that are used to determine the clotting properties of human blood. The tests were listed and a reference concerning their use was given.

Note that the details of conducting the tests are given. The kinds of instruments used are identified by brand and model number. To avoid inconsistencies caused by personal bias, the same person ran the tests. This is similar to asking two cooks to put "some" pepper in a dish. How much is "some"? The amount may vary and so the taste may vary. But the same person is likely to use the same procedure each time.

4 The results obtained by using each of the five tests were given. In (a) no difference was found in the bleeding times of the two groups of subjects.

In (b) there was a difference. In a longer article, all of the bleeding times would be listed. Here, only the mean (average) time of blood clotting was given. The standard deviation (SD) for each group was also noted.

The results were then analyzed statistically and were shown to be significant. This means that there is very little possibility that the same results

Blood coagulation and platelet function in red-haired men*

1 The small study reported here was an attempt to discover evidence for the legend that red-haired subjects bleed more easily than non–red-haired subjects.

Subjects and methods

2 In all, 49 male student volunteers were studied, 25 of whom were objectively classified as having red hair. The following methods have been described for distinguishing red hair from other colours:

(1) Measurement of ash content (Dutcher and Rothman, 1951).
(2) Microscopic observations of structure (Flesch and Rothman, 1945).
(3) Spectrophotometric methods (Reed, 1952; Barnicot, 1956).
(4) Pigment-extraction methods (Arnow, 1938; Rothman and Flesch, 1943).
(5) Fluorescence microscopy (Brunet, Kukita and Fitzpatrick, 1958).

Recently a more detailed examination of the extractable pigments in red hair has been performed by Boldt and Hermstedt (1967), which has shown that the previously described product from extraction of red hair is not homogeneous.

3 We decided to use the fluorescence microscopy method as being the most convenient. One hair from each subject was examined, using a Gillett and Sibert Conference microscope with filters 30–061, 30–062 and 10–285. All the samples of hair from subjects objectively classified as 'red-haired' showed yellowish fluorescence whereas no fluorescence was seen when hair from the remaining 24 subjects was examined.

*Cynthia Reid, M.B., Ch.B., *Lecturer, Department of Anatomy, St. Salvator's College, University of St. Andrews;* C. M. Trotter, B.Sc., Ph.D., *Electron Microscope Unit, Fisons Pharmaceuticals Ltd., Loughborough.*
Published by The Practitioner at 5 Bentinck Street, London W1M 5RN and printed in England by F. J. Parsons Limited London, Hastings and Folkestone.

The following five tests were then performed on each volunteer:

(a) Bleeding time (Duke, 1912).
(b) Whole-blood coagulation time (Lee and White, 1913).
(c) Screening thromboplastin generation test (Hicks and Pitney, 1957).
(d) Platelet count (Feissly, 1959).
(e) Measurement of platelet adhesiveness after addition of ADP (adenosine 5' pyrophosphate) (Eastham, 1963, 1964).

Tests (b) and (c) were always done by the same person (C.M.T.) to avoid technical inconsistencies, and the blood or reagent for these tests was kept at constant temperature by use of a Seroblock heater. For test (e) a Coulter Counter Model A, fitted with a 70μ-orifice tube, was used and all dilutions for this test were performed by means of the Coulter Dual Dilutor.

Results

4 (a) *The bleeding time.* No difference was found in the bleeding times of the two groups of subjects.

(b) *The whole-blood coagulation time.* This test did show a significant difference:

Non–red-haired subjects: mean 8.3 minutes (S.D. 1.15).
Red-haired subjects: mean 9.9 minutes (S.D. 1.25).

(c) *The screening thromboplastin generation test.* There was a wide scatter of results in both groups. It was found that the clotting time in red-haired subjects during the first few minutes of incubation was frequently as long as 180 seconds, whereas the longest time for the non–red-haired subjects was 60 to 70 seconds. Due to the wide variability it was not possible to evaluate these results statistically.

(d) *The platelet count.* There was a wide range of results in both groups of subjects and the values were frequently below those quoted as normal for healthy adults (Dacie and Lewis, 1963), but the results corresponded with those obtained using the Coulter Counter.

Continued.

Blood coagulation and platelet function in red-haired men—cont'd

(e) *Platelet adhesiveness.* The mean percentage loss of platelets after aggregation by use of ADP was 36 per cent in non–red-haired subjects and 35 per cent in red-haired subjects. These results agreed with Eastham's results for this method using siliconized glass.

Discussion

5 The only test that showed any significant difference between the two groups was the whole-blood coagulation time, which was increased in red-haired men although still falling within the normal range. It seems likely that a fuller investigation of clotting factors will yield more specific results. No difference was found in platelet numbers or aggregation between the two groups.

We wish to thank Dr. J. H. Taylor for the idea behind this work; Professor A. E. Ritchie for encouragement and advice; Dr. J. J. Ferguson for assistance in the laboratory, and all the volunteers. This work formed part of the requirement for an Honours B.Sc. Degree in Physiology in the University of St. Andrews (C.M.T.).

References

Arnow, L. E. (1938): *Biochem. J.,* **32,** 1281.
Barnicot, N. A. (1956): *Nature (Lond.),* **177,** 528.
Boldt, P., and Hermstedt, E. (1967): *Z. Naturforsch.,* **226,** 718.
Brunet, P. C. J., *et al.* (1958): in 'The Biology of Hair Growth', edited by W. Montagna and R. A. Ellis, Academic Press Inc., New York, p. 295.
Dacie, J. V., and Lewis, S. M. (1963): 'Practical Haematology', J. & A. Churchill, London, p. 63.
Duke, W. W. (1912): *Arch. Intern. Med.,* **10,** 445.
Dutcher, T. F., and Rothman, S. (1951): *J. Invest. Derm.,* **17,** 65.
Eastham, R. D. (1963): *J. Clin. Path.,* **16,** 168.
— (1964): *Ibid.,* **17,** 45.
Feissly, R. (1959): in 'Blood Platelets', Henry Ford Hospital International Symposium, edited by S. A. Johnson *et al.,* J. & A. Churchill, London, p. 99.
Flesch, P., and Rothman, S. (1945): *J. Invest. Derm.,* **6,** 257.
Hicks, N. D., and Pitney, W. R. (1957): *Brit. J. Haemat.,* **3,** 227.
Lee, K. I., and White, P. D. (1913): *Amer. J. Med. Sci.,* **145,** 495.
Reed, T. E. (1952): *Ann. Eugen. (Lond.),* **17,** 115.
Rothman, S., and Flesch, P. (1943): *Proc. Soc. Exp. Biol. (N.Y.),* **53,** 134.

in the two groups would have been obtained by pure chance. This indicates that there is a real difference in the two groups.

In test (c) the results were given but no conclusions were drawn. The results were so varied that a statistical analysis could not be made. The only result was that the procedure was inconclusive.

Test (d) also showed a wide range of results but did not lead to any conclusions. Finally, test (e) showed a slight difference, but it was not statistically significant. This means that it could be a chance result.

5 The results of all tests were then discussed, indicating both the similarities and the differences in the two groups. As with most scientific studies, this one does not provide a full answer to the original question but leads to more study. It suggests future work and indicates the areas that might prove promising.

6 The list of references was given to enable any other interested person to repeat or verify the results. This aspect of repeatability is an important concept in all of science.

The entire study was then published in a journal easily available to persons interested in the subject. The information was visible for all to see.

Most scientific studies are accomplished by steps that are similar to those used in the above study. The work that results in the award of a Nobel Prize does not just suddenly happen. The sudden flash of brilliance resulting in a major discovery usually comes after many small experiments have provided basic bits of evidence.

SOME SUBDIVISIONS OF BIOLOGY

Biology, which is the "science of living things," was once easily divided into zoology, which deals with the biology of animals, and botany, which deals with the biology of plants. Botany and zoology have grown so extensively that such subdivisions as indicated below are really sciences in themselves and have further specializations. In a poll of its membership, the American Institute of Biological Sciences asked the respondents to indicate their "disciplinary specialties." Examination of the first 2,000 responses resulted in 261 different specialities. Some of these specialities are indicated below.

anatomy* (Gr. *anatemnein*, to cut up) a study of gross structures, especially by dissection.

biogeography (Gr. *bios*, life; *ego*, earth; *graphein*, to write) the science of geographic distribution of organisms in space or throughout a particular region.

cytology (Gr. *kytos*, cell; *logos*, study) a detailed study of cells and their protoplasm.

ecology (Gr. *oikos*, house or home; *logos*, study) a study of the interrelations of living organisms and their living and nonliving environments.

economic biology a study of organisms, which results in the improvement of desirable types or the destruction or control of undesirable ones, including the value of beneficial organisms and the losses because of detrimental ones.

embryology (Gr. *embryon*, embryo; *logos*, study) a study of the formation and development of an embryo.

evolution (L. *e*, out; *volvere*, to unroll or develop) a study of developmental changes undergone by organisms whereby they change throughout time.

heredity or genetics (L. *heres*, heir); (Gr. *genesis*, origin) a study of the inheritance or transmission of characteristics from one generation to another.

histology (Gr. *histos*, tissue; *logos*, study) a microscopic study of tissues.

paleontology (Gr. *palaios*, ancient; *onta*, beings; *logos*, study) the study of the distribution of organisms in time as revealed by their fossil records in the strata of the earth's surface.

pathology (Gr. *pathos*, suffering; *logos*, study) the study of diseases and abnormal structures and functions, including causes, symptoms, and effects.

physiology (Gr. *physis*, function; *logos*, study) a study of the functioning or working of an organism or its parts.

taxonomy (Gr. *taxis*, arrangement; *nomos*, law) the science of systematic classification of organisms.

---

*Derivations are based on *Webster's New International Dictionary*, *Henderson's Dictionary of Scientific Terms*, or *Dorland's Illustrated Medical Dictionary*.

Some early contributors to
NATURAL PHILOSOPHY AND THE BEGINNING OF SCIENCE

Democritus (460-370 B.C.)

A Greek philosopher who, like other early Greeks, enjoyed learning the true nature of things. He developed an atomic theory (atomism) that he applied to natural philosophy. He suggested that all phenomena are to be explained by the incessant movement of atoms, differing only in shape, order, and position. He did this without the benefit of the knowledge of atoms we have today.

Plato (427-347 B.C.)

A Greek philosopher and Aristotle's teacher who interpreted natural phenomena by relying upon intuition or instinct rather than upon Plato reasoning. *(The Bettman Archive, Inc.)*

Aristotle (384-322 B.C.)

A Greek philosopher who stressed the importance of accurate and direct observations in securing facts and data. He drew his conclusions from facts that he secured by direct observations, and thus he initiated the basis for a scientific method of solving problems. Earlier philosophers had a tendency to reach conclusions and then select data and facts that agreed with their conclusions. *(Historical* Aristotle *Pictures Service, Chicago.)*

Johannes Scotus Erigena (815-877 A.D.)

An Irish-Scottish philosopher in France who is thought to have begun a so-called Scholastic Philosophy (scholasticism), the core of which was the doctrine of the continuity and interdependence of the natural with respect to the supernatural order of truth. Scholasticism included the methods and doctrines of the Christian philosophers of the Middle Ages, and its sources were the writings of the church fathers and of Aristotle and his Arabian commentators.

Ibn Roshd Averroes (1126-1198)

An Arabian philosopher and physician who helped to progress the science of his time and has been called "The Aristotle of the Middle Ages." *(Historical Pictures Service, Chi-* Averroes *cago.)*

Roger Bacon (1214-1294)

An English philosopher who was noted for his scholasticism; yet, his precocious explana-tions of various phenomena were not appreciated and he was imprisoned.

Francis Bacon (1561-1626)

An English natural philosopher, who reached his conclusions through the process of induction from facts, thus breaking away from contemporary scholasticism. He stated (1620) that experiments are of fundamental importance in acquiring scientific knowledge, since experiments enable us to establish causes that determine an occurrence and enable us to bring about such occurrences when we wish.

René Descartes (1596-1650)

A Frenchman, was a pioneer in systematic philosophy who, because of his views, was forced to flee to Holland. He stated that analysis is the means of establishing the truth of the first principles of all knowledge. He suggested the wisdom of dividing the problem to be solved into as many parts as possible in order better to solve it. He stated "the purpose of analysis is to find out by means of one single truth, or a particular fact, the principles from which it derives."

Giovanni Borelli (1608-1679)

An Italian philosopher, mathematician, and disciple of Galileo who applied the latter's principles of physics to biology, thus suggesting an experimental approach to the science.

Immanuel Kant (1724-1804)

A German philosopher who believed that there was something in nature that united the mechanistic and teleologic (natural design) views in biology, rather than their being opposed to each other.

Johann Goethe (1749-1832)

A German philosopher who suggested that life is a constant self-destruction and self-recomposition of living matter and that "life is a flame."

John Stuart Mill (1806-1873)

An English philosopher who elaborated on the philosophy of induction still further, propounding as its basis the law of the Uniformity of Nature.

Review questions and topics

1 List reasons why a study of living organisms should be made.
2 Define a scientific method.
3 List and describe each step to be followed in a scientific method, including enough details to ensure that you know the purpose and correct use of each step.
4 Define biology, zoology, and botany.
5 Define and learn the correct derivation and pronunciation of each subdivision of biology as listed in this chapter. Learn the correct pronunciation and derivation of each new term as you encounter it in your study and include a definition to be sure that you understand the meaning of the term.

Selected references

Baker, J. J. W., and Allen, G. E.: Hypothesis, prediction, and implication in biology, Reading, Mass., 1968, Addison-Wesley Publishing Co., Inc.

Beveridge, W. I. B.: The art of scientific investigation, New York, 1960, Random House, Inc.

Butterfield, H.: The scientific revolution, Sci. Amer. **203:** 173-192, 1960.

Cannon, W. B.: The way of an investigator, New York, 1968, Hafner Publishing Co.

Conant, J. B.: On understanding science, New Haven, 1947, Yale University Press.

Conant, J. B.: Modern science and modern man, New York, 1952, Columbia University Press.

Eccles, J. C.: Facing reality: philosophical adventures of a brain scientist, New York, 1970, Springer-Verlag.

Glassman, E., editor: Molecular approaches to psychobiology, Belmont, Calif., 1967, Dickenson Pub. Co., Inc.

Handler, P., editor: Biology and the future of man, New York, 1970, Oxford University Press.

Hardin, G., editor: Population, evolution and birth control, a collage of controversial ideas, ed. 2, San Francisco, 1969, W. H. Freeman and Co. Publishers.

Holton, G.: The twentieth-century sciences; studies in the biography of ideas, New York, 1972, W. W. Norton and Company, Inc.

Lerner, I. M.: Heredity, evolution and society, San Francisco, 1968, W. H. Freeman and Co. Publishers.

McDermott, W.: Air pollution and public health, Sci. Amer. **205:**49-57, 1961.

Westfall, Richard S.: The construction of modern science; mechanisms and mechanics, New York, 1971, John Wiley and Sons, Inc.

2 Characteristics of life

The course in which this text is being used is concerned with living things. However, there should be some understanding about life in general before attempting to study organisms in detail. The statement that "living things look like they are alive" is so obvious as to be ridiculous. We usually have little trouble distinguishing between a living organism and a dead one, or between something that is or was alive and an inanimate object.

However, this is not always the case. We have only to think about the current controversies concerning abortion and transplants of human organs to realize the problems. Such things as seeds, spores, and eggs appear lifeless at times. Studies of viruses, which have the ability to re-produce, to grow, and to undergo abrupt hereditary changes called mutations, emphasize the difficulty. While it appears impossible to establish absolute criteria for distinguishing life from nonlife, the following generalizations are usually considered.

ORGANIZATION

Living things are called organisms because they are organized. They have distinct parts that enable them to function. The arrangement of their cells, tissues, and organs provides an energy-saving efficiency. Organisms consist of single cells or of groups of specialized cells, which may number in the billions in higher plants and animals.

The more complex organisms are composed of coordinated organ systems. Each system consists of a number of organs, each composed of a variety of specialized tissues. These in turn are composed of cells similar in structure and function. For example, the human body is composed of about ten basic systems, each consisting of several organs.

The digestive system consists of the mouth, tongue, teeth, esophagus, stomach, small and large intestines, and associated structures such as the gall bladder and pancreas (Fig. 28-1). An organ such as the stomach is composed of structural tissues, muscles, nerves, and blood vessels.

A complex plant such as an apple tree contains similar groups of organs. For example, roots, stems, leaves, and flowers each perform specific functions contributing to the structure and function of the entire plant.

A particular kind of living organism has rather narrow limits to its form and size, yet we can usually recognize that it is alive. The contents of living cells are far more complex in nature than most nonliving objects. Although there are no chemical elements found only in cells, the arrangement and kinds of molecules are typical of living things.

The parts of organisms are coordinated so that the entire organism acts as a unit or individual. In each living thing there are (1) interdependence and systematic correlations of parts, (2) a variable susceptibility to environmental influences, (3) inherent self-regulatory tendencies, and (4) a centralized control.

METABOLISM

The term metabolism refers to the sum total of the chemical activities that result in maintenance, growth, repair, and coordination of an organism. All of these activities require energy.

Energy and matter

The universe is composed of matter and energy. Energy is the ability to produce change or motion in matter, that is, the ability to perform work. The abilities to produce changes and do work are attributes of living protoplasm. Energy in protoplasm is measured ordinarily by the amount of change or work performed.

Energies are divided into potential and kinetic. Potential (stored) energy is the ability to perform work because of the position or condition of atoms, molecules, or larger bodies. Examples of stored potential energies are coal and wood before they are burned and carbohydrates before

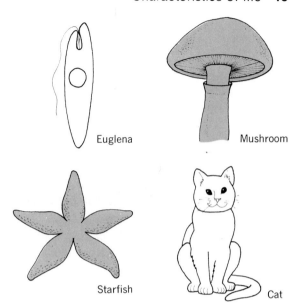

Fig. 2-1. Living things look like they are alive.

they are digested. Chemical digestion of foods results in changing the foods' potential energy into heat, light, electricity, or energy of movement. A stationary ball at the top of an inclined plane has potential energy, but it displays kinetic energy (motion) as it rolls down the incline. Kinetic energy (Gr. *kinein*, to move) is the energy possessed by virtue of motion (active energy). Kinetic energy may become potential, and potential energy may become kinetic. Energy required to form a molecule of substance becomes inactive potential energy when stored in that molecule, but it is converted into active kinetic energy when the molecule is broken down.

All chemical reactions involve changes in energy distribution. Certain chemical reactions require some form of energy, usually heat, whereas others release energy in some form. When a sugar is built, energy is required; when it is broken down, energy is released. The construction and destruction of other foods reveal similar phenomena. The ultimate natural source of energy of foods produced by green chlorophyll-bearing plants is the sun. The energy value of a food is measured by a unit called a kilocalorie, which is the amount of heat required to raise the temperature of 1 kg. (1,000 gm.) of water 1° C. One gram of fat produces about 9 kilocalories of heat; 1 gm. of carbohydrate, about 4 kilocalories; 1 gm. of protein, about 4 kilocalories.

Activity requiring energy in living things must follow the natural laws regarding energy. These are called the *laws of thermodynamics*. In living systems these are designated as bioenergetics.

The first law states that energy can be neither created nor destroyed. This means that living things cannot create *new* energy but that they can transform one type of energy into another. This is shown in organisms by combining energy in chemical work, by light-producing energy, by motion, and in other ways (Table 2-1).

The second energy law states that every energy transformation results in reduction in the usable or available energy in a system. In terms of living things this implies that, in order to remain organized and to function, a constant source of new energy must be available. In plants this takes place by photosynthesis. In animals ingestion of food provides the new energy source.

The total activities of metabolism are often divided into the building up energy–obtaining activities called *anabolism* and the energy releasing activities of *catabolism*.

Anabolism involves many of the nutrition-related activities of a cell or organism. This may then involve *photosynthesis* in plants or *ingestion* of food by animals. Further use of food involves *digestion*, the reduction in size of the food molecules. This is followed by *absorption*, the crossing of cell membranes and circulation of food molecules to all parts of the organism. After this the food may be synthesized into larger more compact molecules and stored or used in growth and repair.

Catabolism involves the "tearing down" or energy releasing activities in the organism. It includes *cell respiration*, resulting in the release of energy in various forms. This may then be used as heat, muscle activity, or the various anabolic activities. Some of this energy may be stored

Table 2-1 Energy transformation in living things

Energy type	Changed to	Transducer (agent of change)
Radiant	Chemical	Plants (photosynthesis)
Chemical	Mechanical	Muscles
Chemical	Light	Fireflies
Chemical	Electricity	Electric eels
Mechanical	Heat	Muscles
Sound	Electricity	Ears

or transferred to complex chemicals. Energy transformation results in the *excretion* of by-products such as water, carbon dioxide, and urea. Breathing provides for an exchange of oxygen and carbon dioxide.

All of the activities of metabolism are influenced by specific catalysts called enzymes. These serve to control the rate at which a chemical activity takes place. Enzymes are produced as one activity of anabolism. Their production and that of other cell components is controlled by complex molecules called DNA and RNA, which will be considered in detail in later chapters.

Use of energy by organisms
PRODUCTION AND USE OF HEAT

When energy is released, there is an accompanying production of heat. In some instances this heat is used to regulate chemical activities or control body temperature, and in others the heat is a waste product that is no longer of use to the organism. In the formation of chemical compounds there is often some heat produced. In some instances, as in the spontaneous combustion of hay, the amount of heat produced is sufficiently great to start a fire. In the destruction of chemical compounds, usually by oxidation, a certain amount of heat is liberated. For example, the oxidation of such foods as carbohydrates, fats, and proteins releases heat for use by the living organism.

A living organism that generates large amounts of heat through its activities is frequently not very efficient in this respect. In such cases much more heat is liberated than is required by that organism. In general, plants are more efficient in this respect than animals. Much of the heat acquired by plants is absorbed from the surroundings. So-called cold-blooded animals maintain a body temperature somewhat similar to that of their environment, while warm-blooded animals generate and conserve heat to maintain a rather constant temperature, regardless of environmental factors. Animals lose heat (1) by conducting it to other objects, (2) by radiating it, (3) by losing it through feces and urine, and (4) by evaporation from the lungs and skin.

BIOELECTRIC PHENOMENA

Protoplasm contains numerous electrolytes. The ions into which acids, bases, and salts dissociate confer charges on surfaces on which they may accumulate. Hence, colloidal particles, each bearing a minute charge, may be changed as chemical changes occur in the protoplasm or as ionizing substances are introduced from the outside. The effects on colloidal particles of protoplasm by inorganic and organic substances brought to it may assist in explaining the many variations in living phenomena.

In certain species of fishes there are modified muscle cells that are arranged in series to serve as electric organs. In such organs the electricity is produced, stored, and discharged into the surrounding water for offensive and defensive purposes. In these electric organs the positive pole of one cell is arranged against the negative pole of the next, so that the voltage produced is determined by the number of cells arranged in the series.

BIOLUMINESCENCE AND LIGHT

Bioluminescence (Gr. *bios*, life; L. *luminescere*, to produce light) is a phenomenon of light production, displayed by certain organisms. In some bacteria and fungi part of the energy released during respiration is used in the production of so-called "cold light" with little heat (1%) being formed. Bioluminescent bacteria are found in decaying wood and leaves, in fish, and in salt water. The production of this type of visible light is controlled by the action of the enzyme luciferase, acting on a substrate called luciferin (L. *lux*, light; *ferre*, to carry), and the production may be continuous if sufficient oxygen and proper foods are available. Luminous bacteria in the eyes of certain fishes in East India emit light constantly. The light emitted by plants is the result of oxidative metabolism, and its usefulness to such plants is an unsolved problem. No green plants have the ability to produce such light.

Luminescence is displayed by such animals as the firefly (beetle), the glowworm, certain squids and fishes, certain jellyfishes and shrimp, and certain species of protozoans. In the firefly, photogenic organs containing localized masses of fatty substances produce light by oxidizing the fatty substances. The photogenic organs are well supplied with oxygen by a copious quantity of tracheal tubes. The greenish yellow light has few nonluminous rays. Its emission is controlled by regulating the oxygen supply. The light seems to be associated with sexual attraction, the female generally producing flashes of a longer duration. In luminous squids and fishes there are organs, lenses, and reflectors to reflect the glow. In the jellyfish *(Pelagia noctiluca)*, the surface of the umbrella is covered with glowing granules. In the protozoan *Noctiluca*, the luminous granules remain inside the cell.

Light affects animals in several ways. The earthworm has no eyes, yet it moves away from light because of light-sensitive cells near the surface. Certain protozoans, planarians, clams, snails, and certain crustaceans are also affected by light. The simple eyes of insects and spiders are influenced by light intensities. The compound eye of arthropods is constructed like a bundle of hollow tubes arranged in the form of a cone. The tubes are isolated from each other by black pigment, and together they produce a reduced image of the object being viewed. The outer end of each tube contains a lens and a facet that are seen on the surface of the compound eye. The inner ends of these tubes possess light-sensitive materials connected with nerves. These eyes also give the organism an interpretation of movement of objects.

The eyes of vertebrate animals act somewhat like a camera. The lens focuses and forms an image on the black, light-sensitive retina on the inside of the eyeball. The retina consists of enormous numbers of nerve cells with chemicals that are changed temporarily by light. Each temporary image produces chemical changes in the nerve cells that vary with the quantity of light on each cell. These chemical changes stimulate other nerve cells that send impulses over the optic nerve to the brain.

PRODUCTION AND RECEPTION OF SOUND

The vibration of some sounding body produces sound waves that are borne to and interpreted by a specialized organ, such as the ear of higher animals. Plants and lower animals do not produce sounds in the accepted sense, although some may be affected by sound waves.

In several higher animals sounds are produced and interpreted in some manner. Almost every insect that has sound-receiving mechanisms also has sound-producing (stridulating) organs. In the common locust there are two types of stridulation. When at rest, certain species draw the femoral joint of the hind leg across a specialized vein of the wing cover to produce sound. When flying, they produce a crackling sound by rubbing wings and wing covers together. Tympanic membranes connected by nerves to the nervous system are assumed to be auditory organs.

GROWTH

It is obvious that living things grow and increase in size and complexity. If a seed is placed in a moist container, we can watch the emergence of a root followed by a young stem and leaves. If a fertilized chicken egg is incubated, it will produce a baby chick. If we watch a developing frog egg, we can see it divide into 2-4-8 and more cells, eventually elongate, form a tadpole, and swim around. If bacteria or yeast cells are placed in a suitable solution, their increase in numbers is indicated as the fluid gets cloudy. While a microscope is necessary to see the individual cells, the increase in cell number is obvious.

All of these examples are indications of life. The entire process of growth and development of complex organisms is usually considered in three somewhat overlapping phases. These are growth, differentiation, and morphogenesis. Strictly speaking, *growth* refers to an increase in the total mass or protoplasm of an organism. Individual cells may increase in size or in number of both (Fig. 2-2). In a growing seed the cells of the young root and stem divide many times. After a cell divides, each resulting cell is only one half the size of the original. Each of these cells increases in size and mass to that of the parent cell. Then they may begin to specialize, that is, take on characteristics of different types, or they may continue to divide. The total development process is recognized by referring to distinct regions of the growing plant as: meristematic (or growth) zone, zone of differentiation (or cell specialization), zone of maturation (or distinct tissue region), and so forth. (See p. 76.)

A dividing starfish egg forms two cells, but they do not increase to the original size. Instead they continue to divide. The number of cells doubles in a pattern of 2-4-8-16-32-64 until a hollow ball is formed. As the number increases, the size of each cell decreases (Fig. 2-3). In a frog egg the same general pattern takes place, but the cells are not all the same size. The first division produces two cells of the same size, each containing half the mass of the original egg. Each of these cells divides to produce four cells of approximately equal size. The next division is unequal and produces eight cells, half of which are smaller than the others (Fig. 2-4). Some difference in the individual cells becomes apparent even at this early stage. The division continues to form a hollow ball stage (although the "hollow" is much less than in the starfish), and then further development takes place. The entire process from egg to

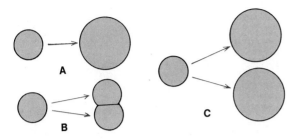

Fig. 2-2. During growth, individual cells may increase: **A,** in size, **B,** in number, or **C,** in size and number.

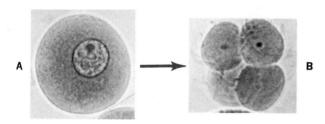

Fig. 2-3. Starfish cleavage. **A,** Egg before fertilization. **B,** Zygote after two divisions; the number of cells has increased and the size of each cell has decreased.

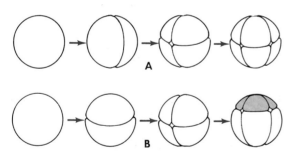

Fig. 2-4. **A,** In starfish cleavage, division produces cells of the same size. **B,** In frogs, the third division produces both large and small cells.

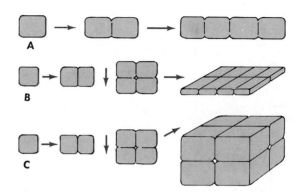

Fig. 2-5. **A,** When cells divide in one direction, a filament forms. **B,** If cells divide first in one direction, then at right angles, a sheet is formed. **C,** When division is in three planes, a cube or other mass is formed.

hollow ball stage is called *cleavage.* (See p. 78.)

The development of form, called *morphogenesis,* takes place as the dividing cells are arranged into rods, tubes, balls, sheets, and other shapes. This follows the plane of division. If a cell always divides in one direction, a filament is formed. If it divides first in one direction (for example, top to bottom) and then in the other (right to left), a sheet of cells will result. If this is followed by a third plane (front to back), then a clump is formed. After this a filament may coil or a sheet may fold or a clump may flatten (Fig. 2-5). Eventually the cells in a particular area begin to differentiate; that is, they change, not only in size and number, but they take on special functions different from other cells. In a plant they may become large globular storage cells, or they may develop chloroplasts and carry on photosynthesis, or they may be conducting cells or many other types. Similarly, animal cells may differentiate into nerve cells or bone or muscle or others. Details of plant and animal development are found in later chapters.

The question of how two cells that started out as apparently equal products of division produce entirely different mature cells has been asked. The answer to this is somewhat complex but does give some indication of being controlled by both biologic and physical factors. It appears that each cell has a complete set of genes typical of its species. During the process of differentiation, some of these genes are active and others are inactive. The particular combination of active and inactive leads to a certain type of cell. The differentiation in turn is influenced by things such as temperature, light, pH, and proximity to other cells.

Light

The fertilized egg of the rockweed *Fucus,* a brown alga, begins development by producing a small protuberance on one side. Cell division proceeds in such a way so that the cell containing the protuberance eventually forms the holdfast, an anchoring device. The other cell divides to form the rest of the plant. D. M. Whitaker showed that the position of protuberance was influenced by the amount of light. The darker side of the zygote was always the site of the formation. In descending order the formation of the protuberance was influenced positively by increased heat, lower pH, nearness to other cells and increased gravity (Fig. 2-6).

In other plants it can be shown that light is also very important. For example, plants grown in the dark show elongated stems and lack of chlorophyll formation. The light influences plant hormones and chemistry.

Nuclear influences

Information about the role of the nucleus in the control of cell activities was obtained by experiments on the single-celled marine alga, *Acetabularia.* This plant (between 5 and 10 cm. long) has basal rootlike rhizoids and a stalk with a caplike "umbrella" at the tip. A single nucleus is present near the base of the stalk. When reproduction occurs, the umbrella divides into reproductive cells.

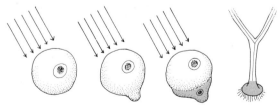

Fig. 2-6. In *Fucus* development, the side away from the light forms the second cell. This cell eventually divides to produce the holdfast attachment organ. This other cell divides to form the rest of the plant.

Two species of *Acetabularia* differ in the shapes of their umbrellas. *A. mediterranea* has a cap with a smooth rim. *A. crenulata* has a deeply indented rim. Hammerling, a German biologist, cut across the stalk and found that the lower, nucleated part would regenerate the distinctive umbrella, while the upper part eventually died without regenerating stalk and rhizoids (Fig. 2-7).

In other experiments on a young plant before the umbrella had formed, Hammerling cut the stalk just above the nucleus and made a second cut just below the tip of the stalk. The growing tip produced both a stalk and an umbrella but neither rhizoid nor nucleus. The stalk section did not grow, and the nucleus-containing rhizoid regenerated an entire new plant complete with umbrella. It was concluded from this that the nucleus supplies a substance that passes up the stalk and initiates umbrella formation. The nucleus apparently produces a substance that regulates cell growth.

From what we know today, we might infer that mRNA had been formed and had led to umbrella formation. However, once this substance was all used, a nucleus was necessary for further regenerative activity.

It is possible to graft the young stalk of one species of *Acetabularia* onto the nucleus-containing rhizoid of another species. When the umbrella regenerates, it is always characteristic of the species that contributes the nucleus, not of the species that contributes the stalk (Fig. 2-8).

The role of the nucleus for cytoplasmic survival can also be shown experimentally by cutting a single-celled protozoan, such as *Amoeba,* into two halves so that one half has a nucleus and the other does not. The nucleated half carries on like a normal ameba, while the enucleated half eventually dies. The ameba without a nucleus may move and feed for a time, but it does not grow or reproduce. Eventually it dies, because the long-range effects of the original nucleus are gone. Here again, we infer that mRNA has been depleted. Just as the cytoplasm depends on the nucleus, the nucleus depends on the cytoplasm,

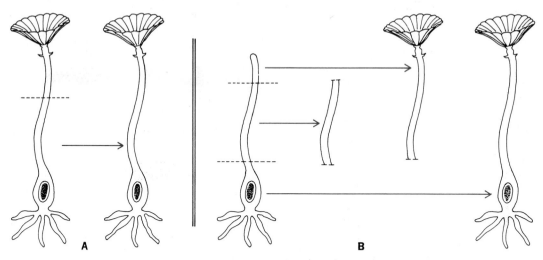

Fig. 2-7. *Acetabularia.* **A,** When cap is removed, a replacement is formed. **B,** If cut while young, the top piece forms a stalk and cap; the midpiece does not grow, and bottom piece with nucleus produces a complete organism.

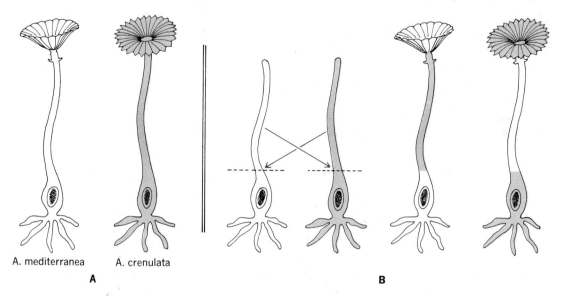

A. mediterranea   A. crenulata

Fig. 2-8. A, *Acetabularia mediterranea* and *A. crenulata.* **B,** When young stalks are transplanted, the resulting cap is characteristic of the nuclear species, not the stalk species.

because the cytoplasm is the site of respiration, food management, and synthesis. A naked, isolated nucleus free from cytoplasm will eventually die, because of a lack of food materials and energy sources. This is an example of the nucleus-cytoplasm interaction necessary for life.

IRRITABILITY

Irritability is an inherent property of all protoplasm. It consists of a sensitivity to stimulation, an ability to transmit the excitation, and to react by various actions. A stimulus is any factor, or change in the environment, that causes an organism to respond in some way. Changes in the direction or intensity of light, in the temperature, in chemical composition, or in moisture content of the surroundings are common stimuli (Fig. 2-9).

Stimuli may be internal or external. A stimulus does not supply energy for a reaction but merely initiates it. The particular response of an organism depends on the quantity and quality of the stimulus and the condition of that organism at the time of stimulation. Organisms have a tendency to restore themselves to their original conditions after responding to a stimulus. This results in *homeostasis,* that is, a steady state, a normal range of living activity.

Even though all cells are irritable, that is, capable of being stimulated, certain cells or parts of cells are particularly sensitive to special kinds of stimuli. In many lower animals the entire body surface is sensitive. In most higher animals, groups of cells are specialized to receive particular stimuli; for example, retinal cells of the eye are light-sensitive, cells of the inner ear are sound-sensitive, and some epithelial cells of the nose and mouth are sensitive to chemicals.

In most animals stimuli are received in one part of the organism but the resulting reaction occurs in some other part. This necessitates a transmission of excitation (conduction) from the place of its origin to the region of reaction. This can be accomplished by means of nerve cells (if they are present) or by means of chemical substances or by a combination of the two.

If a tentacle of a *Hydra* is touched with a glass rod, other tentacles, or even the entire *Hydra,* may contract, showing that there has been conduction, probably by means of its nerve net. Likewise, if a leaf of a sensitive plant, *Mimosa* (Fig. 2-10), is touched, the result of the stimulation may be conducted to adjacent parts of the plant.

In higher animals and plants, substances called hormones (Gr. *hormaein,* to excite) are used. In chemical coordination the action is usually long-lasting, relatively slow, and may be quite extensive. In nervous coordination the actions are usually rapid, of short duration, and somewhat localized. Thus, in higher animals the two systems complement each other in their attempts to adapt to changes in the external and internal environments.

MOTION

One of the most easily observed activities of most living things is motion. It is obvious that animals swim, walk, run, or fly. Microscopic organisms may have hairlike cell processes called cilia or flagella. They use these to propel themselves through fluids such as water or blood.

Reproductive cells of most organisms have cilia, flagella, tails or at least some means of motility. Even the spores and pollen of higher plants show a type of movement. While they do

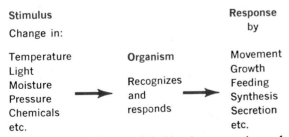

Stimulus Change in:		Response by
Temperature Light Moisture Pressure Chemicals etc.	Organism Recognizes and responds	Movement Growth Feeding Synthesis Secretion etc.

Fig. 2-9. Living things are irritable: they recognize and react to changes in their environment.

Fig. 2-10. Sensitive plant, *Mimosa*. **A,** Normal; **B,** after being touched.

not have cilia or flagella, most of them do show a type of growth movement after spore or pollen germination. This growth in seed plants, for example, is aimed at or attracted to a stimulus given off by the corresponding egg cell.

The human egg cell is propelled toward the uterus by the action of ciliated cells in the fallopian tubes. Ciliated cells are found in many complex animals.

At first examination we may feel that motion is not a real characteristic of the flowering plants. Close observation and slow motion photography, however, shows that leaves twist, curl, and move in response to various stimuli. An obvious example is the Venus fly trap whose leaves close to trap insects. We have already considered the rapid movement of the sensitive plant. The petals of many flowers fold up at night and expand during the day. A sunflower bends to follow the path of the sun across the sky.

Even the contents of cells seem to be in constant motion. If they are observed under a microscope, various streaming patterns can be seen

in the cytoplasm. In some cells this appears to be directly responsive to local stimuli and in others such as *Paramecium* there is a definite path of movement called cyclosis.

If we look at a slide of stained white blood cells, they appear spherical and distinct in outline. These same white blood cells observed when alive are very irregular in shape with a cell membrane that constantly changes form.

REPRODUCTION AND HEREDITY

For any species of plant or animal to successfully maintain its existence, it must possess some means of reproduction or duplication of its kind. There are two general methods of reproduction in the living world: asexual, without the formation of specialized sex cells and sexual, in which distinct sex cells are produced (Fig. 2-11).

In asexual reproduction a new offspring arises from a part of an older individual. This part may be a single-celled spore as in ferns, a many-celled bud as in certain sponges, or an organism may

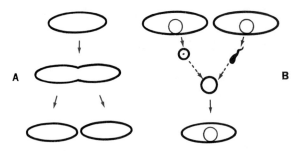

Fig. 2-11. A, Asexual reproduction—one organism divides to form two offspring. **B,** Sexual reproduction —two different organisms produce sex cells in special organs. Egg and sperm unite and develop into offspring.

divide into two parts as in the case of such single-celled animals as *Amoeba* and *Paramecium*. In asexual methods the offspring are very much like the parent.

In sexual reproduction two sex cells unite to form a single fertilized cell from which a new individual develops. In sexual reproduction something more than mere propagation occurs because the new individual has arisen from the union of sex cells from two different parents, and each parent has contributed some of its traits. Hence, the new individual is usually not exactly like either parent but possesses traits from both. When this occurs generation after generation, varieties of offspring are formed, thus making it possible for them to fit into different environmental conditions.

ADAPTATION

Living things are capable of surviving in a great variety of environments. The term adaptation is used in two senses, one to indicate short-term or seasonal changes in response to a change in the environment. The other use of the word indicates the long-term major changes resulting from genetic differences and production of new varieties. These changes lead to evolution.

Short-term adaptations are modifications of activity caused by change of temperature, light, moisture, and other factors of a temporary or seasonal nature. They might include a change in the rate of growth of plants or a change in the ratio of vegetative to flower parts or fruit production. The larger spreading shrub from the south may be a small weed if grown in the north. Animals change the amounts and type of food stored when they are in different parts of their range. Some produce a thick coat of fur in response to colder climates; others change their activity patterns as the climate changes.

In the broader sense adaptation implies all of the hereditary and reproductive changes that result in new varieties. If these new varieties are able to survive and reproduce at the expense of older varieties, then they will become dominant in their area. The organisms are then said to be adapted to land or sea, to fresh water or salt water, to high or low altitudes, to open or vegetation-filled areas, and to warm or cold environments, and so on.

One of the best known and most studied examples of adaptation in the evolutionary sense is in a small group of birds that have come to be known as Darwin's finches. These occur on a group of about twenty small islands called the Galapagos, located on the equator about 600 miles off the coast of Ecuador. The British research ship H.M.S. *Beagle* visited the islands when Charles Darwin was serving as the ship's naturalist.

He noted that although there were only a few kinds of land vertebrates on the islands, there were over a dozen species of a bird called finch. The fourteen species of finches had adapted to the point where they occupied many of the available types of habitats and food sources on the islands. Some were essentially ground-feeders, differing in the type and size of seeds they used for food. Their beaks were adapted to various degrees of seed-crushing activity. Many of the other species were tree-feeders; of these some were plant-feeders and others insect-feeders. Even the insect-feeders showed a difference in the size and type of insect they captured and exhibited corresponding beak modification. One functioned as a woodpecker by pecking a hole in the bark. For the lack of a woodpecker-type tongue, it had adapted by using the spine of a cactus to spear the insect food.

In short, this group of finches had become adapted so that they were able to take advantage of the various kinds of food available on the islands. Their beaks and other characteristics had become modified over a period of time so that when Darwin visited, fourteen distinct species could be identified.

Some contributors to the knowledge of
CERTAIN ACTIVITIES OF PROTOPLASM

Aureolus B. von Hohenheim (Paracelsus) (1493-1541)

A Swiss physician who seemed to gain occasional glimpses of fundamental truth but usually through a haze of astrologic and alchemical absurdities. He may have had some ideas on metabolism and the cycle of elements.

Charles Bonnet (1720-1793)

A Swiss naturalist who discovered parthenogenesis (development of an egg without a male sperm) in certain insects and inferred incorrectly that the females carried many miniature young for succeeding generations in their bodies.

Spallanzani

Lazaro Spallanzani (1729-1799)

An Italian naturalist, philosopher, and priest who described the regeneration of organs in such animals as worms, snails, and frogs. (*Historical Pictures Service, Chicago.*)

Joseph Priestley (1733-1804)

Priestley

An English naturalist and divine who studied the respiratory interdependence of plants and animals. (*The Bettman Archive, Inc.*)

Robert Brown (1773-1858)

A Scotchman who discovered the so-called "Brownian movement" of minute particles (1828), which phenomenon provides a basis for demonstrating molecular motion.

Jacques Loeb (1859-1927)

Loeb

A German-American biologist who experimentally studied sea urchin eggs in which he induced parthenogenesis by artificial means (1910). (*Historical Pictures Service, Chicago.*)

Charles M. Child (1869-1954)

An American biologist who proposed the "Axial Gradient Theory" to explain the differences in rates of metabolism along the axes of living organisms (1921).

C. A. MacMunn

By studying various tissues from echinoderms to man (1886), through the use of microspectrography, he found pigments whose spectra showed a series of absorption bands that varied little from tissue to tissue or from organism to organism.

David Keilin (1887-1963)

A Cambridge University biologist who rediscovered (1925) the same pigments, now called cytochromes ("cellular pigments"), that had been found by MacMunn, thus verifying the latter's contentions.

Otto H. Warburg (1883-   )

A German biochemist who discovered an iron-porphyrin protein known as "cytochrome oxidase" (1930). He was awarded a Nobel Prize (1931) for his contributions to cellular metabolism.

Review questions and topics

1 Discuss each of the activities of living protoplasm, including (1) organization and individuality, (2) metabolism, (3) respiration, (4) growth, (5) movement, (6) irritability, conduction, coordination, and adaptation, (7) reproduction and heredity.
2 Define life in your own words.
3 Contrast, with examples of each: metabolism, anabolism, catabolism, and assimilation.
4 Explain autosynthesis and autocatalysis, with examples of each.
5 Discuss the different types of stimuli, with examples of each.
6 Explain the importance of (1) chemical coordination and (2) nervous coordination; give the attributes of each.
7 Discuss various ways in which organisms may attempt to adapt themselves to meet changing environmental conditions.
8 Explain the methods of reproduction, with the significance of each type.
9 Explain the phenomenon of regeneration and its importance.
10 Discuss viruses and their significance on the borderline between living and nonliving.

Selected references

Chambers, R., and Chambers, E. L.: Exploration into the nature of the living cell, Cambridge, 1961, Harvard University Press.
Comfort, A.: Ageing; the biology of senescence, New York, 1964, Holt, Rinehart & Winston, Inc.
Galston, A. W., and Davies, P. J.: Control mechanisms in plant development, Englewood Cliffs, N.J., 1970, Prentice-Hall, Inc.
Hardin, G., editor: 39 steps to biology. Readings from Scientific American, San Francisco, 1968, W. H. Freeman and Co. Publishers.
Haynes, R. H., and Hanawalt, P. C., editors: The molecular basis of life. Readings from Scientific American, San Francisco, 1968, W. H. Freeman and Co. Publishers.
Johnson, W. H., and Steere, W. C., editors: This is life, New York, 1962, Holt, Rinehart & Winston, Inc.
Spratt, N. T.: Introduction to cell differentiation, New York, 1964, Reinhold Publishing Corp.
Stanley, W. M., and Valens, E. G.: Viruses and the nature of life, New York, 1961, E. P. Dutton & Co., Inc.
Walker, B. S., Boyd, W. C., and Asimov, I.: Biochemistry and human metabolism, ed. 3, Baltimore, 1957, The Williams & Wilkins Co.

3 Cells

One of the basic principles in biology is that all organisms are made of cells, and all phenomena of life are fundamentally cellular in nature. Essentially, the functions of normal plants and animals, as well as those of abnormal, diseased organisms, are but expressions of cell structure and function. Robert Hooke, an Englishman, studied cells as early as 1665. Matthias Schleiden, a German botanist, and Theodor Schwann, a German zoologist, are commonly given credit for the formulation of the cell principle in 1839, although René Dutrochet, a French physiologist, preceded them with similar views in 1824. The cell principle was verified repeatedly by later investigators.

Early investigators used the word *cell* because to them the material looked like the "cells" in a honeycomb. They studied the cell wall and practically ignored the substance within. This emphasis on cells led to the founding of the specialized science of cytology (Gr. *kytos*, cell; *logos*, study).

The cell is considered the basic structural unit of animals and plants; the tissues and organs are made of cells much as the brick is a structural unit in a brick wall. The cells are also functional (physiologic) units because the functions of living organisms are the results of cellular activities. Each cell works somewhat as a unit, but more often groups of cells work together in some common function. There must be coordination and subordination if the organism is to function as a whole with efficiency.

The cell is also a unit of growth and development, as in a complex organism that has matured through a division of its cells, an increase in cell size, and a specialization into tissues. Cells are units of heredity, for it is through them that genes are received from parents, maintained within the embryo and adult, and passed on to future off-spring. During the division of cells, each cell receives genes that enable the organism to express its specific characteristics. Cells are units of regeneration (repair) when tissues or organs are replaced or repaired. Abnormal cell divisions are responsible for growths such as tumors and cancers.

CELL STRUCTURE AND FUNCTION

Many organisms, especially those we call *microorganisms,* are composed of just one cell. Many others, larger and consequently more easily seen, are multicellular, made up of many cells. For example, an adult human male weighing 160 pounds contains approximately 60,000 billion cells. In just one cubic millimeter of human blood, one small drop, there may be 5 million red blood cells.

Detailed studies of cell structure are usually made with either an ordinary microscope or an electron microscope. Before the study the cell or tissue is prepared by making very thin slices and then staining them with special dyes. (Fig. 3-1 is the result of such a process.)

After a number of sections of a particular cell have been studied, it is possible to put all the information together in the formation of a composite cell. You may have seen an example of this technique in a study of the human body. Several transparent sheets are used. Each one has a layer of the body on it. When all of the sheets are held together the skin is visible. When the first sheet is lifted off, the muscle layers are visible. When the second sheet is lifted the internal organs are visible (see Fig. 3-2).

If we cut a thin slice from an ordinary apple, it may have many different features, depending on where the slice was made. A slice made near the surface would be different from a slice across the core or through the stem. If we put all of these slices together, it would be easy to recognize the structure of an apple. It is the same with the study of a plant and its cells.

The rest of this chapter is concerned with a detailed study of cells.

Cell wall

The cell wall is a semirigid, layered covering that is present in most plant cells but absent in most animal cells. It is secreted by the cell and gives protection and support, covering the living plasma membrane beneath (Fig. 3-1). In some plants it contains minute pits or pores between adjacent cells (see Fig. 18-8). In plant cells it is composed primarily of cellulose, a complex carbohydrate material that occurs in long threads called fibrils. Occurring between the fibrils are other complex materials, such as lignin and pectin. Enough space remains so that gases and fluids pass freely through the cell wall.

Two adjacent plant cells are bound together by a shared cell wall layer, the *middle lamella.* Cell walls form the chief component of wood and yield fibers used in the manufacture of such products as paper, flax, cotton, and hemp.

Plasma membrane

The plasma membrane is a living, ultrathin, elastic, porous, semipermeable covering that is present in both plant and animal cells. It is not easily seen with a light microscope. Chemical, physical, and electron microscope studies indicate that it is composed of protein molecules interspersed in a double layer of lipoid or fatlike molecules. Other substances often occur on its surface. Externally it has an irregular contour with deep craterlike infoldings, whereas the inner surface has numerous saclike inpouchings (Fig. 3-3). These invaginations may form *pinocytic* vesicles from which fluid may be ingested by the cytoplasm. The plasma membrane grows as the cell enlarges and has a limited ability to repair itself.

Simple sugars, amino acids, potassium ions, and water may pass through the membrane rapidly. Sodium and other substances, even if con-

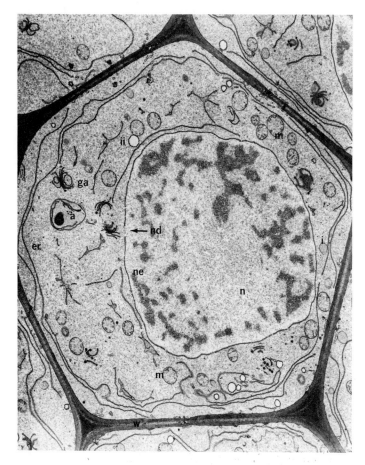

Fig. 3-1. Meristematic cell of the root tip of corn showing the following ultrastructures: **n,** nucleus showing chromatin material; **ne,** nuclear envelope; **nd,** nuclear envelope discontinuity (not a pore); **er,** endoplasmic reticulum; **ga,** Golgi apparatus; **m,** mitochondrion; **a,** amyloplast (starch forming); **i,** unidentified cytoplasmic inclusion body; **ii,** unidentified cytoplasmic inclusion body; **w,** cell wall. Magnification approximately 8,000×.
(From Whaley, W. G. W., Mollenhauer, H. H. M., and Leech, J. H. L.: The ultrastructure of the meristematic cell, Am. J. Bot. 47: 423, 1960.)

structed of small molecules, may not pass readily. The structure and behavior of the membrane, together with the size and character of the entering molecules, are important in the passage of materials through the membrane.

Membrane permeability and osmosis

Permeability is the property of a membrane that determines its penetrability. The permeability of a membrane depends on (1) the size of the pores of the membrane, (2) the size of the particles of the substance attempting to pass through that membrane, and (3) the solubility of the substance in the membrane. A membrane may be permeable to small molecules but impermeable to large molecules. Another membrane may be permeable to ions but impermeable to even the smallest molecules.

PROTOPLASM

The boundary of cells consists of proteins, fatty substances, and other materials that influence its solubility properties, which, in turn, at least partially determine its permeability. Living membranes, such as the plasma membrane of cells, have a selective permeability. Living membranes usually permit the passage of small molecules and certain ions, whereas larger molecules, such as protein molecules, and colloidal particles are restrained. Different cells vary in the permeability of their boundaries. Each has it specific type of permeability, and the plasma membrane

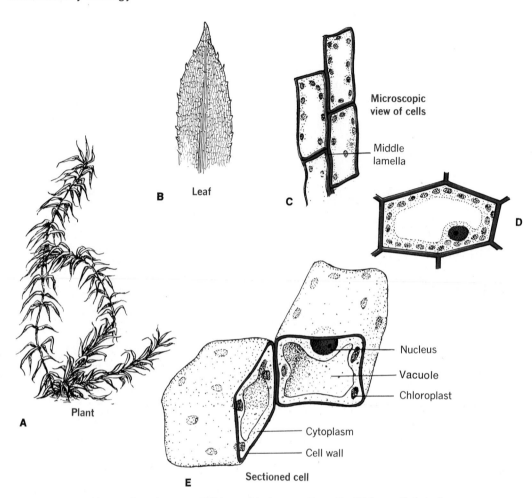

Fig. 3-2. Views of a plant. **A,** With unaided eyes. **B** to **D,** Using a light microscope at increasing magnification. **E,** Reconstruction of a sectioned cell. See Fig. 3-1 for details of an electron micrograph of a plant cell.

of each individual cell plays an important role in regulating the activities of the protoplasm within the cell.

OSMOSIS

The force exerted by the pressure of moving molecules in a solution against a membrane is known as osmotic pressure. The passage of water through a semipermeable membrane is known as osmosis (Gr. *osmos,* push) (Fig. 3-4). The measurable force within living cells is considerable

and usually keeps the cell membrane distended.

A solution with greater concentration (less water) than the protoplasm, which draws water from the protoplasm of the cell, is known as a hypertonic solution. In this case water will pass out of the cell in an attempt to equalize the pressure. Under such circumstances (loss of water) animal cells will tend to shrink because of their delicate cell membrane, whereas the protoplasm of plant cells shrinks away from the rather rigid, resistant cell wall (Fig. 3-5). Such shrinking of protoplasm

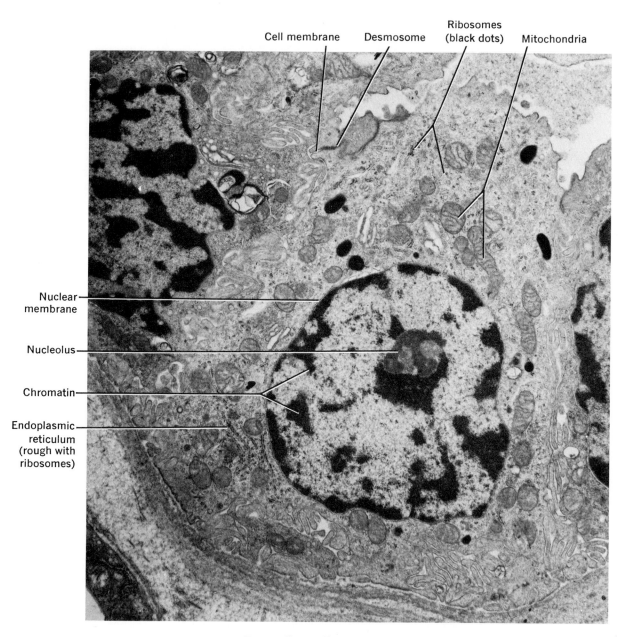

Cell membrane   Desmosome   Ribosomes (black dots)   Mitochondria

Nuclear membrane

Nucleolus

Chromatin

Endoplasmic reticulum (rough with ribosomes)

Fig. 3-3. Human liver cell as seen by electron microscope. *(Courtesy M. A. Hayat, University of Dayton.)*

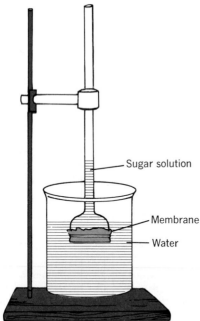

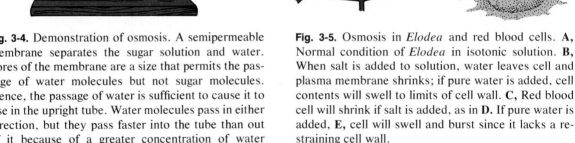

Fig. 3-4. Demonstration of osmosis. A semipermeable membrane separates the sugar solution and water. Pores of the membrane are a size that permits the passage of water molecules but not sugar molecules. Hence, the passage of water is sufficient to cause it to rise in the upright tube. Water molecules pass in either direction, but they pass faster into the tube than out of it because of a greater concentration of water molecules on the outside.

Fig. 3-5. Osmosis in *Elodea* and red blood cells. **A,** Normal condition of *Elodea* in isotonic solution. **B,** When salt is added to solution, water leaves cell and plasma membrane shrinks; if pure water is added, cell contents will swell to limits of cell wall. **C,** Red blood cell will shrink if salt is added, as in **D.** If pure water is added, **E,** cell will swell and burst since it lacks a restraining cell wall.

from the cell wall or membrane during the loss of water is called plasmolysis (Gr. *plasma,* molded; *lysis,* loosening).

A solution with less concentration (more water) than the protoplasm, which places water into the protoplasm of the cell, is known as a hypotonic solution. In this case the addition of water to the protoplasm causes a condition known as turgor (L. *turgere,* to swell). If carried to extreme, the cell may burst. A solution that has the same concentration as the protoplasm and in which there is no net movement of water through the cell membrane is known as an isotonic solution. In this case pressures are equal on

both sides of the cell membrane, and there is no shrinking or swelling.

Passage of materials to or from a cell under such conditions will be by diffusion or by an energy-using process called active transport. It is quite clear that hypertonic and hypotonic solutions around a cell influence the passage of materials out of the cell and into the cell. The absorption of foods and the elimination of wastes probably are accomplished in this way.

Cytoplasmic matrix

Cytoplasmic matrix fills the space between the plasma membrane and the internally located nu-

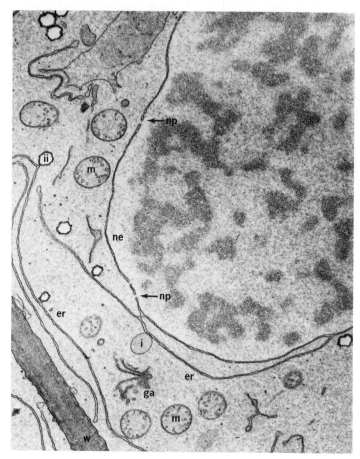

Fig. 3-6. Portion of the meristematic root cap cell of corn showing ultrastructures as follows: **ne,** nuclear envelope (membrane); **np,** nuclear envelope pore (see arrows) showing nuclear membrane continuity with the endoplasmic reticulum; **er,** endoplasmic reticulum; **ga,** Golgi apparatus; **i,** unidentified cytoplasmic inclusions; **ii,** unidentified cytoplasmic inclusions; **m,** mitochondrion; **w,** cell wall. Magnification approximately 17,000×. *(From Whaley, W. G. W., Mollenhauer, H. H. M., and Leech, J. H. L.: The ultrastructure of the meristematic cell, Am. J. Bot. 47:425, 1960.)*

cleus and is composed of a highly organized, intricate meshwork of elongated protein molecules responsible for many of the cell functions. Within the matrix are located the endoplasmic reticulum, mitochondria, the Golgi apparatus, the centrosome, and other cell organelles.

Endoplasmic reticulum

The endoplasmic reticulum, discovered and named by Keith Porter in 1956, extends from the nuclear membrane to the peripheral cytoplasmic meshwork and is composed of delicate, slotlike, roughly parallel membranous or ribbonlike structures.

There are two types of endoplasmic reticulum, rough and smooth. Rough endoplasmic reticulum is found in all cells except erythrocytes and is especially abundant in glandular cells, such as salivary gland cells. Only rough endoplasmic reticulum has direct connections with the nuclear membrane (Fig. 3-6). It has tiny particles of ribonucleic acid (RNA) called ribosomes clinging to its outer surfaces. It is involved in the synthesis, segregation, and accumulation of secretory proteins and in the biosynthesis of membranes.

Smooth endoplasmic reticulum does not show any association with ribosomes and is less common than rough endoplasmic reticulum. It is abundant in the liver and intestinal epithelium and seems to be involved in detoxification mechanisms, in lipid and cholesterol metabolism, and in the biosynthesis of steroid hormones.

Ribosomes and protein synthesis

Ribosomes are tiny spheres consisting of almost equal amounts of RNA and protein. They usually occur in clumps or clusters that are collectively referred to as a polyribosome or, more simply, as a polysome. They are the site of protein formation.

As detailed in Chapter 32, a messenger RNA–ribosome complex is formed. The RNA then moves along the length of the ribosome and attracts transfer RNA (tRNA) to the messenger RNA (mRNA) surface. Amino acids at the end of the tRNA are then brought into contact and peptides are formed. As the ribosome moves along the mRNA, the process is repeated and a polypeptide or protein is formed.

During protein formation, four or five ribosomes are attached to a particular mRNA strand. Apparently, each ribosome is "reading the genetic code" for a particular polypeptide. The attached ribosomes are collectively called the polysome.

Lysosomes

Small membrane-enclosed bodies containing digestive enzymes were first seen by DeDuve in the early 1950s. They were formed on the ribosome-reticulum complex. Because of their digestive function, he referred to them as "a sack full of enzymes" and as "suicide bags." Lysosomes are rich in lytic enzymes (for example, acid hydrolases) and appear to be involved in the engulfment of foreign materials into the cell, in the destruction of dead cells, and in tissue degeneration. They have been found in many cell types, including spleen, kidney, and liver cells, and in meristematic cells of plants.

In lower animals such as *Hydra* the lysosomes in the cell cytoplasm fuse with ingested food vacuoles. The food is then digested. Lysosomes also appear to function in cell breakdown, such as in resorption of the tail of a tadpole, in aging of cells, and in general cell dissolution.

Mitochondria

When a thin section of a plant or animal cell is properly prepared and magnified, numerous round, or rodlike, structures known as mitochondria (Gr. *mitos*, thread; *chondros*, granular) will be seen within the cytoplasm (see Fig. 3-3). Some cells contain 1,000 or more, and their sizes may vary from 0.2 to 3 μm. Mitochondria are surrounded by a thin double membrane of lipoprotein with the inner layer folded to form partitionlike ridges, called cristae, that extend inwardly. Mitochondria may assume different shapes in different organisms and some of the cristae may run crosswise instead of lengthwise. They may even be tubular as in the mitochondrion of *Paramecium*. The outer membrane is elastic and may swell to increase its size. It is composed of protein molecules that can be extended or folded greatly. Mitochondria are called the "power houses" of the cell.

Mitochondria are present in large numbers wherever energy is needed (Fig. 3-7). They contain energy-transfer enzyme systems. The energy released is used for the formation of cell products, the contraction of muscles, and the conduction of nerve impulses. Parts of the mitochondrial enzyme system are formed by such vitamins as riboflavin, nicotinamide, and pantothenic acid. Many different enzyme systems may be present in a mitochondrion, each system composed of many different protein molecules. It is also suggested that many duplicates of each enzyme may exist in the mitochondrial unit at the same time.

Production of energy

Much of our present knowledge of cellular energy and cellular respiration began with the discovery by Otto Warburg of the respiratory enzyme called cytochrome oxidase and the rediscovery of a pigment cytochrome in the cells of living organisms. Cytochrome is related to he-

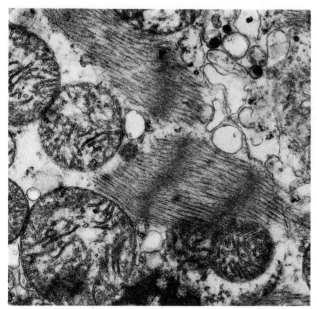

Fig. 3-7. Section of chicken heart showing many mitochondria adjacent to the myofibrils of the ventricle.
(Electron micrograph courtesy M. A. Hayat, University of Dayton.)

moglobin, the oxygen-carrying substance in red blood corpuscles.

When sugar is broken down, it releases hydrogen that is accepted by the cytochrome. In the presence of free oxygen and cytochrome oxidase the cytochrome gives up hydrogen to the oxygen to form water.

Cytochrome oxidase + Oxygen +
(Cytochrome + Hydrogen) → Water

Every cell contains a group of chemicals that are energy trapping and known collectively as adenosine phosphates (Fig. 3-8). Adenosine is a nucleotide and is one of the intermediate reaction products in carbohydrate metabolism. Each cell contains adenosine phosphates. As the name suggests, adenosine triphosphate contains three phosphate groups (H_2PO_3) attached to each adenosine molecule. It is the principal molecule in which energy is stored. Under proper conditions ATP will lose a phosphate group to some other substance and release energy, thus becoming ADP with two phosphate groups. A molecule of ADP can be transformed into ATP by adding a third phosphate group, an addition which of course requires energy. Energy-rich ATP is the chief end product of cellular respiration through the action of the mitochondria of cells. ATP diffuses to all parts of a cell where energy is needed.

Carbohydrates, proteins, and fats are broken down outside the mitochondria into their constituents before passing through the outer member of the mitochondrion. Inside the mitochondria the oxidation reactions remove carbon atoms from these constituents until carbon dioxide and water are produced. The energy thus released by

Cellular respiration

Glucose + Oxygen→Carbon dioxide + Water + Energy
(Sugar)   (Released)

ADP   + Phosphate   + Energy→ATP
(Adenosine   ($-PO_3H_2$)   (Adenosine triphosphate)
diphosphate)   (Energy-rich phosphate)

ATP+Water —— ADP+Phosphate+Energy

Fig. 3-8. Release of energy in living organisms. Adenosine is an organic, phosphate-containing compound that functions in energy transfers within cells.

these successive oxidation steps passes to the ATP by the phosphorylation process.

A sugar, such a glucose, may combine with phosphate, through the process of phosphorylation, to form sugar phosphate (glucose phosphate) and the latter, through a series of steps, eventually ends up as pyruvic acid. These steps are complicated and are considered in more detail elsewhere.

Golgi apparatus

The Golgi apparatus (after Golgi, the Italian histologist) is quite prominent in nerve cells, secretory cells, and germ cells. It has a large surface area, because it consists of a series of membranous structures of variable size and shape. In some cells the apparatus is seen as a pile of joined, platelike, crescentic (curved) layers (Figs. 3-9 and 3-10). It may be observed by both light and electron microscopes and is considered a center of cellular synthesis.

Although it was suggested by Ramon y Cajal in 1914 that the Golgi apparatus secreted the mucus that covers the cells lining the intestine, confirmation of this hypothesis was not made until fairly recently. Electron microscope studies by Neutra and Leblond have shown that globules of mucus do, in fact, arise from the Golgi apparatus.

Further studies by several workers indicate that Golgi apparatus functions in the formation of very large carbohydrate-protein complexes. Apparently, the Golgi apparatus adds the carbohydrate to the protein that has been formed in ribosomes. It appears that most plant and animal cells have active Golgi complexes responsible for the synthesis and secretion of a wide variety of carbohydrate complexes.

Spherosomes

Spherosomes are ovoid or spherical structures that can be seen both with light and electron microscopes in plant cells. They are bound by a single membrane and contain an osmiophilic material. They appear to be involved in fat production or storage, although many other possible functions have been proposed. It is thought that spherosomes are formed by the endoplasmic reticulum.

Centrosome

The centrosome (Gr. *kentron*, center; *soma*, body) is an area of dense protoplasm usually located near the nucleus of many animal cells. It is not commonly present in cells of higher plants, although it may be present in certain lower plants, such as brown algae. It is usually rather inconspicuous in cells except during cell division, when it plays an important role.

After the prophase stage of cell division the centrosome contains two rod-shaped, granular centrioles. The electron microscope reveals them to be short, cylindrical structures placed at right angles to each other. One of the earliest signs of cell division (if the centrosome is present) is the movement of the two centrioles away from each other until eventualy they take positions at opposite sides (poles) of the nucleus. After the nuclear membrane disappears, the centrioles organize cellular proteins into a series of fibers between them that constitute the spindle. More detailed consideration will be given under cell division in Chapter 4.

Fat droplets

Mitochondria are engaged in the breakdown of fats to yield energy and in the build-up of fats when more sugar is available than is needed at that time. Spherical fat droplets are found close to mitochondria in cells, which may explain their role in fat metabolism. Fat is a principal reserve supply of energy in the body.

Cytoplasmic vacuoles

Some of the more interesting cytoplasmic vacuoles occur in *Amoeba, Paramecium,* Protista, and *Euglena*. These contractile vacuoles play an important part in maintaining the water balance of the organism and in expelling waste. Some of them have a series of radiating canals. They can be observed to swell with water, discharge, and

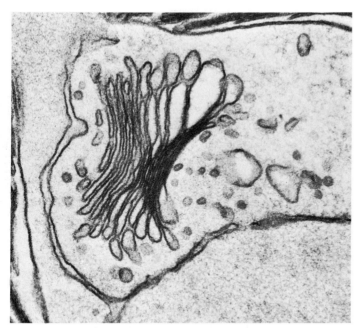

Fig. 3-9. Electron micrograph of the unicellular green alga *Tetracystis excentrica* showing details of a Golgi apparatus. *(From Brown, W. V., and Bertke, E. M.: Textbook of cytology, ed. 2, St. Louis, 1974, The C. V. Mosby Co.)*

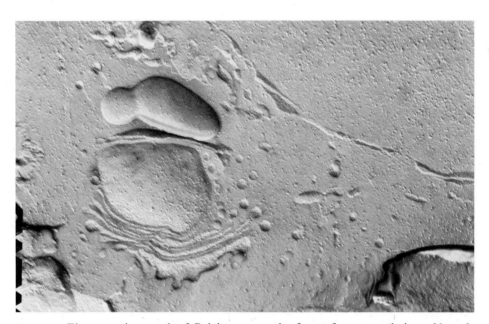

Fig. 3-10. Electron micrograph of Golgi apparatus by freeze-fracture technique. Note the platelike structure and terminal "buds." *(Courtesy Faye D. Schwelitz, University of Dayton.)*

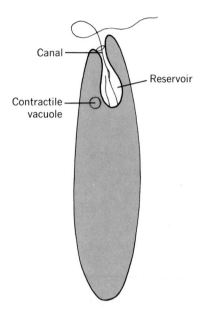

Canal

Reservoir

Contractile
vacuole

Fig. 3-11. Many single-celled organisms, such as
Euglena, have a contractile vacuole, which picks up
excess water from the cell cytoplasm, swells until it
contacts the reservoir membrane, and then discharges
its contents.

fill again (Fig. 3-11 and Chapters 10 and 11).

Most plant cells have a large vacuole bounded
with a single membrane and filled with a watery
cell sap. This fluid may contain high concentrations of pigments, amino acids, sugars, and other
compounds. These plant vacuoles are small in
young cells but are large enough in older cells to
push the cytoplasm to the periphery of the cell
(see Figs. 3-12 and 18-8).

Plastids

Plastids are organized bodies, usually spherical, oval, or ribbon shaped and are present in
certain plant cells (Fig. 3-12). Three types found
in various plant cells include: (1) chloroplasts
(Gr. *kloros,* green), which contain green chlorophyll and absorb radiant energy to photosynthesize foods; (2) chromoplasts (Gr. *chromo,* color),
which are usually yellow, orange, or red and occur in flowers and fruits; (3) leukoplasts (Gr.

leuko, white), which are colorless and may store
energy-rich starch (formed from simple sugars) as
in the potato tuber.

Chloroplasts are especially important in the
process of photosynthesis. (See Chapter 18). The
boundary of a chloroplast is the typical double
membrane. Inside is a structure peculiar to the
chloroplast, called the granum. These structures
are platelike lamellae resembling stacks of coins.
They are interconnected by structures called the
stroma lamellae.

The arrangement of lamellar components is
very precise in that the protein, lipid, and pigment molecules will not support photosynthesis if
disrupted.

Nucleus and nuclear membrane

The nucleus (L. *nucleus,* kernel or nut) contains chromatin materials (gnees) that transmit
hereditary traits from one generation to another.
When chromosomes are formed, they seem to be
helically (spirally) coiled internally and often
appear as two fine, closely twisted strands. This
helical structure of chromosomes resembles the
helical structure of the deoxyribonucleic acid
(DNA) molecules that compose them.

The nucleus is surrounded by a thin, double-
layered nuclear membrane that is composed of
proteins and lipids and supplied with minute,
porelike perforations (Figs. 3-13 and 3-14).

The nucleus exerts regulatory effects on the
cell processes by moving ribonucleic acid (RNA)
from the nucleus to the surrounding cytoplasm.
RNA acquires part of the hereditary pattern from
the DNA of the chromosome, migrates through a
pore of the nuclear membrane, and then acts as a
template (pattern) for protein synthesis in the
ribosomes.

Additional material on nuclear activity will be
found in the section on mitosis and meiosis in
Chapter 4.

Nucleolus

A conspicuous nucleolus (L. *nucleolus,* "little
nucleus") is present within the nucleus when the

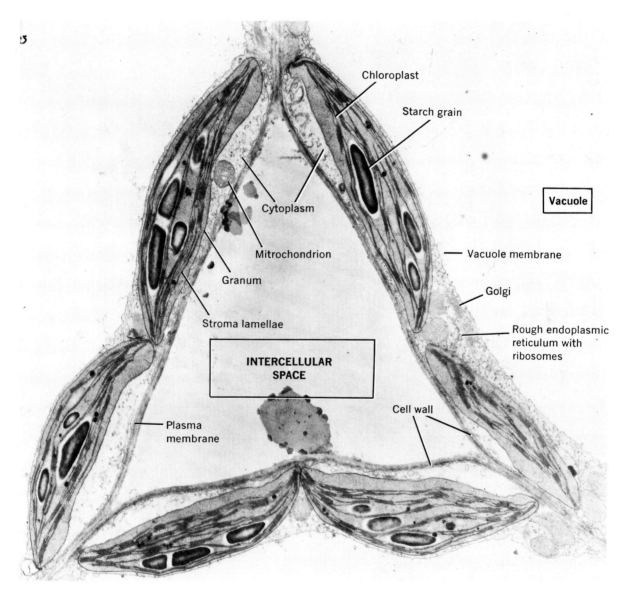

Fig. 3-12. Junction of three plant cells.
(Electron microphotograph courtesy D. R. Geiger and M. A. Hayat, University of Dayton.)

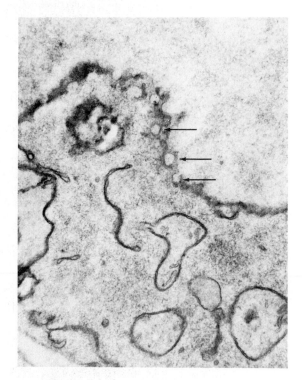

Fig. 3-13. Nuclear pores in cross section.

Fig. 3-14. Nuclear pores in *Euglena*. Freeze-fracture preparation.
(Courtesy Faye D. Schwelitz, University of Dayton.)

cell is not dividing (Fig. 3-15). It is composed of RNA and a high concentration of protein. Sometimes several nucleoli are present in a cell. A nucleolus is formed by a specific region of a particular chromosome, which is called the "nucleolar organizer."

Cilia and flagella

Many cells possess hairlike organelles called cilia or flagella. These are cytoplasmic outgrowths that are surrounded by extensions of the cell membrane. A relatively short extension is called a *cilium*, and a longer one is called a *flagellum*. The distinguishing characteristic is length since other features are the same.

Cilia are usually 10 to 15 μm (micrometers) in

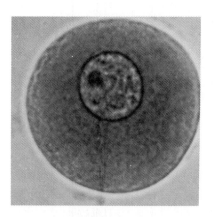

Fig. 3-15. Starfish egg as shown by light microscope. Note the nuclear membrane, nucleolus, and chromatin.

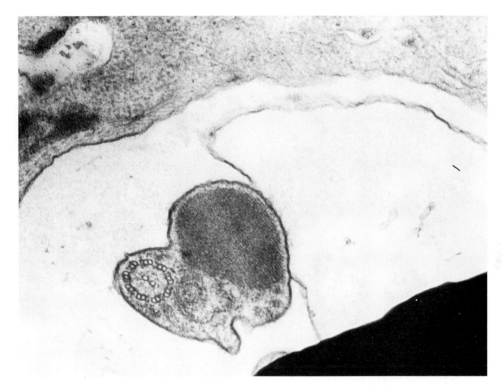

Fig. 3-16. Electron micrograph of a flagellum of *Euglena*. Note the typical 9 + 2 pattern of microtubules. The large striated structure is the flagellar swelling or photoreceptor (see Chapter 10).
(Courtesy Faye D. Schwelitz, University of Dayton.)

length and may occur in tufts, rows, or other patterns, or may cover the entire cell as in *Paramecium* and other ciliates discussed in Chapter 11. Flagella may be several hundred micrometers in length and usually occur singly or in small groups. *Euglena* (Chapter 10) has two flagella of different lengths. One is short and does not emerge from the cell, and the other is elongate and whiplike.

Both cilia and flagella have a characteristic structure of microtubules arranged in a pattern of nine surrounding two in the center (Fig. 3-16).

This 9 plus 2 pattern is found in all nucleated cells that have cilia or flagella. Bacteria may have flagella, but they lack the internal pattern.

Procaryotes

Most of the organelles discussed previously are not found in bacteria and blue-green algae, two groups of organisms classified as procaryotes. They do not have a system of membrane-bound organelles. See Chapter 9 for further details of their structure and function.

Some contributors to the knowledge of
BIOLOGY OF CELLS

Marcello Malpighi (1628-1694)

An Italian scientist and physician, studied the structure of plants and animals and may have referred to cells when he spoke of "globules" and "saccules" (1661). He is considered the "Father of Microscopic Anatomy." He discovered the existence of blood capillaries, whose existence had been hypothesized by William Harvey about thirty years earlier.

Antonj van Leeuewenhoek (1632-1723)

A Dutch microscopist, studied the cellular structure of plants and animals, including bacteria, protozoans, and sperms.

Robert Hooke (1635-1703)

An Englishman, microscopically studied and described many types of natural objects, and from his investigations of cork tissues of plants, he observed minute, boxlike structures that he called "cells" (1665). He gave great impetus to microscopic biology.

Nehemiah Grew (1641-1712)

An English physician, microscopically studied and described (1672) the cells and tissues of plants. He and Malpighi described the microscopic structure of plants so well that few additional contributions were made for over a century.

Robert Brown (1773-1858)

A Scotchman, discovered the general occurrence of the nucleus in plant cells (1831).

Schleiden

Matthias Schleiden (1804-1881)

Schleiden, a German botanist, and Theodor Schwann (1810-1882), a German zoologist, through microscopic studies of many plants and animals, promulgated the "Cell Principle" (1839) in which they stated that all living organisms are composed of cells, or of cells and their products. *(Historical Pictures Service, Chicago.)*

René Dutrochet (1776-1847)

A French physiologist, who preceded Schleiden and Schwann with similar views on cells (1824). He stated that "growth results from both the increase in the volume of cells and from the addition of new little cells."

Edmund B. Wilson (1856-1939)

He was an American cytologist whose text, *The Cell in Development and Heredity* (1924), was an outstanding work in this field. He directed attention to the cellular study and explanation of biologic phenomena.

Review questions and topics

1 State the cell principle, including when and by whom it was formulated.
2 Discuss the various ways in which cells may be considered as units, with examples of each.
3 Discuss the detailed structures and functions of the following: cell wall, plasma membrane, cytoplasmic matrix, endoplasmic reticulum, lysosomes, mitochondria, Golgi apparatus, spherosomes, centrosome, fat droplets, cytoplasmic vacuoles, plastids, nucleus, nuclear membrane, nucleolus.
4 Contrast and give examples of chloroplasts, chromoplasts, and leukoplasts.
5 Discuss such structures and materials as plasmodesmata, ribonucleic acid (RNA), deoxyribonucleic acid (DNA), adenosine triphosphate (ATP), adenosine diphosphate (ADP), phosphorylation, and centriole.

Selected references

Avers, C. J.: Cell biology, New York, 1976, D. Van Nostrand Co.
Brown, W. V., and Bertke, E. M.: Textbook of cytology, ed. 2, St. Louis, 1974, The C. V. Mosby Co.
Capaldi, R. A.: A dynamic model of cell membranes, Sci. Amer. **230:**26-33, 1974.
De Robertis, E. D. P., Nowinski, W. W., and Saez, F. A.: Cell biology, ed. 5, Philadelphia, 1970, W. B. Saunders Company.
Giese, A. C.: Cell physiology, ed. 4, Philadelphia, 1973, W. B. Saunders Company.
Hurry, S. W.: The microstructure of cells, Boston, 1965, Houghton Mifflin Co.
Kennedy, D., editor: Cellular and organismal biology: readings from Scientific American, San Francisco, 1974. W. H. Freeman and Co. Publishers.
Swanson, C. P.: The cell, ed. 3, Englewood Cliffs, New Jersey, 1969, Prentice-Hall, Inc.
Tedeschi, H.: Cell physiology, New York, 1974, Academic Press, Inc.

4 The cell cycle and tissue formation

The body cells of each plant and animal species have a constant number of chromosomes. These occur in pairs; one of each pair is received from the female parent and the other from the male parent. The two members of a pair are called *homologous* chromosomes. The number of chromosomes in different species varies widely (Table 4-1).

Increase in the number of cells of an organism is accomplished by a process called mitosis. This involves the replication of the chromosomes and is usually followed by a division of the cytoplasm called *cytokinesis*. Although mitosis is most easily studied in rapidly growing tissues such as those of an animal embryo or the root tip of a plant (Fig. 4-1), it should be remembered that most of the growth and replacement of cells such as skin and blood, and plant tissues, such as leaves and flowers, is the result of cell division. Special chemical and physical techniques are used to investigate the details of the process.

Table 4-1 Chromosome number

Animal	Chromosomes (pairs)
Horse roundworm (*Ascaris*)	1
Snail (*Helix*)	24
Crayfish (*Cambarus virulus*)	100
Fruit fly (*Drosophila melano-gaster*)	4
Frog (*Rana*)	13
Pigeon (*Columba*)	8
Rhesus monkey (*Macaca mulatto*)	21
Man (*Homo sapeins*)	23

Plant	Chromosomes (pairs)
Green alga (*Spirogyra*)	12
Peat moss (*Sphagnum*)	20
Fern (*Dryopteris cristata*)	164
Pine tree (*Pinus*)	12
Pea (*Pisum*)	7
Wheat (*Triticum vulgare*)	21

THE CELL CYCLE

Until the 1950s most studies of mitosis were concentrated on the actual process of division. The time between cell division was given various designations such as resting phase, metabolic

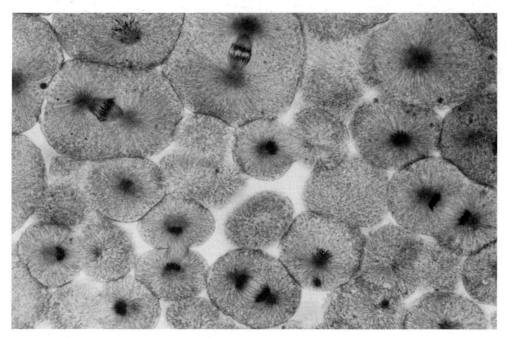

Fig. 4-1. Section of a whitefish embryo showing cells in various stages of mitosis. (*Courtesy General Biological Supply House, Inc., Chicago, Illinois.*)

phase, and interphase. The latter two terms were more correct but still did not really indicate much more than a lack of interest.

About 1950 it was discovered that DNA, the genetic material of the chromosome, doubled during the interphase and separated during the actual mitosis. With the development of more refined techniques, it became evident that there were a number of distinct events in the life cycle of a dividing cell.

In 1953 two researchers, Alma Howard and Stephen Pelc, described the cell cycle in terms of a series of distinct phases. In a cell that is to continue to divide rather than to develop into a specialized tissue, four phases were identified. Immediately after cell division there is a period before DNA replication, called the G_1 *phase*. Then comes the actual period of DNA doubling, the *S phase*. This is followed by a second time interval called the G_2 *phase*, which lasts until the start of mitosis (the actual nuclear and cell divi-

sion), the *M phase*. After mitosis those cells that retain dividing capacities return to the G_1 phase (Fig. 4-2).

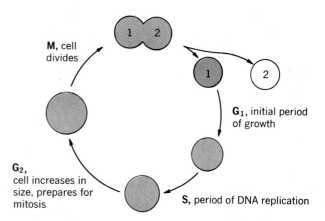

Fig. 4-2. The cell cycle. After division, daughter cells may: *1*, repeat the cycle to form two new cells; or *2*, differentiate as a part of a specialized tissue.

In a general sense, the two cells resulting from mitosis are about half the size of the original. During the G_1, S, and G_2 stages each of these cells usually grows to the normal predividing cell size. In cases such as an early developing embryo, the number of cells increases while the size of each cell decreases. We find that in these cells the growth stages of the G_1 phase are extremely short.

In very general terms, we may assume that during the G_1 phase the cell is producing the enzymes and other components that it will use in the next phase. If the G_1 phase results in increase in cell size, then the cell is also making structural components. This also assumes that it is taking basic materials out of its surrounding and incorporating them into its cytoplasm.

During S phase the cell is using the enzymes and other materials, such as sugars, phosphates, and nucleotides, in the replication of DNA. This chemical doubling of DNA then precedes the physical doubling and separation of the chromosome, which takes place in mitosis.

The events of the G_2 phase are not well known, but we can assume that a preparation is being made for mitosis. The materials necessary for spindle formation and energy production are being synthesized.

The M phase is the period of mitosis. This is treated in detail in the following material.

Mitosis

Although mitosis is a continuous process, it is usually considered during intervals called prophase, metaphase, anaphase, and telophase (Fig. 4-3). These phases are then followed by the G_1, S, and G_2 stages.

The total length of time required for mitosis varies in different cells and in different species. The rate is influenced by the type and age of the tissue involved and the temperature of the surroundings. However, the complex process of chromosome condensation, spindle formation, chromosome splitting, and cytokinesis occur rather quickly when the details are taken into consideration. The time involved varies from less than 10 minutes in some species to several hours in others.

In the following discussion animal cell division will be used to show the general phases of mitosis (Fig. 4-3).

At the end of the G_2 phase the cytoplasm contains mitochondria, ribosomes and other organelles, and the typical chemical contents. A distinct nucleus complete with nucleoli and surrounded by a nuclear membrane is present. Chromosomes as such are not visible, but a diffuse, deeply staining chromatin does exist. It can be shown by chemical analysis that the DNA content has doubled. A pair of centrioles is usually located just outside the nucleus.

PROPHASE

In prophase (Gr. *pro,* before) the first indication of mitotic activity is usually the movement of the centrioles away from each other and toward opposite sides of the cell. During the movement to these opposite poles, distinct raylike fibers, called *asters,* surround the centrioles.

Fine, threadlike fibers, formed by microtubules appear between the two migrating asters and form the mitotic *spindle*. This is broadest at the center of the cell and tapers toward either end. The nuclear membrane begins to disappear, and the spindle occupies the position of the original nucleus. The nucleoli also are no longer visible.

The nuclear materials condense into a darkly staining, coiled network of chromatin threads. The chromatin threads thicken and shorten into distinct chromosomes. At a definite point in each chromosome is a clear area called the centromere (or kinetochore) that seems to assist in the orientation and division of the chromosomes.

It should be noted that each individual chromosome is really two chromosomes produced by the doubling of the DNA and other components during the S phase. This can be seen with the microscope. The doubled condition is usually referred to as two *chromatids* connected by the single centromere.

Individual fibers of the spindle are attached to

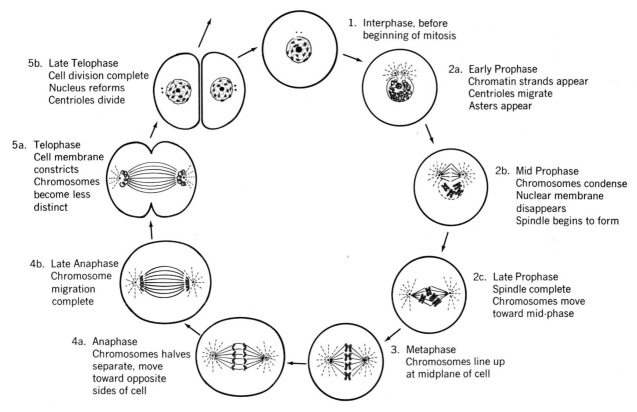

1. Interphase, before beginning of mitosis

2a. Early Prophase
Chromatin strands appear
Centrioles migrate
Asters appear

2b. Mid Prophase
Chromosomes condense
Nuclear membrane disappears
Spindle begins to form

2c. Late Prophase
Spindle complete
Chromosomes move toward mid-phase

3. Metaphase
Chromosomes line up at midplane of cell

4a. Anaphase
Chromosomes halves separate, move toward opposite sides of cell

4b. Late Anaphase
Chromosome migration complete

5a. Telophase
Cell membrane constricts
Chromosomes become less distinct

5b. Late Telophase
Cell division complete
Nucleus reforms
Centrioles divide

Fig. 4-3. Mitosis outline. At end of telophase, daughter cells may specialize or prepare for next mitosis cycle.

the centromeres of all of the chromosomes. The fibers extend toward opposite sides of the cell. The centromeres and attached chromosomes are now moved to a central plane in the cell. This is sometimes called the equator or equatorial plane.

METAPHASE

Once the chromosomes are clustered on the central plane, the cell is said to be in metaphase (Gr. *meta,* after). This usually lasts for a very short time and ends as the centromeres divide. The two chromatids are now referred to as individual chromosomes.

Each new centromere is attached to an individual fiber that moves it away from the central plane. As this movement continues, the cell is in anaphase.

ANAPHASE

During anaphase (Gr. *ana,* up) the chromatids are completely separated, and the newly formed daughter chromosomes move to opposite poles of the spindle. At the same time the spindle seems to elongate, and often the cell itself becomes longer, so that the two groups of daughter chromosomes eventually lie far apart at opposite ends of the cell (Fig. 4-3).

TELOPHASE

In telophase (Gr. *telos,* end) the chromosomes loosen, elongate, and gradually assume a thread-like appearance. The spindle and asters disappear, but the centrioles persist, one with each daughter nucleus. The centrioles later divide to form two per cell. A nuclear membrane again

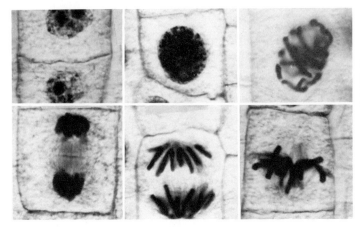

Fig. 4-4. Cell division in onion root tip. Clockwise from top center: 1. Interphase. 2. Prophase. 3. Metaphase. 4. Anaphase. 5. Telophase. 6. Daughter cells.

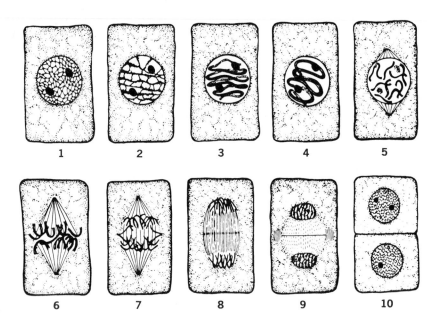

Fig. 4-5. Mitosis in plants. **1, Interphase,** before actual division begins. **Prophase stages: 2** to **5,** chromatin strands thicken and shorten, forming chromosomes consisting of two identical chromatids. In stage **5** the mitotic spindle is forming and the nucleus is disappearing. **Metaphase stage: 6,** chromosomes line up on the equatorial plane of the spindle. **Anaphase stage: 7,** chromatids separate into two identical halves. New chromosomes move along the spindle to opposite poles; **8,** chromosomes at the poles; the cell wall begins to form at minute swellings on the spindle fibers at the equatorial plane. **Telophase stages: 9,** chromosomes become threadlike; the nuclear membrane and nucleoli reappear; the cell wall continues to develop; the spindle disappears; **10,** the division of the cell is complete; two cells similar to the original will grow to normal size.

(Courtesy General Biological Supply House, Inc., Chicago, Illinois.)

forms around the nucleus. Nucleoli reappear, and each daughter nucleus is again in the interphase period.

In animals *cytokinesis,* the actual cell division, usually begins as a furrow or indentation in the plasma membrane. This gradually deepens to divide the cell. Cytokinesis usually begins during telophase, but in some cases it may take place much later.

The obvious result of mitosis is an increase in the number of cells. In a rapidly growing tissue, the original cell forms two cells that form four cells that form eight cells that form sixteen cells, and so on.

Eventually most of the cells stop dividing and take on the characteristics of specialized tissues.

Mitosis in plants

Plant cell division is usually studied in developing seeds, root tips, or other rapidly growing tissues. Most differences between plant and animal mitosis are the result of the structure of the individual cells.

Animals usually have obvious centrioles that lead to formation of asters. Plants, however, lack centrioles and do not form the raylike asters. The cell membrane of plant cells is covered by a semi-rigid cell wall that is not present in animal cells. The actual division into two daughter cells must involve this wall (Fig. 4-4).

A dividing plant cell goes through the DNA and chromosome doubling activities of the G_1, S, and G_2 phases. These are followed by prophase, metaphase, anaphase, and telophase. The doubled chromatids become chromosomes when the centromere splits during anaphase. Each daughter nucleus has a full complement of chromosomes (Fig. 4-5).

New cell membranes begin to appear during telophase. The spindle material in the area of the equatorial plane begins to thicken from the middle toward the edges of the cell. These thickenings enlarge and finally fuse to form a continuous flat plate. New plasma membranes are formed on each side of the cell plate, and two new daughter cells are present.

Fig. 4-6. Three-dimensional view of daughter cells after new cell wall is formed between them.

Cell wall materials are secreted between each plasma membrane and the common cell plate. The original cell plate is now part of the cell wall shared by the daughter cells. The effect may be likened to placing a divider in a drawer (Fig. 4-6). The new cell walls remain somewhat flexible as long as the cell is growing or dividing. Later it may be modified as part of a tissue type. Still later it may become what we call *wood.*

TISSUE FORMATION

An increase in the number of cells of an organism is usually accompanied by the formation of different types of tissues, that is, groups of similar cells with a common function. These tissues in turn may become further specialized as components of an organ, which is then part of an organ system. A cell may be part of an epithelial tissue lining the heart, an organ composed of muscular, epithelial, connective, nervous, and

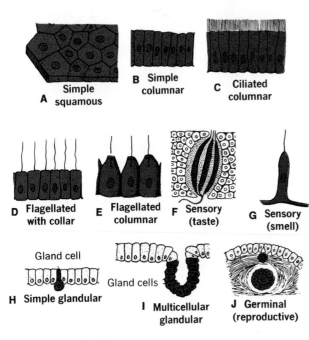

D Flagellated with collar   **E** Flagellated columnar   **F** Sensory (taste)   **G** Sensory (smell)

Gland cell

Gland cells

H Simple glandular   **I** Multicellular glandular   **J** Germinal (reproductive)

Fig. 4-7. Epithelial tissues.

vascular tissues. This heart, together with the blood, veins, and arteries forms the circulatory system.

Animal tissues

There are five major types of animal tissues—epithelial, connective, blood, vascular, and nervous tissues.

Epithelial tissue

Depending upon the type of epithelium (Fig. 4-7), epithelial tissue consists of a layer of flat, cuboidal, or column-shaped cells. Adjacent epithelial cells maintain a contact through delicate strands called protoplasmic bridges. Epithelial tissue forms the outer protective surface of the body, and it lines many internal cavities, such as the digestive and urinary systems. Epithelial

tissue absorbs substances from the exterior and eliminates substances to the outside. It may contain gland cells, which secrete various substances. It may also become modified to produce sex cells within the testes and ovaries; it also forms parts of the sense organs.

SQUAMOUS (PAVEMENT) CELLS

Squamous (L. *squama*, scale) cells are thin, flat, and arranged like stones in pavement. The tissue consists of two types of cells. (1) Simple squamous tissue is made of up of single layers of cells that line internal cavities. *Examples are* endothelium, which lines the heart, blood vessels, and lymph vessels, and mesothelium, which lines the abdominal, lung, and heart cavities. (2) Stratified squamous tissue is composed of more than one layer of cells. It lines the mouth and the esophagus and forms the outer layers of skin of vertebrates.

COLUMNAR CELLS

Columnar (L. *columna*, pillar) cells are column-shaped, and often have tapering ends next to the underlying tissues. In simple columnar tissue there is one layer of cells. This tissue lines the stomach and intestine of most higher animals, where it has a secretory function.

CILIATED, FLAGELLATED, AND COLLARED CELLS

The free surface of the epithelium may bear cilia, flagella, or ''collars.'' Ciliated epithelium is present in gills of clams, the roof of the frog's mouth, and the lining of the air passages of vertebrates, where the cilia move materials from the surface. Flagellated epithelium is found in the inner layer of *Hydra*. Collared epithelium is found in certain canals of sponges.

SENSORY EPITHELIUM

Certain specialized cells help to form sense organs, receptors. Sensory epithelium is found in the retina of the eye, the lining of the nose, the taste buds of the tongue, and the auditory cells of the ear.

Connective tissue

Connective tissue (Fig. 4-8) is common in most parts of the body. This tissue binds the body parts together. Some types form semirigid or rigid structures for the protection and attachment of other tissues and organs. Fibers are usually present. There are large intercellular spaces that may contain nonliving substances, such as fibers, cartilage, and bone.

WHITE FIBROUS (COLLAGENOUS) TISSUE

The white collagenous tissue is composed of long, white or colorless, rather inelastic fibers. This tissue is present in ligaments, which attach bones to bones, and in tendons, which attach muscles to bones. In tendons, there are many white fibers that parallel each other with rows of nuclei between the bundles of fibers. White fibers are called collagenous because when boiled, they produce gelatinous material called collagen.

AREOLAR TISSUE

Areolar (L. *areola,* dim. of *area,* space) tissue is composed of numerous white collagenous fibers and a few yellow elastic fibers, which

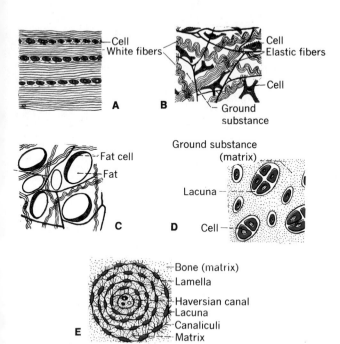

Fig. 4-8. Connective tissues. **A,** White fibrous (from a longitudinal section of a tendon); **B,** areolar (from beneath the skin); **C,** adipose (fat); **D,** hyalin cartilage (from the end of a bone); **E,** bone (cross section of a haversian system).

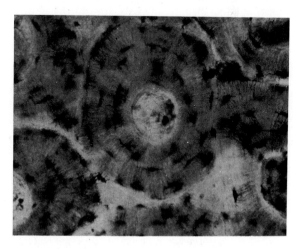

Fig. 4-9. Human bone (cross section); compare with Fig. 4-8, *E.*
(Courtesy General Biological Supply House, Inc., Chicago, Illinois.)

SECRETORY CELLS

Certain glands are unicellular cells, while others are complex and multicellular. Usually, modified columnar epithelial cells produce specific secretions, such as the unicellular goblet cells of the intestine and the multicellular salivary glands.

GERMINAL (REPRODUCTIVE) CELLS

Germinal epithelial cells are modified for the formation of sex cells in the reproductive organs.

branch and form loose networks. This tissue is one of the most widely distributed types of connective tissue. It is found in the fascia (L. *fascia,* band) that surrounds and connects muscles.

ADIPOSE TISSUE

Adipose tissue is a special type of areolar tissue in which the intercellular spaces are filled with fat contained in hollow ''signet-ring'' cells with the nucleus pushed to one side. This tissue is common beneath the skin and around certain organs.

HYALIN CARTILAGE

Hyalin tissue consists of a clear, flexible substance. There are scattered cavities in which rounded cells secrete cartilage. This tissue is found in the ribs, nose, and ends of long bones.

BONE

Bone consists of matrix hardened with calcium salts. Units of bone construction known as the haversian systems (after Havers, an English anatomist) consist of (1) a central canal with an artery, vein, and nerve; (2) lamellae (L. *lamella,* small plate), made of layers of bony plates that form rough-walled canals that are arranged concentrically in circles or ovals around the central canal; (3) lacunae or enlarged spaces associated with lamellae, which contain the irregularly shaped bone cells; (4) tiny, wavy, canal-like canaliculi (L. *canaliculus,* small channel) that radiate from lacunae and connect them with each other and with the central canal; and (5) hard matrix (bone) secreted by the bone cells. Bones support, protect, assist in locomotion, serve for the attachment of muscles and other tissues, and assist in hearing (ear bones).

Blood

Blood or vascular tissue (Fig. 4-10) is sometimes considered as a type of connective tissue. It consists of free cells suspended in an intercellular fluid (plasma). The colorless plasma contains three general types of blood corpuscles. (1) Erythrocytes (Gr. *erythros,* red) are the red blood cells. When mature they are nonnucleated biconcave disks in mammals. The erythrocytes are about 7.6 μm in diameter, and in healthy humans there are approximately 4 to 5 million cells per cubic millimeter of blood. Erythrocytes contain an iron-containing protein, hemoglobin, which carries oxygen. (2) Leukocytes (Gr. *leukos,* white) are the white blood cells. They vary in size, shape, size of the nucleus, and the kinds of granules in the cytoplasm. They are from 7 to 14 μm in diameter, and there are approximately 8,500 per cubic millimeter of blood. Some types destroy foreign materials by phagocytosis; some produce antibodies, and others have different functions. (3) Blood platelets are small, irregular, nonnucleated masses of protoplasm in mammals. They are comparable to nucleated spindle cells of the frog. They are about 3 μm in diameter and there are approximately 300,000 per cubic millimeter. They assist in formation of blood clots.

The vascular tissue carries foods, wastes, oxygen, carbon dioxide, and hormones to all parts of the body. The blood also equalizes temperature

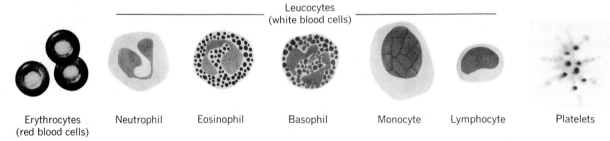

Leucocytes
(white blood cells)

Erythrocytes (red blood cells)   Neutrophil   Eosinophil   Basophil   Monocyte   Lymphocyte   Platelets

Fig. 4-10. Human blood cells and platelets.

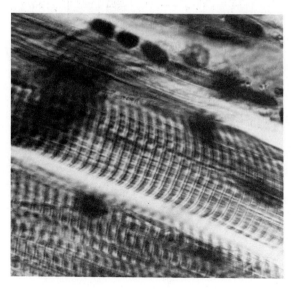

Fig. 4-11. Bas-relief photomicrograph of striated muscles showing the banded condition. (Magnification 1,000×.)
(Courtesy General Biological Supply House, Inc., Chicago, Illinois.)

and maintains the acid-alkaline balance between the various parts of the body. It carries antibodies, which defend against certain diseases. The blood destroys bacteria and other foreign particles by phagocytosis.

Muscular (contractile) tissue

Muscle fibers or cells (Fig. 4-11) are usually elongated and have one or several nuclei per cell, depending upon the type of muscle. There are special internal contractile fibers (myofibrils)

within the cells that cause the muscle to contract when stimulated. Muscle cells move various body parts or the body as a whole.

SKELETAL (STRIATED) TISSUE

Skeletal tissue is associated with the bones and is composed of cylinder-shaped cells, each with several nuclei near the surface of the cell. This tissue makes up the voluntary muscles (under the control of the will). The thin internal myofibrils (Gr. *myos,* muscle) have alternate dark and light bands, thus giving the muscle cell its striated nature. This tissue is found in the muscles of arms, legs, and body walls.

VISCERAL (UNSTRIATED OR SMOOTH) TISSUE

Visceral muscle occurs in the walls of the internal organs. The cells are long and spindle-shaped, and each has one central nucleus. These involuntary muscles are not under the control of the will. The myofibrils are not striated.

CARDIAC TISSUE

Cardiac muscle occurs in the walls of the heart. It is indistinctly striated and has several interior nuclei per cell. Cardiac muscle is involuntary. The internal contractile fibers have alternate dark and light bands that are not as distinct as those in the skeletal muscles. The cells of cardiac tissue are often branched.

Nervous tissue

Nervous tissue (Figs. 4-12 and 4-13) is highly specialized animal tissue consisting of neurons

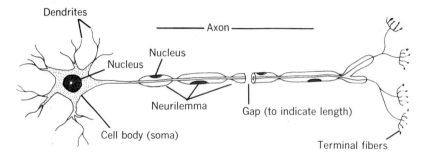

Fig. 4-12. Diagram of a nerve cell, or neuron.

Dendrites

Axon

Nucleus

Nucleus

Neurilemma

Gap (to indicate length)

Cell body (soma)

Terminal fibers

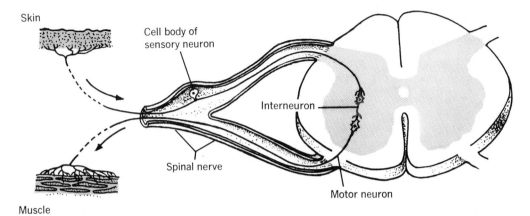

Fig. 4-13. Spinal cord and neurons. Path of impulse is: stimulus on skin → sensory neuron → spinal nerve → spinal cord → interneuron → motor neuron → spinal nerve → muscle → response.

(nerve cells) held in position by a special type of tissue called neuroglia (Gr. *neuron*, nerve; *glia*, glue). Each neuron has a nucleus and processes arising from the cell body. These processes are dendrites (Gr. *dendron*, tree or branched) and axons (L. *axon*, axis). The dendrites are branched and carry impulses to the neurons. Different types may vary in length from 1 mm. to 1 meter. The axons carry impulses away from the neuron, and are usually unbranched. Nervous tissues receive, interpret, and redirect nerve impulses.

The junction of one neuron with the next, a simple contact between two plasma membranes, is called a synapse (Gr. *synapsis*, union). A synapse acts as a polarizing zone, since the impulse will travel across it in only one direction.

The major portion of the brain is composed of the cerebrum and the cerebellum. The cerebrum has an outer cortex region that contains various sizes of pyramid-shaped neurons. The cerebellum consists of an outer cortex of gray matter with an outer layer of scattered cells and fibers; a middle layer of treelike neurons; and an inner granular layer of small neurons.

The spinal cord (Fig. 4-13) is composed of a central, H shaped column of gray matter (seen best in cross section) surrounded by white matter. On each side of the H are two so-called horns: the ventral horn, from which originate the ventral roots of the spinal nerves, and the dorsal horn, from which originate the dorsal roots of the spinal nerves. Motor neurons are found in the larger ventral (anterior) horns. White matter consists of nerve fibers (sensory and motor) that run lengthwise of the spinal cord to transmit nerve impulses up and down.

Organs and organ systems

As animals become more complex, both structurally and functionally, tissues and organ systems develop, with division of labor for the principal life functions (Fig. 4-14). The organ systems and their general functions in man are given briefly as follows.

1. The integumetary system, which covers and protects the body.
2. The skeletal system, which supports, protects, and assists in movement and locomotion from place to place.
3. The muscular system, which together with the skeletal system produces movement and locomotion.

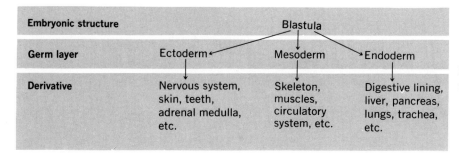

Fig. 4-14. Origin of animal tissues.

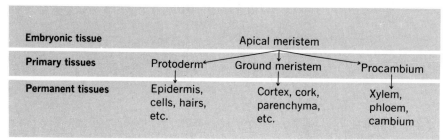

Fig. 4-15. Origin of plant tissues.

4. The digestive system, which takes in foods, changes them chemically into smaller molecules, and absorbs them into the circulatory system.
5. The circulatory system, which transports foods, wastes, oxygen, carbon dioxide, endocrine secretions, and other materials throughout the body.
6. The respiratory system, which permits the exchange of oxygen and carbon dioxide.
7. The excretory system, which excretes the waste products of metabolism.
8. The nervous system, which receives, conducts and redirects nerve impulses throughout the body, thus assisting in the coordination of the activities of the other systems.
9. The endocrine system, which through its ductless gland secretions also assists in coordinating the activities of other systems.
10. The sensory system, which receives stimuli in various parts of the body.
11. The reproductive system, which provides

for the continuation of the individual, and hence the species.

Plant tissues

The formation of tissues in a higher plant begins in growing regions called meristems (Gr. *merizein,* to divide), where active mitosis takes place. (See Chapter 6.) The resulting cells at first form temporary tissues, which may then become modified into the characteristic permanent tissues (Fig. 4-15).

Temporary tissues
MERISTEMATIC

Meristematic tissue cells are small, thin walled, usually cubical, and may divide frequently. They are the basis of the growing regions of the plant and lead to the formation of the other temporary tissues.

PROTODERM

The protoderm (Gr. *protos,* before; *derma,* skin) makes up the outermost layer of cells. It is one cell thick and develops into the epidermis.

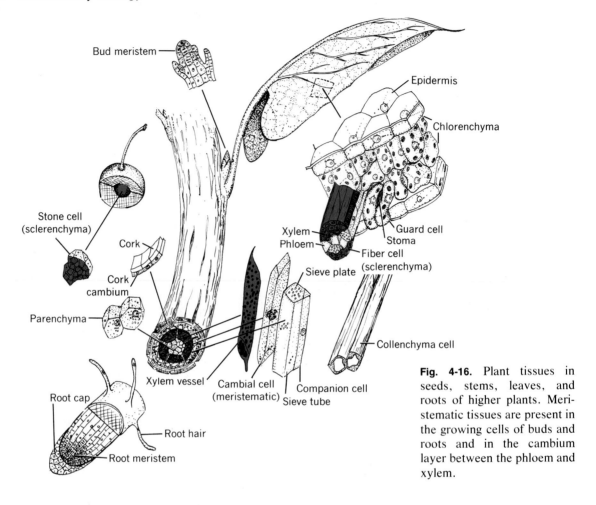

Bud meristem

Epidermis

Chlorenchyma

Stone cell (sclerenchyma)

Cork

Xylem
Phloem

Guard cell
Stoma
Fiber cell (sclerenchyma)

Cork cambium

Sieve plate

Parenchyma

Collenchyma cell

Xylem vessel

Cambial cell (meristematic)
Sieve tube

Companion cell

Root cap

Root hair

Root meristem

Fig. 4-16. Plant tissues in seeds, stems, leaves, and roots of higher plants. Meristematic tissues are present in the growing cells of buds and roots and in the cambium layer between the phloem and xylem.

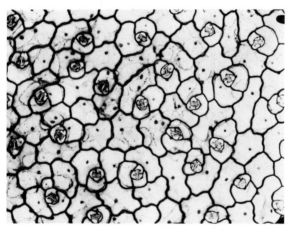

Fig. 4-17. Surface view of the epidermis of a plant, Sedum, showing stomata with surrounding guard cells. *(Courtesy General Biological Supply House, Inc., Chicago, Illinois.)*

Ground meristem forms the greater portion of the developing meristems and may persist for some time as parenchyma or pith cells.

PROCAMBIUM

The procambium cells appear as small patches in the developing stem or root and retain the capacity for cell division. They give rise to the permanent primary vascular tissues.

Permanent tissues

Permanent tissues usually do not change into other kinds of tissues, but in most cases they retain their structural and functional characteristics throughout life.

SIMPLE TISSUES

Simple tissues are composed mostly of one kind of cell and all are constructed similarly.

Epidermis. Epidermis (Gr. *epi,* upon; *derma,* skin) (Figs. 4-16 and 4-17) is usually one cell thick and may contain pigments. The outer cell walls are coated by a waterproof, waxy cutin. Guard cells that control gas movements through epidermal pores called stomata (Gr. *stoma,* opening) possess chlorophyll in chloroplasts.

Epidermal tissues are found on the surface of leaves, flower parts, and younger stems and roots. They function as a protection against mechanical injuries, the effects of parasites, and assist in conserving internal moisture.

Parenchyma. Parenchymal (Gr. *para,* beside; *en,* in; *chein,* to pour) cells are usually ovoid or spherical. There is a large central vacuole, and the cell walls are thin. There are usually numerous intercellular spaces. The protoplasm may remain alive for a long time. Parenchymal cells may form tissues by themselves or may be mixed with other types of cells in complex tissues.

Parenchymal cells are very abundant in higher plants. They are found in the pith of roots, stems, and fruits. They store water and foods in potato tubers, tomato pulp, and watermelon. Green, chlorophyll-bearing parenchyma cells of leaves are called chlorenchyma (Gr. *kloros,* green) tissues.

Sclerenchyma. Sclerenchyma (Gr. *skler-,* hard) cell walls have greatly thickened cellulose and lignin which gives strength and rigidity as in a corn stem. There are two types, (1) fibers that are tough, pliable, strong, elongated cells with tapering ends, and (2) sclereids or stone cells, which give strength and support as in the hull of walnuts.

These tissues give mechanical support and strength; because of the cohesive powers of the fibers, they can be used to make threads, ropes, and textiles; fibers are common in plant stems; sclereids are common in shells of nuts and as gritty masses in pears.

Collenchyma. Collenchyma (Gr. *kolla,* glue) cells frequently are somewhat elongated. The cell walls are thickened at corners or elsewhere, and protoplasm remains alive for a long time. These tissues strengthen both younger and older parts of plants.

Cork. Cork cells frequently are rectangular and regularly arranged. The cell wall contains waterproof, waxy material called suberin (L. *suber,* cork). Protoplasm dies rather soon in these cells.

Cork is present in the outer bark of stems and roots of woody plants where it serves as a protection against mechanical injuries and excessive loss of internal moisture.

COMPLEX TISSUES

Complex tissues consist of several kinds of cells that usually engage in a group of closely related activities.

Xylem. Xylem (Gr. *xylon,* wood) is composed of tracheids and vessels. Tracheids are elongated, tapering cells with short-lived protoplasm. The cell walls usually are strengthened and thickened by spirals or rings of lignocellulose, and often have thin areas or pits. The vessels are long, rather large tubular structures made of series of cells whose walls may be thickened and with pits. Other tissues in xylem may include fibers, xylem parenchyma, and ray cells.

Xylem tissues give strength and conduct water, dissolved mineral salts, and sometimes foods upward through stems (also laterally). Xylem tracheids are the chief conducting tissues in gymnosperms such as pine trees. Xylem vessels are the principal conducting structures in angiosperms such as oak trees.

Phloem. Phloem (Gr. *phloios,* bark) always consists of two types of cells: sieve tubes and phloem parenchyma. The sieve tubes, like xylem vessels, are elongated rows of cylindroid cells whose end walls contain sieve plates containing numerous pores like those in a sieve. The protoplasm remains alive. Phloem parenchyma have the same characteristics described under parenchyma. Other tissues in phloem include fibers and ray cells. In flowering plants there are elongated, living, nucleated cells, known as companion cells, which border sieve tubes and possibly assist in conduction or in food storage.

Phloem tissues give strength and mainly conduct food downward from leaves through stems and roots. Phloem parenchyma stores food.

Review questions and topics

1 Give in detail the distinguishing characteristics and an example of each of the different types of plant tissues.
2 Give the functions of each type of plant tissue.
3 Contrast permanent and temporary plant tissues, with examples of each.
4 Contrast simple and complex plant tissues, with examples of each.
5 Describe in detail each phase in the process of mitosis, explaining the significant events in each.
6 List some factors that might influence the rate of mitosis.
7 Distinguish between mitosis (karyokinesis) and cell division (cytokinesis).
8 What is the origin and significance of the paired condition in chromosomes?
9 Do higher animals necessarily have the greater number of chromosomes?
10 Explain the significance of synapsis and doubling of chromosomes. What significant hereditary phenomenon may occur during synapsis?
11 Contrast haploid (monoploid) and diploid numbers of chromosomes, with examples of each.

Selected references

Brown, W. V., and Bertke, E. M.: Textbook of cytology, ed. 2, St. Louis, 1974, The C. V. Mosby Co.

Esau, K.: Plant anatomy, New York, 1953, John Wiley & Sons, Inc.

Mazia, D.: The cell cycle, Sci. Amer. **230**(1):55-64, 1974.

Rhoades, M. M.: Meiosis. In Brachet, J., and Mirsky, A. E., editors: The cell, vol. 3, New York, 1961, Academic Press Inc.

Sager, R., and Ryan, F. J.: Cell heredity, New York, 1961, John Wiley & Sons, Inc.

Tedeschi, H.: Cell physiology, New York, 1974, Academic Press, Inc.

Windle, W. F.: Textbook of histology, New York, 1969, McGraw-Hill Book Company.

Zeuthen, E., editor: Synchrony in cell division and growth, New York, 1964, Interscience Publishers, Inc.

Some contributors to the knowledge of
TISSUES, ORGANS, AND CELL DIVISION

Marcello Malpighi (1628-1694)

An Italian scientist and physician who studied tissues and organs microscopically. He is considered the founder of microanatomy. He related anatomy and physiology to medicine, and his studies included the detailed structure of lungs, kidneys, spleen, and other organs, and the capillary circulation in frogs (1660).

Antonj van Leeuwenhoek (1632-1723)

A Dutch microscopist who first published an account of sperms.

Jan Swammerdam (1637-1680)

A Dutch microscopic anatomist who studied the anatomy and life histories of insects and also injected blood vessels.

Nehemiah Grew (1641-1712)

An English scientist and physician who microscopically studied cells, tissues, and organs of plants. His book, *Anatomy of Vegetables* (1672), started the work in plant histology.

Marie F. X. Bichat (1771-1802)

A French surgeon and anatomist who recognized that the organs of animals were composed of masses of substance to which the term tissue was applied. He studied without a microscope, and his work is remarkable considering that he died young. He is the "Founder of Histology."

Hugo von Mohl (1805-1872)

He observed cell division in plant root tips and buds (1835), recording the presence of a cell plate between daughter cells.

Rudolph Albert von Kölliker (1817-1905)

A German biologist who demonstrated that sperms and eggs are cellular products of organisms (1845) from which a new organism is derived by cell division. He placed histology on a cellular basis and applied it to embryology.

Rudolph L. K. Virchow (1821-1902)

A German cytolotist who established the law that "all cells arise from pre-existing cells" (*omnia cellula e cellula*). He also stated that pathology was a cellular science (1858), which gave a new emphasis to the explanation of diseases in animals. (*The Bettman Archive, Inc.*)

Virchow

Walther Flemming (1843-1905)

He described cell division in living and fixed cells of salamanders and developed improved methods of fixing and staining. He discovered the longitudinal reduplication of chromosomes (1879) and coined the term "mitosis" (1882).

Eduard A. Strausburger (1844-1912)

He coined such terms as prophase and metaphase in connection with cell division (1884). He stated that the essential factor in both plant and animal fertilization was the fusion of the gametic nuclei of paternal and maternal origin.

Eduard van Beneden (1845-1910)

A Belgian cytologist, demonstrated the constance of chromosome numbers for the species, the reduction of their number during maturation of germ cells, and the restoration of the double number at fertilization.

August Weismann (1834-1914)

A German biologist who postulated that a reduction in chromosome number occurred in the germ cells of plants and animals in such a manner as to separate the diploid chromosomes into two haploid groups without a longitudinal division of each chromosome taking place. His prediction has been verified, and reduction division (meiosis) is found to be coexistent with bisexuality. (*Historical Pictures Service, Chicago.*)

Weismann

M. M. Rhoades (1903-   )

He has contributed much information concerning chromosomes, heredity, meiosis, and crossing-over in numerous publications since 1931.

C. D. Darlington (1903-   )

He demonstrated that synapsis may begin at any of several places along the length of a chromosome in the lily (*Fritillaria*).

C. P. Swanson (1911-   )

He has contributed much information concerning meiosis, chromosomes, heredity, and cellular structure in numerous publications since 1940.

5 Patterns of reproduction

Asexual reproduction
Sexual reproduction
Meiotic cell division.
   Egg and sperm formation
   Meiosis in plants

Organisms produce more of their own kind by a variety of methods. However, when these are examined closely, they fall into just two categories. One of these is based on simple cell division—mitosis—and results in two cells that may continue to divide or to specialize into particular parts of a multicellular organism. In either case this is called *asexual* reproduction. The second type is based on the union of two cells, usually called egg and sperm, and accordingly is called *sexual* reproduction. This process involves a second type of cell division called *meiosis,* which will be considered in detail later.

In the life cycle of many organisms both asexual and sexual types of reproduction may occur. In some organisms one reproductive type is predominant, and the other takes place in response to some seasonal or adverse stimulus.

ASEXUAL REPRODUCTION

When single-celled organisms reproduce, they usually split in half. Each half grows to the usual size of that particular species and eventually reproduces. This is about the most simple type of reproduction.

The resulting halves are now new individuals functioning as entire organisms. This is the result of the presence of hereditary material in the genes and chromosomes. Nuclear division of a single-celled organism is basically mitosis. The total process is often given the descriptive term fission, or splitting. If the plane of division is along a long axis, that is, lengthwise, it is called longitudinal fission. Examples of this are found in such organisms as *Paramecium* and *Euglena* (Fig. 5-1).

Division in a central plane is also common in organisms that are more spherical, such as bacteria. Bacteria such as *Streptococcus,* which grows in a stringlike form, divide in the same direction so that the chain grows in length. Organisms such as the green algae *Spirogyra* grow in filament form. Division of any of the cells results in a lengthening of the filament.

In some cases several nuclear divisions take place before the entire cell divides. This results in a number of very small cells packed inside the original cell membrane. When this bursts, many new organisms are released. This multiple fission is found in some protozoans and algae. In yeasts the mitotic division results in an unequal cleavage

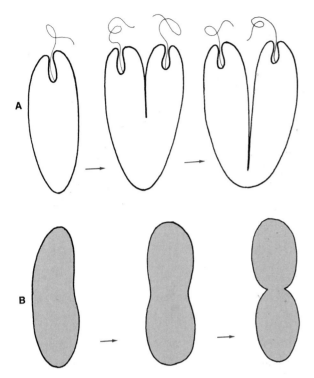

Fig. 5-1. Asexual reproduction by fission. **A,** Longitudinal fission as in *Euglena.* **B,** Transverse fission as in *Paramecium.*

of the cytoplasm so that there is one large cell and a smaller cell attached to it. Often the larger cell may divide again before all of the cells are separated. This is referred to as budding.

The same term, budding, is also used in organisms like *Hydra,* a small coelenterate that resembles a piece of frayed thread. Special cells along the side of the *Hydra* begin to divide, producing more cells. These also divide until a small mound of cells is seen on the side. These cells continue to divide and some begin to specialize. Eventually a complete small *Hydra* will form. When the point of attachment pinches off, there are two complete multicellular organisms.

Many organisms, including molds such as *Neurospora* and *Penicillium* produce special reproductive cells, called spores, at the end of growing tips. As each cell specializes into a spore, the cell

behind it is also becoming a spore. When the spores are released, they settle, germinate, and produce a new strand of the mold. Because of their small size, they are easily dispersed by wind and water currents. Spore formation is also found in many other life cycles, such as those of mushrooms, mosses, ferns, and others.

When a major tissue, usually not associated with reproduction, buds or grows to produce a new plant, the process is called *vegetative reproduction.* This may involve roots, stems, or leaves of various plants. A very common means of increasing the numbers of a plant such as *Coleus* or *Geranium,* is to cut off parts of the stem and then plant them. In a short time these stems will produce roots and then complete plants. There are many variations of this. A natural example of this is found in the strawberry plant. When a strawberry plant grows, it forms elongate stems that touch the ground and produce roots and a new plant. These horizontal stems are called runners or *stolons.* A few plants will produce a strawberry patch in a short time (Fig. 5-2).

A stem that grows underground, producing leaves and flowers that grow upward, is called a *rhizome.* This type of vegetative structure is found in ferns, dandelions, and iris plants. Tubers, such as those found in the potato, are enlarged underground stems. The "eyes" of the potato are really buds, and when these are planted, they produce a new plant. Bulbs, such as found in tulips, lily, and onion, are really clusters of fleshy leaves. When planted, they produce an entire new plant with leaves, stems, and flowers. Other plants, such as *Bryophyllium,* produce new plants at the margins of leaves. When the leaf roots on the ground, many new plants grow from these margins.

A somewhat similar reproductive process takes place in animals. Some of them, such as the *Hydra* mentioned previously, may be cut in half and each half will regenerate the missing portion. Experiments with *Planaria* show that when the worm is cut into several pieces, the missing head

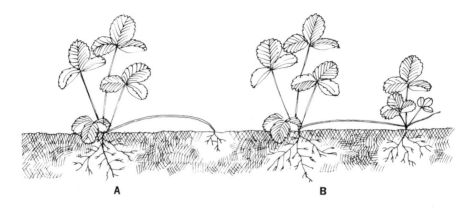

Fig. 5-2. Vegetative reproduction in strawberry plant. **A,** Stolon or runner is formed, touches ground, then takes root. **B,** New plant with new stolon.

or tail will be formed. If the head or tail region is cut lengthwise and not severed, two or more heads or tails may result.

In much the same manner, a starfish may replace a missing arm or the arm may replace a missing starfish! (Fig. 5-3.) Some salamanders are quite adept at replacing a missing tail, some of them even breaking it off to confuse a predator; the tail wiggles, and the salamander runs in another direction. All of the examples indicated here are essentially the basic process of mitosis followed by cell growth and specialization.

SEXUAL REPRODUCTION

Many species, including some of those discussed previously, go through another reproductive process involving cells from two parents. This is sexual reproduction and the cells are usually called eggs and sperms. The function of sexual reproduction is to produce new varieties of the organism by a reassortment of the chromosomes and genes. The number of varieties influences the survival potential of the particular species of organism.

When egg and sperm fuse in the process of fertilization, the nuclei also fuse. The chromosomes of the sperm nucleus are now together with the chromosomes of the egg nucleus. Since we know that the number of chromosomes in a species does not double in each succeeding generation, some means of controlling this amount must exist. The process of reducing the chromosome numbers before fertilization is called meiosis. Details of this will be considered at the end of this chapter.

Sexual reproduction may involve union of sex

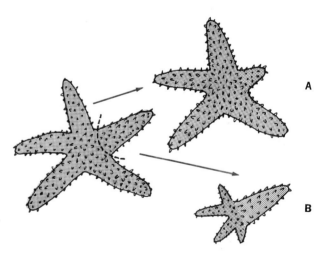

Fig. 5-3. Regeneration in starfish. **A,** When arm is removed, body grows a replacement. **B,** Missing arm may grow a new body.

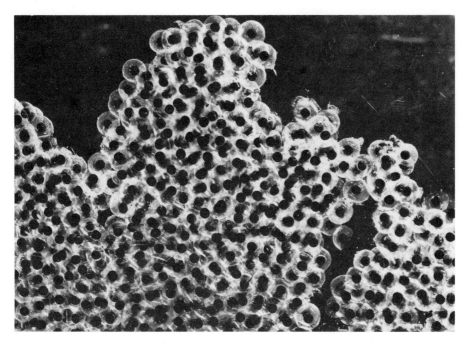

Fig. 5-4. Mass of frog eggs from a pond.
(Courtesy Carolina Biological Supply Company, Burlington, North Carolina.)

cells or gametes of about the same size, or there may be distinct female cells called eggs and male cells called sperms, or more properly spermatazoa. Gametes of equal size or no apparent distinction are called isogametes. In many algae and protozoans either the entire cells may fuse or numbers of gametes of the same size are released. These then fuse and the new organism develops. In some fungi the tips of two filaments may fuse in a sexual type of process.

Two bacteria may form a tube between them and transfer or exchange DNA. After several nuclear rearrangements two paramecia may fuse and exchange nuclear material. Both of these are examples of a process called *conjugation*. The same term is used with the filamentous green algae *Spirogyra*. The process is somewhat different in that the entire cell contents in one filament are transferred to a cell of an adjoining filament.

When cells of unequal size are involved, they are called heterogametes. These are the egg and sperm cells. The resulting fertilized egg is referred to as a *zygote*. Growth of the zygote is by mitosis—simple cell division followed by growth and specialization. Some species develop directly into typical individuals, whereas others produce embryos or larval stages.

Aquatic organisms usually produce a large number of eggs and sperms, which are shed into the water (Fig. 5-4). The resulting fertilization then may be strictly haphazard although some timing mechanism may ensure that egg and sperm have some chance of uniting.

Some fish have distinct behavior patterns that provide that male and female both release their gametes at the same time, thus ensuring fertilization. In crayfish the male inserts sperms into a special receptacle in the female. She then controls the release of both eggs and sperms. The eggs are fertilized as they are deposited.

Land animals usually have some type of internal fertilization where eggs remain inside of the female and sperms are introduced by copulation. Even here there are many variations. After internal fertilization some insects, reptiles, and birds shed the eggs to the outside. Some animals, such as birds and turtles, may have a specialized nest. In some amphibians, fish, and reptiles, the fertilized egg is retained inside the female, and after hatching inside, the young animals are released. In still other cases, such as humans and other mammals, there is a distinct internal protective relationship inside the mother by means of placenta formation.

Other variations of both asexual and sexual reproduction will be encountered in the section dealing with specific plants and animals.

MEIOTIC CELL DIVISION

A second type of cell division occurs in organisms that reproduce by the fusion of egg and sperm. This process is called meiosis (Gr. *meioun,* to reduce) and is accompanied by a reduction in the number of chromosomes in the daughter cells.

The time at which meiosis occurs in the life cycle varies with different organisms but is constant for each particular species. Three times when meiosis can occur are considered: (1) during the formation of spores, as is common in plants; (2) during the formation of gametes (eggs or sperms), which is common in animals (and in a few lower plants); or (3) immediately after the fertilization of an egg by a sperm (at the beginning of the cleavage of the egg), which occurs in certain lower plants.

Regardless of the particulars, sexually reproducing species have a process that ensures that each generation has the same number of chromosomes and, therefore, the same genetic information. In typical cells of higher plants and animals, two of each kind of chromosome are usually present. These are called *homologous chromosomes,* and when they are present the cell is in the *diploid* condition. The process of meiosis reduces the total number of chromosomes so that one member of each homologous pair is present in the resulting egg or sperm cell.

A cell whose nucleus contains the number of pairs of chromosomes typical of the species is referred to as a 2N or *diploid* cell. After the reduction division, the resulting cells are called N or *haploid* cells. Diploid and 2N refer to cells with pairs of chromosomes; haploid and N refer to cells with single members of a pair of chromosomes.

Except for the kind of cell resulting from the process of meiosis, that is, egg or sperm, the phenomenon is similar in both sexes, following the same phases or stages in each. The sequence of events is similar to, but not quite the same as, mitosis.

Reduction in chromosome number takes place by two nuclear divisions preceded by only one replication of chromosomes. This results in the formation of four cells whose nuclei contain the haploid number of chromosomes. The phases (prophase, metaphase, anaphase, and telophase) occur twice in meiotic division (Fig. 5-5).

INTERPHASE

The cell that will eventually divide twice to form reproductive cells was formed as a result of a prior division by mitosis. This cell, then, increased in size, synthesized the usual cellular components, and in short is a normal, metabolic cell. As we noted earlier, a cell that undergoes mitosis first doubles the DNA content and the chromosome materials. The same is true in the first division by meiosis.

PROPHASE I

In meiosis, the same pattern of spindle formation, nuclear disappearance, and chromosome appearance takes place. Each chromosome can be seen to consist of two chromatids joined at a centromere. Next, in a process quite different from mitosis, the members of homologous pairs of chromosomes come together. There is actual contact between the centromeres and the joined chromatids. This active joining is called synapsis

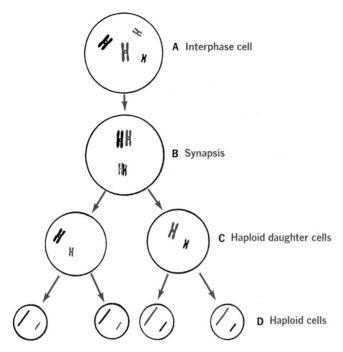

Fig. 5-5. Meiosis. **A,** Diploid cells (2N) with two pairs of homologous chromosomes. **B,** Diploid cell (2N) with tetrad formation during synapsis. **C,** Haploid (N) daughter cells with one double-stranded member of each homologous pair of chromosomes. **D,** Haploid (N) cells, from second division, each with one single-stranded member of each pair of homologous chromosomes.

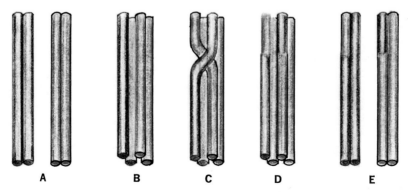

Fig. 5-6. Synapsis and crossover (exchange of parts) during meiosis. **A,** Homologous pairs of chromosomes; **B,** synapsis (coming together); **C,** crossover; **D,** exchange of chromatid parts; **E,** separation.

(Gr. union). Since each homologous chromosome consists of two chromatids, there is a total of four chromatids in the union. Sometimes this is called a *tetrad*. The chromatids may twist around each other, fuse at various places, and exchange parts in a process called crossover (Fig. 5-6). As in mitosis the chromosomes move toward a central plane.

METAPHASE

A second important distinction is now apparent. The entire tetrad lines up in the same area. A spindle fiber extends from the centromere of each original chromatid pair to the opposite ends of the cell. Remember that in mitosis each centromere of a chromatid pair is on a separate fiber leading to both ends of the cell. When the centromeres split, the chromatids become independent chromosomes and are moved to opposite ends of the cell. In meiosis I, however it is the tetrad that is divided, not the centromere.

ANAPHASE I

The centromeres in meiosis do not split. Since the two homologous chromosomes (four chromatids) are in contact at the centromeres and are on the same fiber, the centromeres are moved apart. This means that the original doubled chromatids are still joined by their original centromere. They have not split and are not yet independent chromosomes.

TELOPHASE I

The chromatid pairs, joined at their common centromere, are moved to opposite ends of the parent cell. An important distinction from the process of mitosis is that one entire chromosome (consisting of two chromatids) from each homologous pair is present at either end of the cell.

After telophase in both processes the total number of "chromatids" at each end of the dividing cell is the same. The difference is that in mitosis they are separate and distinct and originated from different parent chromosomes. In meiosis the chromatids exist as joined pairs,

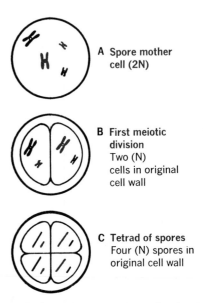

A Spore mother cell (2N)

B First meiotic division
Two (N) cells in original cell wall

C Tetrad of spores
Four (N) spores in original cell wall

Fig. 5-7. Meiosis during spore formation in plants, for example, moss, fern, pollen, and so on.

and they are the entire original chromosome.

In mitosis the chromatids separate; in meiosis the homologous chromosomes separate. Mitosis is complete after this first division. A second division may take place, but the result is simply more cells of the same type.

MEIOSIS II

A second meiotic division *must* take place before the process is complete. It is significant that there is no further doubling of DNA or of total chromosome material during this meiotic interphase.

The joined chromatid pairs move to the central plane. The centromeres split and the newly independent chromosomes move to opposite poles—complete cell division takes place. The result of the two meiotic divisions is that one diploid cell with homologous chromosomes (2N) has divided twice to produce four haploid (N) cells, each with one member of each original homologous pair. Spore formation in plants is a

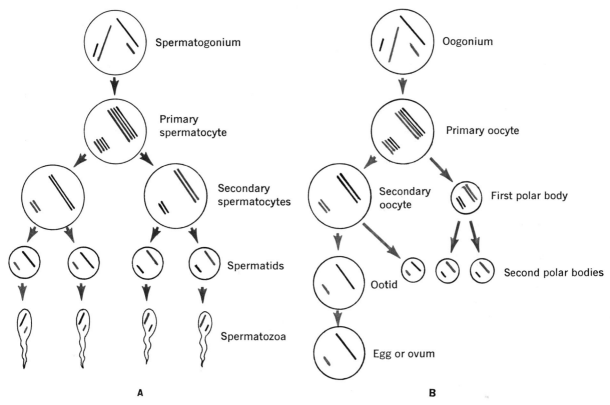

Fig. 5-8. Gametogenesis. **A,** Spermatogenesis. **B,** Oogenesis. Note that four sperms—but only one functional egg—are formed from each parent cell.

good example of this (Fig. 5-7) since the four new haploid cells are retained in the original cell wall until dispersed.

Egg and sperm formation

The reduction of chromosome numbers during meiosis in many animals, including humans, takes place in the male testes and the female ovaries. The sperm-producing process is called spermatogenesis, and the egg-producing process is oogenesis (pronounced OH-OH-genesis).

Spermatogenesis

In males, a cell that will become a sperm cell is called a *spermatogonium*. Each one of these cells goes through the two divisions of meiosis to pro-

duce four sperm cells (Fig. 5-8, *A*). In what corresponds to the beginning of the meiotic division, the chromosome content of a spermatogonium is doubled. The cell is now called a *primary spermatocyte*. Each primary spermatocyte goes through synapsis and resulting tetrad formation. The members of homologous chromosome pairs separate, and two cells are formed. These *secondary spermatocytes* contain one member of each pair of homologous chromosomes, and each chromosome contains two chromatids joined at a centromere.

Each of the secondary spermatocytes goes through the second meiotic division. As in other types of meiosis, this second division is *not* preceded by a doubling of DNA or chromosomes.

During anaphase of this division, the centromeres split, the chromatids now are independent chromosomes, and the entire cell divides. The newly formed cells are called *spermatids*. Each spermatid quartet results from one original pri-

mary spermatocyte. They now mature, change form somewhat, and become tailed spermatozoa or sperm cells.

Each spermatogonium has one pair of each kind of chromosome. After the chromosome con-

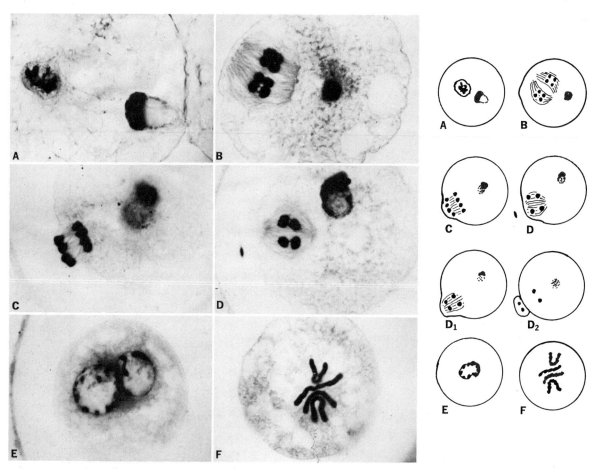

Fig. 5-9. Oogenesis and fertilization in *Ascaris*. **A,** Unfertilized egg with a central nucleus and the entrance of a sperm; **B,** metaphase of first meiosis showing the diploid number of chromosomes doubled by longitudinal splitting; **C,** telophase of first division showing half the number of chromosomes being extruded into the first polar body; **D,** second division during metaphase, showing formation of the second polar body; **D**$_1$ and **D**$_2$, diagrams showing completion of formation of second polar body; **E,** fertilization, sperm and egg nuclei fuse; **F,** mitosis during cleavage of the zygote.

(Courtesy Carolina Biological Supply Company, Burlington, North Carolina.)

tent doubles, each cell is now a primary spermatocyte. Each primary spermatocyte has one pair of each kind of *doubled* chromosomes. After the first division of meiosis, each resulting secondary spermatocyte contains only doubled chromosomes, that is, two chromatids joined by a common centromere but not pairs of doubled chromosomes. When the secondary spermatocytes divide, the centromeres also divide and each resulting spermatid contains independent chromosomes. The number of these is one half of the original number in the spermatogonium before chromosome doubling took place. The original diploid (2N) spermatogonium has produced four haploid (N) spermatids.

Oogenesis

Egg production is similar to sperm production in that an original diploid 2N cell produces a haploid (N) egg. It is different in that only one functional egg results from the process of oogenesis (Fig. 5-8, *B*). Three nonfunctional polar bodies are also formed. The original cell in the ovary is called the oogonium. After chromosome doubling and the beginning of meiosis, it is called a primary oocyte. During meiosis there is tetrad pairing and synapsis. There is also cell division resulting in two cells that have joined chromatids.

These two cells are different from what might be expected since one is quite large and the other quite small. This disparity in size results from a very unequal division of the cytoplasm during cytokinesis. In effect, a small portion of the original primary oocyte is pinched off with one set of chromosomes. The larger portion contains most of the cytoplasm and the other set of chromosomes. This larger portion is the secondary oocyte. The other cell is called the first *polar body*.

The secondary oocyte now goes through a second meiotic division. Here, again, the chromosomes do not double prior to cell division. The centromeres, however, do split during anaphase. The actual cell division is again unequal, resulting

in one large cell with most of the cytoplasm and one set of chromosomes and one smaller cell with the other set of chromosomes. This larger cell is the egg or ovum. The smaller cell is a second polar body. If the first polar body divides, then three polar bodies, in varying states of disintegration, may be found on the surface of the egg.

In chromosome terms, a diploid 2N oogonium has doubled its chromosomes, divided twice, and produced four haploid (N) cells. One of these is a functional egg; the other three are the polar bodies.

The preceding account applies equally well to human gametogenesis, and essentially the same thing happens in other animals and in many plants. In many cases the female meiosis stops before formation of the polar bodies. The penetration of the egg membrane by a sperm stimulates the formation of polar bodies. After this the remaining female nuclear material fuses with the sperm nucleus (Fig. 5-9). Oogenesis in the roundworm *Ascaris* is often used to demonstrate this phenomenon.

Meiosis in plants

The time at which meiosis occurs in the life cycle of plants varies in different groups of plants. In flowering, angiospermous plants it immediately precedes the formation of microspores (pollen grains) and megaspores (embryo sacs). Such spores are produced in fours, by two divisions of a spore mother cell.

Life cycles in plants

Life cycles are characteristic of living organisms. An organism begins its development and, after passing through stages of growth and development, eventually reaches the same state again in the next generation. Organisms that reproduce sexually go through a cycle, such as egg, embryo, adult, and egg. In most plants that reproduce sexually this cycle is complicated by having two parts, a gamete-producing *gametophyte* generation and a spore-producing *sporo-*

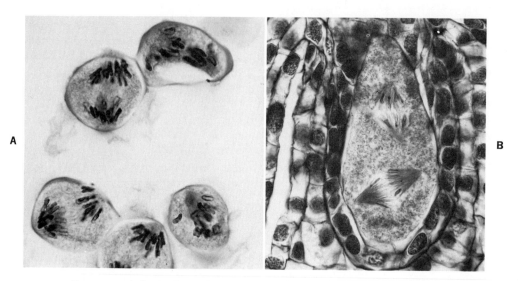

Fig. 5-10. Sporogenesis in the lily *(Lilium)*. **A,** Several microspore mother cells dividing. **B,** Megaspore formation in the embryo sac.

phyte generation. The alternation between these two generations corresponds with the change in the number of chromosomes in the two stages of the life cycle, as described earlier.

Alternation of generations is found in certain algae, liverworts, true mosses, horsetails, club "mosses," ferns, gymnosperms, and angiosperms (flowering plants). In many colonial green algae the plant body is a haploid (N) gametophyte that gives rise to haploid gametes that fuse to form the sporophyte generation. In certain brown algae the diploid generation is rather conspicuous and the haploid one is small. In most red algae and some brown algae the two generation are much alike and independent.

Gametogenesis and sporogenesis in plants

Many plants reproduce by specialized sex cells called gametes. Fusion of pairs of gametes is called fertilization, and the resulting cell is a zygote. When gametes are similar in size and structure, the process is called isogamy; if gametes are unlike, it is called heterogamy. Sometimes the smaller gamete is called a sperm and the larger, an egg. The formation of gametes is called ga-

metogenesis. In many plants minute specialized cells called spores are produced for reproduction by a process called sporogenesis.

In the life cycle of most plants there are two phases, each of which produces the other, a process called alternation of generations. One generation, the gametophyte, produces gametes. The fertilized egg does not develop into another gamete-producing individual (as it does in most animals) but into another generation, known as the sporophyte. The sporophyte produces spores, each of which develops eventually into a new gametophyte.

In seed-producing flowering plants, as in a few of the more advanced spore-bearing lower plants, two kinds of spores are produced: the microspores, which produce male gametophytes, and the megaspores, which produce female gametophytes (Fig. 5-10). In higher plants the megaspore is retained within the spore case, closely attached to the mother plant, where it germinates into a small female gametophyte. This whole structure, with the protective integument (covering), is known as an ovule (immature seed). Usually only a few ovules are formed, in contrast to the great

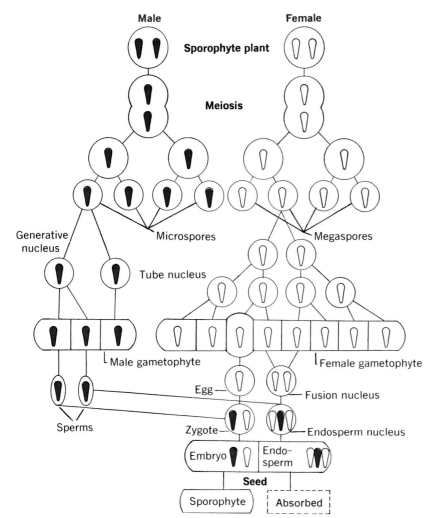

Fig. 5-11. Life cycle of a flowering seed plant, showing sporogenesis, including meiosis, and gametogenesis. The double number of chromosomes (2N) of the sporophyte is reduced by meiosis to a single number (N) when spores are produced. The microspore nucleus divides by mitosis to form a generative nucleus and a tube nucleus. In the pollen tube the generative nucleus divides by mitosis to form two nuclei that become sperms (male gametes). These two nuclei and tube nucleus constitute the microscopic, male gametophyte. The megaspore nucleus gives rise, by a series of mitotic divisions, to eight nuclei that constitute the female gametophyte. One of these eight nuclei becomes the nucleus of the egg (female gamete); two others fuse to form the fusion nucleus. The union of the egg and one sperm forms the zygote; the other sperm and the fusion nucleus form the endosperm nucleus. Note the phenomenon of double fertilization. (See next chapter for further details.)

numbers of spores produced by lower plants (Fig. 5-11).

Microspores (pollen grains) are liberated into the air, but instead of falling to the ground and germinating there, they are carried to the ovule (or near it), where each microspore produces two sperms (male gametes). One sperm fertilizes the egg to form an embryo, which obtains its nourishment from the mother plant, thus acting like a parasite. Around the embryo is placed a supply of stored food in the endosperm, which results from the union of another sperm and the female fusion nucleus. The embryo grows and forms an embryonic root and one or more embryonic leaves. The embryo and endosperm are enclosed by the protective integument of the ovule, thus forming a seed. See Chapter 6 for details.

Review questions

1 As a biologic process, reproduction is usually considered under two major categories. What are they, and what is the basis for the division?
2 How do single-celled organisms reproduce? List several types and give examples of each.
3 How is reproduction of single-celled organisms different from, or similar to, the division of single cells in a multicellular organism?
4 What is meant by "budding"? Compare this process in simple and in more complex organisms.
5 Compare bud formation with spore formation.
6 Discuss vegetative reproduction. Can this process also apply to animals?
7 In terms of chromosomes, how are eggs and sperms similar to and different from buds and spores?
8 Discuss the varied mechanisms used by aquatic organisms to ensure fertilization.
9 Discuss the types of egg development in land organisms.
10 What is meiosis? List the stages in this process.
11 In terms of chromosomes, how do the egg and sperm compare with the original cell that initiated their formation?
12 What is synapsis? What is tetrad?
13 How does spermatogenesis compare with oogenesis?
14 Define the following: haploid, diploid, 2N.
15 Define the following: oogonium, oocyte, ovum, polar body, spermatogonium, spermatocyte, spermatid, and spermatozoan.

Selected references

Asdell, S. A.: Patterns of mammalian reproduction, ed. 2, Ithaca, N. Y., 1964, Cornell University Press.

Barnes, R. D.: Invertebrate zoology, ed. 2, Philadelphia, 1968, W. B. Saunders Co.

Berrill, N. J.: Sex and the nature of things, New York, 1953, Apollo Editions Inc.

Bloom, W., and Fawcett, D. W.: A textbook of histology, ed. 8, Philadelphia, 1962, W. B. Saunders Co.

Bold, H. C.: The plant kingdom, ed. 3, Englewood Cliffs, N. J., 1970, Prentice-Hall Inc.

Grenlach, V. A.: Plant function and structure, New York, 1973, The Macmillan Co.

Michelmore, S.: Sexual reproduction, Garden City, N. Y., 1964, Natural History Press.

Rhoades, M. M.: Meiosis. In Brachet, J., and Mirsky, A. E., editors: The cell, vol. 3, New York, 1961, Academic Press, Inc.

Salisbury, F. B., and Ross, C.: Plant physiology, Belmont, Cal., 1969, Wadsworth Publishing Co.

6 Growth and development

Changes in form during the growth and development of an organism result from several distinct processes. The total number of cells may increase by simple mitosis. The size of individual cells may increase while the number remains the same or also increases, and finally, the position of one cell with reference to another may change. The same number of cells may be present as a flat sheet, a solid mass, a hollow ball, or a mound. All types of increase in size and shape occur during the development of an organism.

PLANT DEVELOPMENT

If we carefully look at an apple, some interesting observations may be made. At one end is a stem and at the other is a ring of dried up hairlike structures. We usually do not spend much time remembering that this fruit was once a flower attached to a branch by its stem. If we cut

the apple, the seeds can be observed. In the tradition of Johnny Appleseed we have some confidence that if the seeds are planted they will produce a tree.

This, in general, is what plant development is all about. It involves mature plants that produce flowers, seeds, and fruit. The seeds develop to produce a new flowering plant. Some general outline of this has already been considered. Now we will study the details.

A flower is said to be complete when it consists of four floral organs: sepals, petals, stamens, and pistil (Fig. 6-1). Sepals are the outermost floral organs. They are usually green and may give some protection, especially when young. Petals are the floral organs just inside the sepals and are usually colored. All the sepals together constitute the calyx (Gr. *kalyx,* cup), and all the petals together constitute the corolla (L. *corolla,* crown). Sepals and petals together constitute the perianth (Gr. *peri,* around; *anthos,* flower). A stamen is composed of an enlarged, pollen-producing anther at the tip of a stalklike filament. A pistil is composed of a pollen-receiving stigma and a style leading to the enlarged ovary at the base.

Carpels (Gr. *karpos,* fruit) are the units of which a pistil is composed. They bear and enclose the ovules (immature seeds). An ovule de-

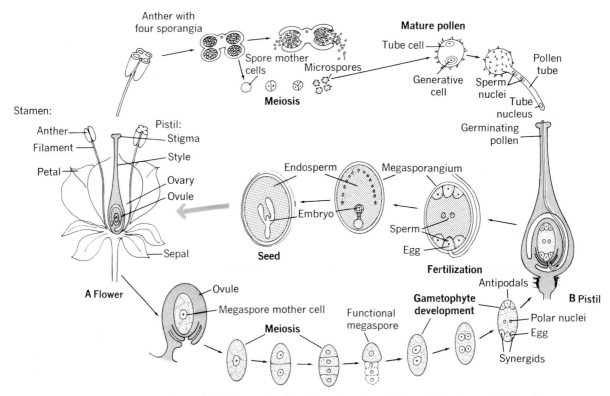

Fig. 6-1. Reproduction and seed formation in a flowering plant. **A,** Detail of a flower. **B,** Detail of the pistil at time of fertilization.

velops into a seed after fertilization. A mega-sporangium is the central part of an ovule. It encloses the megagametophyte and is covered by the integument.

If the flower consists of both male and female parts, it functions as follows. Within the micro-sporangium each microspore mother cell pro-duces four haploid microspores or pollen grains. Before the pollen is shed, the microspore nucleus divides to form a tube nucleus and a generative nucleus. If the pollen alights on the stigma of the proper plant, a pollen tube is formed that extends through the style and ovary to the megagame-tophyte in the ovule. By the time the generative nucleus passes through the pollen tube, it has divided into two microgametes or sperms. The microgametophyte is very small and is reduced to

three nuclei (one tube nucleus and two sperms).

Within the megasporangium each diploid megaspore mother cell forms four megaspores by meiosis. However, three of the four megaspores degenerate. The nucleus of the remaining megaspore divides to form two, one moving to each end of the young megagametophyte. Each of these nuclei divides into two, and these two into four, so there are two groups of four nuclei each (Fig. 6-2). One from each group moves to the center of the megagametophyte where the two polar nuclei fuse. Thus at the micropyle end of the megagametophyte there are typically three cells, one egg and two nonfunctional synergid cells (Gr. *synergos,* working together). At the opposite end of the megagametophyte are three nonfunctional antipodal cells. The endosperm

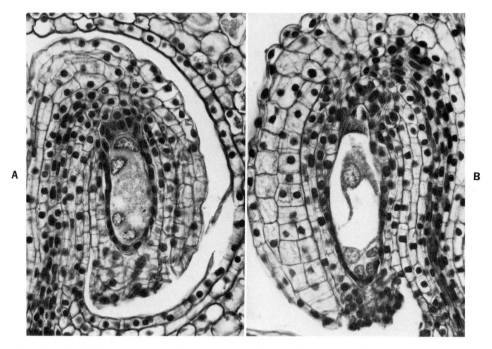

Fig. 6-2. Lily *(Lilium)* ovule. **A,** Embryo sac at four-nucleate stage. **B,** Mature embryo sac at eight-nucleate stage.
(Courtesy General Biological Supply House, Inc., Chicago, Illinois.)

nucleus, resulting from previous fusion of two polar nuclei, is in the center.

The pollen tube grows toward the ovule, being nourished by the tissues of the style through which it passes. The pollen tube usually enters the ovule through a small opening called the micropyle. The two sperms that come down the pollen tube are discharged into the megagametophyte (Fig. 6-3). One sperm fertilizes the egg to form a zygote, which will develop into the embryo, and the other sperm fertilizes the endosperm nucleus (two polar nuclei) to form the unique triploid number of chromosomes (3N). This triploid nucleus gives rise to the nutritive endosperm tissue, by mitosis.

Hence, double fertilization occurs. One sperm (N) fuses with one egg (N) to form a zygote (2N) and eventually the embryo. The other sperm (N) fuses with the endosperm nucleus (2N) to form the endosperm tissue (3N).

The embryo and the seed

After endosperm formation, the zygote divides several times, forming a narrow stalk. The topmost cell will continue to divide and form the embryo, while the rest of the stalk becomes an anchoring and absorptive device called a suspensor.

As the cells of the embryo continue to divide, several regions become apparent. The cells closest to the suspensor will form the future root, and the opposite end forms the shoot, that is, stem and leaves. As further development takes place, fingerlike growths appear. These growths form the *cotyledons,* which will serve as the food source for the germinating seed.

At this stage of development a seed is complete. It now consists of an embryo and cotyledons, surrounded by protective seed coats. The embryo itself is a somewhat rod-shaped column whose lower end, the future root, is called an

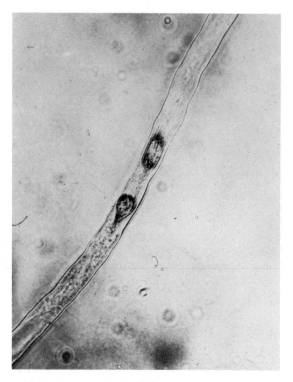

Fig. 6-3. Germinating pollen tube of *Amaryllis*. Note the migrating sperms' nuclei.
(Courtesy R. A. Joly, University of Dayton.)

hypocotyl. The upper end, the future stem, is called the epicotyl and bears the plumules (primitive leaves). Growing points called *meristems* are present at each end.

Germination and seedling formation
Seed germination

A seed usually undergoes a period of dormancy before it may germinate. Usually it dries out and must encounter some physical change before dormancy is broken. Eventually it will absorb water, swell, and begin to show signs of growth. This results from enzymatic hydrolysis providing energy for cell metabolism and from repeated division.

Seedling formation

In a bean seed, for example, after the seed absorbs water, the coat cracks, the cotyledons become soft and somewhat wrinkled, and the hypocotyl becomes elongated by repeated cell division. The hypocotyl is the first part of the plant to emerge from the seed (Fig. 6-4). Its lower part, called the radicle, now grows downward into the soil as the primary root. This downward growth is controlled by plant hormones and takes place regardless of the orientation of the seed. (If this did not happen, farmers would have to make sure that each seed was planted with the tip down). As the new primary root elongates, root hairs develop followed by secondary roots. The new plant now has an anchoring device and an absorptive surface. Further growth in the other plant parts now becomes more apparent.

The upper part of the hypocotyl enlarges, pulling the rest of the shoot out of the seed coats and out of the ground. In some seeds the cotyledon may remain underground and the epicotyl grows to emerge from the ground. At any rate a tiny plant or seedling is now apparent.

The cotyledons become wrinkled and reduced in size as they are used as food for the developing plant. The area above the cotyledon, the epicotyl, has a growing point at its tip. This contains the primary leaves and an apical meristem. Now that the leaves are exposed to light, they turn green as chlorophyll is produced. Photosynthesis takes place, and food is produced for further growth and development of the plant. At this stage a functioning plant consisting of roots, stem, and leaves is present. Further growth takes place in special areas of the root and stem called meristems. These areas retain the property of continued cell division and do not specialize into mature tissues.

Plant development
Root growth and development

If we examine the young root that developed from the hypocotyl, it becomes evident that growth takes place by both cell division and elongation of cells. New cells are formed primarily in an apical meristem region at the root tip. This is

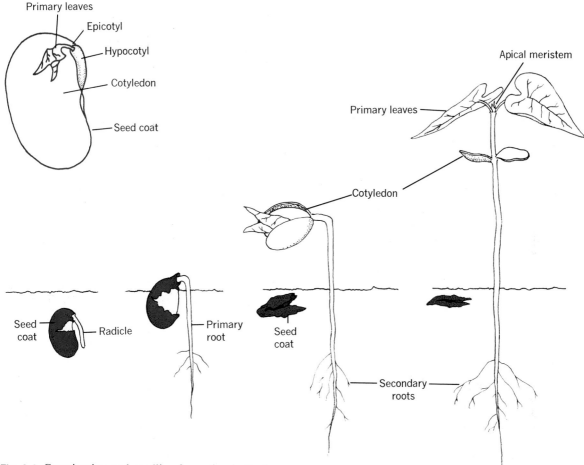

Fig. 6-4. Germination and seedling formation in the bean.

sometimes called the *zone of cell multiplication*. These cells go through repeated mitosis forming a *root cap* (Fig. 6-5), which is worn away as the root enlongates. As cells are lost here, they are replaced by further division. Some of the young cells do not divide further but begin to elongate. To understand the process of growth here, consider a single cell that divides, forming two cells. The one closest to the root tip divides again. The other cell begins to elongate. Now there are three cells: one elongating, the second beginning to elongate, and the third beginning to divide again. Elongation is influenced by plant hormones called auxins.

As a number of elongating cells are formed, a definite *zone of elongation* can be distinguished. And as these cells begin to mature and specialize, a *zone of maturation and differentiation* can be seen.

Root differentiation

If a slice of the root is taken at the zone of differentiation, several regions can be observed as distinct types of cells are formed. This slice will show three concentric areas of specialization. The innermost is called the *procambium* or *provascular* cylinder. It will develop into the conducting tissues of the *stele*. The procambium

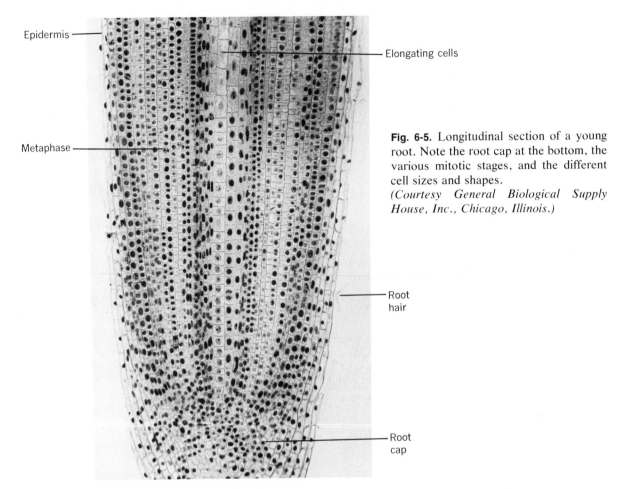

Epidermis

Elongating cells

Metaphase

Root hair

Root cap

Fig. 6-5. Longitudinal section of a young root. Note the root cap at the bottom, the various mitotic stages, and the different cell sizes and shapes.
(Courtesy General Biological Supply House, Inc., Chicago, Illinois.)

tissue is surrounded by relatively unspecialized *ground* meristem, which becomes the *cortex* and *endodermis*. This is very often a storage region in a mature plant. The outermost cylinder called the *protoderm* is very thin and will form the epidermis of the mature plant (Fig. 6-6, *A*). The younger cells of the region are usually specialized as absorptive *root hairs*.

Shoot growth and differentiation

As in the root, development in the stem takes place in apical meristems. After initial formation from the seed, the shoot consists of a stem and the primary leaves. The apical meristem is pres-

ent at the apex of the stem. The process of growth is similar to that in the root except for the absence of a cap. Thus, at an early level of development the young stem consists of zones of cell multiplication, elongation, and differentiation and maturation.

Although these differentiating cells do develop into primary tissues similar to the protoderm, ground meristem, and procambium of the root, the arrangement is somewhat different. Thus, a slice through a young region of a stem might show four concentric cylinders, an innermost *pith*, then a procambium or vascular area, arranged eventually into vascular bundles, the *cortex*, and the outermost *epidermis* (Fig. 6-6, *B*).

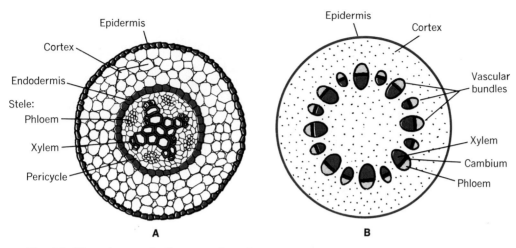

Fig. 6-6. Plant tissues. **A,** Cross section of a young root. **B,** Section of a young stem to emphasize the pattern of vascular bundles. Dotted area represents undifferentiated pith cells.

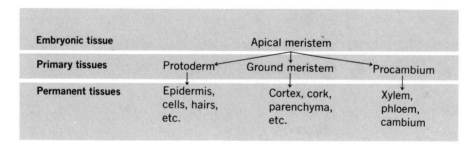

Embryonic tissue		Apical meristem	
Primary tissues	Protoderm	Ground meristem	Procambium
Permanent tissues	Epidermis, cells, hairs, etc.	Cortex, cork, parenchyma, etc.	Xylem, phloem, cambium

Fig. 6-7. Origin of plant tissues.

Summary of differentiation

Plant growth in general takes place in apical meristems, which develop to produce primary tissue called protoderm, ground meristem, and procambium (Fig. 6-7). These in turn produce the mature tissues. *Protoderm* forms the outermost layer, the epidermis. Procambium forms the vascular tissues, which contain different kinds of xylem and phloem cells, as well as cambium. Ground meristem produces the cells of the cortex and pith in primary food and water storage areas. Cambium remains a meristematic tissue and leads to the formation of secondary tissues that account for the increase in width of stems.

Further details in cells and tissues are in earlier chapters.

ANIMAL GROWTH AND DEVELOPMENT

Three distinct stages in the formation of a new animal are recognized; these are (1) cleavage, during which the fertilized egg divides repeatedly, resulting in an increase in the number of cells and a decrease in the size of cells: (2) gastrulation, which is the formation of the primitive tissues; and (3) morphogenesis, which is the organization of distinct organs and tissues characteristic of the particular animal.

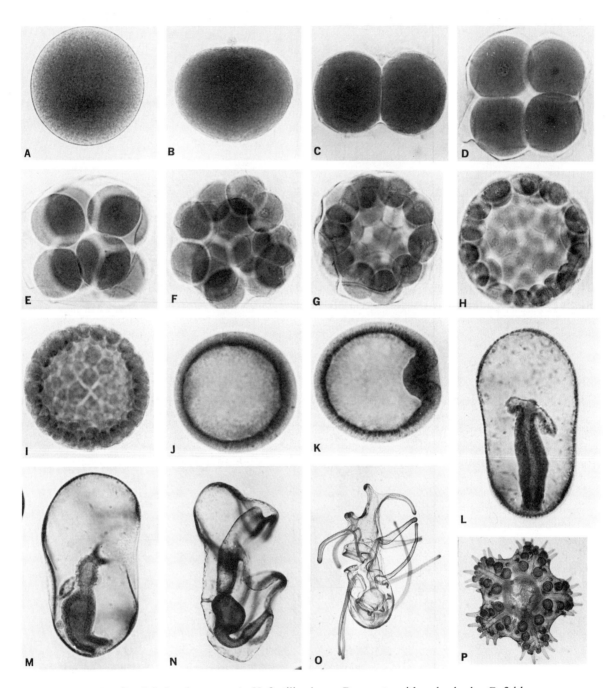

Fig. 6-8. Starfish development. **A,** Unfertilized egg; **B,** zygote with polar body; **C,** 2 blastomeres; **D,** 4 blastomeres; **E,** 8 blastomeres; **F,** 16 blastomeres; **G,** 32 blastomeres; **H,** 64 blastomeres; **I,** nonmotile blastula; **J,** motile blastula; **K,** early gastrula; **L,** late gastrula (ventral view); **M,** early bipinnaria (lateral view); **N,** late bipinnaria (lateral view); **O,** brachiolaria; **P,** young starfish.

(Courtesy Carolina Biological Supply Company, Burlington, North Carolina.)

Development of the starfish

The stages in cleavage and gastrulation in the starfish zygote are quite easy to follow because of its small size. The entire series may be studied on one slide.

In general, after fertilization, the zygote goes through a number of equal divisions resulting in 2, 4, 8, 16, 32 cells, and so on (Fig. 6-8). Although the number of cells increases, the total mass or bulk does not change. The repeated divisions result in the formation of a hollow ball of cells called the blastula (Gr. *blastos,* bud).

After this stage some of the cells divide faster than others, resulting in an invagination in one side of the blastula. This invagination represents the beginning of the gastrula stage (Gr. *gastros,* digestion).

Eventually, a distinct two-layered embryo is formed. The outer layer is the ectoderm, and the inner is the endoderm. The cells of the endoderm form a third layer, the mesoderm. These three primitive tissues form the distinct structures of the larval stages of the starfish.

Embryology of the frog

The early development of many animals is similar. The frog has been chosen to illustrate the general principles because the frog has a rather typical method of development and the materials used in the study are available and rather inexpensive.

Frog eggs (and also eggs of fish, reptiles, and birds) have a large amount of yolk concentrated toward the lower or vegetal pole, while the cytoplasm is at the upper or animal pole.

Early cell stages

In 2 or 3 hours after fertilization the zygote divides by mitosis to form the 2-cell stage. A second division occurs by mitosis in about 1 hour and at right angles to the first plane of division, thus forming the 4-cell stage. These four more or less equal cells are called blastomeres (Gr. *meros,* part).

The next plane of cleavage is horizontal and

slightly above the middle or equator, thus dividing each of the four previous cells to form the 8-cell stage (Fig. 6-9). Of these eight cells, four are pigmented, smaller, located at the animal pole, and known as micromeres (Gr. *mikros,* small); the other four are unpigmented, larger, located at the vegetal pole, and known as macromeres (Gr. *makros,* large).

Morula and blastula stages

These cells continue to divide until there is a large number of cells, packed together in a somewhat solid mass known as the morula stage (L. *morum,* berry). The micromeres continue to divide more rapidly than the macromeres at this time. It is evident that growth cannot continue indefinitely in this manner, or the animal would be solid, without cavities in which tissues and organs could be placed. Consequently, at a certain stage in the cleavage process, the cells of the morula stage line up to form a one-layered, hollow sphere known as the blastula stage. This sphere has a central, fluid-filled cavity known as the blastocoel (Gr. *koilos,* hollow). The blastula stage consists of (1) an outer, transparent, jellylike capsule, (2) a dark, pigmented animal hemisphere, composed of smaller and more numerous cells, and (3) a light-colored, unpigmented vegetal hemisphere, which is composed of larger and fewer cells. The vegetal cells are quite large and contain yolk or food that is supplied to the cells of the animal hemisphere. This arrangement makes the blastula wall on the vegetative side much thicker than that on the upper or animal side. The active growth in this stage occurs primarily in the animal hemisphere region.

Gastrula stage

The gastrula (yolk plug) stage is formed as follows. At a certain point between the animal and vegetal hemispheres, the vegetal cells turn in toward the blastocoel. The pigmented animal cells (ectoderm) grow over the lighter, unpigmented vegetal cells (endoderm) and fold in with them to some extent at that point. Thus, an inner layer of

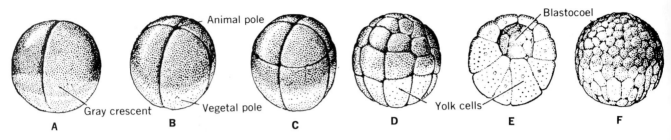

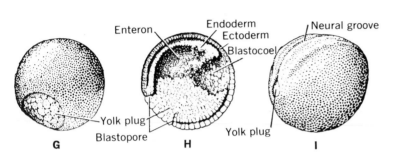

Fig. 6-9. Embryology of the frog. **A** to **C**, Two-, four-, and eight-cell stages of the dividing egg; **D**, early blastula; **E**, section of **D**; **F**, late blastula; **G**, early gastrula with very small ectodermal cells overgrowing other cells; **H**, section of **G**, showing germ layers, blastocoel, etc.; **I**, neurula stage with the neural groove; **J**, neural tube stage with the neural groove closed and assuming a tadpole form; **K**, section of **J**.
(From Woodruff, L. C.: Animal biology, New York, The Macmillan Company.)

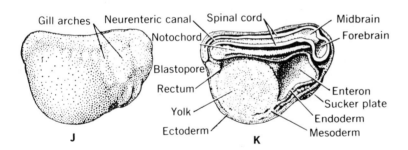

cells is continuous with the outer layer. Because of more rapid mitosis, the animal hemisphere continues to grow almost entirely over the vegetal hemisphere, leaving a small, light yolk plug exposed. The space between the boundaries of the infolded layers of cells, which surround the yolk plug, is the blastopore, or primitive mouth. The ingrowth of the latter is shown on the surface by a thin, crescent-shaped fold or groove. The outer layer of cells, the ectoderm, is continuous with the inner inturned layer or endoderm. The point at which the endoderm cells of the vegetal region turn in is one side of the yolk plug and is known as the dorsal lip of the blastopore. This inturned endoderm forms a cavity, known as the archenteron (Gr. *arche*, beginning; *enteron*, gut), primitive intestine (primitive gut). The blastocoel now appears as a reduced cavity at the opposite side and is gradually being crowded out by the developing archenteron and endoderm. The indentation on the side of the yolk plug opposite the dorsal lip of the blastophore is known as the ventral lip of the blastopore.

Neurula stage

The neurula (Gr. *neuron*, nerve) stage follows the gastrula stage. The neural groove, which is the forerunner of the future nervous system, begins as a small depression on the dorsal side of the blastopore and grows anteriorly along the

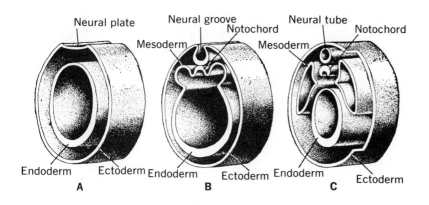

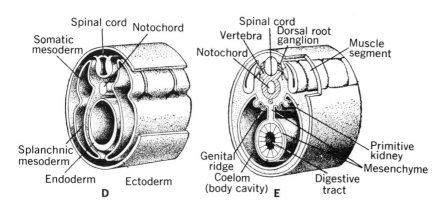

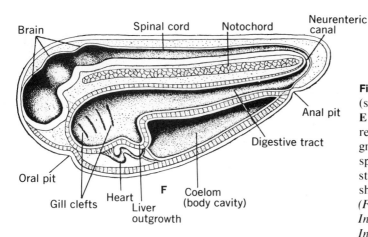

Fig. 6-10. Development of a vertebrate (somewhat diagrammatic). Stages **A** to **E** are a cross section of the mid-body region. **A,** Neural plate stage; **B,** neural groove stage; **C,** neural tube stage; **D,** spinal cord stage; **E,** spinal cord stage still later; **F,** embryo cut lengthwise to show the internal structures.
(From Guyer, M. F.: Being well born, Indianapolis, The Bobbs-Merrill Co., Inc.)

dorsal side of the embryo as a thickened neural plate (medullary plate) in the ectoderm (Fig. 6-10). During this time the embryo grows longer and develops definite anterior and posterior ends. A thickened fold at each margin of the original neural plate forms a neural fold (medullary fold). These folds at first are flat and far apart; later they arch toward the median dorsal line and unite to form the future neural tube. At this time the neural plate sinks to form a definite neural groove along the middorsal side of the embryo. The neural groove is composed of ectoderm cells.

An elongated mass of cells dorsal to the archenteron forms the long, rodlike notochord (Gr. *noton*, back; *chorde*, rod), which still may be connected with the endoderm from which it originates. The mass of cells at either side of the neural groove is known as the mesoderm (middle germ layer).

Neural tube stage

The neural tube stage closely follows the neurula stage. The two neural folds on the dorsal surface of the embryo at this time have met and fused into an elongated neural tube. At this stage the neural tube probably will be free from the outer ectoderm from which it originated. The anterior part of the neural tube constricts and enlarges by well-regulated mitosis to form the future fore-, mid-, and hindbrains. The notochord is now free from the archenteron and is just below the neural tube.

The mesoderm completely surrounds the archenteron ventrally, and near the middorsal part there appears a small split or break that is the forerunner of the coelom (body cavity). This break continues ventrally, thus forming the body cavity between the two layers of the mesoderm. The inner layer of the mesoderm, known as the splanchnic layer (Gr. *splancnon*, entrail), lies next to the endoderm, whereas the outer layer, known as the somatic layer (Gr. *soma*, body), lies next to the ectoderm. The cells of both ectoderm and endoderm are now quite distinct.

Larval stages

The larval stage with external gills follows the neural tube stage. A pair of oval, thick-lipped suckers on the ventral side of the tadpole serve for attachment purposes. The stomodeum (primitive mouth) appears as an oval pit in front of the suckers. The olfactory pits are a pair of small depressions above the anterior to the stomodeum. The three pairs of external gills are finger-

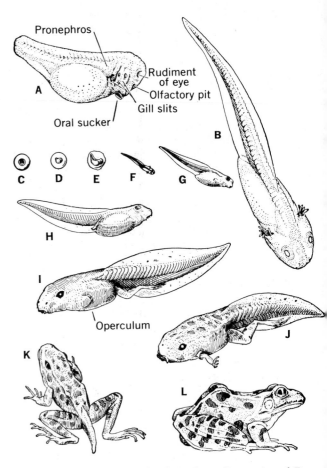

Fig. 6-11. Metamorphosis of the frog. Stages **A** and **B** closely follow stage **K** of Fig. 6-9.
(From Woodruff, L. C.: Animal biology, New York, The Macmillan Company.)

like processes on either side of the head that act as specialized organs of respiration. The proctodeum (primitive anus) is located on the dorso-posterior part of the tadpole. A tail and a pair of eyes are also present.

The larval stage with internal gills (Fig. 6-11) follows the stage with external gills. The external gills are covered by a fold of skin known as the operculum (gill cover), which has a single opening, the spiracle. As the three pairs of external gills are resorbed, four pairs of internal, fishlike gills are formed. In this stage the suckers are small projections just behind the mouth. The mouth is surrounded by a number of small projections known as the circumoral papillae and has a pair of horny jaws. The intestine shows through the transparent ventral body wall as a long, coiled tube. This great length of intestine suggests a typical vegetarian animal, which the tadpole really is at this stage. The hind limb buds appear as small outgrowths on either side of the anal opening. These buds will continue to grow by mitosis into the real hind limbs.

The later stages of development follow. The front limb buds appear and develop into typical front legs. The tail gradually is resorbed and disappears, the materials being taken to the liver and stored. The internal gills are resorbed, and their place taken with rapidly growing lungs. The adult frog is not aquatic but has lungs similar to other landliving (terrestrial) animals. The coiled intestine gradually shortens, suggesting the typical carnivorous (flesh-eating) animal that the frog has become.

Review questions and topics

1 Discuss the stages of embryonic development (embryogenesis) in animals.
2 Explain the structures and functions of the embryonic development of a frog and man.
3 Discuss the tissues that arise from the ectoderm, mesoderm, and endoderm.
4 Review cleavage and gastrulation in the starfish.
5 How does cleavage in the starfish differ from cleavage in the frog? How is it similar?
6 Explain the relationship between the Latin or Greek origins of the name and the actual morphology of the organism in the following: morula, blastula, gastrula, archenteron, neurula.

Selected references

Alston, R. E.: Cellular continuity and development, Glenview, Ill., 1967, Scott, Foresman and Co.
Balinsky, B. I.: An introduction to embryology, ed.3, Philadelphia, 1970, W. B. Saunders Company.
Barth, L. J.: Development; selected topics, Reading, Mass., 1964, Addison-Wesley Publishing Co., Inc.
Berrill, N. J.: Developmental biology, New York, 1971, McGraw-Hill Book Co.
Edwards, R. G.: Mammalian eggs in the laboratory, Sci. Amer. 215:72-81, 1966.
Etkin, W.: How a tadpole becomes a frog, Sci. Amer. 214: 76-88, 1966.
Frieden, E.: The chemistry of amphibian metamorphosis, Sci. Amer. 209:110-118, 1963.
Gray, G. W.: Human growth, Sci. Amer. 189:65-76, 1953.
McElroy, W. D., and Glass, B., editors: The chemical basis of development, Baltimore, 1958, The Johns Hopkins Press.
Moore, J. A.: Heredity and development, New York, 1972, Oxford University Press.
Sussman, M.: Developmental biology: its cellular and molecular foundations, Englewood Cliffs, N. J., 1973, Prentice-Hall, Inc.
Tanner, J. M.: Earlier maturation in man, Sci. Amer. 218: 21-27, 1968.
Waddington, C. H.: Principles of development and differentiation, New York, 1966, The Macmillan Company.
Waddington, C. H.: Principles of embryology, London, 1956, George Allen & Unwin, Ltd.
Whittaker, J. R.: Cellular differentiation, Belmont, Calif., 1968, Dickenson Pub. Co., Inc.
Willier, B. H., Weiss, P. A., and Hamburger, V., editors: Analysis of development, Philadelphia, 1955, W. B. Saunders Co.

Early contributors to the study of
GROWTH AND DEVELOPMENT

Marcello Malpighi (1628-1694)

An Italian scientist who studied the development of organs of chick embryos and erroneously concluded that the adult was preformed in a miniature form in the egg.

Jan Swammerdam (1637-1680)

Swammerdam, a Dutch naturalist, described the cleavage (division) in frogs' eggs.

Charles Bonnet (1720-1793)

Bonnet, a Swiss naturalist, proposed the "Preformation Theory," which stated that an organism carried in its body many preformed individuals, in miniature form, for succeeding generations.

Kaspar F. Wolff (1733-1794)

Wolff, a German physiologist, disproved Bonnet's preformation theory by demonstrating that new embryonic structures develop where none had existed before. He stated that the particles that constitute all animal organs in their earliest inception are little microscopic globules.

von Baer

Lazaro Spallanzani (1729-1799)

Spallanzani, an Italian biologist, attempted to induce frogs' eggs to develop artificially by using vinegar and lemon juice (1785).

Karl Ernst von Baer (1792-1876)

A German embryologist who developed the science of comparative embryology. He studied eggs and tissue formation, which led to the formulation of the "Germ-layer Theory" (1828). He observed the resemblance between embryos of lower and higher animals and discovered mammalian eggs (1827). (*Historical Pictures Service, Chicago.*)

Balfour

Francis M. Balfour (1851-1882)

An English embryologist who wrote a comprehensive book *Comparative Embryology*, thus ushering in this phase of the science (1880). (*Historical Pictures Service, Chicago*).

Jacques Loeb (1859-1927)

A German-American biologist who caused frog eggs to develop into mature frogs (1899) by artificial parthenogenesis (development of an egg without a sperm).

Eduard van Beneden (1846-1910)

Van Beneden, a Belgian cytologist, studied the behavior of chromosomes during meiosis (reduction division) and helped to clarify their role in heredity and development (1883).

Thomas Hunt Morgan (1866-1945)

An American biologist who made many valuable contributions to experimental embryology.

Oskar Hertwig (1849-1922)

A German biologist who first observed (1875) all the steps in fertilization, including the union of egg and sperm chromosomes in sea urchins.

Spemann

Hans Spemann (1869-1941)

A German zoologist who proposed the "Organizer Concept" in embryology (1921), which stated that certain parts of a developing embryo act as organizers and influence developmental patterns of the organism. (*Historical Pictures Service, Chicago.*)

Alexis Carrel (1873-1944)

A French surgeon and biologist who came to the United States in 1905, cultured tissues outside the body (in vitro) in 1912, and established a culture of chick embryo connective tissue that grew continuously for 25 years under laboratory conditions.

Ross V. Harrison (1870-1959)

An American biologist who devised a technique for culturing cells outside the body (1907) by mounting a fragment of tissue in clotting lymph fluid on a cover slip of a hanging drop (concave) slide.

7 Nutrients and basic materials

Nutrients
Chemical composition of organisms
  Water and minerals
  Organic components

The preceding chapters have been concerned with the general structure and function of living organisms. Some of the structures appear to be quite simple and others are highly complex. Similarly, some activities are easy to understand and others are more difficult.

This chapter is concerned with the basic materials of cells and organisms. It considers the basic raw materials and how these materials become more complex structures involved in living activities.

NUTRIENTS

Any consideration of raw materials in a living organism is necessarily a study of its food and basic nutrients. This can be done in several ways. One method is simply to grind up the organism and then to make a chemical analysis of its contents. The results of such an examination are similar to those shown in Tables 7-1 to 7-3.

Another method of study is to determine whether or not the organism can grow and de-

Table 7-1 Chemical analysis of cells

Element	Percent (by weight)
Oxygen (O)	76.0
Carbon (C)	10.5
Hydrogen (H)	10.0
Nitrogen (N)	2.5
Phosphorus (P)	0.3
Potassium (P)	0.3
Sulfur (S)	0.2
Chlorine (Cl)	0.1
Sodium (Na)	0.04
Calcium (Ca)	0.02
Magnesium (Mg)	0.02
Iron (Fe)	0.01

Table 7-2 Trace elements in various cells

Copper (Cu)	Chromium (Cr)
Cobalt (Co)	Tin (Sn)
Manganese (Mn)	Vanadium (V)
Zinc (Zm)	Flourine (F)
Iodine (I)	Silicone (Si)
Molybdenum (Mo)	Nickel (Ni)
Selenium (Se)	

velop by using particular combinations of foods which have been analyzed previously. This is often done by feeding animals, such as rats, a variety of special diets each of which is slightly

Table 7-3 Percentages of compounds in protoplasm

Organic compounds (percent by weight)		Inorganic compounds (percent by weight)	
Protein	15	Water	80
Fats	3	Inorganic salts	1
Carbohydrates	1		

different from the other. The rates of growth and development are then compared.

Plants can be studied by growing them in water solutions containing various chemicals. They are also often studied by planting seeds in the soil and by adding different fertilizers or nutrients. Bacteria and other microorganisms are grown either in test tubes or in other special containers. The growth medium is varied as each new nutrient is studied.

Some of these substances and how they are involved in the structure and function of organisms will now be considered.

CHEMICAL COMPOSITION OF ORGANISMS

As indicated above, basic information about chemical composition can be obtained by analyzing body fluids and cell extracts. Such analysis shows that about one quarter of the 92 naturally occurring chemical elements can be found in living things (Table 7-1). Presumably they are of some importance in the metabolism and survival of the particular organisms in which they are found.

About 98% of the contents of most living organisms is made up of just six chemical elements: oxygen, carbon, hydrogen, nitrogen, calcium, and phosphorus. The exact amount will vary somewhat in a specific organism. Approximately 20 other elements can be found in very small amounts, usually in hundredths of a percentage. These elements are called trace elements since only a "trace" of them may be found in a particular analysis (Table 7-2).

The small amounts do not necessarily imply, however, that they are of little importance. For example, only four thousandth (0.004) of a percent of the human body is iron. Iron is, however, a necessary component of hemoglobin, the oxygen-carrying pigment of the blood. When less iron is present, less hemoglobin is manufactured and less oxygen is available to the cells. This results in anemia, lack of energy, and serious illness.

The amount of iodine in the human body is even less; however, it is vital in the formation of the thyroid hormone *thyroxine,* which regulates the basic metabolism of the body. Others are just as important in humans and in other organisms such as plants and bacteria.

Water and minerals

Water is the most abundant substance in living organisms. The amount differs in each species of organisms. Some species have as little as 10% and others may have up to 90% water. A dormant seed contains very little water. In contrast, an average human may be two thirds water and some jellyfish may be 95% water.

Life is absolutely dependent on water. Water is the fluid other materials are dissolved in. It is a solvent, a carrier, a lubricant, a temperature regulator, and a protectant. It is the major component of blood, digestive fluids, and cytoplasm. It takes part in many chemical reactions in cells, and it carries away waste products in the form of urine and sweat. Oxygen dissolved in water enters the lungs, and the materials that we taste and smell are also first dissolved in water.

The minerals found in organisms are important in both structural components and in metabolic activity. Shells, bones, and teeth are composed largely of calcium and phosphorus compounds. Lack of sufficient amounts of either of these during childhood may result in a bone-deforming disease called rickets. Other human deficiency diseases may result from insufficient amounts of several other minerals. These include iron and iodine, mentioned above, as well as others.

In the blood and tissue fluids, sodium and chlorine make up the 0.9% salt content. This is important in maintaining the osmotic balance and the acid-base ratio. Chlorine is a major component of the acidic stomach juices and of the digestive fluids of the pancreas and intestine.

Extensive sweating caused by heat and physical exertion may result in salt losses sufficient to cause cramps and heat exhaustion. This is the reason salt tablets are used during heavy work. Many other minerals are important in such activities as blood clotting, muscle contraction, nerve conduction, and numerous other functions. Some of these will be considered later in this chapter and in other chapters.

Mineral requirements similar to the above are found also in other organisms. The familiar N, K, and P on the plant fertilizer label refer to nitrogen, potassium, and phosphorus, all necessary elements for growth. The percentage of trace elements refers to other minerals that may not always be present in the soil and must be added for growth.

Organic components

Organic components were first identified by analyzing either living organisms or their waste products. Because of this they were named *organic*. They include the carbon compounds in these living organisms. Examples of these important biological compounds are carbohydrates, lipids, proteins, and nucleic acids. Human requirements for these important nutrients are considered in greater detail in connection with digestion and metabolism in Chapter 28.

Carbohydrates

Carbohydrates are organic compounds consisting of carbon, hydrogen, and oxygen in the ratio of $(CH_2O)n$. As the name indicates, they may be considered as "hydrates" of carbon, although this is not correct chemically. They exist in many different forms, the most common of which are various sugars or *saccharides* (Table 7-4).

Simple sugars or *monosaccharides* contain 3 to 7 carbon atoms and the corresponding H_2O units.

Table 7-4 Some common carbohydrates

Type	Sugar	Molecular formula
Monosaccharides		
Pentose (5 carbon)	Ribose	$C_2H_{10}O_5$
	Deoxyribose	$C_5H_{10}O_4$
Hexose (6 carbon)	Glucose	$C_6H_{12}O_6$
	Fructose	$C_6H_{12}O_6$
	Galactose	$C_6H_{12}O_6$
Disaccharides	Lactose	$C_{12}H_{22}O_{11}$
	Maltose	$C_{12}H_{22}O_{11}$
	Sucrose	$C_{12}H_{22}O_{11}$
Polysaccharides	Starch	60 to 80 glucose units
	Glycogen	12 to 16 glucose units
	Cellulose	100's of glucose units

Five-carbon sugars, *ribose* and *deoxyribose* (which has 1 less oxygen atom), are important as components of RNA—ribonucleic acid—and DNA—deoxyribonucleic acid. These form the structure of the genes and ribosomes of the cell.

Probably the most significant naturally occurring sugar is the 6-carbon *glucose* (Fig. 7-1). It is the primary blood sugar and is important in the energy relations of both plants and animals. Its 6 carbon atoms may occur as a straight chain or as a ring. This is also true for other 6-carbon sugars that have the same general formula, $C_6H_{12}O_6$, but differ in the placement of H and O atoms. Unless we know the structural arrangement, the same $C_6H_{12}O_6$ may refer to glucose, fructose, or ga-

Fig. 7-1. Structure of common sugars.

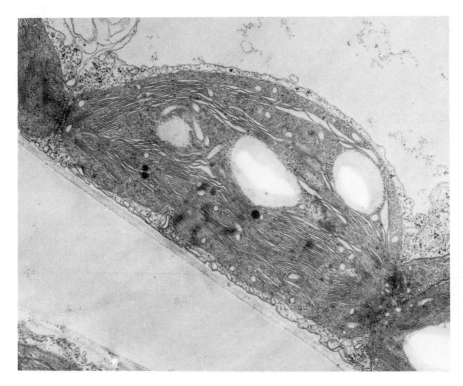

Fig. 7-2. Three large starch granules formed in a spinach chloroplast. *(Courtesy Dr. Faye Schwelitz, University of Dayton.)*

lactose, which are different monosaccharides and have different chemical properties.

The combination of 2 monosaccharides produces a double sugar or *di*saccharide. Table sugar is the double sugar *sucrose,* formed by the combination of glucose and fructose.

Combinations of 3 monosaccharides produce *tri*saccharides such as raffinose. Multiple units of simple sugars combine to produce polysaccharides.

Polysaccharides are among the important biologic products. The animal carbohydrate *glycogen* is composed of 12 to 18 glucose units. It is an important storage form of animal carbohydrate. The primary plant storage form is *starch* composed of 300 or so glucose units (Fig. 7-2). *Cellulose,* an important component of plant cell walls and fibers, may contain 2000 or so glucose units.

Cellulose is an important component of wood.

It is of interest that some termites, insects that feed primarily on wood, cannot digest cellulose since they lack the appropriate digestive enzyme, *cellulase.* They do harbor, however, a population of cellulose-digesting protozoans in their intestines. The termite chews wood, the protozoan digests it, and both gain a nutrient. If humans produced the proper enzyme, we could eat trees!

The final decomposition or digestive product of any of these saccharides is glucose, the primary energy yielding source of most organisms. Carbohydrates also form many important compounds necessary for living organisms.

Lipids

Lipids are organic substances also containing C, H, and O but in a different ratio and arrangement than in the carbohydrates. The most important of the lipids are the fats.

Fats are formed by the combinations of an al-

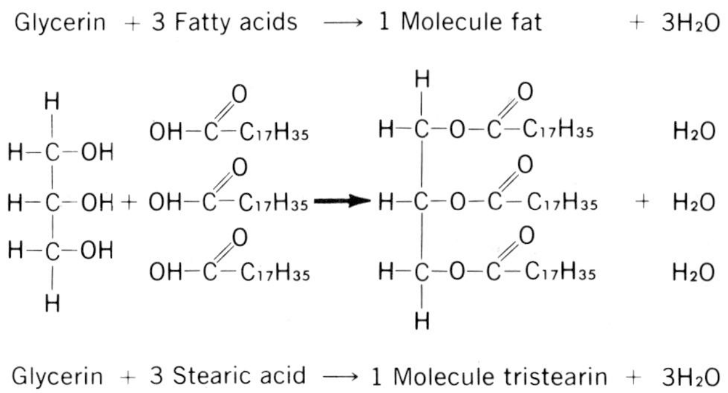

Glycerin + 3 Fatty acids ⟶ 1 Molecule fat + $3H_2O$

$$
\begin{array}{c}
H \\
| \\
H-C-OH \\
| \\
H-C-OH \\
| \\
H-C-OH \\
| \\
H
\end{array}
+
\begin{array}{c}
OH-C\!\!\stackrel{O}{\diagup}\!\!-C_{17}H_{35} \\[4pt]
OH-C\!\!\stackrel{O}{\diagup}\!\!-C_{17}H_{35} \\[4pt]
OH-C\!\!\stackrel{O}{\diagup}\!\!-C_{17}H_{35}
\end{array}
\longrightarrow
\begin{array}{c}
H \\
| \\
H-C-O-C\!\!\stackrel{O}{\diagup}\!\!-C_{17}H_{35} \\
| \\
H-C-O-C\!\!\stackrel{O}{\diagup}\!\!-C_{17}H_{35} \\
| \\
H-C-O-C\!\!\stackrel{O}{\diagup}\!\!-C_{17}H_{35} \\
| \\
H
\end{array}
+
\begin{array}{c}
H_2O \\[8pt]
H_2O \\[8pt]
H_2O
\end{array}
$$

Fig. 7-3. Formation of a fat molecule.

Glycerin + 3 Stearic acid ⟶ 1 Molecule tristearin + $3H_2O$

cohol—usually *glycerol*—and *fatty acids*. Fatty acids are composed of long C—C chains that have a carboxyl group (COOH) at one end. One molecule of glycerol combines with 3 fatty acids to form 1 molecule of a fat. In the process 3 molecules of water are split out.

Tristearin is a fat found in beef cattle. It is formed when glycerol combines with 3 molecules of stearic acid ($C_{17}H_{35}COOH$), as in the above reaction (Fig. 7-3).

The fatty acids found in living things usually contain an even number of carbon atoms. If the bonds linking the carbon atoms are single bonds, the fat is said to be *saturated*. If there are many double bonds linking the carbon atoms, it is unsaturated. Unsaturated fats are usually liquid at room temperature and are called oils. Saturated fats are usually present as solids or tallow forms. The kind of fat an organism produces and stores may be related to its environment and body temperature. Some organisms that live in cold water, for instance, have a high oil content.

Waxes are lipids formed by fatty acids combined with alcohols other than glycerol. Some examples are beeswax, which is secreted by abdominal glands of worker bees, and the waxy cuticle that protects plant epidermis.

One or two of the fatty acids in a fat may be replaced by other molecules. This happens in the formation of *phospholipids,* which are important components of cell membrane.

Lipids, especially fats, are an important form of energy storage in an organism (Fig. 7-4). After the carbohydrate stores are used up, the lipids are metabolized for energy release and for other cell uses. It is of some interest that in animals such as camels, whose hump is largely fat, water is released during the fat decomposition. This water release may account for the famous lack of thirst of the camel.

Proteins

Proteins are organic compounds containing carbon, hydrogen, oxygen, nitrogen, and usually sulfur. They are important metabolic and structural compounds in all living organisms. Proteins are formed by combination of amino acids, of which there are about 20 to 24 that occur in nature. Ten of these are nutritionally essential in the diet since they cannot be made by metabolic reaction. An amino acid is formed by the combination of an amine group—NH_2 to an organic acid C=COOH to form the basic structure.

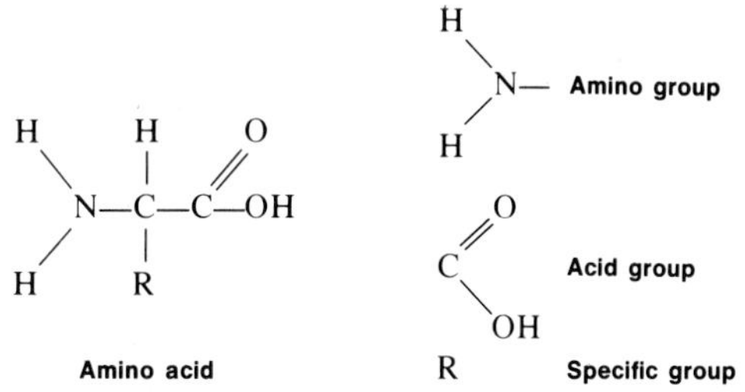

Examples of amino acids are glycine where R is H, alanine where R is CH_3 and cysteine where R is CH_2—SH. Two amino acids combine to form a dipeptide. The point of combination is between the amino end of one and the acid end of the other. In the formation of the resulting *peptide*

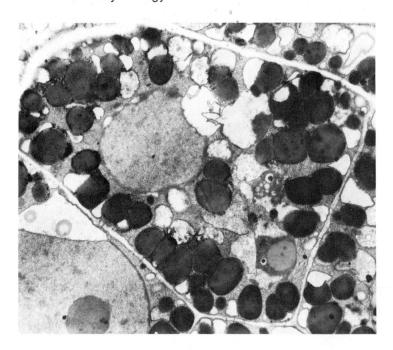

Fig. 7-4. Electron micrograph of cell with many dark-staining lipid granules. *(Courtesy M. A. Hayat, University of Dayton.)*

bond —C——N—H, a molecule of water is split off.

Peptides are combinations of two or more amino acids held together by the peptide bond. They occur as dipeptides, tripeptides, and multiple polypeptides. Many important biologic products are simple polypeptides.

Two hormones important in human beings are each composed of nine amino acids. One of these, oxytocin, stimulates uterine contractions during childbirth and later initiates milk secretion. The other, vasopressin, is important in water metabolism and kidney function. The antibiotics *gramicidin* and *tyrocidin* are both 10-amino-acid polypeptides. Insulin, important in sugar metabolism, consists of two polypeptide chains linked together.

Proteins are polypeptides composed of long chains of amino acids. Some examples are: egg albumin, 50 amino acids; wheat protein, 196 amino acids; and hemoglobin, 574 amino acids.

Protein composition may vary quite widely. Proteins vary in the total number of amino acids present. They differ in the amounts of the total possible amino acids, although most proteins do have 17 to 18 different kinds. Ribonuclease, a protein enzyme that breads down RNA, has a total of 124 amino acids of 19 different types. Of these only two are leucine, whereas 15 are serine.

Probably most importantly proteins differ in the sequence of the amino acids present. It is interesting to note that the serious blood disease, sickle cell anemia, is caused by a change in just one amino acid in the hemoglobin molecule. At

Normal
Val·His·Leu·Thr·Pro·Glu·Glu·Lys·

Sickle Cell
Val·His·Leu·Thr·Pro·Val·Glu·Lys·

Fig. 7-5. The first six amino acids in normal and sickle cell hemoglobin. Note that the substitution of only one amino acid leads to a disease-producing condition.

the sixth position in the molecule, one amino acid, glutamic acid, has been replaced by another, called valine (Fig. 7-5).

When this does occur, very serious illness results and may lead to death, if all of the hemoglobin contains the altered sequence. If only some of the hemoglobin is affected, this *sickle cell trait* may result in disease sometimes, but not as often and as seriously as when all of the hemoglobin is affected. In addition, the sickle cell trait also confers a type of resistance to malaria. Because of the distortion of the cells, the malaria parasite is unable to penetrate the red blood cells.

To consider the different kinds of possible proteins, let's use a protein composed of 124 amino acids of which there are 19 different kinds, as in ribonuclease. Let us imagine that we are filling in columns of letters, with each column to contain 124 positions, each position to be occupied by any one of 19 letters. The possible number of combinations is astounding!

As we have seen with the example of sickle cell hemoglobin, the replacement of just one amino acid in a combination may change the biologic activity of the protein.

PROTEIN STRUCTURE

Proteins occur in different shapes, such as fibers, spheres, or other three-dimensional forms. The type of arrangement is largely the consequence of the type of bonding of the amino acids and other atoms involved.

The *primary structure* is the covalent peptide structure resulting from amino acid combinations. This "backbone" may form a *secondary*

structure by coiling in a springlike fashion to form what is called an *alpha*-helix. The coils are held together by hydrogen bonds. These coils of protein may further bend in various fashions to form cubes, spheres, or "globs," which are the *tertiary* structure. And finally, as we have seen in hemoglobin, several of these "globs" may bind together, forming a *quatenary* structure.

The various shapes are important in the formation of protein fibers, spheres, and globular forms as in enzymes and antibodies. These are important and form the unique structure (or specificity) of different types of proteins. This specificity is the basis of blood types, allergy, and transplant failure.

Proteins are sensitive to changes in their immediate environment. For example, changes in temperature, pH, or pressure may influence protein structure and cause a clumping or coagulation.

Enzymes

Enzymes are proteins that form a special class of biologic catalysts. They function to control the rate at which a chemical reaction takes place. Enzymes influence the rate of reaction in the same manner as an increase in temperature. That is, since molecules can react only when they are in contact, heat increases the speed of molecular movement and therefore the possibility of collision. Within limits, then, an increase in temperature results in an increase in chemical reaction.

Enzymes provide a mechanism that brings molecules together, thus increasing the opportunity for contact and reaction. They function as organic catalysts, speeding the rate at which a reaction takes place without being consumed in the process.

One idea about how they function is the lock and key theory. Since enzymes are large globular protein molecules, they have many projections and depression—nooks and crannies formed by the various types of folding of the protein chains. These shapes are similar to the ridges of a key that the lock or chemical reactants fit into. Once

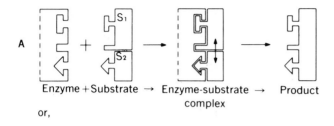

A

Enzyme +Substrate → Enzyme-substrate → Product
complex

or,

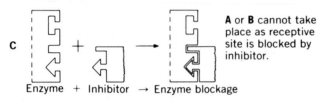

B

Enzyme + Substrate → Enzyme-substrate → Products 1
complex and 2

Fig. 7-6. Lock-and-key theory of enzyme activity.

C

Enzyme + Inhibitor → Enzyme blockage

A or **B** cannot take place as receptive site is blocked by inhibitor.

the key is in the lock, the reaction takes place (Fig. 7-6). The particular chemical substances involved are referred to as the "substrate" of the enzyme. During a reaction an enzyme-substrate complex forms, which results in the products of the action.

Enzyme + substrate → E-S complex → products

Sometimes the reaction is blocked by a similar type of substrate that fits the enzyme surface but cannot complete the reaction. These *inhibitors* prevent the normal substrate from contacting the enzyme, and the reaction does not take place.

Since they are proteins, enzymes are subject to all of the factors influencing this type of compound. They may be influenced or inactivated by temperature, pH, other ions, and pressure.

The enzymes that influence digestion are a good example. Starch digestion starts in the saliva, which has a pH of about 6.8. Salivary *amylase* works at this pH but not at that of the stomach, which is highly acid. Protein digestion

is influenced by the enzyme *pepsin* at pH 2. As food passes into the small intestine, the pH changes to about pH 8. This is a suitable pH for the protein-digesting *trypsin,* the fat- or lipid-digesting *lipase,* and the enzymes influencing sugar digestion as indicated above.

Enzymes are named according to specific rules related to their activity. Thus, the enzyme assisting the splitting of carbonic acid (H_2CO_3) into water and carbon dioxide is called *carbonic anhydrase* (Gr. *an,* away; *hydr,* water). Similarly, the enzyme linking the subunits of the DNA molecule is called *DNA polymerase.* The "ase" ending is a general suffix denoting an enzyme. Another, older system of naming enzymes refers to the substrate; for example, *sucrase* is an enzyme that splits table sugar or sucrose into glucose and fructose.

Even though sucrase acts to split the double sugar sucrose, it will not catalyze the breakdown of the very similar sugars lactose and maltose. These are split by *lactase* and *maltase.* This prop-

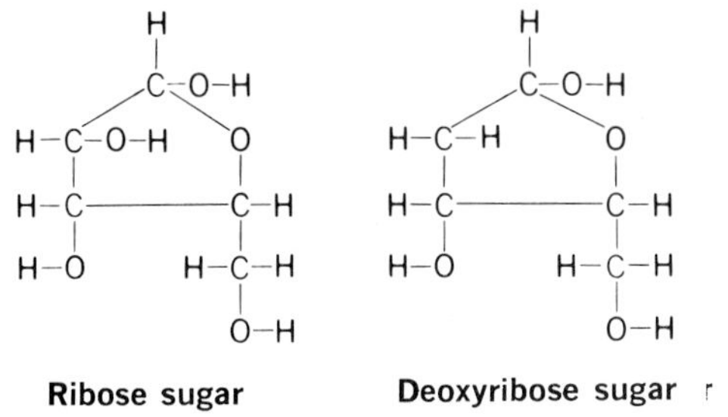

Fig. 7-7. Sugars (pentose or 5-carbon types) that form basic units of nucleic acids. Ribose sugar is found in ribonucleic acid (RNA) and deoxyribose sugar is found in deoxyribonucleic acid (DNA).

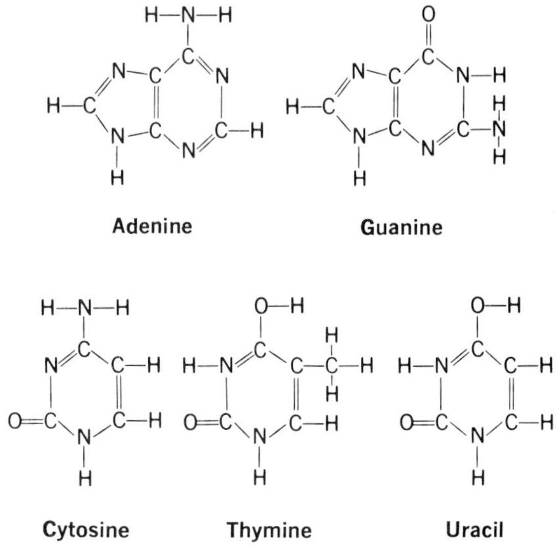

Fig. 7-8. Purine bases (adenine and guanine) and pyrimidine bases (cytosine and thymine) that form basic units of nucleic acids. Uracil is a pyrimidine base found only in RNA in place of thymine, which occurs in DNA. The other three bases are common to both DNA and RNA.

erty of influencing only one reaction (or a few at most) is called enzyme specificity.

In the breakdown of a large molecule, which proceeds in several steps, each reaction has a particular enzyme. Starch is hydrolyzed to maltose by *amylase,* and maltose to glucose by *maltase.* The complete breakdown of a protein may involve several enzymes.

As we will see later, most of the chemical reactions of living organisms are catalyzed by specific enzymes.

Nucleotides and nucleic acids

Nucleotides and nucleic acids are so named because they were first isolated from the nuclei of various cells. They are very important to a number of living activities.

Some nucleotides are associated with the energy release necessary for metabolism. The energy required during active transport of materials through a cell membrane comes from the splitting of adenosine triphosphate (ATP) into adenosine diphosphate (ADP) and inorganic phosphate Pi. This is accomplished by the release of energy as follows:

$$ATP \rightarrow ADP + \text{\textcircled{P}}i + energy$$

Some nucleotides combine with enzyme molecules in a joint function. These are called coenzymes. Others are important in regulating cell processes. Large numbers of four specific kinds

of nucleotides combine to form the complex polymers deoxyribonucleic acid (DNA) and ribonucleic acid (RNA), which are the basis of heredity and cell metabolism.

NUCLEOTIDES

Nucleotides are combinations of phosphate, sugar, and a nitrogen base. The phosphate, PO_4, is a derivative of phosphoric acid, H_3PO_4. It combines with one of the two 5-carbon sugars, ribose or deoxyribose, which has one less oxygen atom (Fig. 7-7). This in turn is linked to a nitrogen base, so called because of its nitrogen content and the fact that chemically it is a base (as opposed to an acid).

Five different nitrogen bases occur in nucleotides of biologic importance. Two of these, adenine and guanine, are double-ring structures called *purines.* The other three, thymine, cytosine, and uracil, are single-ring structures called pyrimidines (Fig. 7-8).

Table 7-5 Relationships between DNA and RNA

DNA (deoxyribonucleic acid)	RNA (ribonucleic acid)
Contains pentose (5-carbon) sugar called deoxyribose	Contains pentose (5-carbon) sugar called ribose
Contains bases adenine, guanine, cytosine, thymine	Contains bases adenine, guanine, cytosine, and uracil
Contains phosphoric acid (phosphate) that connects various sugars with one another	Contains phosphoric acid (phosphate) that connects various sugars with one another
DNA is genetic material of life	RNA present in large amounts in nucleoli
DNA always associated with chromosomes (genes); each set of chromosomes seems to have fixed amount of DNA	RNA is found mainly in combination with proteins in ribosomes in the cytoplasm as messenger RNA and as transfer RNA

The entire combination, then, may be called by the name of the base and the term nucleotide, thus, adenine nucleotide, guanine nucleotide, and so on. Strict chemical nomenclature refers to these as adenylic acid or adenosinemonophosphate. When the base adenine is combined with ribose, it is called adenosine. If combined with deoxyribose, it is called deoxyadenosine.

ADENOSINE PHOSPHATES

As indicated above, the combination of adenine, ribose, and phosphate forms adenosine monophosphate (AMP). In living things a second phosphate is joined to the molecule to form adenosine diphosphate (ADP), and similarly, a third phosphate forms adenosine triphosphate (ATP).

ATP is referred to as a high energy compound. This indicates that more energy is present than normally expected. The "extra" energy is present in the bonds between the phosphate groups, and for this reason this high energy bond is designated by a squiggle (undulating line) $\sim$ rather than the dash — used to indicate bonds. ATP, then, is written adenosine —Ⓟ$\sim$Ⓟ$\sim$Ⓟ. The energy released by splitting off the last $\sim$Ⓟis four times as much as that in the second to first —Ⓟ. It is this extra energy that drives the reactions of metabolism. Although other types of nucleotides occur in cytoplasm, ATP is the most common energy source.

Of course we realize that the ultimate source of energy is the sun. Plants utilize light energy during the process called photosynthesis. This works to form various sugars. These in turn are used in the plant or eaten by animals. At any rate the energy in the sugar is released in the formation of the high energy $\sim$Ⓟbonds.

NUCLEIC ACIDS

Nucleotides in living organisms combine to form long chains called nucleic acids. According to the type of sugar in the nucleotides, they are called ribose nucleic acid (RNA) or deoxyribose nucleic acid (DNA). DNA forms the substance of the hereditary material called genes and directs the assembly of RNA. RNA is found in the nucleolus and cytoplasm of the cell and controls the actual synthesis of polypeptides (Table 7-5).

In the bonding of one nucleotide to another, the combination is between the sugar of one and the phosphate of the other with the nitrogen bases bonded to the sugar at a different point. The "backbone" of a string of nucleotides is composed of sugar-phosphate-sugar-phosphate-S-P-S-P, and so one. The "spines" of nitrogen bases project away from the sugar (Fig. 7-9).

Both double ring nitrogen bases, adenine and guanine, and one single ring, cytosime, form nucleotides with ribose or deoxyribose. Thymine is found only in deoxyribose molecules and uracil only in ribose nucleotides.

Considering only the formation of a DNA molecule, four nucleotides are present: ade-

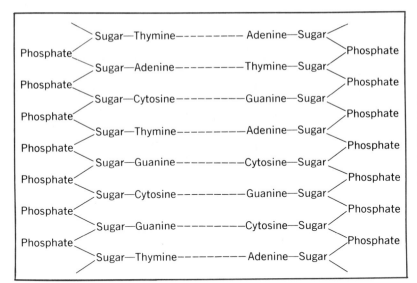

Fig. 7-9. Structure of DNA (deoxyribonucleic acid), showing the complementary structure of two chains. Sugar is deoxyribose. Two complementary chains twisted around one another are connected by hydrogen bonding (–––––––).

nine—A, guanine—G, cytosine—C, and thymine—T. A portion of the DNA may then be represented by A-G-C-T-A-G-C-T, and so on. It is important to note that any one of these positions may be occupied by any one of the four nucleotides. The molecule may also be T-G-C-A-T-G-C or any other combination.

As we have seen before, this presents an immense number of possible combinations and therefore different nucleic acids. It is these different sequences that determine the hereditary properties of the nucleic acid. See Chapter 32, Genes and gene action, for further details.

Review questions and topics

1 Define matter, atom, ion, element, compound, and molecule, with examples of each.
2 Discuss the chemical composition of protoplasm, including the common elements present.
3 Distinguish between organic and inorganic compounds, with examples.
4 Discuss the important characteristics of carbohydrates, fats, and proteins, and the roles of each in the metabolic activities of protoplasm.
5 Define kilocalorie.
6 Discuss the functions of inorganic salts in the living processes of organisms.
7 Discuss the importance of water as a constituent of living protoplasm.
8 Discuss the general characteristics and properties of enzymes, including their importance in living processes.

Selected references

Baker, J. J. W., and Allen, G. E.: Matter, energy and life, ed. 2, Reading, Mass., 1970, Addison-Wesley Publishing Co., Inc.
Barrington, E. J. W.: The chemical basis of physiological regulation, Glenview, Ill., 1968, Scott, Foresman and Company.
Bohinsky, R. C.: Modern concepts in biochemistry, Boston, 1973, Allyn & Bacon, Inc.
Bronk, J. R.: Chemical biology; an introduction to biochemistry, New York, 1973, The Macmillan Co.
Calvin, M., and Jorgenson, M. J., editors: Bio-organic chemistry. Readings from Scientific American, San Francisco, 1968, W. H. Freeman and Co., Publishers.
Christensen, H.: pH and dissociation, Philadelphia, 1964, W. B. Saunders Company.
Crick, F. H. C.: Nucleic acids, Sci. Amer. **197:**188-200, 1957.

Epstein, H. T.: Elementary biophysics, Reading, Mass., 1963, Addison-Wesley Publishing Co., Inc.

Glassman, E., editor: Molecular approaches to psychobiology, Belmont, Calif., 1968, Dickenson Pub. Co., Inc.

Hanawalt, P. C., and Haynes, R. H., editors: The chemical basis of life. Readings from Scientific American, San Francisco, 1973, W. H. Freeman and Co., Publishers.

Ingram, V. M.: The biosynthesis of macromolecules, New York, 1965, W. A. Benjamin, Inc.

Jellinck, P. H.: The cellular role of macromolecules, Glenview, Ill., 1967, Scott, Foresman and Company.

McElroy, W. D.: Cell physiology and biochemistry, Englewood Cliffs, New Jersey, 1964, Prentice-Hall, Inc.

Neal, Arthur L.: Chemistry and biochemistry; a comprehensive introduction, New York, 1971, McGraw-Hill Book Co.

White, E. H.: Chemical background for the biological sciences, Englewood Cliffs, New Jersey, 1964, Prentice-Hall, Inc.

Whittingham, C. P.: The chemistry of plant processes, New York, 1964, Philosophical Library, Inc.

Some contributors to the knowledge of
CHEMICAL PROPERTIES OF LIFE

Purkinje

Johannes Purkinje (1787-1869)

A Bohemian physiologist, applied the ▮ protoplasm (Gr. *protos,* first; *plasma,* liq▮ to the jellylike substance (1840), although▮ meaning of the term was probably somew▮ different from the meaning in later usage. ▮ *Bettman Archive, Inc.)*

von Mohl

Felix Dujardin (1801-1860)

A French zoologist, observed the jellylike ▮ terial in animal cells (1835) and applied ▮ term sarcode (Gr. *sarx,* flesh) to it. This ▮ stance was later found in living plant cell▮

Hugo von Mohl (1805-1872)

A German botanist, found that plant ▮ were composed of the living substance (▮ and used the term protoplasm much as we▮ today. *(Historical Pictures Service, Chica▮*

Max Schultze (1825-1874)

A German cytologist, enunciated the P▮ plasmic Theory and stated that the jelly▮ protoplasm was similar in plant and an▮ cells (1861). He concluded that "the cell ▮ accumulation of living substance or p▮ plasm definitely delimited in space and ▮ sessing a cell membrane and nucleus."

Huxley

Thomas H. Huxley (1825-1895)

An English biologist, referred to protoplasm as "the physical basis of life" (1868). *(Historical Pictures Service, Chicago.)*

Otto Bütchli (1848-1920)

A German protozoologist and physiologist, made experimental observations on nonliving substances (mixtures of olive oil, potassium carbonate, and glycerin) that simulated some of the movements and appearance of living protoplasm.

Max Verworn (1862-1921)

A German physiologist, suggested that chemical particles with special chemical actions are what we really mean by life.

8 The kinds of living things

People have always had a natural curiosity about the identity of living things. There is a definite satisfaction in being able to distinguish a robin from a bluejay or to note that the Christmas tree is a pine and not a spruce. The oft-repeated saying about life in a drop of pond water takes on a new significance with the awareness of water pollution. The increase in backpacking is one sign of the renewed interest in environmental studies. Handbooks for the identification of various plants and animals have become more numerous and more popular. It is not at all unusual to find a series of "How to Know the Plants," "Pictorial Nature Guide to Animals," or "Field Guide to the Insects" in a neighborhood drug store or supermarket. A list of selected readings on the indentification of living things is given at the end of this chapter. Thousands of people take part in the annual Christmas count of birds conducted by the Audubon Society.

All of this activity is not new. We know from the paintings on prehistoric cave walls that we carry on the interest of early mankind. When cavemen noticed that some animals had horns and others did not, that some fed by grazing and others by eating flesh, that some fruits made him sick while others were edible, he left a record of this in the wall paintings.

CLASSIFICATION

Beyond the normal curiosity and common sense, knowledge of different kinds of living things is the work of the scientists called taxonomists or systematists. These biologists study the similarities and the differences in various groups of living things. They use the results of their studies to indicate evolutionary relationships and to determine the relationships of various plants and animals.

Any system of grouping or classification may have several objectives. For example, your college is organized into departments or majors. The student body is classified, according to year in school, as senior, junior, sophomore, and freshman. A senior English major both resembles and differs from a senior history student or a freshman biology major. The similarities and differences enable the staff to keep track of things as well as to serve the student body. The same general concepts are used in biologic classification to keep track of, and to show the similarities and differences between, living organisms.

Tiger (common)   Persian   Siamese

A   B   C

Fig. 8-1. Varieties of the domestic cat, *Felis domesticus*.

The kinds of information used to determine relationships of various organisms were and are dependent on the tools and techniques used to examine the organisms. No special equipment is needed to determine that two flowering plants have much in common. Similar observations can be made for many animals.

More detailed information can be obtained by a close examination of the same structure in a variety of animals. For example, a comparative study of the anatomies of the forelimb of the frog, bird, cat, horse, and human shows that they are constructed on the same general plan. Studies of these same structures in the embryo show that they develop in the same general way. The same thing is true for many other structures in these organisms. Many other examples are discussed later in the text in Chapter 33.

As smaller and structurally simpler organisms are studied it is necessary to use different techniques. In the study of microorganisms an electron microscope is useful. Even more common are biochemical analyses of the organism in order to determine its ability to produce a certain enzyme or to use a particular substance as food.

Classification of organisms

The basic unit in nature, and in biologic classification, is the species. A species is considered to be a group or population of similar organisms having many characteristics in common and differing from all other groups in some significant ways. Members of a species are capable of interbreeding to produce fertile offspring.

Because language barriers may present problems of communication to scientists of different nationalities who are studying the same organism, the scientific name of the species is used. The scientific name consists of the genus and species and, hence, is called *binomial* (two name) nomenclature. Some examples are: man, *Homo sapiens;* housefly, *Musca domestica;* and corn plant, *Zea mays.* In strict practice the binomial is followed by the name of the person who first described and named the organism. Thus, the full scientific name of the sparrow is *Passer domesticus* Linnaeus (often abbreviated as Linn. or L.).

All of the domestic cats are placed in the same species, *Felis domestica* (Fig. 8-1), since they can freely interbreed and produce fertile offspring. The proud and aloof Persian or Siamese cats

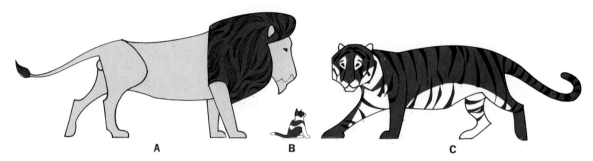

Fig. 8-2. Members of the cat family Felidae. **A,** *Panthera leo;* **B,** *Felis domesticus;* **C,** *Panthera tigris.*

maintain their distinct characteristics only by the efforts of the cat breeders' association. They can and do, if not prevented, freely mate with the alley cat and produce fertile kittens.

The large cats such as the panther, lion, and tiger are placed in the genus *Panthera.* Among their characteristics is a voice box that enables them to roar. The panther, sometimes called leopard, is *Panthera pardus;* the lion is *Panthera leo,* and the tiger is *Panthera tigris.*

Different species in nature do not normally interbreed. In experiments in zoos, offspring of the mating of a lion and a tiger have occasionally been produced. These "ligers" or "tiglons" have invariably been sterile.

The domestic cats in the genus *Felis,* the big cats in genus *Panthera,* and some others such as the short tailed bobcat or wildcat *Lynx rufus* are all members of the cat family **Felidae** (Fig. 8-2). Cats, the dog family **Canidae,** and other flesh eating animals are placed in the order **Carnivora.** These in turn are grouped with other milk-producing, hair-bearing animals in the class **Mammalia.** The mammals, birds, amphibians, snakes, fish, and a few lower groups form the phylum **Chordata.** Most of these have a vertebral column and other related structures. All of the kinds of animals are placed in the kingdom **Animalia.**

The kingdom is the largest and most inclusive category of organisms. Each level below this has fewer species since more differences are considered. The entire hierarchy of classification is: kingdom, phylum, class, order, family, genus, and species, as shown in the classification of cats:

KINGDOM	Animalia
PHYLUM	Chordata
CLASS	Mammalia
ORDER	Carnivora
FAMILY	Felidae
GENUS	*Felis*
	Panthera
	Lynx
SPECIES	*F. domestica*—domestic cats
	P. pardus—leopard
	P. leo—lion
	P. tigris—tiger
	L. rufus—bobcat

In strict practice the term *division* is used in plants in place of phylum.

The starting point in modern biologic nomenclature is generally accepted to be the system used by Linnaeus in the tenth edition of his *Systema Naturae,* published in 1758. Some plant classifications date from his 1753 work, *Species Plantarum.*

THE KINGDOMS OF LIFE

Until the invention of the microscope, people had no trouble in grouping living organisms. If it was green and had roots, it was a plant. If it moved around and ate things, it was an animal. Biologists worked with the Kingdom Plantae and the Kingdom Animalia.

Many microscopic organisms are green and they move about freely. The problem that this presented was either ignored by calling them both plants and animals or solved by lumping the problems in a new kingdom called Protista.

With the widespread use of the electron microscope it became evident that the bacteria and blue-green algae differ from all other organisms in several fundamental ways. The most significant distinctions involve the lack of membranes around a nucleus and the absence of other organelles that are found in higher plants and animals.

These differences led to the establishing of two major categories of organisms, the *procaryotes* and *eucaryotes,* based on the absence or presence of a distinct membrane-enclosed nucleus. Because of this, bacteria and blue-green algae were placed in a separate kingdom called *Monera.*

In 1969 R. H. Whittaker published an article in *Science,* detailing reasons for establishing the fungi as a separate kingdom. Since then several people have made modifications of this five kingdom scheme. Since many schemes of classification have been proposed to deal with larger categories such as kingdom and phylum, Table 8-1 gives some indication of the various interpretations of the living organisms.

In a comment on the different schemes, one writer has suggested (jokingly, we hope) that they tell more about the originators of classifications than about the organisms themselves.

The five kingdom scheme of classification

Although there are some problems with this as with any other scheme of classification, it seems to come close to being a suitable system. As we will see in detail later, it is based on differences and similarities in nuclear organization, mode of nutrition, and tissue complexity.

Procaryotes—Kingdom Monera

Those organisms that do not show much internal organization and are without an organized system of membranes around a nucleus are called *procaryotes,* meaning first, or primitive, nucleus. They also lack organized mitochondria and plastids and complex flagella. They obtain food primarily by absorption. These organisms include the bacteria and blue-green algae and are placed in the kingdom *Monera.*

Among the five phyla are the Cyanophyta, commonly called the blue-green algae, and the Eubacteriae or true bacteria. Although the characteristics of the kingdom Monera would seem to stress structures lacked, the organisms classified in this kingdom are living and have the necessary structures and functions to carry on all of the life activities. For example, they do have chromosomes and DNA in spite of lacking a nuclear membrane. They do carry on all sorts of metabolic activities even though their enzymes are not enclosed in mitochondria or plastids, and the motile forms do move even though they lack a "typical" flagellar structure. Details of these will be studied in the following chapter.

Eucaryotes

All other organisms do have the membrane enclosed nucleus, which is the basis of the name Eucaryote (true nucleus). Membrane enclosed mitochondria are also present. In addition, many

Table 8-1 The kingdoms of life*

Basic	3 Kingdom	3 Kingdom	4 Kingdom	5 Kingdom
Plantae	Protista	Monera	Monera	Monera
Animalia	Plantae	Plantae	Protista	Protista
	Animalia	Animalia	Plantae	Plantae
			Animalia	Fungi
				Animalia

*The phyla included in these kingdoms may vary with the author of the scheme.

eucaryotes have distinct plastids and other organelles. Those that have flagella show an internal flagellar structure of fibrils in a 9 + 2 arrangement; that is, there are two central fibrils surrounded by a ring of nine more, all making up the elongate whiplike structure. Some are unicellular, whereas others have varying degrees of complexity. They have varied types of nutritive patterns. Some make their food by photosynthesis; some absorb it from the surroundings, and others directly ingest the nutrients.

Kingdom Protista

Organisms classified in the Kingdom Protista are primarily single cells or colonies of single cells with nuclear membranes and mitochondria. Many have plastids and other organelles. If present, the flagella show the 9 + 2 arrangement of fibrils. The mode of nutrition is diverse—some are photosynthetic, some absorptive, some ingest, and some have combinations of these. They have varied means of reproduction, including both asexual division as well as true meiosis and sex cell union.

Included among the many phyla are: the Euglenophyta, the euglenalike organisms; the Chrysophyta, the golden algae; the Sporozoa, the entirely parasitic sporozoans; Zoomastigina, the animal flagellates; and the Ciliophora, the ciliates.

There are more than 100,000 species in this large and diverse group. In older systems of classification the Kingdom Monera and the first group of Protista are included with the Plants. The remaining phyla are called protozoa and are grouped with the animals. Lynn Margolis suggests that the red, brown, and green algae and the slime molds should also be included here.

Kingdom Fungi

The eucaryotic organisms placed in the Kingdom Fungi lack photosynthetic pigments and plastids. Their mode of nutrition is by absorption from a suitable food supply. They usually have many nuclei enclosed in a tubular threadlike mycelium. Mycelial walls are present and sometimes inner dividing walls. Tissue formation, other than in reproductive areas, is missing or primitive. Reproduction may include both asexual and sexual modes that in higher forms take place in a complex life cycle.

The phyla of the fungi include: the Myxomycota or slime molds, the Zygomycota or conjugation fungi, the Ascomycota or sac fungi, and the Basidiomycota or club fungi. The approximately 50,000 species include common molds, yeasts, mushrooms, and related forms.

Kingdom Plantae

Plants are characterized by multicellular eucaryotic cells with distinct cell walls. They frequently have vacuoles in the cell. Photosynthetic pigments are contained in plastids. The chief mode of nutrition is by photosynthesis. Plants show increasing structural complexity, forming organs for photosynthesis, anchorage, and support. Higher forms have specialized photosynthetic, vascular, and covering tissues. Reproduction is primarily sexual with an alternation of haploid and diploid generations. The haploid forms are progressively less complex in the higher members of the kingdom (Table 8-2).

The kingdom comprises those algae not listed with the Protista and also the higher green plants (Table 8-3). Representative algal phyla include: Rhodophyta or red algae, Phaeophyta or brown algae, and Chlorophyta or green algae. The other progressively more complex phyla are: the Bryophyta or liverworts and mosses and the Tracheophyta or vascular plants.

The algal phyla are simply constructed; some species are unicellular (one cell); some consist of a linear series of cells or in some cases spheres or sheetlike masses. They occur in water, both fresh and salt. They appear to have at most a primitive level of tissue structure. Some more complex "seaweed" types may grow to a length of over 100 feet. These may show some specialization in anchoring and floatlike devices. Reproductive spores are formed in sporangia (spore cases),

Table 8-2 Some characteristics of plants

Plant	Chloro-phyll	Multi-cellular embryos	True leaves, stems, and roots	Conducting tissues (phloem and xylem)	Flowers	Seeds exposed (naked)	Seeds enclosed
Algae	+	−	−	−	−	−	−
Liverworts and true mosses	+	+	−	−	−	−	−
Club "mosses," horsetails, and ferns	+	+	+	+	−	−	−
Gymnosperms (conifers and allies)	+	+	+	+	−	+	−
Angiosperms (flowering plants)	+	+	+	+	+	−	+

Table 8-3 The number of plant species

Plant	Number of species*
Green algae	7,000
Brown algae	1,500
Red algae	4,000
Liverworts	9,000
True mosses	14,000
Club mosses	11,000
Horsetails	25
Ferns	10,000
Gymnosperms (conifers)	500
Angiosperms (flowering plants)	300,000
TOTAL	350,000

*This is a rough figure since numbers are not exact and some groups are omitted.

which are usually of simple structure. When sex cells are present, a gametangium (gamete case) is formed. Fertilization results in a zygote, which grows directly into the mature form. No embryo is produced.

The bryophyte and tracheophyte plants are much more complex. They show increasing specialization with the formation of complex vascular tissues, reproductive organs, and supporting tissues. The reproductive cycle includes an alternation between a multicellular spore-producing plant and a multicellular gamete-producing plant. Fertilization results in formation of a multicellular embryo that is parasitic for some time. Different parts of the embryo produce the different parts of the mature plants. The higher plants are mostly terrestrial although some types may live in water. Over two thirds of the Kingdom Plantae are flowering plants. These produce the flowers, seeds, fruit, and vegetation that we associate with the green world.

Kingdom Animalia

The animals are eucaryotic organisms that do not have photosynthetic pigments, plastids, or cell walls. Most of them ingest their food into an internal digestive cavity. Some forms lack this cavity and absorb nutrients. Higher forms have an extremely complex level of organization with sensory and motor systems. Reproduction is primarily sexual with only a few lower groups showing haploid forms (Tables 8-4 and 8-5).

Except for the primitive Mesozoa and Porifera or sponges, the remaining phyla are relatively complex multicellular organisms with considerable tissue differentiation. These groups are placed in a subkingdom Eumetazoa. Metazoan animals are multicellular with bodies usually composed of distinct tissues, organs, and often with several organ systems. The cells contain centrioles and lack cell walls. They possess multicellular reproductive organs and generally pass through distinct embryo and larval stages. Adult tissues are formed from either two or three embryonic tissue layers.

It is thought that metazoans evolved from

Table 8-4 Some characteristics of animals

Phylum	Germ layers	Symmetry	Segmented body	True body cavity (coelom)	Skeleton	Vertebral column
Porifera	2	− or radial*	−	−	Spicules, fibers of spongin*	−
Coelenterata	2	Radial	−	−	Perisarc, mesoglea, limy*	−
Ctenophora	3	Biradial*	−	−	−	−
Platyhelminthes	3	Bilateral	−	−	−	−
Nematoda (Aschelminthes)	3	Bilateral	−	−	−	−
Annelida	3	Bilateral	+	+	−	−
Mollusca	3	− or bilateral*	−	+ (small)	Calcareous shells*	−
Arthropoda	3	Bilateral	+	+	Chitin	−
Enchinodermata	3	Radial, biradial*	−	+	Calcareous plates*	−
Chordata	3	Bilateral	+	+	Cartilage, bone*	+*

*In general, the characteristics are typical for a majority within the phyla.

Table 8-5 Numbers of animal species

Phylum	Number of species*	Phylum	Number of species
Porifera (po -rif′ er a) (Gr. *poros*, pore; *ferre*, to bear)	10,000	Mollusca (mo -lus′ ka) (L. *mollis*, soft)	80,000
Coelenterata (*Cnidaria*) (se len ter -a′ ta) (Gr. *koilos*, hollow; *enteron*, digestive cavity) (ni -da′ ri a) (Gr. *knide*, nettle)	9,000	Arthropoda (ar -throp′ o da) (Gr. *arthron*, joint; *pous*, appendage or foot)	1,000,000
Platyhelminthes (plat i hel -min′ thez) (Gr. *platys*, flat; *helmins*, worm)	13,000	Echinodermata (e ki no -dur′ ma ta) (Gr. *echinos*, spiny; *dermos*, covering or skin)	5,500
Nematoda† (Aschelminthes) (Gr. *askos*, cavity; *helmins*, worm)	10,000	Chordata (kor -da′ ta) (L. *chorde*, chord or string)	40,000
Annelida (a -nel′ i da) (L. *annellus*, little ring; *eidos*, like)	7,000		

*Does not include all phyla, groups, or species.
†Classified as Nematoda in this textbook. Phylum Aschelminthes includes many groups.

Table 8-6 Summary of the origins of mesoderm

Superphylum	Phylum	Origin of mesoderm	Characteristics
Radiata	Coelenterata (Cnidaria)	Ectoderm	Mesoderm absent in some species; when present, it is not well developed but consists largely of a mass of jellylike materials with few cells embedded in it; often called mesoglea (Gr. *mesos*, middle; *gloia*, jelly)
Acoelomata	Platyhelminthes Nemertea	Ectoderm	Mesoderm remains as solid layer
Pseudocoelomata	Acanthocephala Aschelminthes (Nematoda)	Ectoderm	Mesoderm does not form solid layer, but tissues collect in limited areas between ectoderm and endoderm
Schizocoelomata	Bryozoa Mollusca Annelida Arthropoda	Early mesoderm from ectoderm; later mesoderm from endoderm	Mesoderm eventually splits into two layers, and space between them forms true coelom
Enterocoelomata	Echinodermata Chordata	Early mesoderm from endoderm	Mesoderm forms pouches that fill space between ectoderm and endoderm; pouches separate off and form true coelom

some ancestral unicellular form of life and then became multicellular, retained and improved their ability to move about, and lost any photosynthetic capacity. In general, metazoan animal bodies possess a structure that promotes a way of life based on ingestion and locomotion.

Their bodies show a characteristic symmetry: radial in the phyla Cnidaria or coelenterates and Ctenaphora or comb jellies and either bilateral or biradial in the rest.

The animals with complex embryos are further characterized by the absence of, or type of formation of, a body cavity (Table 8-6). The phylum Platyhelminthes does not have a body cavity and is placed in the group of *acoelomates* (Gr. *a-*, without; *coeloma*, true body cavity). The roundworms, phylum Aschelminthes, and others have a cavity but it is not lined; that is, it is "false," and these groups are called *pseudocoelomates*.

The remaining phyla do have a coelom and are grouped according to the way they are formed. The *schizocoels* include the phyla Mollusca or mollusks, the Annelida or segmented worms, and the Arthropoda or arthropods.

The final group, the *enterocoela*, includes the Echinodermata, or echinoderms, and the Chordata, or chordates. All of the groups listed previously contain other phyla.

Some contributors to the knowledge of
CLASSIFICATION

Theophrastus (370-285 B.C.)

A Greek who was a student of Aristotle and is called the "Father of Botany." He described about 500 plants that he classed as herbs, undershrubs, shrubs, and trees. He wrote a *History of Plants*.

Pliny the Elder (23-79 A.D.)

A Roman literary man and general who described nearly a thousand plants, many of which were useful for medicinal purposes. His *Natural History* of thirty-seven volumes contained half-true data on natural history collected from his predecessors and is not considered too important in the history of botany.

Konrad von Gesner (1516-1565)

A Swiss naturalist and zoologist who published *Historia animalium* and founded the first zoologic museum.

Andrea Cesalpino (1519-1603)

An Italian physician and herbalist who classified plants into fifteen classes, largely on the basis of fruits and flowers (1583).

Gaspard Bauhin (1560-1624)

A Swiss naturalist who published excellent descriptions of nearly 6,000 species of plants (1623). He used a system of binominal nomenclature (a genus name and a species name) and, by means of his accurate descriptions, attempted to clear up the confusion of plant classification that had existed for years.

Mathias de Lobel (Lobelius) (1538-1616)

He pointed out that leaves are valuable in plant classification and divided plants into two groups on this basis. His great work *Kruydtboeck* (1581) contained many excellent wood engravings.

Linnaeus

Cuvier

John Ray (1628-1705)

An English biologist who wrote the *Historia Planatarum* (1686-1704) and used the terms dicotyledons and monocotyledons (embryonic seed leaves). He grouped animals on the basis of opposite traits. His catalog of plants was standard reference and laid the foundation for Linnaeus.

Joseph P. de Tournefort (1656-1708)

A French botanist who was the first to provide the genera of classification systems with careful descriptions, thus setting genera apart from species quite definitely.

Carolus Linnaeus (1707-1778)

A Swedish biologist who is considered the "Father of Modern Classification" because of his revisions and reorganizations of biologic nomenclature. He used the binary (two name) system in the descriptions of hundred of species much better than previous workers. He wrote *Systema Naturae* (1735) and *Genera Planatarum* (1737). (*Historical Pictures Service, Chicago.*)

Johann F. Blumenbach (1752-1840)

A German anthropologist and anatomist who classified animals into five classic varieties (1775) and called attention to the taxonomic problem of man himself.

Georges L. Cuvier (1769-1832)

A French naturalist who attempted to show the relationship between structure and function, not only in living animals, but also from fossil remains. (*The Bettman Archive, Inc.*)

Augustin P. de Candolle (1778-1841)

A Frenchman who published his *Théorie élémentaire de la botanique* (1813) in which he laid down the laws of plant classification so definitely that the natural system was permanently established.

Constantine S. Rafinesque (1784-1840)

A Frenchman who made a classification of medical plants. He came to the United States in 1802 and in 1815 was Professor of Botany at Transylvania College, Kentucky.

Louis Agassiz (1807-1873)

A Swiss naturalist and teacher at Harvard University who did much to focus attention on living organisms.

Adolphe T. Brongniart (1801-1876)

He proposed (1843) a system of classifying plants, dividing them into *Cryptogamae* (without flowers) and *Phanerogamae* (with flowers). The latter contained monocotyledonous and dicotyledonous types.

Asa Gray (1810-1888)

Gray

An American who, using European systems of classification, discovered and described the plants of the United States in the nineteenth century. He was the first widely known botanist in the United States. He improved the system and wrote *Gray's Manual of Botany* (1848). The Gray Herbarium of Harvard University is named in his honor. *(Historical Pictures Service, Chicago.)*

Adolph Engler (1844-1930)

A German who proposed a system of classifying plants that formed a basis for modern systems of taxonomy. He was assisted by K. Prantl at the University of Berlin.

Libbie H. Hyman (1888-1969)

An American zoologist who spent years in research and in writing her extensive treatise of several volumes, *The Invertebrates*. Special reference is given to the anatomy, embryology, physiology, ecology, and taxonomy of invertebrate animals.

Review questions and topics

1 Contrast the subkingdoms Thallophyta and Embryophyta in as many ways as possible.
2 Define (a) sporangia, (b) gametangia, (c) archegonium, (d) antheridum, (e) gametophyte, (f) sporophyte, (g) egg, (h) sperm, and (i) zygote.
3 Give reasons for using scientific names and classification rather than common names and personal systems of classification.
4 What is meant by binominal nomenclature, and what is its importance?
5 In the construction of a scientific classification, explain each of the following: kingdom, subkingdom, phylum, class, order, family, genus, and species.
6 What is meant by "true" leaves, stems, and roots? Why are certain structures that resemble the foregoing not considered to be true stems, leaves, and roots?
7 What are the chief characteristics that distinguish each of the following groups from each other: (a) algae and fungi, (b) liverworts and true mosses, (c) club "mosses," horsetails, and ferns, and (d) gymnosperms and angiosperms?
8 How many species are there in each phylum? What is the total number of species of animals? Explain how this number may vary.
9 Can you give some explanations why there are so few species in one phylum and so many thousands in another?
10 Why are the phyla of animals listed in a particular sequence? Could this arrangement be changed? Explain why this might be possible.
11 Contrast acoelomate and coelomate; unsegmented and segmented; invertebrate and vertebrate; diploblastic and triploblastic; radial, biradial, and bilateral symmetry.
12 What do you think the status of biology might be today if we did not have a scientific method of naming, classifying, and identifying animals?

Selected references

Barnes, R. D.: Invertebrate zoology, ed. 3, Philadelphia, 1973, W. B. Saunders Co.

Barrington, E. J. W.: Invertebrate structure and function, Boston, 1967, Houghton Mifflin Co.

Blackwelder, R. E.: Classification of the animal kingdom, Carbondale, Ill., 1963, Southern Illinois University Press.

Blair, W. F., and others: Vertebrates of the United States, New York, 1957, McGraw-Hill Book Company.

Bold, H. C.: The plant kingdom, Englewood Cliffs, New Jersey, 1964, Prentice-Hall, Inc.

Boughey, A. S., editor: Population and environmental biology, Belmont, Calif., 1967, Dickenson Pub. Co., Inc.

Cronquist, A.: The divisions and classes of plants, Bot. Rev. **26:**425-482, 1960.

Dittmer, H. C.: Phylogeny and form in the plant kingdom, Princeton, New Jersey, 1964, D. Van Nostrand Co., Inc.

Fingerman, M.: Animal diversity, New York, 1969, Holt Rinehart and Winston, Inc.

Gottlieb, J. E.: Plants: adaptation through evolution, New York, 1968, Reinhold Publishing Corp.

Hyman, L. H.: The invertebrates; vol. I, Protozoa through Ctenophora, 1940; vol. II, Platyhelminthes and Rhynchocoela, 1951; vol. III, Acanthocephala, Aschelminthes, and Entroprocta, 1951; vol. IV, Echinodermata, 1955; vol. V, Smaller coelomate groups, 1959, New York, McGraw-Hill Book Co.

Jensen, W. A., and Kavaljian, L. G., editors: Plant biology today, Belmont, Calif., 1963, Wadsworth Publishing Co., Inc.

Margulis, L.: Five kingdom classification and the origin and evolution of cells, Evolutionary Biology **7:**45-78, 1974.

Meglitsch, P. A.: Invertebrate zoology, New York, 1967, Oxford University Press, Inc.

Moore, J. A., editor: Ideas in modern biology, Garden City, New York, 1965, Natural History Press, Division Doubleday & Company, Inc.

Ross, H. H.: Biological systematics, Reading, Mass., 1974, Addison-Wesley Publishing Co.

Scagel, R. F., and others: An evolutionary survey of the plant kingdom, Belmont, Calif., 1965, Wadsworth Publishing Co., Inc.

Simpson, G. G.: Principles of animal taxonomy, New York, 1961, Columbia University Press.

Whittaker, R. H.: New concepts of kingdoms of organisms, Science **163:**150-156, 1969.

Selected readings on identification of living things

Abbott, I.: How to know the seaweeds, Dubuque, Iowa, 1974, William C. Brown Company, Publishers.

Baerg, H. J.: How to know the western trees, Dubuque, Iowa, 1973, William C. Brown Company, Publishers.

Helfer, J. R.: How to know the rocks and minerals, Dubuque, Iowa, 1970, William C. Brown Company, Publishers.

Jaques, H. E.: How to know the insects, Dubuque, Iowa, 1947, William C. Brown Company, Publishers.

Jaques, H. E.: How to know the land birds, Dubuque, Iowa, 1947, William C. Brown Company, Publishers.

Jaques, H. E.: How to know the living things, Dubuque, Iowa, 1946, William C. Brown Company, Publishers.

Jaques, H. E.: How to know the plant families, Dubuque, Iowa, 1948, William C. Brown Company, Publishers.

Prescott, G. W.: How to know the aquatic plants, Dubuque, Iowa, 1969, William C. Brown Company, Publishers.

Prescott, G. W.: How to know the freshwater algae, Dubuque, Iowa, 1954, William C. Brown Company, Publishers.

Raun, G.: How to know the snakes, Dubuque, Iowa, 1974, William C. Brown Company, Publishers.

PART TWO

Microbiology

Viruses, blue-green algae, and bacteria
Kingdom Protista—simple algae
Kingdom Protista—protozoans
Kingdom Fungi—molds and mushrooms

9 Viruses, blue-green algae, and bacteria

Viruses
Monera
    Phylum Cyanophyta (blue-green algae)
    Phylum Schizophyta (bacteria)

Part two of this text deals with a variety of organisms usually studied by microbiologists. The particular specializations and the organisms they study are: bacteriology, bacteria; virology, viruses; phycology, algae; protozoology, protozoa; and mycology, fungi. Depending on where these microorganisms are found and their possible importance, microbiologists may study either the organisms of medical importance or those found in water and sewage, food, milk, air, soil, or even outer space. Industrial and commercial aspects are studied by industrial microbiologists.

The topics considered in this chapter comprise a rather strange collection about which there is much doubt and discussion.

The viruses are macromolecular complexes that appear to have only the process of reproduction in common with truly living things. Bacteria and blue-green algae are simple living organisms that lack an organized nucleus and do not have membrane-bound organelles.

VIRUSES

A virus is a submicroscopic structure consisting of a nucleic acid core surrounded by a protein coat. Either DNA or RNA may be found in the viral core, although DNA is more common. The coat is composed of one or more kinds of proteins, usually forming a symmetrical pattern of various types. Some viruses appear rather simple (Fig. 9-1), while others have a more complex structure with several types of projections. Since viruses range in size from less than 10 to more than 200 nm (nanometers), they cannot be seen with an ordinary microscope. Information about their structure has been obtained through the use of the electron microscope and through such techniques as x-ray diffraction.

The question of whether viruses are living or nonliving is still subject to discussion, although the controversy has not stopped virology from bcoming one of the fastest growing areas of biology. Researchers in the field of virology do not agree completely on the use of standard biological taxonomy for viruses. There is, however, a growing use of "families" to designate particular groups of viruses. For example, members of 16 distinct viral "families" are known to infect vertebrates.

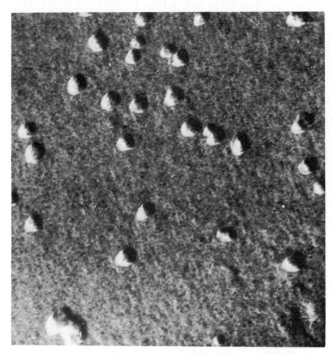

Fig. 9-1. Internal cork virus (ICV) of sweet potato. *(Electron micrograph courtesy M. A. Hayat, University of Dayton.)*

About all that can really be said about "living" viruses is that they are the smallest macromolecules capable of reproducing themselves. Although viruses possess DNA or RNA, any reproduction must take place in the presence of components of a living cell. Viruses, then, may be called obligate parasites; that is, they exist only in the presence of a living cell.

Viruses may be crystallized, yet they are capable of assuming typical activity later. If they are disrupted chemically or physically, they can reassemble various parts into an intact virus.

Viruses produce many kinds of diseases in plants, animals, and human beings. Many viruses that cause animal diseases show an affinity for specific tissues and organs, such as nerves, skin, respiratory organs, and internal organs.

Among the human viral diseases are measles, mumps, smallpox, chickenpox, influenza, rabies, poliomyelitis, yellow fever, warts, cold sores, shingles, virus pneumonia, infectious hepatitis, and infectious mononucleosis. Effective vaccines have been developed against many human virus diseases, such as smallpox, yellow fever, and poliomyelitis.

Among viral diseases of animals are foot-and-mouth disease of cattle, rabies, swine flu, hog cholera, fowl pox, parrot fever (psittacosis) of birds (and man), cattle plague, and dog and cat distemper. Effective vaccinations have been developed for many of these diseases. Other viruses attack birds, fish, insects, bacteria, and blue-green algae.

Among the viral diseases of plants are tobacco mosaic disease, tomato mosaic, cucumber and melon mosaic, the yellows disease of peach, aster yellows, curly top of beets, and many others.

Bacteriophage

Bacteriophage (bacteria; Gr. *phagein*, to destroy) is a type of virus that may rapidly destroy certain bacterial cells (Figs. 9-2 and 9-3). Some bacteriophages seem to attack only one species or strain of bacteria, whereas others attack several. The phage particles enter the bacterial cell, multiply, and eventually destroy the cell. Even

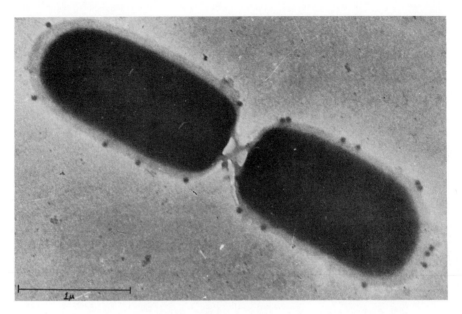

Fig. 9-2. Bacteriophage attached to bacterial cells *(Escherichia coli)* as shown by an electron micrograph. Bacterial cells are shown dividing by fission, and nineteen phage particles are shown adsorbed to the cell walls.
(Courtesy Society of American Bacteriologists; by permission of Dr. S. E. Luria.)

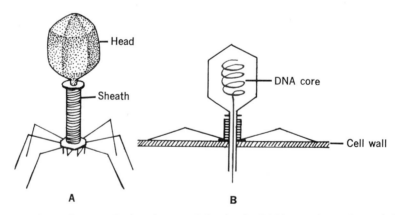

Fig. 9-3. Phage virus. **A,** Entire virus. **B,** Injecting its DNA core into a bacterial cell.

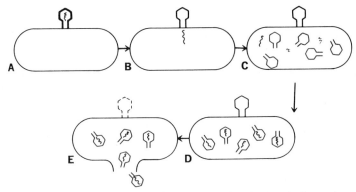

Fig. 9-4. Bacteriophage life cycle. **A,** Virus particle attached to the bacterium; **B,** nucleic acid core migrates into the cell; **C,** DNA replicates and induces the formation of new protein coats from the cell contents; **D,** new virus particles are assembled; **E,** cell bursts, releasing a new generation of the virus.

bacteria have their destructive, disease-producing pathogens!

A virus attaches itself to the bacterium, enters quickly, and minutes later the bacterium ruptures, releasing many new viruses, each an exact replica of the original invading virus. By the use of radioactive tracers it is possible to find out how the virus makes these replicas of itself from the materials available. By labeling with radioactive atoms either the substances of the virus or the medium in which it grows, experimenters can trace the materials and events resulting in the construction of new virus offspring.

The virus of the T_2 strain that infects the intestinal bacterium *Escherichia coli* has a hexagonal head and a tail and is approximately 0.0007 of an inch long. The outer layer consists of protein that has the ability to attach itself to the surface of the bacterium of the proper species. Inside the head the core contains DNA (Fig. 9-3).

Some of the most fascinating experiments with viruses are studies of their heredity by means of tracer techniques. Our knowledge of how bacterial viruses reproduce may be summarized briefly as follows: (1) the virus can attach itself to the surface of a bacterial cell by means of its protein coat; (2) the virus then pours its DNA into the cell, the emptied protein coat remaining outside and serving no further purpose; (3) the DNA within the cell makes replicas of itself, using as raw materials the nucleic acids of the bacterium and the fresh substances absorbed by the bacterium from its surroundings; and (4) the DNA induces the synthesis of new protein in the cell, and the units of protein combine with DNA replicas to form about 200 exact copies of the original (parent) virus (Fig. 9-4). About 40% of the parent virus DNA is saved and appears in the descendants.

These facts give only a brief outline of the process, but they assist in solving the problem of how organisms build structural copies of themselves and pass on their heredity from generation to generation.

MONERA

Many of the characteristics of cells described in earlier chapters should have been prefaced by the phrase "except in blue-green algae and bacteria." These two groups of organisms are procaryotes (Gr. *pro,* before; *karyon,* kernal); that

Table 9-1 Comparison of procaryote and eucaryote* cells

Cell component	Procaryotic cells	Eucaryotic cells
Cell wall	Contain amino sugars and muramic acid	Amino sugars and muramic acid not present
Nuclear membrane	Absent	Present
Chromosomes	Contains DNA only	Contain DNA and protein
Mitochondria	Absent	Present
Golgi apparatus	Absent	Present
Lysosomes	Absent	Present
Endoplasmic reticulum	Absent	Present
Chlorophyll	Contain chlorophyll and other pigments, but not in chloroplast	Chlorophyll, if present, is in chloroplasts
Flagella	Lack 9-2 fibrillar structure	Have characteristic 9-2 fibrillar structure

*With a complex nucleus.

is, they lack an organized nucleus, since they do not have a nuclear membrane or complex chromosomes (see Table 9-1).

They also lack other membrane-bounded structures, such as mitochondria, lysosomes, plastids, Golgi apparatus, and endoplasmic reticula. However, it is probable that the enzymatic functions of these membranes may take place along identations of the cell membrane.

If flagella are present in bacteria, they do not have fibrils arranged in a pattern of 2 in the center, surrounded by a circle of 9 more. However, this 9-2 pattern is present in flagella of most other organisms.

Because of these missing structures, the blue-green algae and the bacteria are placed in the kingdom Monera. The latest edition of *Bergey's Manual of Determinative Bacteriology,* the most widely used reference for bacterial taxonomy, recognizes this grouping but chooses to call it *Kingdom Procaryotae.*

Phylum Cyanophyta (blue-green algae)

The blue-green algae are placed in the phylum Cyanophyta (Gr. *kyanos,* blue; *phyta,* plant). In addition to the missing features indicated above, they have certain special characteristics.

They are simple, unicellular forms, often forming a colony of similar cells with little differen-

tiation among the cells. The cells do not have an organized nucleus, although the DNA content may be concentrated in one area. The cells are often surrounded by a slimy, gelatinous sheath. No flagellated cells are present. Furthermore, the cells do not contain the large watery vacuoles found in higher plant cells.

Like the photosynthetic plants, blue-green algae cells contain chlorophyll a and beta carotene. Two other carotenoid pigments, myxoxanthine and myxoxanthophyll, are often present. Two special phycobilin pigments are also present in the cells of the blue-green algae—phycocyanins and phycoerythrins. These pigments appear to function in the energy transfers that take place during photosynthesis. While plastids do not appear to be present within the cell, organized photosynthetic lamellae, similar to the grana of green plants, can be seen with the electron microscope.

Most blue-green algae live in fresh water, although many are marine and others are found on wet rocks and in other moist places. At times, they become so abundant that they produce ''water blooms,'' resulting in foul odors and the murky appearance and strange taste of the water.

Some species grow in hot springs with temperatures of over 80° C., where they precipitate calcium and magnesium salts to produce traver-

Table 9-2 Some characteristics of blue-green algae and bacteria

Characteristic	Blue-green algae	Bacteria
Photosynthesis and pigments	By chlorophyll a that, like chlorophyll b of certain other plants, absorbs light from the red portion of the spectrum; oxygen given off Other pigments include carotenoids and phycobilins	Some species, for example, photosynthetic purple-sulfur bacteria and green-sulfur bacteria, contain chlorophyll-like pigments that differ only in minor structural detail from chlorophyll a and b Obligate anaerobic organisms that require reduced sulfur (usually H_2S) for photosynthetic metabolism Bacterial chlorophylls principally absorb longer wavelengths (outside visible light spectrum in near infrared region); no oxygen given off
Nucleus	Not organized (amorphous)	Not organized (amorphous)
Cell wall	Distinct	Distinct
Number of cells	Unicellular or in colonies	Unicellular or in colonies
Slimy, sheathlike covering	Often present	Present in some species
Flagella	Absent	Present in certain species
Fixation of atmospheric nitrogen	By some species	By some species
Average size	Usually larger than bacteria	Coccus types (about 1μm diameter) Rod types (about 0.5×2 μm) Spiral types (2×25 μm)
Reproduction	Asexual: by (1) fission; (2) spores; (3) fragmentation of filaments or colonies Sexual: not known	Asexual: by (1) fission; (2) endospores Sexual: by conjugation

tine, a chalky material with brilliant colors. Calcium carbonate in lakes may be precipitated by the algae to form marl, an earthy material formerly used on lime deficient soils.

One colorless species of blue-green algae, *Beggiatoa*, which is sometimes called a bacterium, is a sulfur oxidizing form. The chemical process relative to this oxidation may be illustrated as follows:

$$2 \ H_2S + O_2 \rightarrow 2 \ S + H_2O$$

Minute granules of elemental sulfur accumulate in the filaments of *Beggiatoa* as a result of this process.

Several species of blue-green algae, including *Nostoc* and *Anabena* (Figs. 9-5 and 9-6) are able to fix atmospheric nitrogen as do certain species of bacteria. Possibly these species also play a role in maintaining the fertility of certain soils.

Asexual reproduction occurs by (1) fragmentation of filaments or colonies, (2) spores, or (3) fission. Filamentous types may have occasional thick-walled cells, known as heterocysts (Gr. *heteros,* different; *kystis,* sac), in which the protoplasm becomes colorless. Fragmentation of the filament usually occurs in the heterocysts. In other filamentous types fragmentation occurs at points where two vegetative cells are separated by separation disks of gelatinous material. The parts of the filament that are separated (by disks or heterocysts) are called hormogonia (singular, hormogonium) (Gr. *hormos,* chain; *gonos,* off-

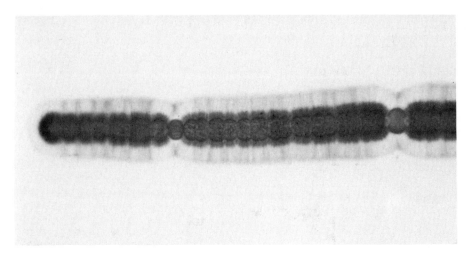

Fig. 9-5. *Anabena,* a filamentous, blue-green alga showing the external watery sheath and heterocysts.
(Courtesy General Biological Supply House, Inc., Chicago, Illinois.)

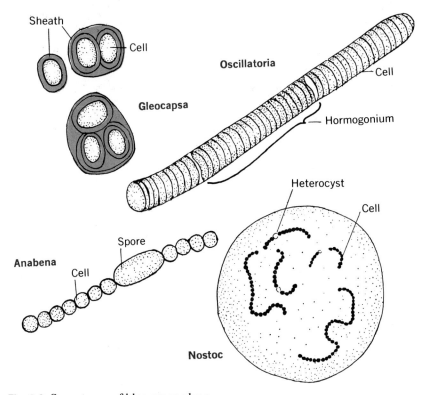

Fig. 9-6. Some types of blue-green algae.

spring). Colonies may also fragment because of the failure of the gelatinous sheath to hold them together.

Spores may be of two kinds. Numerous endospores (Gr. *endon,* within; *spora,* spore) may be formed by the repeated division of the cell contents. These endospores are released by a rupture of the cell wall. An akinete (Gr. *a-,* not; *kinein,* to move) is a resting cell in which the cell wall is enlarged and thickened before germination occurs under favorable conditions. Certain species reproduce asexually by fission that results from simple cell division. It is not known whether sexual reproduction occurs at all in blue-green algae.

It has been suggested that in the future, blue-green algae may play an important part in waste disposal. They would be used to capture nitrogen from the air or water by using solar energy. The resultant increases in the biomass of the algae could be used as fertilizer.

Examples of blue-green algae

Gleocapsa (Gr. *gloios,* jelly; L. *kapsa,* box) (Fig. 9-6) has a bluish green region, the color resulting from the diffused pigments (chlorophyll and phycocyanin). Several individual cells may be associated together and surrounded by a jellylike sheath. *Gleocapsa* is common on wet rocks and in other moist places. Reproduction is by cell division and colony fragmentation.

Oscillatoria (L. *oscillare,* to swing) (Fig. 9-6) consists of a series of independent plants associated in an unbranched, filamentous colony. The chlorophyll and phycocyanin are located in the outer region of the cell. Living filaments may swing back and forth, or oscillate, hence the name *Oscillatoria.* Soft gelatinous areas, called separation disks, may develop between cells and break the filament into hormogonia. A hormogonium may form a new colony. *Oscillatoria* is common in damp earth, on moist stones, and on flowerpots.

Nostoc (Fig. 9-6) is composed of a chainlike colony of individual cells arranged like a twisted string of beads. At intervals there are thick-walled, transparent heterocysts that serve to break the filament into hormogonia, as in *Oscillatoria.* A heterocyst may germinate to form a new filament. Numerous filaments may aggregate into a spherical colony surrounded by a sheath. *Nostoc* is common in fresh water and in soils.

Anabena (Figs. 9-5 and 9-6) resembles filaments of *Nostoc* but differs in that certain thick-walled, elongated spores (akinetes) may separate to form a threadlike colony. Heterocysts may be formed as in *Nostoc. Anabena* occurs chiefly as free-floating colonies in ponds and lakes. The gelatinous sheath is quite watery.

Phylum Schizophyta (bacteria)

Bacteria are placed in the phylum Schizophyta because they reproduce by splitting cells. Their numbers increase very rapidly as a result of this process, doubling as often as every 20 minutes during active growth. *Bergey's Manual,* mentioned previously, refers to this group simply as "the bacteria," and recognizes 19 major groups.

Most people are aware of bacteria primarily in relation to disease. Although it is true that there are many pathogenic (disease-producing) species, most are indifferent to man, and others are even beneficial.

Studies of bacteria began with the work of Leeuwenhoek, an early Dutch inventor of microscopes. With his primitive microscope (see Fig. B-3, appendix B), which was really not much more than a hand lens, he observed many types of bacteria and reported their activities to the scientific community of his day.

Today bacteria are studied with more complex light and electron microscopes, by biochemical tests, and by various physiologic and growth methods.

Morphology

Bacteria are quite small, ranging in size from about $5\mu m$ (1 micrometer is 1/250,000 of an inch) to about $0.1\mu m$. Most bacteria have one of three basic shapes: spherical, cylindrical or rod-shaped, or helical (spiral). Cells with these shapes are called coccus, bacillus, and spirillum, respectively. Bacterial cells may also occur in colonies, such as in clusters (as in *Staphylococcus*), in chains *(Streptococcus),* or in pairs, fours, and other groupings (Figs. 9-7 to 9-10).

As indicated earlier, bacteria do not have a very complex internal organization. They do, however, contain DNA, RNA, various proteins, enzymes, sugars, lipids, and amino acids as well as more complex molecules.

The bacterial cell is surrounded by a wall composed of various kinds of mucopeptides, amino acids, and carbohydrate derivatives. These substances add to the strength of the wall and assist in other functions. Because of their presence, bacterial cells react differently to various stains. The well-known Gram stain, named after its developer, takes advantage of this selective staining property of bacteria. Many activities of a particular bacterium can be inferred by its reaction to this stain.

Some bacteria secrete a complex polysaccharide or peptide that forms a gelatinous capsule around the cell wall. Many of the more virulent types of bacteria form this capsule. Studies of one of these types, a *Pneumococcus* that causes pneumonia, helped determine that DNA was the genetic substance of a cell (see Chapter 32).

Motility

Movement in bacteria is by means of slender flagella. The number and position of these flagella vary with the species (Fig. 9-10). Although the flagella are extensions of the cytoplasm, they differ structurally from the more complex organelles in other organisms, for bacteria lack the internal 9-2 arrangement of fibrils.

Metabolism

Most bacteria, like higher plants, require free oxygen for their normal activities and are known as aerobes (a′ er ob) (Gr. *aer*, air; *bios*, life). Other types can thrive only in the absence of free oxygen, and hence are known as anaerobes (an

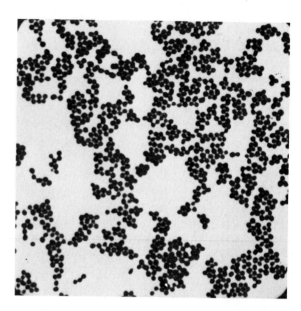

Fig. 9-7. *Staphylococcus:* Note the characteristic grapelike clusters.

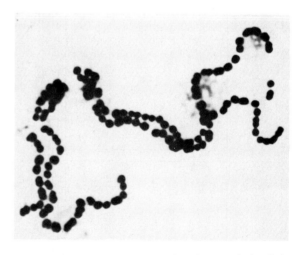

Fig. 9-8. *Streptococcus:* Note the characteristic chain formation.

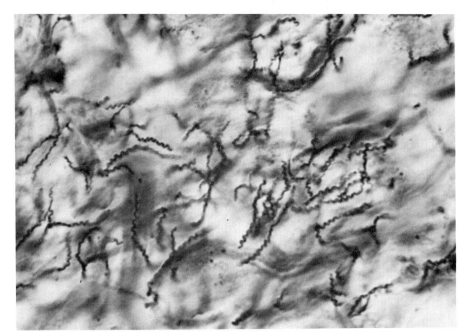

Fig. 9-9. *Treponema pallidum*, the spiral shaped organism that causes syphilis. *(Courtesy General Biological Supply House, Inc., Chicago, Illinois.)*

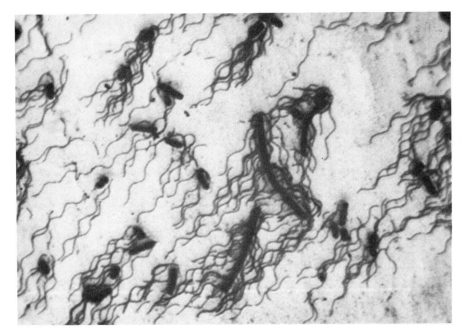

Fig. 9-10. Typhoid organism *(Salmonella typhosa)* stained to show flagella extending from the surface of rod-shaped bacteria. *(Courtesy General Biological Supply House, Inc., Chicago, Illinois.)*

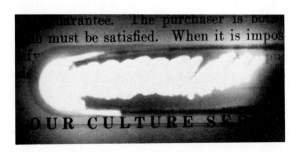

Fig. 9-11. Luminescent bacteria *(Photobacterium fischeri)* photograph in total darkness using only the light produced by a growing culture of rod-shaped organisms. *(Courtesy Carolina Biological Supply Company, Burlington, North Carolina.)*

-a′ er ob) (Gr. *an*, without). The anaerobes secure their oxygen from oxygen-containing foods, such as carbohydrates. Respiration in cells is similar to that considered in Chapter 28.

Various types of bacteria grow at different temperatures. Those growing best at temperatures of about 20° C. or below are known as psychrophiles (Gr. *psychros,* cold; *philein,* to love); those having an optimum temperature between 20° and 40° C. are mesophiles (Gr. *mesos,* middle), while those that grow best above 45° C. are called thermophiles (Gr. *thermo,* heat). In spite of the high temperatures, numbers of thermophilic bacteria are found in a great variety of places, including hot springs. Their cytoplasms seem to be especially constructed to live under such temperatures. Psychrophilic organisms are common in cold, deep waters, where they exist as saprophytes. Psychrophiles may decompose foods in cold storage. Most bacteria, however, are mesophiles. Saprophytic mesophiles are common in soils, water, and decomposing materials.

A few species of bacteria (Fig. 9-11) are able to emit light, which may continue for a time after they die. They may be present in decaying leaves, wood, meat, fish, and in salt water. They are commonly called luminescent or photogenic bacteria. Light production is thought to be caused by an oxidizing enzyme called luciferase (L. *lux,* light; *ferre,* to bear), which acts on its substrate material known as luciferin. The reaction is thought to be one in which the reduced luciferase combines with oxygen to emit light. When certain species are grown in a large volume, they produce light in such quantities that they may be photographed by their own light. One may read in a dark room with light produced under such conditions.

Bacteria vary widely in their methods of securing foods and energy. Most bacteria lack the chlorophyll that characterizes higher plants and, in general, cannot carry on photosynthesis. Some species, however, contain chlorophyll-like pigments by means of which they manufacture their foods. One species, *Halobacterium halobium,* contains the pigment *bacteriorhodopsin* in its cell membrane. It is able to convert light energy into metabolic energy.

From food and energy standpoints bacteria may be classified as follows.

AUTOTROPHIC BACTERIA

Autotrophic (Gr. *autos,* self; *trophe,* nourishment) bacteria require inorganic substances only, from which they manufacture their foods. They are usually classified as chemosynthetic or photosynthetic bacteria.

Chemosynthetic organisms are without photosynthetic pigments and obtain energy from chemical reactions (oxidations).

Among the chemosynthetic organisms are certain iron bacteria that utilize carbon dioxide of the atmosphere as a source of carbon and derive their energy from the oxidation of ferrous iron to basic ferric sulfate or insoluble ferric hydroxide. Some of the chemical processes relative to such iron bacteria may be illustrated as follows:

$$4\ FeCO_3 + O_2 + 6\ H_2O \rightarrow$$

Ferrous carbonate

$$4\ Fe(OH)_3 + 4\ CO_2 + Energy$$

Ferric hydroxide

Another group of chemosynthetic bacteria oxidize certain sulfur compounds and utilize some of the released energy in the manufacture of their foods. Two of the chemical processes relative to such sulfur bacteria may be illustrated as follows:

$$2\ H_2S + O_2 \rightarrow 2\ S + 2\ H_2O + \text{Energy}$$

Hydrogen   **Elemental**
sulfide   **sulfur**

$$2\ S + 2\ H_2O + 3\ O_2 \rightarrow 2\ H_2SO_4 + \text{Energy}$$

Sulfuric
acid

Photosynthetic organisms contain chlorophyll-like pigments and obtain energy from sunlight.

Among the photosynthetic organisms are certain sulfur bacteria that contain bacteriochlorophyll pigments and grow in environments containing sulfur compounds. Yellow and red carotenoid pigments may also be present in some species. In the presence of light energy, they photosynthesize foods, but do not liberate oxygen as is true of higher chlorophyll-bearing plants. Some of the chemical processes relative to such sulfur bacteria may be illustrated as follows:

$$2\ H_2S + CO_2 + \text{Light energy} \rightarrow$$

Hydrogen
sulfide   Carbohydrate $+ 2\ S + H_2O$

                              Elemental
                              sulfur

$$2\ S + 3\ CO_2 + 5\ H_2O + \text{Light energy} \rightarrow$$
$$2\ \text{Carbohydrate} + 2\ H_2SO_4$$

                              Sulfuric
                              acid

HETEROTROPHIC BACTERIA

Heterotrophic (Gr. *heteros,* different) bacteria require an organic source of carbon as food. They usually exist as saprophytes or parasites.

Saprophytes (Gr. *sapros,* dead or rotten; *phyta,* plants) generally live entirely on nonliving organic matter and are not ordinarily involved in disease production.

Fig. 9-12. Colonies of bacteria growing on an agar plate. Each group started with one cell.

Parasites or pathogens (Gr. *para,* beside; *sitos,* food) (Gr. *pathos,* suffering; *gen,* origin) can and do live in or upon other living organisms and may cause disease but may or may not be able to live as saprophytes.

Reproduction

The primary method of reproduction in bacteria is asexual fission (cell division), in which the parent cell divides crosswise into two unicellular organisms. In coccus types fission may occur in various planes, depending upon the species. Often after fission many newly formed cells remain together to form a colony. Different species produce colonies of different sizes, shapes, and even colors. Each species forms a definite type of colony (Fig. 9-12) whose specific characteristics are used for identification purposes. Under favorable growth conditions, fission may occur in approximately 20 to 30 minutes. Theoretically, a bacterial cell dividing at its maximum rate would, with continuous division, produce 4,700,000 quadrillion offspring in 24 hours. This mass would weigh approximately 2,000 tons. This never happens because sufficient food is not available in any one place, and the accumulation

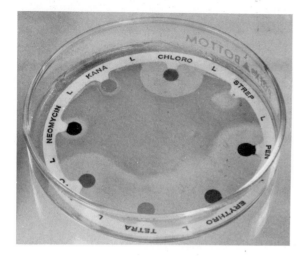

Fig. 9-13. Test for the sensitivity of bacteria to various antibiotics. The clear zones indicate an inhibition of bacterial growth as the antibiotic diffuses from the disk; darker areas are colonies of bacteria.

of waste products of metabolism prevents such reproduction.

Some bacteria under such environmental conditions as extremes of temperature and lack of moisture may form internal endospores. Bacterial spores may occur at the end of a rod (polar), at the center of a rod (central), or between these two points (subpolar or subterminal). Spores may be smaller than the diameter of the rod or they may be larger than the diameter of the rod at the time spores are formed, in which case the rods may appear spindle shaped, or club shaped. The size and location of a spore within a cell are definite for a particular species.

Spores are formed by condensing the cytoplasm into a spherical or ovoid mass that is surrounded by a resistant wall. A loss of water accompanies spore formation. When a spore encounters favorable environmental conditions, it absorbs water and germinates to form an active cell.

SEXUAL REPRODUCTION

Three types of genetic recombinations are known in bacteria—conjugation, transformation, and transduction. Conjugation is related to the sexual process in other organisms in that the parental cells actually come together.

Conjugation has been studied most in *Escherichia coli,* a common intestinal bacterium. This bacterium has a single chromosome with genes arranged in a definite order.

In 1946, Joshua Lederberg and Edward Tatum studied mutants of *E. coli* that were unable to synthesize certain amino acids. When two different mutants were grown on the same medium, some new combinations with characteristics of both ''parent'' strains were seen. Since this only happened when two bacteria were in contact, some sort of a sexual process was thought to have occurred.

Studies with the electron microscope in the 1950s showed that when bacteria from two different sexes (or mating types) come together, a cytoplasmic bridge is formed between them, the chromosome of the ''male'' is transferred to the ''female,'' and, presumably, the process of genetic recombination takes place. The male cell dies after the transfer. In many cases only part of the chromosome is transferred, and the new traits are governed by whatever genes are on that part.

In transformation a bacterium picks up strands of DNA that have come from dead cells. In transduction the new DNA is carried from one cell to another by viruses.

Harmful activities of bacteria
BACTERIAL DISEASES OF MAN

Louis Pasteur is generally credited with advancing the idea that bacteria cause disease. The work of Robert Koch, who studied the causes of tuberculosis and anthrax, helped this concept to gain acceptance.

Koch developed a procedure to illustrate that a particular organism was related to a particular disease. These methods have survived to this day in the form of Koch's postulates. In general, they are as follows:

1. The bacterium in question must always be found in cases of the particular disease.

2. It must be isolated from the diseased organism and grown in culture.

3. When injected into a healthy organism, it must produce the disease.

4. It must be isolated from the second organism and compared with the bacteria from the first organism.

The use of this procedure has helped in establishing the causes of the diseases shown in Table 9-3.

BACTERIAL DISEASES OF OTHER ANIMALS

Many bacterial diseases of man seem to have their counterparts in lower animals, although certain animal diseases do not affect human beings. Among the bacterial diseases of animals might be included such forms as tuberculosis of cattle and hogs; anthrax of sheep; chicken cholera; glanders of horses, sheep, and goats; pneumonia; septicemia in cattle; botulism in chickens and other animals; tularemia in rabbits; rat plague; and Bang's disease (brucellosis or undulant fever) in cattle.

BACTERIAL DISEASES OF PLANTS

It may seem surprising that there are many plant diseases caused by bacteria, but the list is growing constantly as further research is done in this highly important field of bacteriology. By his research on the fire blight of pears in 1879, Burrill of the University of Illinois proved that bacteria could cause plant diseases. Among the symptoms by which some of the bacterial plant diseases are recognized are blights, in which there is a rapid death of blossoms, young leaves, and stems, usually when they are not fully developed; rots, which are soft discolorations in which cell walls of the affected tissues are destroyed; wilts, in which the aerial parts of plants wilt and dry out, sometimes caused by the interference with the conduction of water in the xylem tissues; and cankers, which are dead depressions in the surface tissues caused by bacterial activities. Examples of plant diseases caused by specific bacteria are soft rot of carrots, root rot; pineapple

Table 9-3 Bacterial diseases of man

Bacterial disease	Causal agent
Boils, abscesses, pus	*Micrococcus (Staphylococcus) pyogenes*
General infections	*Streptococcus pyogenes*
"Sore throat" (certain types)	*Streptococcus pyogenes (hemolyticus)*
Meningitis	*Neisseria meningitidis*
Gonorrhea	*Neisseria gonorrhoeae*
Pneumonia	*Diplococcus pneumoniae*
Anthrax	*Bacillus anthracis*
Diphtheria	*Corynebacterium diphtheriae*
Typhoid fever	*Salmonella typhosa*
Paratyphoid	*Salmonella paratyphi*
Tuberculosis	*Mycobacterium tuberculosis*
Leprosy	*Mycobacterium leprae*
Undulant fever	*Brucella abortus*
Plague ("black death")	*Pasteurella pestis*
Tularemia ("rabbit fever")	*Pasteurella tularensis*
Whooping cough	*Hemophilus pertussis*
Tetanus ("lockjaw")	*Clostridium tetani*
Gaseous gangrene	*Clostridium perfringens (welchii)*
Botulism (toxic food poisoning)	*Clostridium botulinum*

rot; wilt diseases of tomato, potato, cantaloupe, squash, cucumber, and corn; citrus canker; fire blight of pears and apples; crown galls of apples, grapes, and raspberries; bacterial blight of beans; and bacterial blight of walnut trees. Annual losses of crops because of bacterial diseases reach millions of dollars.

SPOILAGE OF FOODS

Bacteria cause spoilage of foods as illustrated by rotting of meats, spoilage of butter, souring of milk and other dairy products, and spoilage of fresh and canned wines, fruits, and vegetables. These damages can result in rendering the foods unfit for use or in the production of detrimental substances in the foods as a result of bacterial activities. As an example of this, certain anaerobic bacteria *(Clostridium botulinum)* may attack some types of foods (for example, meat, canned corn, and beans) and produce a very potent toxin which may cause death if even a small quantity is ingested. This disease is known as botulism (L. *botulus,* sausage) because it was originally discovered in sausage.

Beneficial activities of bacteria

Probably a preponderance of the common knowledge of the roles of bacteria pertains to their harmful effects. However, their activities can be utilized quite beneficially, as will be shown.

MANUFACTURE OF FOODS

Through their products and metabolic activities bacteria may play important roles in the production of sauerkraut and pickles, the manufacture of butter and certain kinds of cheese, the production of vinegar by the oxidation of alcohol to acetic acid, and the curing of coffee, cocoa beans, and black tea.

INDUSTRIAL AND AGRICULTURAL ACTIVITIES

Through increased bacteriologic research the roles of bacteria in a great variety of industrial and agricultural activities are being realized. Among such applications are the "tanning" of skins to produce leather; the removal of flax fibers from the stems of flax plants; the fermentation of shredded green plants for the production of ensilage for animal feed; bacterial actions on sugars, proteins, and other organic materials that form such commercially important products as acetone (used in the manufacture of photographic films), and explosives; the production of butyl alcohol used in the manufacture of lacquers; the production of lactic acid used in the "tanning" of skins; the production of citric acid used in lemon flavoring; and the production of vitamins used in food and medicinal preparations.

INCREASE OF SOIL FERTILITY

Bacteria may decompose proteins, fats, carbohydrates, and other organic compounds from the bodies of animals and plants, or from their waste products, thereby removing much of the organic debris from the earth and returning the simple products thus formed to the soil to help maintain its fertility. The anaerobic bacterial decomposition of nitrogenous organic compounds usually results in the formation of malodorous materials, chiefly sulfur compounds. This process is known as putrefaction (L. *putrefacere,* to make rotten). The aerobic bacterial decomposition of organic compounds in the presence of oxygen and without the development of malodorous materials is called decay (L. *de,* down; *cadere,* to fall). In the decomposition of organic substances several species of bacteria, some aerobic and some anaerobic, may carry the process through numerous successive stages until there is a complete breakdown into simpler materials that will be usable by future plants.

The various bacteria involved in the important nitrogen transformations to maintain soil fertility include ammonifying bacteria, nitrifying bacteria (both nitrite and nitrate bacteria), and nitrogen-fixing bacteria. These important phenomena will be considered in greater detail elsewhere in the text.

The sulfur bacteria in soils may convert hydrogen sulfide, a product of protein destruction, through a series of stages to eventually form sulfuric acid. This is transformed into sulfates, which are an essential, major source of sulfur for the metabolism of green plants. Thus the supply of sulfates is maintained.

DIGESTIVE AND OTHER PHYSIOLOGIC ACTIVITIES IN ANIMALS

Some species of bacteria perform certain beneficial physiologic activities in the intestines of animals. Cellulose-digesting enzymes secreted by bacteria assist in the digestion of cellulose in herbivorous animals, such as cattle and horses. Bacteria in the large intestine and in the lower part of the small intestine of man may synthesize vitamins. About one-third of the dry weight of human feces consists of the bodies of microorganisms, primarily bacteria.

Some contributors to the knowledge of
VIRUSES*

Louis Pasteur (1822-1895)

He artificially attenuated (lowered the patho-genicity) rabies virus (1885) but preserved its immunizing property. *(Historical Pictures Service, Chicago.)*

Martinus Beijerinck (1851-1931)

A Dutch microbiologist, substantiated the evidence of filtrable viruses in living orga-nisms (1899). He discovered the root-nodule bacteria that fix free atmospheric nitrogen in certain plants (1888). He also studied the fixa-tion of free atmospheric nitrogen by bacteria *(Azotobacter)* in pure culture (1908). He sug-gested a *"contagium vivum fluidum"* as the cause of the mosaic disease of tobacco plants (1899).

Walter Reed (1851-1902)

Reed and members of the American Army Commission in Havana, Cuba, described the viral cause of yellow fever (1900).

*See Chapter 8 for contributors and refer-ences to blue-green algae.

Pasteur

Wendell M. Stanley (1904-   )

An American biochemist who isolated from the juice of tobacco plants infected with mo-saic disease, a protein of great molecular size. It displayed the properties of a virus and from it he isolated a crystalline form by sedimen-tation in ultracentrifuges (1935).

John E. Enders (1897-   )

An American microbiologist and virologist, has given us a powerful tool to produce im-munization against certain diseases. He first demonstrated how to grow the dangerous po-lio virus in a place other than nerve tissue, for which he received a Nobel Prize. Jonas Salk has been credited for developing a polio vac-cine; this was made possible because of the studies of Enders and his associates.

Joshua Lederberg (1925-   )

An American biologist and winner of a Nobel Prize (1958), showed that viruses can alter the heredity of bacteria. With his teacher, Edward Tatum of Yale University, he demonstrated that bacteria may have sex.

Review questions and topics

1 What are viruses? Why are they considered to be on the borderline between living and nonliving?
2 List several diseases of plants and animals caused by vi-ruses.
3 What are bacteriophages and what are some of their more important characteristics?
4 Define bacteria in your own words.
5 Why should bacteria be classified in the phylum Schizo-phyta?
6 Why are bacteria placed in the kingdom Monera?
7 Describe all the ways in which bacteria may be (1) beneficial and (2) harmful.
8 Contrast and give examples of heterotrophic and auto-trophic nutrition.
9 Do any bacteria photosynthesize food even though they do not possess chlorophyll? Explain in detail.
10 Do bacterial cells possess nuclear materials? Do any of them possess an organized nucleus?
11 Explain the methods of reproduction of bacteria.

12 Describe the common types of true bacteria, giving ex-amples of each type.
13 Explain the following terms: (1) bacterial locomotion, (2) flagella, (3) chromogenic bacteria, (4) cytochromes, (5) luminescent or photogenic bacteria, and (6) luciferase.
14 Contrast and give examples of parasitic and saprophytic bacteria.
15 Explain the physiologic differences between aerobic and anaerobic bacteria.
16 Explain and give examples of the two groups of autotro-phic bacteria.
17 Explain these terms: psychophiles, mesophiles, and ther-mophiles.
18 Discuss the characteristics of the blue-green algae.
19 In what ways are bacteria and blue-green algae alike? Dif-ferent?

Some contributors to the knowledge of
BACTERIA

Antonj van Leeuwenhoek (1632-1723)

A Dutch microscopist, was probably the first to see bacteria (1676), and his descriptions were quite good considering the low magnifications (300 diameters) of the 247 microscopes that he constructed. *(Historical Pictures Service, Chicago.)*

Louis Pasteur (1822-1895)

A French bacteriologist and chemist, proved that fermentations and decompositions of substances resulted from the activities of microbes. He proved that certain diseases were caused by bacteria, and he is often called the "Father of Bacteriology."

Robert Koch (1843-1910)

A German bacteriologist and physician, devised a plate method for growing bacteria on solid media (1881). He proved that specific bacteria cause the diseases of anthrax (1877) and tuberculosis (1822). He formulated Koch's Postulates for proving that a particular microbe causes a specific disease. *(Historical Pictures Service, Chicago.)*

Leeuwenhoek

Koch

William H. Welch (1850-1934)

An American and a student of Koch, discovered the "gas bacillus" in infected wounds after his return to America. Welch is called the "Dean of American Medicine." Based on Koch's methods he gave the first course in bacteriology in the United States at Johns Hopkins University, Baltimore, Md.

Sergius Winogradsky (1856-1913)

A Russian microbiologist, studied the physiology of sulfur bacteria (1899). He identified the bacteria of the soil that oxidize ammonia to nitrites and nitrates (1890).

Alexander Fleming (1881-1955)

An English bacteriologist, discovered the antibiotic penicillin (1929). This led other investigators to discover many additional antibiotics with which to treat numerous diseases today.

Selman A. Waksman (1888-   )

An American, discovered the antibiotic streptomycin (1944) and received a Nobel Prize (1952).

Selected references

Brieger, E. M.: Structure and ultrastructure of microorganisms, New York, 1963, Academic Press Inc.

Echlin, P.: The blue-green algae, Sci. Amer. **214**:75-81, 1966.

Edgar, R. S., and Epstein, R. H.: The genetics of a bacterial virus, Sci. Amer. **212**:70-78, 1965.

Fritsch, F. E.: The structure and reproduction of the algae, New York, 1945, Cambridge University Press.

Horne, R. W.: The structure of viruses, Sci. Amer. **208**:48-56, 1963.

Pelczar, M. J., and Reid, R. D.: Microbiology, ed. 3, New York, 1972, McGraw-Hill Book Co.

Salle, A. J.: Fundamental principles of bacteriology, New York, 1973, McGraw-Hill Book Co.

Stanier, R. Y., Doudoroff, M., and Adelberg, E. A.: The microbial world, ed. 3, Englewood Cliffs, New Jersey, 1970, Prentice-Hall, Inc.

Stent, G. S.: The multiplication of bacterial viruses, Sci. Amer. **188**:36-39, 1953.

Umbreit, W. W.: Modern microbiology, San Francisco, 1962, W. H. Freeman and Co., Publishers.

10 Kingdom Protista—simple algae

Phylum Euglenophyta (euglenoids)
  Euglena
Phylum Chrysophyta (yellow-green algae, golden
  brown algae, and diatoms)
  Diatoms
  Vaucheria

As indicated in Chapter eight, the organisms placed in the kingdom Protista are mostly single cells with distinct nuclei and membrane-bound organelles. They obtain their food in a variety of ways. Some are photosynthetic, some ingest food directly, and some absorb predigested or decaying food particles.

In general the Protista includes the simple algae and the protozoa. Some specialists would also include the more complex green, brown, and red algae. In this text these complex algae are included in section three along with the plants.

PHYLUM EUGLENOPHYTA (EUGLENOIDS)

Almost all of these algae are found in fresh water. Most of them contain chlorophyll combined with other pigments, such as carotenoids and xanthophylls. Euglenoids have an organized nucleus, and the chlorophyll is localized in chloroplasts (Gr. *chloros,* green). They store food as paramylon, a starchlike carbohydrate.

Most species are single-celled organisms, lacking a cell wall and moving by one or two whiplike flagella. Reproduction occurs by longitudinal cell division. Some species may form cysts, thick-walled resting cells that protect them from unfavorable conditions, such as drought. Encysted cells germinate into swimming cells. Euglenoids form an important part of plankton (Gr. *planktos,* wandering), which consists of free-swimming or floating aquatic organisms. It is an important part of the diet of fishes and other aquatic organisms. *Euglena* is a typical example.

Euglena

Euglena and its related forms are classified by some botanists as algae, since most of them possess chlorophyll with which to photosynthesize foods, as do higher green plants. However, they are classified by zoologists as single-celled protozoans because they may have tubular gullets. Furthermore, some of them may ingest solid foods, as do most animals. Euglenoids might be considered to be plants that possess animal-like methods of securing foods or to be animals that show some plant characteristics. This is one reason for using the kingdom Protista.

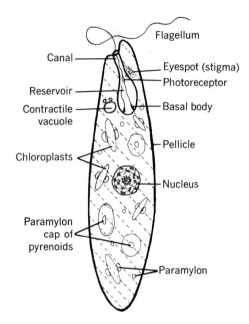

Fig. 10-1. *Euglena.*

Euglena is spindle shaped and has a large, centrally located nucleus (Fig. 10-1). An anterior flagellum propels it through the water. Numerous small, green chloroplasts carry on photosynthesis. Paramylon granules are scattered throughout the cytoplasm. Pyrenoids (colorless plastids) may be present in some instances, assisting in food production.

At the anterior end of the *Euglena* is a tabular reservoir. The basal bodies of two flagella lie at the base of this reservoir. One flagellum is very short and does not emerge from the cell. The emergent flagellum has a swelling just inside the reservoir. This may serve as a light receptor. Near the reservoir is a light-sensitive, red eyespot and a contractile vacuole that maintains osmotic balance (Fig. 3-16).

During reproduction the nucleus undergoes mitosis, after which the cell divides into two new individuals by longitudinal fission. During mitosis the flagellum, photoreceptor, and eyespot are duplicated and the reservoir divides. The daughter reservoirs open into the still single canal, but each now has its own contractile vacuole, eyespot, and two flagella.

A thick-walled, resistant cyst may be formed around the cell during adverse conditions (Fig. 10-2). A few species lack chloroplasts and are probably saprophytes, absorbing materials by diffusion through the cell surface.

Euglena gracilis is about 100μm (1/250 of an inch) in length. *Euglena* may move in a spiral rotary manner by the vibratile movements of the flagellum. It may also move by a twisting of the body known as euglenoid movement. Various species of *Euglena* may be found. They occur in a wide variety of habitats including acid mine streams, shaded soft-water, woodland pools, stagnant polluted streams, and ocean beach pools. Needless to say, not every species is found in every habitat. Because of their abundance in different areas, they are probably an important part of food for various organisms, such as ciliates, insect larvae, and flatworms. Some species, especially *Euglena gracilis,* serve as important experimental organisms. This is especially true in studies of photosynthesis.

PHYLUM CHRYSOPHYTA (YELLOW-GREEN ALGAE, GOLDEN BROWN ALGAE, AND DIATOMS)

The phylum Chrysophyta (Gr. *chrysos,* gold) contains algae in which there is a greater proportion of yellow or brown carotenoid pigments than there is chlorophyll, hence their color. The pigments and chlorophyll are contained in plastids. Stored foods are oils or leucosin, a carbohydrate. Flagella may be present or absent, depending upon the species. The nucleus is organized. These algae may be unicellular, colonial, or even multicellular. Asexual reproduction occurs by cell division, motile zoospores, or by nonmotile spores (aplanospores). Sexual reproduction may occur by oogamy.

Yellow-green algae are primarily freshwater

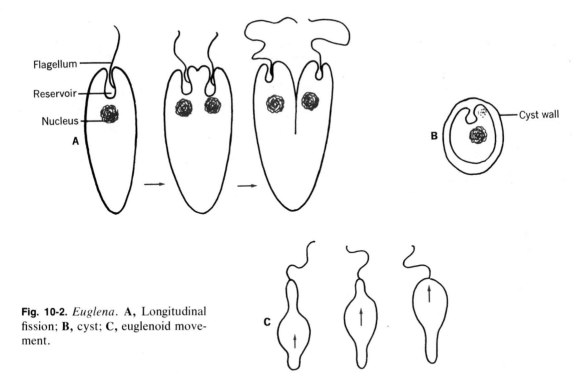

Fig. 10-2. *Euglena.* **A,** Longitudinal fission; **B,** cyst; **C,** euglenoid movement.

forms. However, some live in soils, on trees, or in damp places. Golden brown algae are predominately freshwater types and may contaminate water supplies. Diatoms occur in fresh and salt waters, in soils, on other algae and plants, and even in hot springs. Since diatoms are so widely distributed, they will be considered in more detail here.

Diatoms

Diatoms (Gr. *dia,* across or two; *tome,* to cut) are usually unicellular. The cells of various species may be shaped like rods, disks, triangles, or boats. Their innumerable shapes and delicate beauty label them as the "jewels of the plant world" (Fig. 10-3). The cell walls are usually composed of two overlapping valves (halves) and are often impregnated with silica, making them glasslike, fragile, and transparent. They do not decay after the plant dies. The plastids, in a layer of cytoplasm just within the cell wall, impart dif-

ferent colors to different species. One organized nucleus is present. Stored foods are oils and/or leucosin.

The siliceous cell walls are beautifully patterned with tiny dots or perforations. These are so symmetrically arranged that the efficiency of lenses may be tested by viewing them. The markings on the cell walls of the round types are usually radially arranged, whereas on the elongated types the markings assume a bilateral pattern. In the cell wall a longitudinal slit, called the raphe (Gr. *rhaphe,* seam), may be observed. Vegetative cells do not possess flagella.

Reproduction occurs by cell division, each newly formed cell remaining in the two original valves (Fig. 10-4). Each new cell then secretes a new valve inside each of the old valves. This results in two new cells. One is about the size of the original and the other is slightly smaller than the parent cell. Hence, some of these cells in succeeding divisions will become smaller and

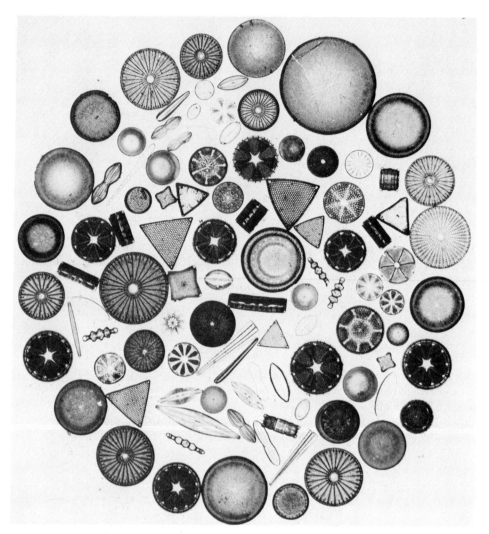

Fig. 10-3. Diatoms in a pattern.
(Courtesy General Biological Supply House, Inc., Chicago, Illinois.)

smaller. Finally, these smaller cells regain their original size by the production of rejuvenescent cells called auxospores. In the sexual method, diatom cells form gametes, and the fusion of two gametes forms a zygote that acts as an auxospore. This zygote enlarges to maximum cell size for that species, and two new valves are then formed.

Diatoms are important food supplies for aquatic animals. Large accumulations of their siliceous valves are known as diatomaceous earth, which is frequently used in manufacturing polishes and tooth powders and pastes, for insulation, and for filtration purposes. Diatomaceous earth is used in dynamite as an absorbent for the liquid nitroglycerin. When it is added to

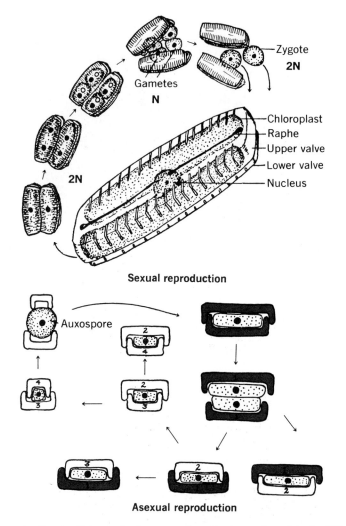

Sexual reproduction

Asexual reproduction

Fig. 10-4. Reproduction in diatoms. In the asexual method, numbers **1** to **4** within the valves show decreases in sizes in successive generations (divisions).

concrete, strength and workability are improved. Bricks may also be sawed from the earth deposits that in certain parts of the world are hundreds of feet thick. Geologists suggest that diatoms may have assisted in the formation of petroleum. Diatomaceous earth can be molded into hollow cylinders to be used in making bacteriologic filters.

Vaucheria

Vaucheria (after Vaucher, a Swiss botanist), a golden brown filamentous alga (Fig. 10-5) is present in fresh water, but some are often found as a feltlike mass on moist soils. The plant (thallus) is long, tube-like, and normally not divided by cross cell walls. Numerous chloroplasts and nuclei are present in the cytoplasm that sur-

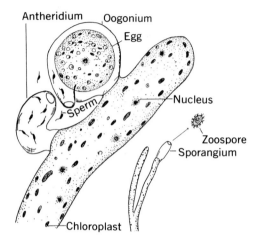

Fig. 10-5. Tubular type of golden brown alga *(Vaucheria)*, showing methods of reproduction. Observe the numerous nuclei and chloroplasts in the tubular filament, which is usually not divided by cross cell walls. Sexual reproduction by oogamy is shown above, and asexual reproduction by zoospores is shown in lower right.

rounds a central vacuole. Reserve food is stored as oil. The multinucleated filament is known as a coenocyte (Gr. *koinos,* shared; *kytos,* cell) and develops by the elongation of the cell in which the nucleus divides repeatedly without forming cross cell walls.

Asexual reproduction may occur in any one of three ways, depending on the habitat. In aquatic species a sporangium is produced by the formation of a cross wall near the tip of a branch. The protoplast of this cell becomes a motile, multi-flagellated, multi-nucleated zoospore that develops into a new filament. In species on moist soil, asexual nonmotile spores may be formed. In other terrestrial species the coenocytic protoplasts may divide into numerous segments, around which heavy walls are secreted to form thin-walled, asexual resting cells.

Sexual reproduction is accomplished by oogamy, in which sex cells of unequal sizes unite, the egg being nonmotile. Both sexes are present on the same filament. The antheridium is a hook-like branch in which numerous biflagellated sperms, each with a single nucleus, are produced. The oogonium is an enlarged structure containing an egg supplied with food and a single nucleus. The fertilized egg develops into a thick-walled resistant oospore that later germinates to form a new filament.

Review questions and topics

1 Describe in your own words that group of plants commonly referred to as simple algae.
2 Why are these algae placed in the kingdom Protista?
3 List the distinguishing characteristics by which the following phyla may be differentiated: Euglenophyta and Chrysophyta. Give examples of each phylum.
4 What are the relationships among algae, chlorophyll, and photosynthesis?
5 Define: plankton, epiphyte, and parasite.
6 Discuss the economic importance of these algae, including beneficial and harmful conditions.
7 List the asexual methods of reproduction found in algae, describing each.
8 List the sexual methods of reproduction found in algae, describing each.
9 What is meant by a life cycle? Make a diagram of a typical life cycle of an alga of each phylum.
10 Define, giving an example of each: isogamy, heterogamy, and oogamy.
11 List the various types of pigments found in the various algae, giving the specific alga in which each is found.
12 List the various types of stored foods found in various algae, giving the specific alga in which each is found.

Selected references

Alexopoulos, C. J., and Bold, H. C.: Algae and fungi, New York, 1967, The Macmillan Company.
Chapman, V. J.: The algae, London, 1962, Macmillan & Co., Ltd.
Dawson, E. Y.: How to know the seaweeds, Dubuque, Iowa, 1956, William C. Brown Company, Publishers.
Fritsch, F. E.: The structure and reproduction of the algae, vol. 1 and 2, Cambridge, 1935 and 1945, Cambridge University Press.
Gibor, A.: Acetabularia: a useful giant cell, Sci. Amer. **215:** 118-24, 1966.
Guberlet, M. L.: Seaweeds at ebb tide, Seattle, 1956, University of Washington Press.
Jackson, D. F., editor: Algae and man, New York, 1964, Plenum Press.
Lamb, I. M.: Lichens, Sci. Amer. **201**(4):144-156, 1959.
Lewin, R. A., editor: Physiology and biochemistry of algae, New York, 1962, Academic Press Inc.

Some contributors to the knowledge of
ALGAE

A. L. de Jussieu (1748-1836)

A French botanist who proposed a natural system of plants (1789) from which the term algae dates in a somewhat modern sense.

Friedrich Kützing (1807-1893)

A German botanist who first distinguished between diatoms and desmids (1833).

W. H. Harvey (1811-1866)

Harvey

An Irish botanist who divided the algae into three groups that correspond quite well with present-day groups of blue-green, green, brown, and red algae (1836). *(The Bettmann Archive, Inc.)*

Karl W. von Naegeli (1817-1891)

A Swiss botanist who distinguished the blue-green from the green algae (1853).

Gottlob L. Rabenhorst (1806-1881)

A German botanist who recognized the green algae group (Chlorophyta) in approximately its present-day sense (1863).

C. J. Friedrich Schmitz

He made important contributions to our knowledge of the structure and classification of red algae (1883-1889).

W. G. Farlow

He made extensive studies and published a volume on the marine algae of the south coast of New England (1873).

Georg Klebs (1857-1918)

A German botanist who suggested that several groups of flagellated organisms are related to different groups of algae and are not all to be included with the unicellular animals as they had been. This proposal caused **Adolph Pasher (1887-1945)**, a German botanist, to work out a system classifying the major groups of algae along present-day lines (1910-1931). He proposed the term Chrysophyta (1914).

Harald Kylin (1879-1949)

A Swedish botanist who first clearly recognized the major groups of brown algae based primarily on life cycles (1933). He also modified Schmitz' classification of red algae (1923-1932).

Milner, H. W.: Algae as food, Sci. Amer. **189**:31-35, 1953.

Prescott, G. W.: How to know the fresh-water algae, Dubuque, Iowa, 1954, William C. Brown Company, Publishers.

Prescott, G. W.: The algae: a review, Boston, 1968, Houghton Mifflin Company.

Smith, G. M.: Cryptogamic botany, vol. 1, New York, 1955, McGraw-Hill Book Company.

Tiffany, L. H.: Algae, the grass of many waters, Springfield, Ill., 1958, Charles C Thomas, Publisher.

Weiss, F. J.: The useful algae, Sci. Amer. **187**:15-17, 1952.

11 Kingdom Protista—protozoans

Protozoans are primarily unicellular organisms. Several genera have species, however, in which colonies are formed by aggregates of cells. Since chlorophyll is not present, protozoan nutrition depends on either ingesting smaller organisms and food particles or absorbing partially decayed products of other organisms. Protozoans that live in other organisms may cause serious harm by their feeding activities.

The protozoan cell is either naked or surrounded by a nonrigid cuticle. In some instances shells of various inorganic materials are secreted as external skeletons. Many protozoans are free living, while several in each class are symbiotic and, particularly, parasitic. Members of the phylum Sporozoa are all parasitic.

There are probably more than 100,000 species of protozoans. Most adult forms are motile. Three types of locomotion are common: by (1) fingerlike pseudopodia, or "false feet," (2) whiplike flagella, and (3) short, hairlike cilia. Because the sporozoan types are parasitic in cells, their active locomotion is greatly restricted. Most protozoans may form encapsulated cysts as temporary phases in their life cycles. In free-living flagellated and ciliated types the mouth-like gullets are usually well developed. In ameboid types the fingerlike pseudopodia are used as ingesting structures. Contractile vacuoles for waste elimination and water balance are present in most protozoans.

The organisms considered in this chapter are extremely varied. Although often described as simple, single-celled animals, they are very complex and have many differences as well as similarities. Many biologists call attention to this diversity by placing the protozoa in a separate kingdom called Protista.

Most people become aware of Protista's existence only by hearing of such problems as the increase of "red tide" along the seacoasts, and the outbreak of dysentery among foreign travelers. However, in addition to their place in nature, protozoans are important research organisms. Our knowledge of cell metabolism has been greatly increased by studying them.

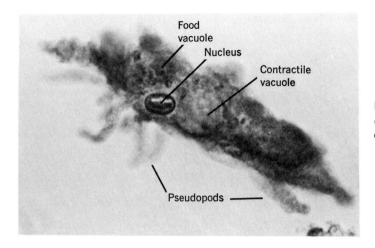

Fig. 11-1. *Amoeba proteus.* Note the nucleus, contractile vacuole, food vacuole, and pseudopodia.

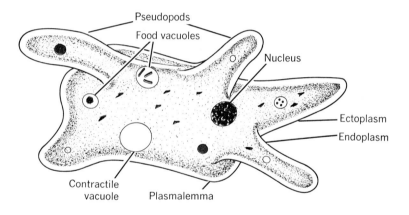

Fig. 11-2. Diagram of *Amoeba.*

CLASSIFICATION OF PROTOZOA

Phylum Sarcodina (Gr. *sarx,* protoplasm or flesh)
   Locomote by protoplasmic projections called pseudopodia ("false feet"). Example: *Amoeba.*
Phylum Zoomastigina (Gr. *mastix,* whip).
   Locomote by one or more whiplike flagella. Examples: *Peranema* and *Trypanosoma.*
Phylum Ciliophora (L. *cilium,* hairlike cilia)
   Locomote by hairlike cilia. Example: *Paramecium.*
Phylum Sporozoa (Gr. *spora,* spore; *zoon,* animal)
   Have no locomotor organelles but may change body shapes. Reproduce by forming spores. Example: *Plasmodium.*

PHYLUM SARCODINA

Amoeba

   Amoeba (Gr. *amoibe,* change) is a common fresh-water protozoan, averaging about 500 μm (1/50 of an inch) in length. It appears to be a colorless, jellylike, granular mass that frequently changes shape by forming fingerlike pseudopodia. A disk-shaped nucleus is present but is not easily seen in living specimens (Figs. 11-1 and 11-2). The protoplasm displays a streaming movement. Some details are observed more easily in stained specimens. There is an outer,

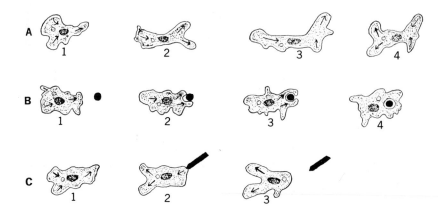

Fig. 11-3. *Amoeba* behavior. **A,** Locomotion; **B,** ingesting a solid particle of food; **C,** reaction to contact with a glass rod. Arrows represent the direction of the flow of protoplasm.

delicate plasma membrane (plasmalemma), which has a clear ectoplasm beneath it. Internally, there is a granular endoplasm, which is made up of a rather viscous plasmagel and a more fluid plasmasol, which exhibits the flowing movements.

Locomotion is accomplished by forming temporary pseudopodia (Fig. 11-3, *A*). At the beginning a pseudopodium is a fingerlike projection of ectoplasm; then the granular plasmasol flows into the projection as it lengthens. In this process the posterior part of the moving *Amoeba* is changed from the plasmagel to the plasmasol, while just the reverse is occurring at the anterior end where the pseudopodium is forming. In the gel state the protoplasm may contract somewhat and squeeze the sol, thus pushing in into the pseudopodium, the tip of which is not a gel. When the fluid sol comes near the tip of the pseudopodium, it moves to the sides and is converted into the gel state. Hence, at the point where a pseudopodium forms, the ectoplasm liquefies as a result of local chemical action, thus changing to the sol state. The endoplasm flows through this liquefied area and, as it does, the surface gels to form a tube of ectoplasm. Elsewhere the ectoplasm is converted to endoplasm. An *Amoeba* may form pseudopodia from any part of the cell. This type of activity, known as ame-

boid movement, is also found in some cells of complex animals, such as in the white blood cells of man.

Nutrition of an *Amoeba* is maintained by the ingestion of food particles. Amebae feed on other protozoans, rotifers, bacteria, and algae. Ingestion may occur at any part of the *Amoeba* by a thrusting out of pseudopodia to engulf the food. By this process the ectoplasm surrounds the food, and the structure eventually becomes a food vacuole that flows along in the endoplasm (Fig. 11-3, *B*). Enzymes from the cytoplasm move into the food vacuole, and the process of digestion begins. The digestive enzymes assist in breaking down complex foods, such as proteins and fats, into simpler molecules that can pass through the food vacuole boundary into the cytoplasm. Here these simple molecules are used as food or recombined into more complex substances. Other enzymes assist in recombining these simpler molecules into complex foods. Food vacuoles, when first formed, give an acid reaction but later become alkaline. As digested foods pass from the food vacuole into the protoplasm, the vacuole becomes smaller, and finally only the indigestible materials are egested (eliminated) through the cell surface.

Respiration occurs by diffusing oxygen, which is dissolved in the water, through the body surface. In order to secure necessary energy an *Amoeba* oxidizes foods, which results in such waste materials as water, urea, and carbon dioxide. Wastes may be eliminated through the body surface, but some are excreted by a special spheroid organelle, called a contractile vacuole. As this vacuole slowly collects water, it increases in size, contacts the cell surface, and expels its contents. It then repeats the process.

An *Amoeba* responds to various stimuli in different ways, as do other animals, but its behavior pattern is simpler than that of higher organisms (Fig. 11-3, *C*). The responses of amebae and other protozoans are often called taxes (Gr. *taxis,* locomotion), which are orientations of the organism, either toward the source of a stimulus (positive taxis) or away from the stimulus (negative taxis). Responses are influenced by the quality and intensity of the stimulus, as well as by the physiologic status of the organism at that particular time.

Reproduction occurs asexually by binary fission, in which the nucleus divides by mitosis, and the cytoplasm elongates and divided to form two daughter amebae (Fig. 11-4). Usually an *Amoeba* will divide every two or three days, although this rate may be influenced by many factors.

Several kinds of amebae are parasitic on such things as *Hydra,* annelids, and vertebrates. One of these, *Entamoeba histolytica,* is an intestinal parasite in man, which causes the disease amebic dysentery and leads to intestinal abcess, liver damage, and other complications.

Several other protozoans of the class Sarcodina are shown in Fig. 11-5.

PHYLUM ZOOMASTIGINA

Peranema

Peranema (Gr. *pera,* sac; *nema,* thread) is a single-celled, flagellated protozoan found in stagnant fresh water and in ponds (Fig. 11-6). When not swimming, the body form may assume many different sizes (25 to 100 μm wide) and temporary shapes, which suggest that its external pellicle is flexible. One large, stiff flagellum and a smaller one are present on the anterior end. The posterior end of the body is blunt but tapers anteriorly. An oval nucleus contains chromatin. Nutrition is holozoic (ingestion and digestion of solid organic materials) and saprozoic (absorption of soluble organic materials). Chlorophyll is absent so food cannot be photosynthesized. A contractile waste vacuole is located near a saclike reservoir that is connected to a gullet. The gullet is supported by rods so that it opens and closes. A slitlike mouth (cytosome or pharynx) opens into the cytoplasm near the base of the flagellum. Food vacuoles are formed in which foods are digested by enzymes. Foods are stored as paramylon (starchlike material). Asexual reproduction occurs by longitudinal binary fission.

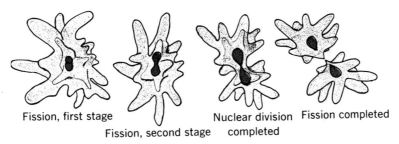

Fission, first stage   Nuclear division   Fission completed
Fission, second stage   completed

Fig. 11-4. Reproduction of *Amoeba* by binary fission. Note how the nucleus divides and how the cells eventually separate to form two amebae.
(From Braungart, D. C., and Buddeke, R.: An introduction to animal biology, ed. 7, St. Louis, 1968, The C. V. Mosby Co.)

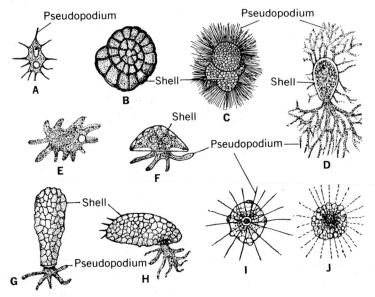

Fig. 11-5. Representative Sarcodina, showing various types of pseudopodia. **A,** *Protomonas;* **B,** *Rotalia;* **C,** *Globigerina;* **D,** *Allogromia;* **E,** *Amoeba;* **F,** *Arcella;* **G,** *Difflugia;* **H,** *Centropyxis;* **I,** *Actinophrys;* **J,** *Thalassicola.*

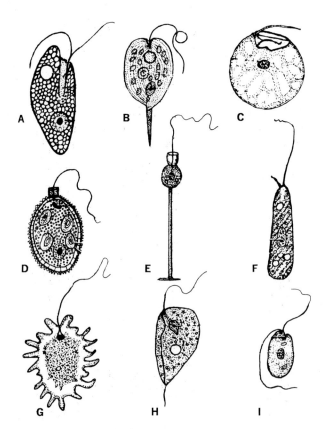

Fig. 11-6. Representative flagellates. **A,** *Chilomonas;* **B,** *Phacus;* **C,** *Noctiluca;* **D,** *Trachelomonas;* **E,** *Monosiga;* **F,** *Peranema;* **G,** *Mastigamoeba;* **H,** *Cercomonas;* **I,** *Bodo.*

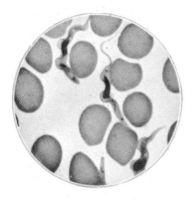

Fig. 11-7. *Trypanosoma gambiense* is the cause of sleeping sickness in central and western Africa. *(Courtesy General Biological Supply House, Inc., Chicago, Illinois.)*

Trypanosoma

Trypanosoma (tri pan o -so' mah) (Gr. *trypanon*, awl; *soma*, body) is a genus containing elongated, ribbonlike, mononucleated, monoflagellated protozoans that are parasitic in certain vertebrate animals. The flagellum is attached to the trypanosome body by a thin undulating membrane, and the actions of the two propel the organism.

Trypanosoma gambiense (Fig. 11-7) (Gambia, a country on the west coast of Africa) causes African sleeping sickness. The trypanosomes are injected into the human bloodstream by infected, blood-sucking, tsetse flies. Normally, the trypanosomes may live in the blood plasma of certain vertebrates. The tsetse flies bite infected animals and suck up the trypanosomes, which multiply and migrate to the salivary glands in about 3 weeks. If an infected fly bites a man or other mammal, the saliva and reorganized trypanosomes are injected. The multiplication of trypanosomes in the human blood may liberate fever-producing toxins. Sometimes these organisms may invade the nervous system, resulting in damage to the brain, which leads to lethargy that becomes sleeping sickness and ends in death.

Efforts to control the tsetse fly throughout Africa are hindered by poverty, poor transportation, and war. Both care of the sick and control of disease are interrupted. The tsetse fly is increasing and crosses national borders at will. The World Health Organization reports large increases in sleeping sickness in man and animals.

Flagellated protozoa in termites and roaches

The digestive tracts of certain termites and wood roaches contain many kinds of flagellates. These protozoans perform an essential function in the digestion of wood. The insect host eats the wood, the flagellates digest the cellulose it contains, and both absorb the digested products. This type of mutual collaboration between two different kinds of organisms in which each depends on the other for the accomplishment of a process beneficial to both is called mutualism (L. *mutuus*, exchanged).

PHYLUM CILIOPHORA

Paramecium

Paramecia (Gr. *paramekes*, oblong) are ciliated protozoans, which are sometimes called "slipper animalcules" because of their shape. Paramecia are common in fresh water that contains an abundance of decaying organic matter. Paramecia have holozoic nutrition, living on bacteria, algae, and other organisms. A common type, *Paramecium caudatum* (L. *cauda*, tail), has a tuft of tail-like cilia on the posterior end and may be up to 0.3 mm. long, whereas *Paramecium aurelia* (o -re' lia) (L. *aurum*, brown or gold) may be less than 0.2 mm. long (Fig. 11-8). A depression on the ventral side called the oral groove runs obliquely backward, ending just posterior to the middle of the body.

Covering the entire surface is a clear, elastic membrane, the pellicle, which is covered with fine cilia. The pellicle is divided by tiny, elevated ridges into may small hexagonal areas. Just beneath the pellicle is the clear ectoplasm, which encloses the granular endoplasm. Embedded in

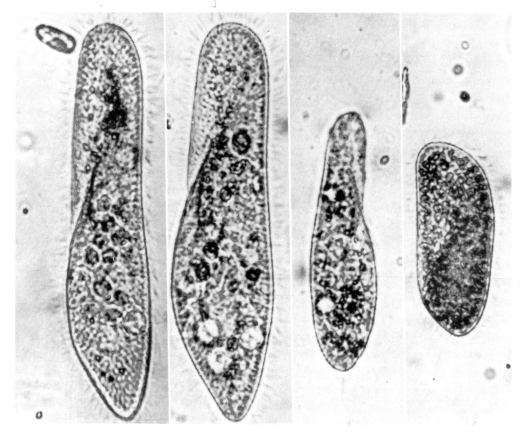

Fig. 11-8. Four species of *Paramecium*. Left to right: *P. multimicronucleatum, P. caudatum, P. aurelia,* and *P. bursaria.* The first two are about 0.3 mm. long, the others about 0.2 mm.
(Courtesy Carolina Biological Supply Company, Burlington, North Carolina.)

the ectoplasm, just below the surface, are spindle-shaped cavities called trichocysts (Gr. *thrix,* hair; *kystis,* sac), which are filled with a semifluid that may be discharged as long threads, presumably for defense purposes. There are two kinds of nuclei the larger macronucleus and a smaller micronucleus (Fig 11-9). In *Paramecium aurelia* there are typically two micronuclei.

Cilia in the oral groove sweep food in the water into the mouth (cytostome). Along the gullet (cytopharynx) are rows of cilia fused into an undulating membrane to pass food inwardly from the mouth to the posterior part of the gullet where the food is collected into a food vacuole. This, when of proper size, disconnects from the gullet to circulate through the endoplasm. Enzymes digest the food in the food vacuoles, which gradually become smaller. Food vacuoles at first have an acid reaction and later become alkaline. Indigestible fecal materials are ejected through the anus, which may be seen as a pore posterior to the oral groove (Fig. 11-10).

Respiration occurs when the oxygen dissolved in the water diffuses into the cell and the carbon

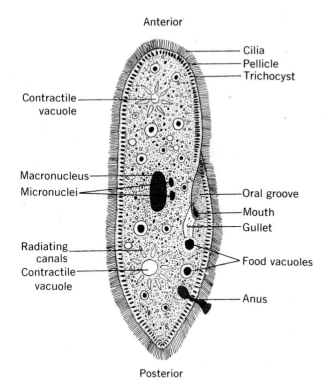

Anterior

Cilia
Pellicle
Trichocyst

Contractile
vacuole

Macronucleus
Micronuclei

Oral groove

Mouth

Gullet

Radiating
canals

Contractile
vacuole

Food vacuoles

Anus

Posterior

Fig. 11-9. *Paramecium aurelia,* a ciliated protozoan of class Ciliata.

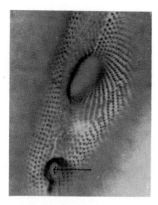

Fig. 11-10. *Paramecium,* showing anal pore, enlarged. *(Courtesy Dr. Elizabeth E. Powelson, Wittenberg University.)*

dioxide passes out. Toward each end, near the dorsal surface, is a contractile vacuole, composed of a central hollow area and several radiating canals. These canals conduct excess water and possibly wastes to the vacuole, which increases in size. When a certain size is reached, the contractile vacuole contracts and discharges its contents to the outside through a pore. The effect of this is to maintain water and ion balance.

By special fixation and staining it can be observed that the cilia are attached to basal granules that are connected to each other by interciliary fibrils. This comprises the noncellular neuromotor mechanism for the coordination of ciliary action. The cilia vibrate either forward or backward, so that the animal can move in either direction. By beating obliquely the cilia cause a rotation on the long axis. The longer cilia of the oral groove vibrate more rapidly so that the anterior end swerves. A combination of such actions causes the animal to move along a helical path.

When a *Paramecium* encounters an undesirable stimulus, such as certain chemicals, it performs an avoiding reaction by reversing its cilia to move backward. The anterior end is then swerved aborally (away from the mouth), thus pivoting on the posterior end (Fig. 11-11). This permits a sampling of new areas by the oral groove, which continues until a condition is encountered that again permits forward movement. This might be called a "trial-and-error" type of adjustment.

The specific type of response of a *Paramecium* even to the same stimulus, varies from time to time; a great number of conditions influence the reactions. The responses of protozoans are attempts to orient themselves, either toward or away from a stimulus, and are called taxes, which were mentioned earlier in this chapter. The names of taxes are based on the type of stimulus suggested by the following: (1) thermotaxis (Gr. *therme,* heat; *taxis,* locomotion), response to heat; (2) chemotaxis, response to chemicals, (3) phototaxis, response to light; (4) thigmotaxis (Gr.

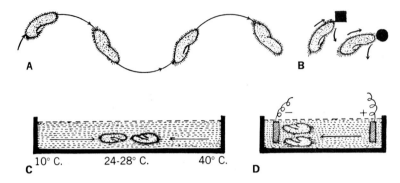

Fig. 11-11. Paramecia behavior. **A,** Locomotion (helical movement); **B,** reaction to contacts with solid objects (avoiding reaction); **C,** reaction to temperature; **D,** reaction to electric current.

thigema, touch), response to contact; (5) geotaxis, response to gravity; (6) rheotaxis (Gr. *rhien,* to flow), response to water or air currents; and (7) galvanotaxis (L. Galvani, Italian physiologist), response to electric currents.

Paramecium has an average optimum temperature of about 26° C. and tends to be negatively thermotaxic to temperatures much above or below this temperature; it reacts negatively to most chemicals, yet may react positively to certain concentrations of mild acids; it avoids either bright light or no light at all; it usually avoids objects; it responds negatively to gravity; it usually swims against water currents and will point its anterior end toward the negative pole (cathode).

A *Paramecium* reproduces by transverse binary fission, whereby the micronucleus divides by mitosis into two micronuclei, which migrate to opposite ends of the cell. The macronucleus divides into two parts, without mitosis, and the two take positions in opposite ends of the cell. The gullet buds off a new gullet, and two new contractile vacuoles are formed. A transverse constriction deepens across the middle of the cell, thus dividing the cytoplasm. Hence, two smaller paramecia are visible in about 20 minutes. Many generations may be produced annually by this process. All individuals produced from a single individual by binary fission have a

similar heredity and constitute a group called a clone (Gr. *klon,* twig). Such groups of organisms can be used experimentally where numbers of individuals with similar heredity are required.*

At certain times two paramecia may attach themselves temporarily at their oral groove regions. They then build between them a protoplasmic conjugating tube through which they will mutually exchange micronuclear materials by a process called conjugation (Fig. 11-12). The behavior of the micronucleus and macronucleus is quite complex during this process. After the mutual exchange of the micronuclear materials the two paramecia separate again and may continue to divide by fission. Conjugation may not be necessary for rejuvenation, since cultures have been kept for years without its taking place. However, in those cases a related process called autogamy usually took place. Conjugation is more common in *Paramecium caudatum* than in *Paramecium aurelia.*

Autogamy (Gr. *autos,* self; *gamos,* marriage) is another method of reproduction in which there is a nuclear reorganization within a single *Paramecium.* The macronucleus disintegrates, and the

---

*One researcher, Ralph Wichterman, reported that he had maintained cultures of *Paramecium bursaria* for over 11 years. They were then discarded because they were no longer needed.

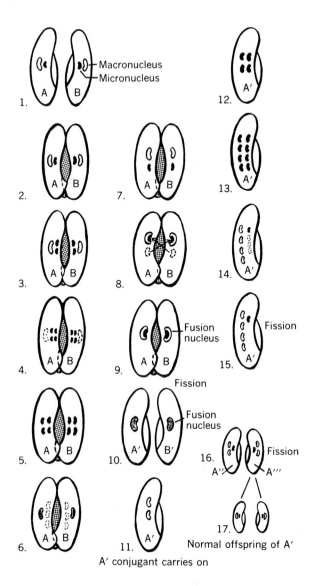

Fig. 11-12. *Paramecium caudatum* conjugation. **2,** Two partners meet and fuse at the oral groove. **3** to **5,** Macronucleus in each disappears and micronucleus divides twice. **6** and **7,** Three of the four micronuclei disappear and the fourth one divides unequally. **8** and **9,** Paramecia exchange nuclei, which then fuse. **10,** Partners separate and the nucleus in each now has genes from two sources. **11** to **13,** Fusion nucleus divides three times, resulting in a total of eight nuclei. **14,** Three of the new nuclei disappear, four of them enlarge, and one remains a micronucleus. **15,** Micronucleus *and organism* divide. **16,** New micronucleus and new organism(s) divide again. **17,** Process is now complete. Each of the eight new paramecia (four from each original parent) has a micronucleus and a macronucleus.

(Modified from Braungart, D. C., and Buddeke, R.: An introduction to animal biology, ed. 7, St. Louis, 1968, The C. V. Mosby Co.)

micronuclei divide. Two of the micronuclei (each with one half the normal chromosome number) then fuse and restore the original chromosome number. The *Paramecium* may then divide by fission. This process involves nuclear reorganization but differs from conjugation in that only one individual is involved.

PHYLUM SPOROZOA

Plasmodium

Plasmodium is a protozoan belonging to the class Sporozoa. There are no well-defined organelles for locomotion, although certain immature stages may be motile. There are no contractile

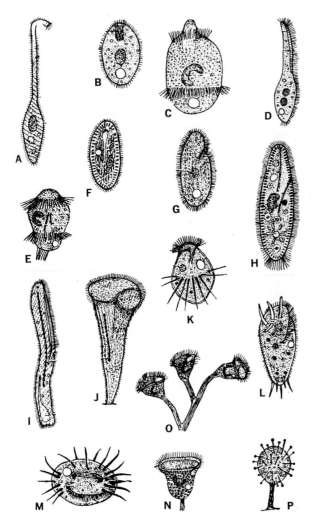

Fig. 11-13. Representative ciliates. **A,** *Lacrymaria;* **B,** *Prorodon;* **C,** *Didinium;* **D,** *Lionotus;* **E,** *Urocentrum;* **F,** *Frontonia;* **G,** *Colpoda;* **H,** *Paramecium caudatum;* **I,** *Spirostomun;* **J,** *Stentor;* **K,** *Halteria;* **L,** *Stylonychia;* **M,** *Euplotes;* **N,** *Vorticella;* **O,** *Carchesium;* **P,** *Podophyra.*

When an infected female *Anopheles* mosquito bites (actually this "bite" is really a puncture) a human, she injects minute, slender sporozoites along with saliva from the salivary glands (Fig. 11-14). These sporozoites pass through the bloodstream and are filtered out in the liver. They undergo multiple divisions in the liver and finally enter the red blood cells of the general circulatory system.

Inside the red blood cells they grow and divide into many merozoites, which are released as the red blood cells burst.

Because of the great numbers of parasites and the toxins that they produce, the characteristic fevers and chills develop. The time between the fever and the chill varies with the type of malaria.

After the asexual stages some merozoites become sexual forms called gametocytes. When the gametocytes are sucked into the mosquito's stomach, they become microgametocytes and macrogametocytes. Two gametes of opposite sex unite to form a zygote, which enters the stomach walls. Later, these zygotes enlarge to form oocysts, which produce thousands of sporozoites. The sporozoites escape from their enclosing cyst and travel to the salivary glands from which the female mosquito transfers them by biting a human being. The developmental stages in the mosquito require 7 to 18 days or longer. In man the symptoms usually appear in 10 to 14 days after the parasites enter his body.

Although malaria has been eliminated as a major disease in the United States, its incidence increased tremendously among servicemen in the Far East during the Vietnam War. It is still present in many areas of the world. People who donate blood are always asked where they have traveled. If they have been in an area where the disease is present, then their blood will not be used for transfusion.

ECONOMIC IMPORTANCE OF PROTOZOA

Most protozoans are neither harmful nor beneficial to man. Some types contribute to the for-

vacuoles and no digestive organelles, since food is absorbed. Many species of *Plasmodium* infect man and other vertebrates, producing different types of marlaria. The parasites are transmitted by the female mosquito of genus *Anopheles*.

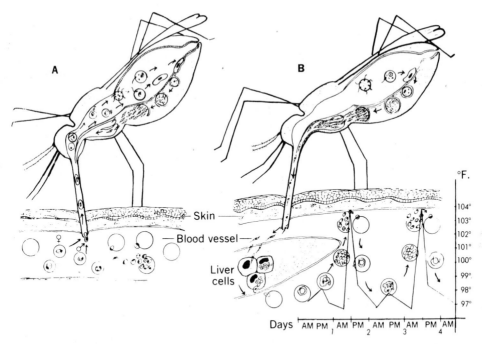

Fig. 11-14. Malarial parasite *(Plasmodium vivax)* life cycle. **A,** Female mosquito *(Anopheles)* ingests parasites from blood; **B,** sporozoites are injected into man. Fever cycle is shown in lower right corner. Note the two high points at certain times when certain stages are present.
(From Hickman, C. P.: Integrated principles of zoology, ed. 3, St. Louis, 1966, The C. V. Mosby Co.)

mation of soils and earth deposits, such as chalk and rocks formed by *Radiolaria* and *Foraminifera*. When they die, certain shelled types form an ooze at the bottom of the ocean that may solidify into rocky formations. Water supplies may become undesirable because of bad tastes or odors produced by such protozoans as *Bursaria, Uroglena, Dinobryon,* and *Synura*.

The cellulose eaten by termites is digested in their intestines by certain flagellated protozoans, thus changing it into a form that is usable by the termites. In turn the termites give protection and distribution to the protozoans. Many types of protozoans live in the intestines of animals and man without causing harm. *Entamoeba gingivalis* and

Trichomonas tenax may live in the mouth. Many types of protozoans have been studied in laboratories, and much has been learned about living protoplasm, the inheritance of traits, nutritional requirements, behaviors of living organisms, osmosis, and permeability.

Among the many human protozoan diseases are: intestinal diarrhea, which may be caused by a variety of protozoans, including *Giardia lamblia* (see box on p. 147); African sleeping sickness, caused by *Trypanosoma gambiense* and transmitted by the tsetse fly; amebic dysentery, caused by *Entamoeba histolytica;* and several types of malaria, caused by different species of *Plasmodium*.

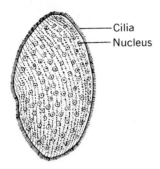

Fig. 11-15. *Opalina ranarum,* a flagellated protozoan that is parasitic in frogs, worms, and mollusks.

Among protozoan diseases of animals are the parasitism of frogs, worms, and mollusks by *Opalina* (Fig. 11-15); Nagana diseases of game animals of Africa, caused by *Trypanosoma brucei;* Texas fever of cattle, caused by *Babesia* and transmitted by a tick; coccidiosis of poultry, rabbits, and other animals, caused by *Coccidia;* and destruction of sperms and seminal vesicles in earthworms, caused by *Monocystis.*

Review questions and topics

1 From your studies of protozoans list the characteristics that they have in common.
2 From your studies of protozoans make a table listing the differences among them.
3 Describe the following for each protozoan studied: (1) integument (covering), (2) locomotion, (3) ingestion and digestion, (4) respiration, (5) circulation, (6) excretion, (7) coordination, nervous, and sensory equipments, and (8) reproduction.
4 Why do unicellular (acellular) organisms not have organs and tissues? What is the function of organelles?
5 What characteristics of living protoplasm are displayed by protozoans?
6 Do you consider the various types of protozoans to be simple or complex? Explain.
7 Do individual protozoans die of old age? Of what significance is the fact that protozoans simply divide when they reach a certain stage in their life? Explain the relationship between the environmental conditions and the death of protozoans.
8 Discuss the economic importance of protozoans from beneficial and detrimental standpoints.
9 Explain the complex life cycle of *Plasmodium,* including the stages in proper sequence, the different hosts, and the detrimental effects.
10 How many studies of certain types of protozoans contributed to our knowledge of cytoplasmic inheritance? Explain what is meant by "killers" and "sensitives" in paramecia.

Selected references

Barnes, R. D.: Invertebrate zoology, ed. 3, Philadelphia, 1974, W. B. Saunders Company.
Cheng, T. C.: General parasitology, New York, 1973, Academic Press, Inc.
Corliss, J. O.: The ciliated Protozoa, New York, 1961, Pergamon Press, Inc.
Jahn, T. L.: How to know the protozoa, Dubuque, Iowa, 1949, William C. Brown Company, Publishers.
Jeon, K. W., editor: The biology of amoeba, New York, 1973, Academic Press.
Jones, A. R.: The ciliates, New York, 1974, St. Martins Press Inc.
Levine, N. D.: Protozoan parasites of domestic animals and of man, ed. 2, Minneaspolis, 1973, Burgess Publishing Company.
Manwell. R. D.: Introduction to protozoology, New York, 1961, St. Martins Press, Inc.
Meglitsch, P. A.: Invertebrate zoology, ed. 2, New York, 1972, Oxford University Press, Inc.
Schmidt, G. D. and Roberts, L. S.: Foundations of parasitology, St. Louis, 1977, The C. V. Mosby Co.
Vickerman, K., and Cox, F. E. G.: The protozoa, Boston, 1967, Houghton Mifflin Company.

Giardiasis—California, Colorado*

The report by a local California physician of laboratory confirmed giardiasis in a family in July led to an investigation which uncovered an outbreak of giardiasis traced to exposure in Estes Park, Colorado.

On July 23, 1976, the Los Angeles County Health Department notified the state health department that a local physician had reported laboratory-confirmed giardiasis in a patient; 2 other family members had also been ill. Illness was also suspected in relatives from 4 other states who, along with this family, attended a reunion near Estes Park in the Rocky Mountain National Park in late June. The Colorado Department of Health was alerted, and it initiated and investigation.

Altogether, 9 of 17 reunion members were subsequently found to have become ill. The group had stayed at summer cabins supplied by water from a small reservoir on the Fall River; that water was chlorinated but not filtered.

Colorado officials studied 2 additional groups: 48 other persons who had stayed at the same cabins in the period June 1-July 28, and a control group of 42 who had stayed for about 4 days during this same period, but at a nearby lodge which received filtered, chlorinated city water. Symptoms of giardiasis (diarrhea plus any 2 of the following: foul-smelling stools, bloating, abdominal cramps, loss of appetite, or weight loss) were reported in 37% of those in the first group but in none of the controls.

Filtrates of reservoir water were found positive for cysts. However, 100 non-human, mammalian fecal specimens collected upstream from the reservoir revealed no *Giardia* organisms on microscopic examination. Eleven water samples collected from sources upstream revealed no fecal coliform counts above .5/100 ml.

Editorial note: The carrier rate of *Giardia lamblia* in the United States ranges between 1.5% and 20%, depending on the community and age group surveyed. An ongoing CDC intestinal parasite survey shows *G. lamblia* to be the most frequently diagnosed intestinal parasitic pathogen in public health laboratories. Reports of epidemics are increasing: at Aspen during the 1965-1966 winter season, at Boulder, Colorado, in 1972, at Rome, New York, in 1975, in Utah in 1975, at a Colorado resort lodge in 1976, and in Camas, Washington, in March 1976 *(1,2,3)*. A survey taken during the Rome outbreak indicated that more than, 4,800 of the city's 46,000 residents became ill *(4)*.

Campers ingesting untreated water from mountain streams are at increased risk of acquiring giardiasis. At least one non-human species (beaver) has been implicated in *Giardia* cyst contamination of community water supplies—at Camas Washington. *Giardia* cysts are not destroyed by chlorination at dosages and contact times commonly used in water treatment. Negative coliform counts, therefore, do not assure safety from *G. lamblia;* filtration is needed.

Reported by Colorado Disease Bulletin, No. 42, 1976, and California Morbidity, No. 46, 1976.

References

1. Moore, G. T., Cross, W. M., McGuire, D., et al.: Epidemic giardiasis at a ski resort. N. Engl. Med. **281**(8):402-427, 1969
2. Wright R. A., Vernon T. M.: Epidemic giardiasis at a resort lodge. Rocky Mt. Med. J. **73**(4):208-211, 1976
3. Barbour, A. G., Nichols, C. R., Fukushima, T.: An outbreak of giardiasis in a group of campers. Am. J. Trop. Med. Hyg. **25**(3):384-389, 1976
4. MMWR **24**(43):366-371, 1975

*From Morbidity and Mortality Weekly Report, Feb. 18, 1977.

Some contributors to the knowledge of
PROTOZOA

Konrad von Gesner (1516-1565)

A Swiss naturalist who described a foraminifera type of protozoan in 1565.

Antonj van Leeuwenhoek (1632-1723)

A Dutch microscopist who studied many types of cells and organisms. Through his accurate accounts and observations he discovered free-living and parasitic protozoans (1674-1703) and is known as the "Father of Protozoology."

Christian G. Ehrenberg (1795-1876)

A German microbiologist who was an enthusiastic student of protozoans, especially infusorians. He postulated that they are highly organized unicellular organisms and classified them. (*Historical Pictures Service, Chicago.*)

Felix Dujardin (1801-1860)

A French microscopic zoologist, recognized living substance in protozoans and applied the name "sarcode" to them (1835).

Ehrenberg

Siebold

Carl T. E. von Siebold (1808-1896)

A German parasitologist who recognized protozoans as single-celled animals (1845) and classified them. (*Historical Pictures Service. Chicago.*)

Émile François Maupas (1842-1916)

A French protozoologist who demonstrated sexuality in protozoans, which later stimulated much work in protozoology and parthenogenesis (reproduction without fertilization).

Lösch

He described a protozoan (*Entamoeba histolytica*) in the feces and intestinal ulcers in a fatal case of dysentery (1875).

12 Kingdom Fungi—molds and mushrooms

Phylum Myxomycota (slime molds)
The true fungi
    Phylum Zygomycota
    Phylum Oomycota
    Phylum Ascomycota
    Phylum Basidiomycota

As indicated in Chapter eight, fungi are organisms with membrane-bound nuclei and other organelles. Their means of obtaining nutrition is by absorption of pre-digested food from a suitable source. The approximately 50 thousand species include common molds, yeasts, mushrooms, and related forms.

PHYLUM MYXOMYCOTA (SLIME MOLDS)

The organisms in the phylum Myxomycota (Gr. *myxos*, slime; *mykes*, fungus) are so named because of their slimy appearance. The slime molds are primarily saprophytes and absorb foods from non-living organic materials in soils and from decaying plant materials. A few species are parasitic, some even pathogenic (disease-producing). They resemble certain true fungi in their methods of spore formation. In addi on they resemble certain lower animals because of their slimy, ameboid bodies, their ameboid

methods of locomotion by forming pseudopodia (Gr. *pseudo*, false; *pous*, foot), and their ingestion of solid foods. The mold is a thin mass of naked, slimy protoplasm called a plasmodium (Gr. *plasma*, liquid; *eidos*, form) (Fig. 12-1). It

Fig. 12-1. Plasmodium of slime mold *(Physarum). (Courtesy General Biological Supply House, Inc., Chicago, Illinois.)*

149

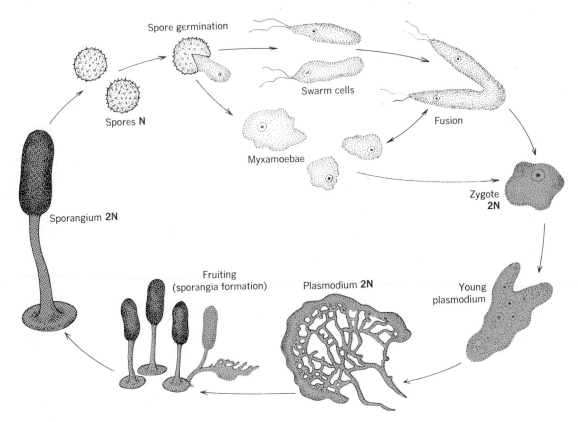

Fig. 12-2. Life cycle of a slime mold.

contains numerous nuclei and moves by a flowing of the protoplasm that forms the pseudopodia.

The plasmodium may be colorless, red, yellowish, violet, or some other color. The plasmodium produces a number of spore cases, called sporangia, which may vary in size and form in different species. The sporangium may be colorless, orange, brown, or purple, depending upon the species. In some species the sporangia are borne on stalks; in others they are stalkless. They may be spherical, ovoid, or some other shape. As a sporangium matures, the internal protoplasm forms a network of fine fibers called a capillitium (L. *capillus,* hair) in the meshes of which are formed numerous unicellular, nonmotile spores. Germinated spores form motile, flagellated cells known as isogametes (Fig. 12-2). Two isogametes unite to form a motile, ameboid

zygote, which later loses the flagella. The zygote moves over the substratum, engulfing bacteria and organic particles that are digested in vacuoles. The multinucleated plasmodium does not contain a cellulose wall. Among the common slime molds are *Stemonitis, Lycogala,* and *Physarum* (Fig. 12-3).

THE TRUE FUNGI

True fungi are diverse in size, form, physiology, and methods of reproduction. Nutrition for all true fungi is heterotrophic and, since none of them contain chlorophyll, they are either saprophytes or parasites. The bodies of most true fungi consist of filaments called hyphae (Gr. *hyphe,* web). In some species the hypha is unicellular; in

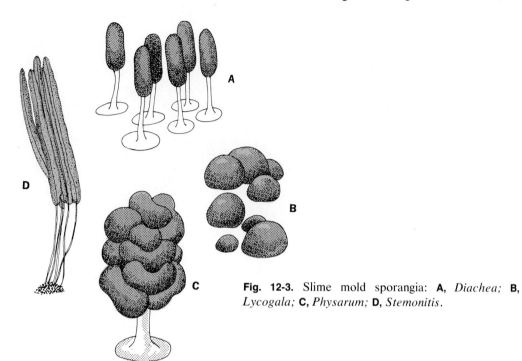

Fig. 12-3. Slime mold sporangia: **A,** *Diachea;* **B,** *Lycogala;* **C,** *Physarum;* **D,** *Stemonitis.*

Table 12-1 Comparison of slime molds and true fungi

| | | True fungi | | |
Characteristic	Slime molds (Myxomycota)	Algalike fungi (Zygotomycota)	Ascus fungi (Ascomycota)	Club fungi (Basidiomycota)
Plastids and chlorophyll	−	−	−	−
Filamentous hyphae	−	+	+	+
Septate hyphae	−	−	+	+
Ameboid plasmodium	+	−	−	−
Locomation (adult stage)	Pseudopodia	Nonmotile	Nonmotile	Nonmotile
Asexual reproduction	Sporangiospores Motile swarm spores (myxamoebae)	Sporangiospores Motile swarm spores (zoospores) (aquatic species)	Ascospores Conidiospores* Budding*	Conidiospores* Chlamydospores* Uredospores* Telispores* Pycnicspores* Aeciospores*
Sexual reproduction	Isogamy	Isogamy* Heterogamy*	Ascospores* (by fusion)	Basidiospores (by fusion)

*Certain species.

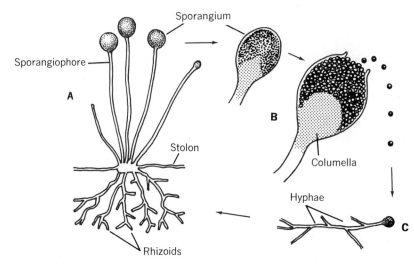

Fig. 12-4. Asexual cycle of black bread mold, *Rhizopus stolonifer*. **A,** Mycelium with sporangiophores; **B,** detail of a sporangium; **C,** spore germination.

others it is multicellular. In some species certain hyphae contain yellow, orange, red, or other types of pigments. A mass of hyphae is called a mycelium (my -se' li um) (Gr. *mykes,* fungus). All true fungi reproduce by means of some type of microscopic spores, and other methods of reproduction may also occur in various species (Table 12-1). Spores are usually dispersed by air currents, water, animals, or in other ways. True fungi require abundant moisture, a favorable temperature, and a sufficient supply of proper organic foods. Most of them are aerobic, but a few are anaerobic.

Phylum Zygomycota

True fungi that produce a zygospore are placed in this phylum. Many of these true fungi are threadlike or filamentous, like many of the algae, except that the fungi lack chlorophyll. The characteristics of this phylum will be shown in the following examples.

Black bread mold

Rhizopus stolonifer (Figs. 12-4 to 12-8) is a common saprophytic mold that lives on moist,

organic materials, such as bread, fruits, animals, and animal dung. The hyphae are whitish or grayish and form a weblike mycelium. Young hyphae are branched, lack cross walls (nonseptate), and contain numerous nuclei. Older hyphae, especially when reproducing sexually, may have cross walls (septate). The rootlike hyphae, known as rhizoids (Gr. *rhiza,* root; *eidos,* like), absorb nourishment from the substratum and anchor the plant. Other hyphae grow over the surface and are called stolons from which arise spore-forming sporangiophores (Gr. *spora,* spore; *angios,* case; *phorein,* to bear). Each sporangiophore bears a globular spore case (sporangium) at its tip that becomes darker as it matures (Fig. 12-6). An enlarged, basal, internal structure within the sporangium is called the columella (L. *columella,* small column), which may be visible when the sporangium ruptures. The airborne, asexual, nonmotile sporangiospores germinate to form branching hyphae and eventually form a new mycelium.

Sexual reproduction might be considered isogamous. Small branches, known as gametangia, form between two adjacent hyphae, which fuse at

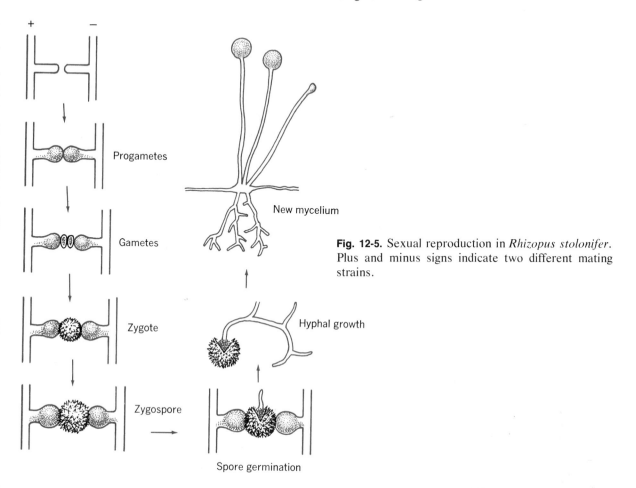

+   −

Progametes

Gametes

Zygote

Zygospore

New mycelium

Hyphal growth

Spore germination

Fig. 12-5. Sexual reproduction in *Rhizopus stolonifer*. Plus and minus signs indicate two different mating strains.

Fig. 12-6. Mass of sporangia of black bread mold *(Rhizopus stolonifer)*. Each tiny dot is a sporangium containing hundreds of spores. The culture was started by placing a few spores on a nutrient agar medium.

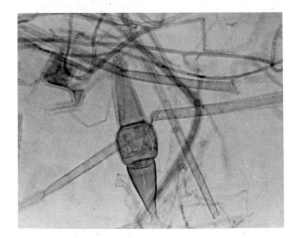

Fig. 12-7. Young zygospore of *Rhizopus stolonifer;* note the strands of mycelium in the background.

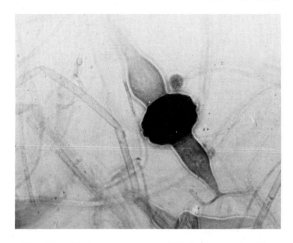

Fig. 12-8. Mature zygopore of *Rhizopus stolonifer.*

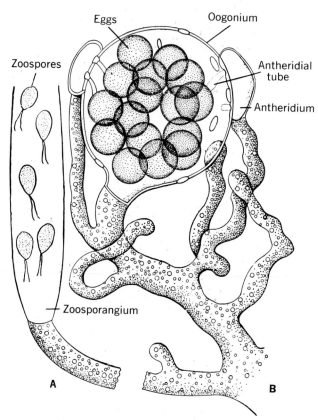

Fig. 12-9. Common water mold *(Saprolegnia).* **A,** Reproduction by asexual, motile zoospores; **B,** sexual reproduction by gametes produced in the antheridia and oogonia. Antheridia form antheridial tubes through which the male gametes pass.
(From Fuller, H. J., and Tippo, O.: College botany, New York, Holt, Rinehart & Winston, Inc.)

their tips. *Rhizopus stolonifer* is said to be heterothallic (Gr. *heteros,* different; *thallos,* shoot); that is, two types of hyphae, called "plus" and "minus," are necessary for this sexual fertilization. Each gametangium forms a multinucleated "gamete." The two "gametes" unite to form a thick-walled, multinucleated zygospore, which under proper conditions will germinate to form a new hypha. Some forms of bread mold are homothallic (Gr. *homos,* same) since both types of gametes are produced on the same plant.

Phylum Oomycota
Water mold

The fungi of the water mold group are primarily aquatic saprophytes (Fig. 12-9), securing foods from organic matters, especially from dead plants and animals, thus having a heterotrophic nutrition. A few species cause serious damage by parasitizing fishes, amphibia, and turtles. Fishes in aquaria may become affected by whitish mycelia of water molds.

Penicillium
    Fungus (hypha) (N) → Conidiophores (N) → Conidiospores   (N) → Fungus (hypha) (N)

OR

    Fungus (hypha) (N) → Ascus (2N) → Ascospores (N) → Fungus (hypha) (N)
        (dikaryotic*)   (by karyogamy)† (by meiosis)

Yeast
    Yeast cell (N)
                    ↘
                      Zygote (2N) → Ascus (2N) → Ascospores (N) → Yeast cell (N)
                    ↗   (by meiosis)
    Yeast cell (N)
                    ↘
                      Buds (2N) → Yeast
                                  cell (2N)

Fig. 12-10. Life cycles of certain fungi.

Saprolegnia (Gr. *sapros,* rotten; *legnon,* edge) has slender hyphae with tapering ends that penetrate the substratum and also branched hyphae whose tips bear enlarged zoosporangia. The zoospores (swarm spores) with two terminal cilia are liberated by the zoosporangia and swim ("swarm") in the water, eventually losing their terminal cilia and surrounding themselves with a wall. Later, each escapes from this wall, but now has two lateral cilia. Each of these zoospores forms a new hypha.

Sexual reproduction is heterogamous whereby unicellular, enlarged oogonia and clublike male antheridia are developed. Each oogonium produces several eggs. The antheridia develop antherical tubes that penetrate the oogonia. Male nuclei (unorganized sperms) are discharged through the antheridial tubes into the oogonia. An egg fertilized by a male nucleus forms a thick-walled oospore that develops into new hyphae. In some species eggs develop without fertilization. *Saprolegnia* produces both antheridia and oogonia on the same plant, hence is said to be homothallic. In older systems of taxonomy both *Rhizopus* and *Saprolegnia* are placed in the class Phycomycetes.

Phylum Ascomycota

True fungi that at some time in their life cycle produce ascospores in a saclike ascus (Gr. *askos,* sac) (Fig. 12-10) are placed in this phylum. Other methods of reproduction may be present, depending upon the species. The Ascomycota are filamentous except for certain types of yeasts. When filaments are present, they have cross walls (septa), and organized nuclei. Additional characteristics of this phylum are shown in the following representatives.

Penicillium

Penicillium (L. *penicillus,* painter's brush) is a bluish green mold whose loosely arranged hyphae grow in or on such materials as damp leather, foods, and fruits (Fig. 12-11). *Penicilium* produces ascospores in saclike asci (Fig. 12-12). In addition the sporebearing hyphae, called conidiophores, bear chains of small, colored spores (conidia) at their tips. Masses of these conidiophores resemble tiny brushes, hence the name *Penicillium.*

Various species of *Penicillium* cause the spoilage of such foods as bread, apples, pears, grapes, and citrus fruits, and cause the destruction of

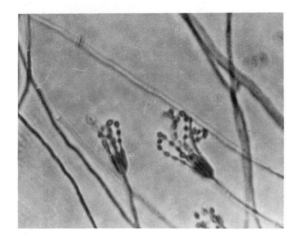

Fig. 12-11. Conidia and conidiospores of blue-green mold *(Penicillium)*.

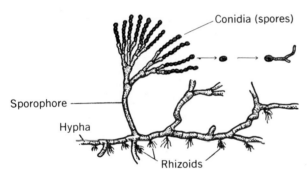

Fig. 12-12. Blue-green mold *(Penicillium)*. Ascospores are seldom observed under ordinary conditions.

paper, leather, and lumber. A few members of this group of fungi may cause skin infections in man. Certain types are used in the manufacture of Camembert cheese *(Penicillium camemberti)* and of Roquefort cheese *(Penicillium roqueforti)*. The characteristic flavors and odors of these cheeses result from the activities of these molds. The bluish green areas in the cheese are masses of conidia. Certain species of *Penicillium* are used in the manufacture of the antibiotic penicillin.

Antibiotics (Gr. *anti*, against; *bios*, life) are organic substances that are synthesized by one type of organism and inhibit or destroy another type of organism. This antagonistic inhibition between two species is called antibiosis. In 1940 the medicinal value of penicillin and other antibiotics was established.

Yeasts

Yeasts are typically unicellular fungi, usually without hyphae, although a few species may develop short ones. Each cell is usually ovoid in shape and contains a nucleus (Fig. 12-13). Asexual reproduction is often accomplished by budding, during which small protuberances (buds) are projected from the cell. The bud may free itself or remain attached and produce more buds, eventually forming a chainlike arrangement of individual cells.

Under certain conditions a yeast cell may become an ascus, in which ascospores (usually four) are formed. In other instances two yeast cells may fuse before ascospores are formed. Certain yeasts produce an enzyme, zymase, which ferments sugars to form ethyl alcohol. In dough certain yeasts ferment sugars to release carbon dioxide whose bubbles cause the bread to "rise" (become porous). Some yeasts manufacture vitamins, while others synthesize protein from molasses and ammonia. The common yeast cake contains yeast cells and a quantity of starch. Yeasts may be classified as (1) *Saccharomyces* (sugar fungi), which are harmless or even beneficial, and (2) *Blastomyces* (germ fungi), which are pathogenic. Many sugar fungi are of commercial value, but many germ fungi produce a variety of common diseases (Table 12-2).

Pathogenic Ascomycotes

Some Ascomycotes cause diseases in plants. In *ergot* of rye the fungus turns cereal grains and wildgrasses purple and makes them poisonous to man. *Chestnut blight* has destroyed most of the chestnut trees in the United states. The mycelium kills the cambium and inner bark, eventually stopping the food supply. *Dutch elm disease* (introduced to the United States in 1930) causes

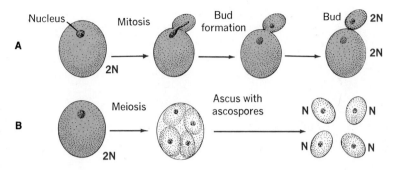

Fig. 12-13. Bread yeast *(Saccharomyces cerevisiae).* **A,** Mitosis and bud formation; **B,** meiosis and ascospore formation.

Table 12-2 Diseases of man and animals caused by pathogenic fungi

Disease	Causal agent and symptoms
North American blastomycosis (Gilchrist's disease)	Fungus causing chronic infection with suppurative and granulomatous lesions in body, especially in skin, lungs, and bones
Moniliasis	Fungus causing lesions in mouth, skin, nails, lungs, or vagina; creamy white, ulcerlike patches in mouth constitute disease called thrush
Coccidioidomycosis	Fungus causing acute but benign respiratory infection; or progressively chronic, malignant infection of skin, bones, or internal organs
Sporotrichosis	Fungus causing nodular lesions in skin and lymph nodes that may soften and break to form ulcers
Dermatomycoses	
(a) Ringworm of feet ("athlete's foot" or *Tinea pedis)*	Fungus causing infection of skin of feet
(b) Ringworm of body *(Tinea corporis)*	Fungus causing infection of skin of body
(c) Ringworm of scalp *(Tinea capitis)*	Fungus causing infection of scalp and hair, and scaly, red lesions and sometimes deep ulcers
Actinomycosis (lumpy jaw)	Anaerobic, pathogenic fungus causing chronic, systemic infection with lesions and abscesses

great damage to many elm trees. As with the chestnuts, the American Elm is rapidly disappearing from many areas of the country. This disease is not to be confused with phloem necrosis of elms, caused by a virus. (4) In *peach leaf curl* the yeastlike fungus causes the leaves to curl and become yellow. (5) In *apple scab* the fungus reduces the yield of apples by millions of bushels by producing raised, discolored areas on the leaves and scabby spots on the fruits, which may be small and distorted. (6) In *powdery mildews* the parasitic hyphae appear as grayish, powdery areas on the leaves and stems of flowering plants, such as lilacs, roses, apples, grapes, cherries, cereals, clovers, and dandelions. Slender hyphae, the haustoria, absorb food from the host plant.

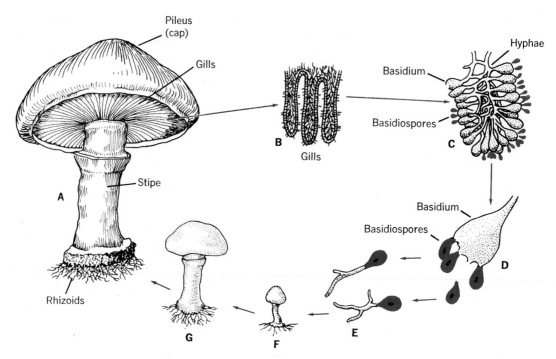

Fig. 12-14. Mushroom cycle. **A,** Mushroom; **B,** cross section of three gills; **C,** detail of a gill to show compact mass of hyphae, club-shaped basidia, and basidiospores; **D,** a basidium and basidiospores; **E,** germinating spore; **F and G,** young mushrooms.

Phylum Basidiomycota

Fungi in the phylum Basidiomycota (Gr. *basidium*, club; *mykes*, fungus) produce basidiospores on club-shaped basidia. Closely packed basidia are arranged in a flat sheet called the hymenium. The filamentous hyphae have cross walls, and the cells possess nuclei. In some types the hyphae are closely compacted and the interhyphal spaces are filled with rather solid materials, as in the bracket fungi. Each basidiospore is attached to the tip of a stalk called a sterigma (Gr. *sterigma*, support). Basidia typically develop from binucleate, terminal cells of hyphae. As these cells enlarge, the two nuclei fuse to form one nucleus, which divides twice to form four nuclei, one of which passes into each basidium as part of a basidiospore. The mature basidiospores may be dispersed by winds, insects, or other agencies. When proper temperature, moisture,

and foods are encountered, a spore germinates, forming a hypha with mononucleate cells. After some growth these cells become binucleate. this binucleate condition persists throughout the growth of these hyphae, the nuclei dividing as new cells are formed. Terminal cells of some hyphae produce the basidia, in which two nuclei fuse.

Mushrooms

Mushrooms are saprophytic fungi (Fig. 12-14) that derive their foods from decomposing organic materials in the soil, dead leaves, bark, and wood. The vegetative body is composed of masses of septate hyphae, some of which penetrate the substratum. Each sporophore (Gr. *spora*, spore; phorein, to bear) typically consists of a broad, caplike or umbrella-shaped pileus (L. *pileus*, cap) and a stalklike stipe. On the undersurface of the pileus are sheet-like gills. The gills are composed of plates of compacted hyphal tissues that bear the club-shaped basidia. The common, edible mushroom *(Psalliota campestris)* may produce nearly two billion spores, each of which may germinate to form a new hypha.

There are several hundred species of mushrooms, and only a few of them are poisonous. Most poisonous forms belong to the genus *Amanita*. Unless the collector is familiar with the mushrooms, he should take no chances with the deadly species. It is better to forego the mushrooms than to suffer fatal illness.

Smuts

Smuts are caused by basidiomycote fungi that parasitize flowering plants. Masses of septate hyphae penetrate the tissues of the host plant. They are called smuts because the fungi produce heavy-walled, dark-colored, ill-smelling spores called chlamydospores (klam' i do spores) (Fig. 12-15). The resistant smut spores may lie dormant until the following spring, when each ger-minates into a tubelike basidium (three to four cells). Each basidium produces a basidiospore (sporidium) The basidiospores attack the host plant and form hyphae, which eventually form smut spores.

In some types of smuts chainlike groups of spores (conidia) may be formed on the parasitized plant. Thus in certain life cycles as many as three kinds of spores may be produced. Smuts primarily parasized members of the grass family, including corn, oats, wheat, rye, rice, and barley. They can produce enormous crop losses.

Corn smut, *Ustilago zeae,* consists of tumor-like, black masses of smut on any part of the corn plant, but especially on the flower parts. When these tumors mature in the summer or fall, their hyphae contain masses of black chlamydospores that usually germinate the next spring or summer to infect new corn plants. The basidiospores, formed on basidia, produce germ tubes capable of infecting any part of the plant. The resulting hyphae mass together in certain areas and break out as smut tumors. The tumors are white at first but become black as the chlamydospores mature.

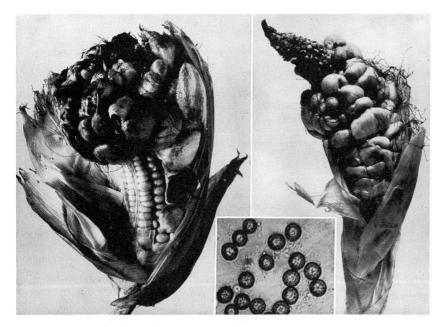

Fig. 12-15. Corn smut *(Ustilago zeae).* Unbroken tumors are at right and broken and disseminating spores are at left; the inset shows chlamydospores of corn smut. *(From Hill, J. B., Overholts, L. O., and Popp, H. W.: Botany, New York, McGraw-Hill Book Company.)*

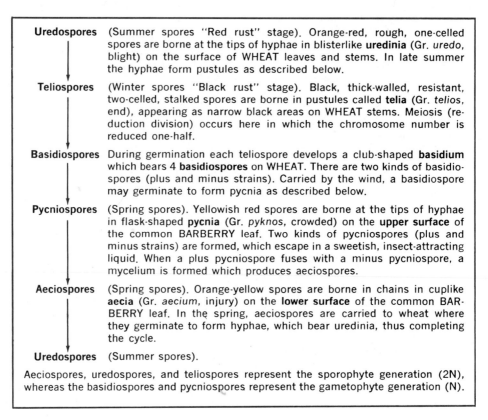

Uredospores ↓	(Summer spores "Red rust" stage). Orange-red, rough, one-celled spores are borne at the tips of hyphae in blisterlike **uredinia** (Gr. *uredo*, blight) on the surface of WHEAT leaves and stems. In late summer the hyphae form pustules as described below.
Teliospores ↓	(Winter spores "Black rust" stage). Black, thick-walled, resistant, two-celled, stalked spores are borne in pustules called **telia** (Gr. *telios*, end), appearing as narrow black areas on WHEAT stems. Meiosis (reduction division) occurs here in which the chromosome number is reduced one-half.
Basidiospores ↓	During germination each teliospore develops a club-shaped **basidium** which bears 4 **basidiospores** on WHEAT. There are two kinds of basidiospores (plus and minus strains). Carried by the wind, a basidiospore may germinate to form pycnia as described below.
Pycniospores ↓	(Spring spores). Yellowish red spores are borne at the tips of hyphae in flask-shaped **pycnia** (Gr. *pyknos*, crowded) on the **upper surface** of the common BARBERRY leaf. Two kinds of pycniospores (plus and minus strains) are formed, which escape in a sweetish, insect-attracting liquid. When a plus pycniospore fuses with a minus pycniospore, a mycelium is formed which produces aeciospores.
Aeciospores ↓	(Spring spores). Orange-yellow spores are borne in chains in cuplike **aecia** (Gr. *aecium*, injury) on the **lower surface** of the common BARBERRY leaf. In the spring, aeciospores are carried to wheat where they germinate to form hyphae, which bear uredinia, thus completing the cycle.
Uredospores	(Summer spores).

Aeciospores, uredospores, and teliospores represent the sporophyte generation (2N), whereas the basidiospores and pycniospores represent the gametophyte generation (N).

Fig. 12-16. Life cycle of the black stem rust of wheat *(Puccinia graminis).* (See also Fig. 12-18.)

Annual losses in the United States because of corn smut amount to millions of dollars.

Rusts

Rusts are fungi that are parasitic on flowering plants and ferns; they are so named because of reddish brown spores on the surface of leaves and stems. Hyphae penetrate the tissues of the host plant. A rust may parasitize two unrelated species of plants, alternating between the two hosts. Examples of two-host rusts include the "cedar-apple" rust, which alternates between cedars and apple trees; white pine blister rust, which alternates between white pine trees and wild gooseberries or currants, and the black stem

Table 12-3 Reduction of crop yield in the United States by plant diseases

Disease	Estimated reduction in yield
Wheat stem rust	85,452,000 bushels
Barley stem rust	12,046,000 bushels
Oat loose smut	32,728,000 bushels
Field corn smut	95,087,000 bushels
Field corn—all diseases	335,826,000 bushels
Potatoes—all diseases	58,673,000 bushels
Tobacco downy mildew	81,502,000 pounds
Cherry leaf spot	16,222,000 pounds
Apple scab	12,774,000 bushels
Strawberries—all diseases	746,000 crates
Grape black rot	31,830,000 pounds

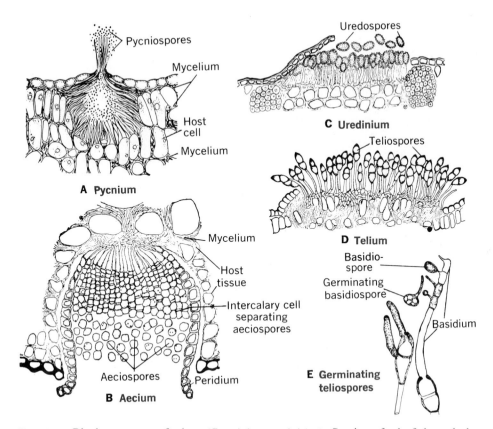

Fig. 12-17. Black stem rust of wheat *(Puccinia graminis)*. **A,** Section of a leaf through the pycnium; **B,** section through the aecium in a barberry leaf; **C,** section through the uredinial sorus, showing unicellular, rough uredospores on slender stalks; **D,** section through the telial sorus, showing two-celled teliospores on long pedicles; **E,** germinating teliospores—left, both cells of the spore germinating; right, only an apical cell germinating; the germ tube has been transformed into a basidium, from each cell of which a basidiospore has been or is being formed. A germinating basidiospore is also shown.
(From Hill, J. B., Overholts, L. O., and Popp, H. W.: Botany, New York, McGraw-Hill Book Company.)

rust of wheat, which alternates between wheat and the common European barberry (not the Japanese barberry). Other rusts may appear on corn, oats, rye, pears, plums, cherries, cone-bearing trees, garden vegetables, and cultivated flowers.

Black stem rust of wheat is a destructive disease caused by a basidiomycete fungus called *Puccinia graminis* (puk -sin′ i a; gram′ in is; after Puccini, an Italian anatomist; L. *graminis,* grass). The life cycle is described briefly in Figs. 12-16 to 12-18.

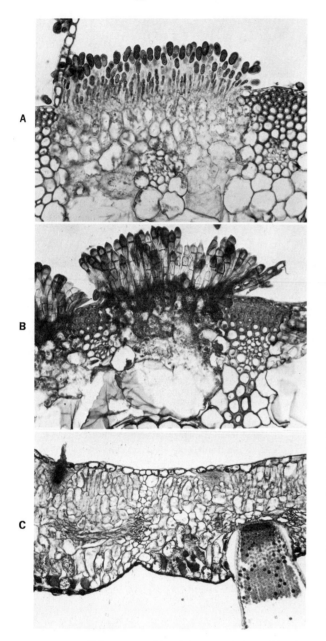

Fig. 12-18. Wheat rust, *Puccinia graminis*. **A,** Uredospores in blister-like uredinia on wheat leaves; **B,** teliospores on wheat stem; **C,** aecium and aeciospores on lower surface of barberry leaf. Note flask-shaped pyenium on upper surface. (See also Fig. 12-17.)

Some contributors to the knowledge of
SLIME MOLDS AND TRUE FUNGI

**Clusius (Charles de la Cluse)
(1529-1609)**

He described and illustrated many edible and poisonous fungi (1601).

Gaspard Bauhin (1560-1624)

He classified about one hundred species of fungi and lichens.

Joseph P. de Tournefort (1656-1708)

He recognized six genera of fungi in his *Elements de Botanique* (1964).

Sebastian Vaillant (1669-1722)

He illustrated many fungi in his *Botanicon Parisiense* (1727)—studies that were not equaled for over 100 years.

Pier Antonio Micheli (1679-1737)

An Italian botanist who was an outstanding student of the microscopic anatomy of fungi and their classification.

Carolus Linnaeus (1707-1778)

A Swedish botanist who in his *Species Plantarum* recognized 24 classes of plants, one of which included fungi.

Christian H. Persoon (1755-1837)

He made great contributions to fungi classification and the roles of spores in this respect.

Elias M. Fries (1794-1878)

A Swedish mycologist who contributed greatly to fungi classification, especially the larger types. He recognized smuts and rusts as a natural group (1821). He is the "Father of Systematic Mycology." *(Historical Pictures Service, Chicago.)*

Fries

Buchner

August C. J. Corda (1809-1849)

A Czechoslovakian mycologist who first recognized the sac fungi (Ascomycetes) as a group (1842) only a few years after the ascus had been described by **Joseph Henri Léveillé (1796-1870)**, the French mycologist. Léveillé discovered the paddle-shaped basidium of Basidiomycetes (1837).

Louis Tulasne (1815-1885) and Charles Tulasne (1816-1884)

Two French mycologists who beautifully illustrated works on the structures of the ascus fungi (1861-1865). They contributed much to the knowledge of the morphology and classification of ascus fungi.

Louis Pasteur (1822-1895)

A French bacteriologist who demonstrated that yeasts cause fermentations (1871). He classified ferments into "organized ferments," such as yeasts and certain bacteria, and "unorganized ferments," such as digestive enzymes like pepsin.

Heinrich Anton De Bary (1831-1888)

A German mycologist who first recognized the alga-like fungi (Phycomycetes) as a group (1873). He worked out many of the life cycles of fungi (1887), including the one for the black stem rust of wheat (*Puccinia graminis*). He thought that slime molds were more closely related to animals than to plants.

Ernst A. Bessey (1877-1963)

An American who advanced the studies of fungi, especially those of the United States.

Pier Andrea Saccardo (1845-1920)

An Italian mycologist who published the *Sylloge Fungorum,* in which he reprinted in standardized form most of the original descriptions of species of fungi. The first complete set of eight volumes was published from 1822 to 1889, and seventeen supplements were issued up to 1931.

Eduard Buchner (1860-1917)

A German chemist and Nobel Prize winner who discovered that cell-free yeast extracts caused fermentation of sugars (1897). He called this water-soluble, heat-labil principle "zymase." (*Historical Pictures Service, Chicago.*)

Bernard O. Dodge (1872-1960)

An American mycologist who pointed out the usefulness of the pink bread mold (*Neurospora*) as a tool for basic genetic studies (1930), and it has been widely used since.

G. W. Martin (1886-   )

Martin, of the State University of Iowa, an authority on slime molds, believes that fungi originated from a protozoan-like ancestor. He includes slime molds in his Division Fungi and calls them Myxomycetes (1949).

Review questions and topics

1 Define and distinguish between slime molds and true fungi.
2 Why are slime molds and true fungi classed in the kingdom Fungi?
3 Describe the life cycle of a typical slime mold.
4 List the distinguishing characteristics of the following Phyla of the true fungi; Zygomycota, Ascomycota, and Basidiomycota, including examples of each.
5 Discuss the ways in which fungi may be of importance to man.
6 Describe each of the asexual methods of reproduction that are found in the slime molds and in the true fungi, giving an example of each.
7 Describe the process of conjugation as found in certain fungi. In which fungi does conjugation occur? In what ways does this process resemble sexual reproduction?
8 Describe how sunlight and lack of moisture may be detrimental to fungi.
9 Explain how fungi obtain their nourishment and oxygen.
10 Diagram the life cycle of *Rhizopus stolonifer*.
11 Explain how bread molds develop inside the loaf. How do they get into the loaf?
12 Diagram a typical life cycle of each class of true fungi.
13 Why are yeasts classed as Ascomycota? In what ways do yeasts differ from other Ascomycota?
14 Why is *Penicillium* classed as an Ascomycota? Of what economic importance is it?
15 Describe the life cycle of the black stem rust of wheat, including the host plants, various stages, types of spores, damages.
16 Which types of fungi possess septate hyphae? Which types possess filamentous hyphae?

Selected references

Alexopoulos, C. J.: Introductory mycology, ed, 2, New York, 1962, John Wiley & Sons, Inc.

Alexopoulos, C. J., and Bold, H. C.: Algae and fungi, New York, 1967, The Macmillan Company.

Batra, S. W. T., and Batra L. R.: The fungus gardens of insects, Sci. Amer. 217(5):112-120, 1967.

Bonner, J.: The cellular slime molds, ed. 2, Princeton, New Jersey, 1967, Princeton University Press.

Bonner, J.: How slime molds communicate, Sci. Amer. 209: 84-93, 1963.

Cochrane, V. W.: Physiology of fungi, New York, 1958, John Wiley & Sons, Inc.

Cook, A. H.: The chemistry and biology of yeasts, New York, 1958, Academic Press Inc.

Duddington, C.: Beginner's guide to the fungi, New York, 1972, Drake Publishers, Inc.

Gray, W. D.: The relation of fungi to human affairs, New York, 1959, Holt, Rinehart & Winston, Inc.

Lamb, I. M.: Lichens, Sci. Amer. 201(4):144-156, 1959.

Maio, J. J.: Predatory fungi, Sci. Amer. 199(1):67-72, 1958.

Moore-Landeker, E.: Fundamentals of the fungi, Englewood Cliffs, New Jersey, 1972, Prentice-Hall, Inc.

Pramer, D.: Nematode-trapping fungi, Science 144:382-388, 1964.

Rose, A. H.: Yeasts, Sci. Amer. 202:136-146, 1960.

Snell, W. H.: A glossary of mycology, Cambridge, Massachusetts, 1957, Harvard University Press.

Sparrow, F. K.: Aquatic phycomycetes, Ann Arbor, 1960, University of Michigan Press.

PART THREE

Plant biology

13 Thallophytes—complex algae

Phylum Chlorophyta (green algae)
  Chlamydomonas
  Volvox
  Ulothrix
  Spirogyra
Phylum Phaeophyta (brown algae)
  Kelp (Laminaria)
  Rockweed (Fucus)
Phylum Rhodophyta (red algae)
  Nemalion
  Polysiphonia

The kingdom Plantae is sometimes considered as two subkingdoms—Thallophyta, the more complex algae, and Embryophyta, those plants which produce a distinct embryo following sexual reproduction. The thallophytes all have chlorophyll and other pigments and carry on photosynthesis. Many of them are microscopic and become obvious only following massive "blooms" due to addition of nutrients caused by pollution of ponds or streams.

Many of the thallophytes are large seaweed plants that grow over 100 feet in length. The green, brown, and red algae are considered here. Some systems of classification place them with the simple algae in the kingdom Protista instead of with the Plantae as is done in this text.

PHYLUM CHLOROPHYTA (GREEN ALGAE)

Green algae are placed in the phylum Chlorophyta. The chlorophyll is present in chloroplasts. Additional orange-reddish pigments known as carotenoids (L. *carota,* carrot) may also be present. The cell wall contains cellulose, and starch is the stored food. Starch is formed by special structures known as pyrenoids that are associated with the chloroplasts (Table 13-1). Green algae vary in structure; they may be unicellular, colonial, or multicellular, depending upon the species. When the vegetative body cells or the reproductive cells are motile, each bears two to eight anterior flagella, usually of equal lengths.

Reproduction occurs (1) asexually by fission, by fragmentation, by motile spores, or by nonmotile spores; or (2) sexually by isogamy (Gr. *isos,* equal; *gamos,* marriage) with the union of gametes of equal size; by heterogamy (Gr. *heteros,* different) with the union of gametes of unequal size; or by oogamy (Gr. *oon,* egg), which is a special type of heterogamy in which the female gamete is a nonmotile egg. The particular methods of reproduction vary with the species. The gametangium that produces the sex cells is unicellular, hence it cannot be called a sex

Table 13-1 Some characteristics of algae

Plants	Pigments	Cell wall	Stored food	Flagella
Green algae (Chlorophyta)	Chlorophyll a, b; carotenoids	Cellulose	Starch $(C_6H_{10}O_5)_n$; oils (sometimes)	When present, 2 to 8 anterior
Brown algae (Phaeophyta)	Chlorophyll a; carotenoids;* fucoxanthin (yellow brown); xanthophylls†	Cellulose; algin	Laminarian carbohydrate; sometimes mannitol or fats	When present, 2 lateral, un-equal
Red algae (Rhodophyta)	Chlorophyll a, d; carotenoids; phycoerythrin (reddish); phyco-cyanin (bluish)	Cellulose; pectin	''Floridean starch''; floridoside sugar	Absent

*Carotene $(C_{40}H_{56})$; varies from creamy white to orange red.
†Xanthophylls $(C_{40}H_{56}O_2)$, producing yellowish colors.

"organ" in the true sense, and it does not have sterile jacket cells surrounding it.

Green algae as well as other types supply foods for freshwater and marine animals. Most green algae live in fresh water, but a few species live in the ocean. Others live in soils, on rocks and trees, and in ice and snow. Certain species live in salt waters in which the concentration of salt is much greater than that in ocean water. Some species live in other plants or animals. Some marine green algae in association with red algae may form lime salts that contribute to the formation of reefs.

Chlamydomonas

In addition to the general characteristics of green algae, the following details apply to *Chlamydomonas* (Gr. *chlamys,* cloak; *monas,* one) (Fig. 13-1). Each ovoid plant contains a single, cup-shaped chloroplast with a pyrenoid; a pigmented stigma (eyespot); and two anterior flagella of equal length. There are two contractile vaculoes near the anterior end for excretion. Some investigators classify this organism with the single-celled animals.

At times there are formed within the cell two, four, or eight flagellated zoospores. These are motile, or swarm spores that look like miniature

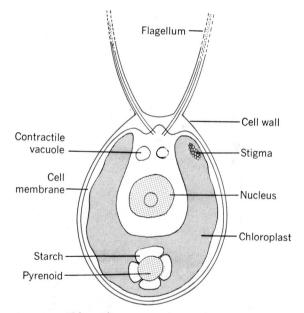

Fig. 13-1. *Chlamydomonas.*

parent cells and that swim out to form a new generation. In some instances the parent cell divides into eight, sixteen, or thirty-two isogametes. In the water two isogametes unite to form a single-celled, tetraflagellated zygote. When this is surrounded by a thick, resistant wall, it is called a zygospore. The single nucleus of the

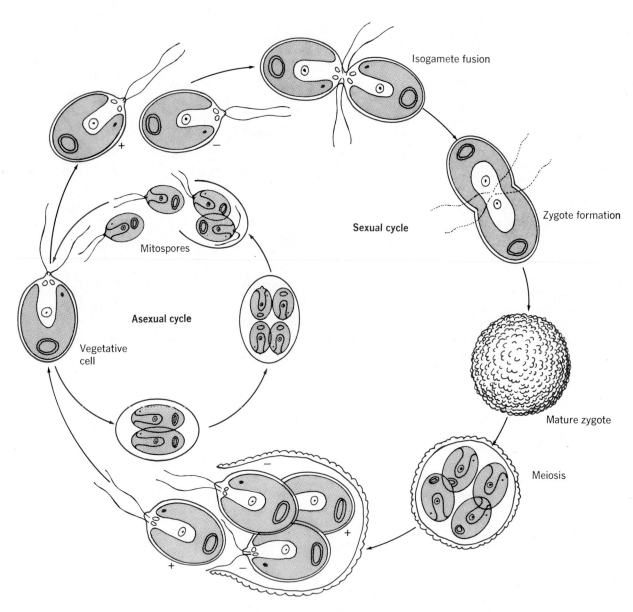

Fig. 13-2. *Chlamydomonas* life cycle. *Asexual cycle:* Cell wall functions as a cyst wall. Protoplast divides twice to form four mitospores, which escape and grow to normal size. *Sexual cycle:* Cells from different mating types (+ and −) fuse, form a zygote, then undergo meiosis forming four spores, which are released to form normal sized vegetative cells, two plus (+) and two minus (−).

Fig. 13-3. *Volvox,* a colonial alga, showing many colonies with daughter colonies inside. *(Courtesy General Biological Supply House, Inc., Chicago, Illinois.)*

zygospore produces four nuclei, one for each of the four zoospores produced (Fig. 13-2). As these are formed, the number of chromosomes is reduced by the process of meiosis from the double (2N) number to the haploid (N) number. The zoopores and gametes look somewhat alike except that the former are larger. *Chlamydomonas* may occur in fresh water, in damp soil, and even in snow.

Volvox

Volvox (L. *volvere,* to roll) is a colonial green alga in which hundreds to thousands of individual cells are arranged in a single layer. These cells are held together by a gelatinous secretion and joined by protoplasmic strands that establish physiologic continuity between them. The individuals form a hollow sphere that is 1 to 2 mm. in diameter and is composed of 500 to 50,000 cells (Fig. 13-3). The individual cells have much the same structure as *Chlamydomonas.*

Division of labor is accomplished by two kinds of cells, the biflagellated body or vegetative cells and the reproductive cells. Most of the cells are vegatative. Each body cell has two flagella, chloroplasts, a contractile vacuole, and a light-sensitive eyespot or stigma. The latter may assist in orientation for photosynthetic purposes.

A young colony consists primarily of somatic or body cells, but a mature colony contains, in addition, a number of cells capable of forming new colonies, either asexually or sexually. In asexual reproduction certain body cells redivide until a mass of cells is formed. These cells are arranged into a hollow sphere that forms a new colony.

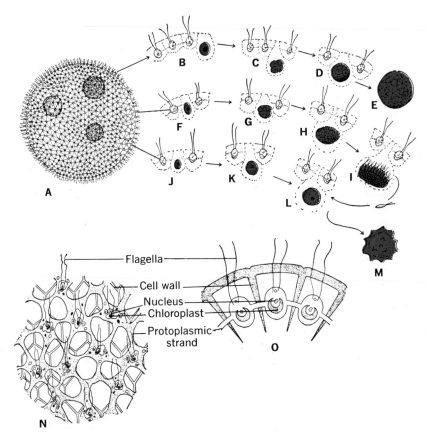

Fig. 13-4. Morphology and methods of reproduction in *Volvox*. **A,** Colony with hundreds of somatic cells, showing three immature daughter colonies within; **B** to **E,** formation of a new colony by cell losing its flagella and dividing repeatedly to form a new colony; **F** to **I,** formation of a bundle of flagellated sperms; **J** to **L,** formation of an egg and its fertilization by a sperm to produce a resistant zygospore, **M,** from which a new colony develops; **N,** surface view of the colony; **O,** side view of the colony.

Colonies are usually of different sexes, but in some species (for example, *Volvox globator*) both types of gametes (sperms and eggs) are formed by different cells of one colony. Numerous biflagellated sperms are produced by certain cells called antheridia or male gametangia (Fig. 13-4). Other cells, called oogonia or female gametangia, produce single eggs, which are large, without flagella, and contain a large amount of stored food.

A sperm swims to the egg and fertilizes it. This fusion, involving a smaller, motile sperm and a larger, nonmotile egg, is called oogamy. The resulting zygote develops a thick wall, thus forming a resistant zygospore. The zygospore undergoes meiosis or reduction division, after which the protoplast forms a new colony, or else a single zoospore (motile spore) is formed that makes a new colony. *Volvox* occurs in freshwater pools or lakes.

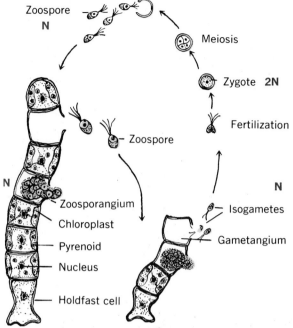

Zoospore
N

Meiosis

Zygote **2N**

Fertilization

Zoospore

N

— Zoosporangium

— Chloroplast

— Pyrenoid

— Nucleus

— Holdfast cell

Isogametes

Gametangium

N

Fig. 13-5. Structure of vegetative cells and methods of reproduction in *Ulothrix*. The numbers of chromosomes in the cells are represented by diploid (2N) and haploid (N).

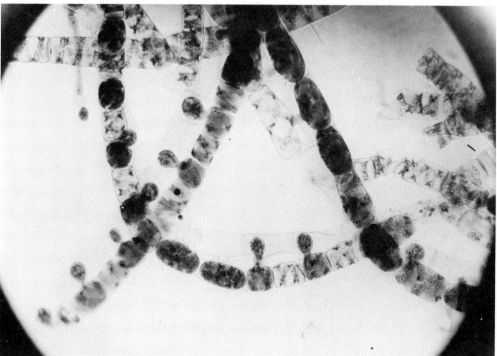

Fig. 13-6. *Spirogyra,* a green alga. Transfer of protoplast (protoplasm of cell) through the conjugation tube, from one filament to the other, during the process of conjugation. *(Courtesy General Biological Supply House, Inc., Chicago, Illinois.)*

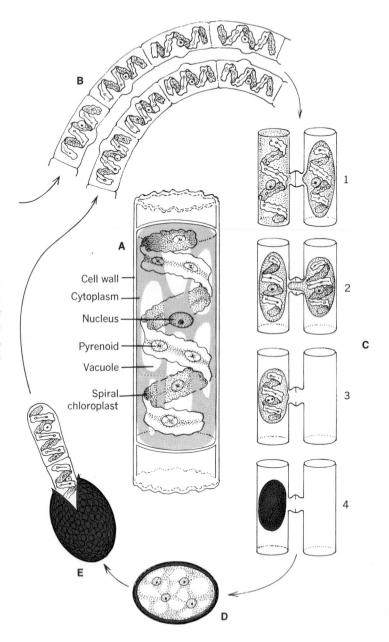

Fig. 13-7. *Spirogyra* life cycle. **A,** Single cell in a filament. **B,** Filaments beginning conjugation. **C,** Conjugation: **1,** Conjugation tube formation; **2,** cell contents move through tube; **3,** protoplasts fuse, one "cell" is empty; **4,** zygospore is formed. **D,** Meiosis in zygospore. **E,** Germination and filament formation.

Cell wall
Cytoplasm
Nucleus
Pyrenoid
Vacuole
Spiral chloroplast

Fig. 13-8. *Spirogyra.* Zygospores are formed; note the empty cells.

Ulothrix

Ulothrix (Gr. *oulos,* woolly; *thrix,* hair) (Fig. 13-5) is a freshwater, filamentous, unbranched, multicellular green alga that has a basal holdfast cell for attachment. Each vegetative cell has a nucleus and a chloroplast, which resembles an open band or ring and which contains pyrenoids. Each cell has a large central vacuole.

Unicellular reproductive structures, known as zoosporangia, produce from two to 32 large, motile zoospores, each with four flagella. Each zoospore forms a new filament by cell division. Other unicellular reproductive structures called gametangia produce from eight to 64 gametes, each of which bears only two flagella and is smaller than the four-flagellated zoospores. The gametes arise from different filaments. The union of isogametes produces a single-celled zygote. Each zygote divides by meiosis to produce four zoospores, each of which will form a new filament through repeated cell divisions. Asexual reproduction may occur by fragmentation or by the formation of nonmotile aplanospores (Gr. *aplanes,* not moving) in which the cell contents round up and secrete a new wall. The vegetative cells of *Ulothrix* are differentiated, interdependent, and may be considered to be above a true colony type of green alga in the evolutionary scale.

Spirogyra

Spirogyra (Gr. *speira,* spiral; *gyros,* curved) (Figs. 13-6 to 13-8) is common in fresh water where, because of its somewhat slimy apperance, it is called "pond scum" or "water silk." Each unbranched filament is composed of a linear series of cells and is covered with a mucilaginous sheath. One or more spiral, ribbon-shaped chloroplasts with several pyrenoids may be present in each cell, depending upon the species. An organized nucleus near the center of the cell is surrounded by a layer of cytoplasm. Strands of cytoplasm also extend to the pyrenoids that are located on the chloroplasts (Fig. 13-7).

Reproduction is accomplished asexually by fragmentation and by cell division; sexually it takes place by conjugation. In conjugation cells in two adjacent filaments form a conjugation tube between them, and the contents of one cell pass through the tube to the cell of the other filament. This union, or fertilization, forms a resistant zygospore (zygote) from which a new filament can develop. Even though the nonmotile gametes are of equal size, the one that migrates might be considered as male and the other as female. Sometimes the cells in a single filament unite.

PHYLUM PHAEOPHYTA (BROWN ALGAE)

The brown algae belong to the phylum Phaeophyta (Gr. *phaeo-,* brown) (Figs. 13-10 to 13-12). They are multicellular, some types of kelps being more than 100 feet long. The plant body may show considerable differentiation, with large rootlike holdfasts for attachment, a stemlike stipe, and one or more leaflike blades. Gas-filled bladders (floats) are often present. Brown algae are marine and are usually present along the shores in colder waters. The golden brownish pigment called fucoxanthin (fu ko -zan' thin) masks the chlorophyll. Usually there are several

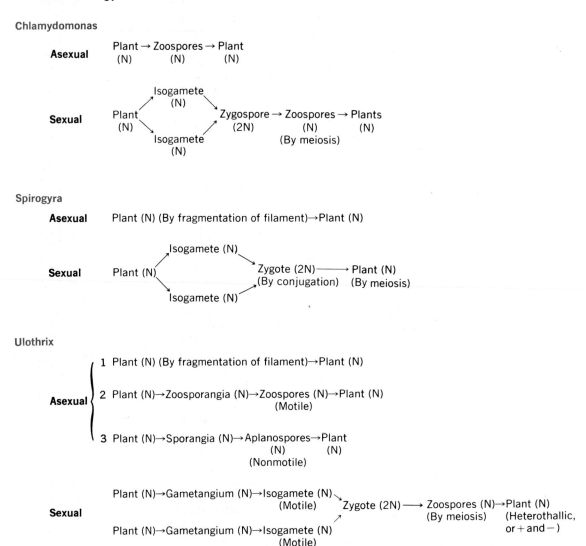

Fig. 13-9. Life cycles of certain green algae. Numbers of chromosomes are represented by (N) and (2N).

plastids per cell, but no pyrenoids are present. Each cell has an organized nucleus and may contain vacuoles. Some cells are so highly organized that they resemble cells of higher plants.

Brown algae show alternation of generations, in which a gamete-producing gametophyte alternates with a multicellular, spore-producing sporophyte. The brown alga, *Sargassum,* re-produces by asexual fragmentation, and its sexual reproduction is much like the life cycle of the rockweed *(Fucus).* The so-called Sargasso Sea of the Atlantic Ocean, between the West Indies and Africa, has great masses of these plants.

Depending upon the species of brown algae, sexual reproduction may occur by isogamy, heterogamy, or oogamy. When present, the pear-

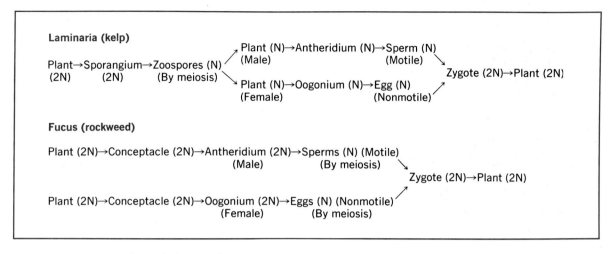

Fig. 13-10. Life cycles of certain brown algae.

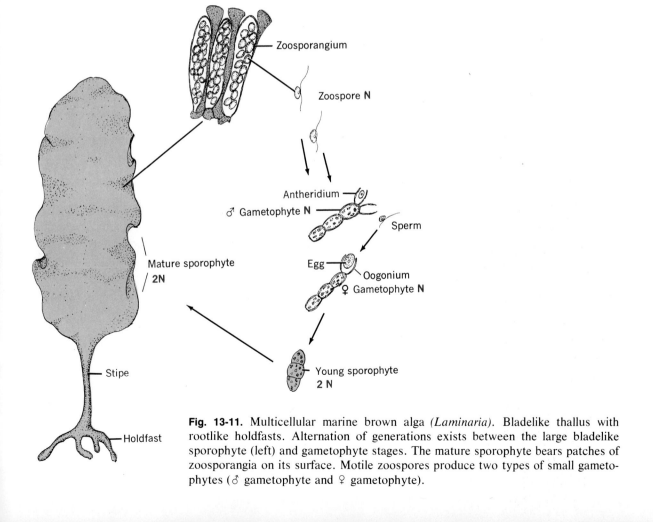

Fig. 13-11. Multicellular marine brown alga *(Laminaria).* Bladelike thallus with rootlike holdfasts. Alternation of generations exists between the large bladelike sporophyte (left) and gametophyte stages. The mature sporophyte bears patches of zoosporangia on its surface. Motile zoospores produce two types of small gametophytes (♂ gametophyte and ♀ gametophyte).

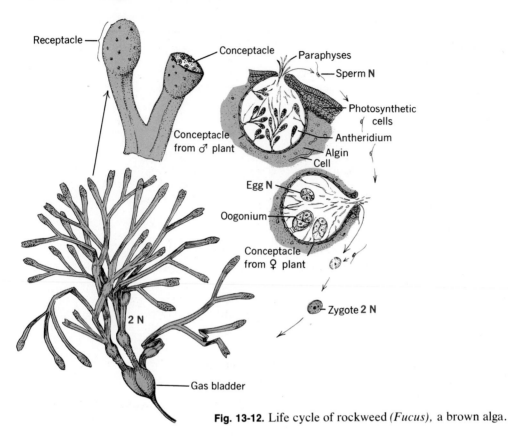

Receptacle

Conceptacle

Paraphyses

Sperm N

Conceptacle
from ♂ plant

Photosynthetic
cells

Antheridium

Algin
Cell

Egg N

Oogonium

Conceptacle
from ♀ plant

2 N

Zygote 2 N

Gas bladder

Fig. 13-12. Life cycle of rockweed *(Fucus),* a brown alga.

shaped reproductive cells bear two lateral flagella of unequal length.

Certain brown algae are sources of iodine, potassium, fertilizers, and foods for animals and man.

Kelp (Laminaria)

The kelps (Fig. 13-11), which belong to the genus *Laminaria* (L. *lamina,* thin plate or blade), are called ''devil's aprons'' and occur on both coasts of the United States. In addition to the general characteristics of brown algae already given, the following details apply more specifically to *Laminaria.* The large sporophyte plants bear patches of zoosporangia on the blade surfaces. The biflagellated zoospores formed in the zoosporangia produce two types of microscopic

gametophytes. The microscopic male gametophyte is a simple, branched filament, bearing one-celled antheridia that produce biflagellated *sperms* with flagella of unequal length. The small, less-branched female gametophyte is also a short filament, bearing oogonia that produce nonmotile eggs. Fertilization by oogamy produces a zygote that eventually develops into a mature sporophyte. Hence, there is alternation of generations between the more conspicuous sporophyte and the microscopic gametophyte.

Rockweed (Fucus)

The brown alga called the rockweed or *Fucus* (fu' kus) (Figs. 13-10 and 13-12) is commonly attached to rocks along sea coasts. The sporophyte plant is leathery, frequently branched, and at-

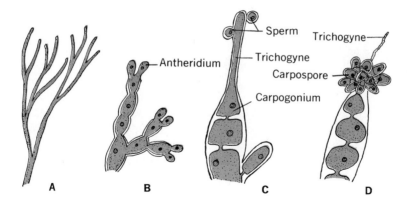

Fig. 13-13. Red alga *Nemalion*. **A,** Portion of forked, cylindrical body (thallus) that grows attached to rocks along the seacoast; **B,** portion of a branch bearing brushlike filaments whose tips divide into antheridia, each of which contains a nonmotile sperm (spermatium); **C,** portion of a branch bearing a basal carpogonium (with an egg) and an elongated trichogyne; a nonmotile sperm, carried by water, descends the trichogyne to unite with the egg to form a zygote; **D,** portion of a carpogonium showing asexual, nonmotile carpospores produced by a cluster of filaments; released carpospores develop into new *Nemalion* plants.

tached by a disk-shaped holdfast. Some branches bear gas-filled bladders (floats) for buoyancy in the water. The enlarged tips of the branches are called receptacles and are covered with small openings (ostioles) leading into cavities called conceptacles, within which are the sex organs. In some species the antheridia are borne on one plant and the oogonia on another plant. In other species the male and female organs are located on the same plant, even within the same conceptacle.

When mature, each oogonium contains eight eggs and is surrounded by a series of multicellular, branched, hairlike paraphyses (Gr. *para,* beside; *physis,* growth). The *antheridia* produce numerous pear-shaped sperms, each with two unequal, lateral flagella. The egg is fertilized by the motile sperm in the water. A zygote eventually produces a new *Fucus* plant by numerous cell divisions (Fig. 13-12).

Unfertilized eggs may be induced to develop artificially, without fertilization, through treatment with various substances. *Fucus* zygotes have been used experimentally in studies of en-

vironmental factors and development. Asexual reproduction may occur by fragmentation.

PHYLUM RHODOPHYTA (RED ALGAE)

Red algae are placed in the phylum Rhodophyta (Gr. *rhodon,* red). They are primarily marine, especially in warmer waters, although a few species live in fresh waters. The chlorophyll is associated with a red pigment called phycoerythrin (Gr. *phykos,* seaweed; *erythros,* red) and sometimes with a blue pigment called phycocyanin. Most species are sessile (attached), multicellular, and may be relatively simple or branched in the form of a ribbon or a sheet. Sizes vary from a few inches to several feet in length. Each cell contains a nucleus, central vacuoles, and one or more plastids, some of which possess pyrenoids. Broad cytoplasmic strands that connect adjacent cells are characteristic of red algae. Stored foods are "starches," which are carbohydrate intermediates between glucose and true starch.

Most red algae reproduce sexually. None of

the reproductive cells bear flagella. Many red algae show an alternation of generations between a free-living sporophyte and a free-living gametophyte. The specific methods of reproduction of typical forms are considered later in this chapter.

Some of the red algae grow on other algae as epiphytes, within the bodies of plants, or even as parasites. One species is associated with a green alga among the hairs of the three-toed sloth. Some species are able to grow in the sea to a depth of 600 feet, probably because the phycoerythrin acts as a photosynthetic pigment in the blue light of deep sea water.

Certain red algae are sources of agar, which is used as a solid medium for the cultivation of bacteria as well as for a medicine. A jellylike food may be obtained from the Irish "moss" (*Chondrus*) (Fig. 13-14). Some types become encrusted with lime and thus help to form reefs, atolls, and even islands. Other red algae are sources of jellylike substances used in shoe polishes, hair dressings, shaving creams, cosmetics, and various lubricating jellies. Some species serve as foods for fish, cattle, sheep, and even man.

Nemalion

Nemalion (Gr. *nema,* thread) is a cylindrical, forked, marine red alga found attached to rocks along the sea coast (Fig. 13-13). The body (thallus) is composed of a mass of interwoven, branched, threadlike structures surrounded by a gelatinous material. Some branches bear rushlike filaments whose tips are divided into short antheridia (spermatangia), each containing one nonmotile sperm (spermatium). The tips of other branches bear female structures, each with an enlarged, basal carpogonium (Gr. *karpos,* fruit; *gone,* offspring) that bears an egg and an elongated, hairlike, tubular trichogyne (Gr. *thrix,* hair; *gyne,* female) to receive the sperm that is brought by water. The sperm nucleus descends within the trichogyne to the carpogonium where it fuses with the egg nucleus to form a zygote. From the carpogonium is formed a cluster of short filaments whose tips produce the asexual, nonmotile carpospores. These carpospores germinate to form a new *Nemalion* plant.

There are a few freshwater red algae that are sometimes found in clear, colder waters, which are similar to marine types in many ways but are usually smaller.

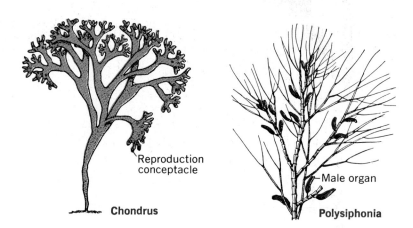

Reproduction conceptacle

Chondrus

Male organ

Polysiphonia

Fig. 13-14. Red algae. *Chondrus* is sometimes erroneously called Irish "moss." *Polysiphonia* is also shown in Fig. 13-15.

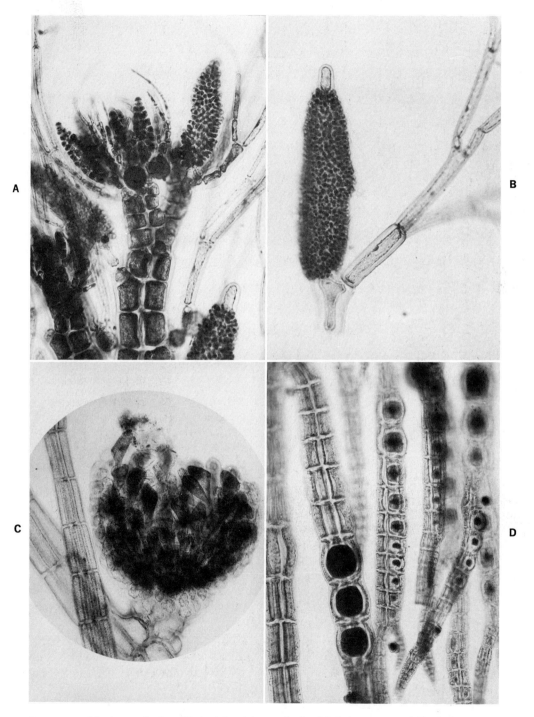

Fig. 13-15. Marine red alga *(Polysiphonia)*. **A,** Antheridial (spermatangial) clusters; **B,** antheridial branch; **C,** cystocarp (carpogonium) with an enclosed gonimoblast; **D,** tetraspores.

(Courtesy General Biological Supply House, Inc., Chicago, Illinois.)

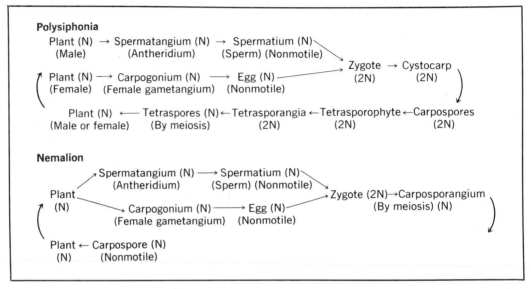

Fig. 13-16. Life cycles of certain red algae.

Polysiphonia

Polysiphonia (Gr. *poly-*, many; *siphon*, tube) is a multibranched red alga that grows on rocks of the sea coast (Figs. 13-14 and 13-15). The main axis and larger branches consist of a central core made of a single row of elongated core cells and surrounded by a layer of jacket cells. The core cells are connected with each other by strands of cytoplasmic connectives that form tubelike siphons, hence the name *Polysiphonia*. Each cell has a nucleus and numerous disk-like red plastids that contain phycoerythrin and mask the chlorophyll.

This red alga is diecious (Gr. *di-*, two; *oikos*, house), the male gametes being produced by one plant and the female gametes by another plant. The lateral branches of the male plants bear clusters of antheridia (spermatangia) that produce numerous nonmotile sperms (spermatia). The side branches of other plants bear female carpogonia. Each carpogonium (female gametangium) bears an elongated trichogyne to receive the sperm that is brought by water. The nucleus of the sperm travels down the trichogyne to the carpogonium where it fuses with the egg. After numerous cell divisions, many filaments are produced whose tips form carpospores. Other filaments of the plant form an urn-shaped cystocarp around the carpospores.

When the carpospores are released, they produce new plants that form tetrasporangia, each with four nonmotile tetraspores. The tetrasporangia are borne on the central core and lie just beneath the jacket cells. When liberated, each tetraspore produces a *Polysiphonia* plant that bears either antheridia or carpogonia. This complex life cycle consists of (1) male and female plants, (2) zygote and carpospores, and (3) tetrasporophyte plants with tetrasporangia that produce the tetraspores (Fig. 13-16).

Some contributors to the knowledge of
ALGAE

A. L. de Jussieu (1748-1836)

A French botanist who proposed a natural system of plants (1789) from which the term algae dates in a somewhat modern sense.

Friedrich Kützing (1807-1893)

A German botanist who first distinguished between diatoms and desmids (1833).

W. H. Harvey (1811-1866)

An Irish botanist who divided the algae into three groups that correspond quite well with present-day groups of blue-green, green, brown, and red algae (1836). *(The Bettmann Archive, Inc.)*

Harvey

Karl W. von Naegeli (1817-1891)

A Swiss botanist who distinguished the blue-green from the green algae (1853).

Gottlob L. Rabenhorst (1806-1881)

A German botanist recognized the green algae group (Chlorophyta) in approximately its present-day sense (1863).

C. J. Friedrich Schmitz

He made important contributions to our knowledge of the structure and classification of red algae (1883-1889).

W. G. Farlow

He made extensive studies and published a volume on the marine algae of the south coast of New England (1873).

Georg Klebs (1857-1918)

A German botanist who suggested that several groups of flagellated organisms are related to different groups of algae and are not all to be included with the unicellular animals as they had been. This proposal caused **Adolph Pasher (1887-1945)**, a German botanist, to work out a system classifying the major groups of algae along present-day lines (1910-1931). He proposed the term Chrysophyta (1914).

Harald Kylin (1879-1949)

A Swedish botanist who first clearly recognized the major groups of brown algae, based primarily on life cycles (1933). He also modified Schmitz's classification of red algae (1923-1932).

Review questions and topics

1 Describe in your own words that group of plants commonly referred to as algae.
2 Why are the algae placed in the subkingdom Thallophyta?
3 List the distinguishing characteristics by which the following phyla may be differentiated: Chlorophyta, Phaeophyta and Rhodophyta. Give examples of each phylum.
4 What are the relationships among algae, chlorophyll, and photosynthesis?
5 Define: plankton, epiphyte, and parasite.
6 Describe the special kind of plants known as lichens.
7 Discuss the economic importance of each phylum of algae, including beneficial and harmful conditions.
8 List the asexual methods of reproduction found in algae, describing each.
9 List the sexual methods of reproduction found in algae, describing each.
10 What is meant by a life cycle? Make a diagram of a typical life cycle of an alga of each phylum.
11 Describe the increase in complexity of structures methods of reproduction, and functions from the simpler to the higher types of algae.
12 Define, giving an example of each: isogamy, heterogamy, and oogamy.
13 List the various types of pigments found in the various algae, giving the specific alga in which each is found.
14 List the various types of stored foods found in various algae, giving the specific alga in which each is found.

Selected references

Alexopoulos, C. J. and Bold, H. C.: Algae and fungi, New York, 1967, The Macmillan Company.

Chapman, V. J.: The algae, London, 1962, Macmillan & Co., Ltd.

Dawson, E. Y.: How to know the seaweeds, Dubuque, Iowa, 1956, William C. Brown Company, Publishers.

Fritsch, F. E.: The structure and reproduction of the algae, vol. 1 and 2, Cambridge, 1935 and 1945, Cambridge University Press.

Gibor, A.: Acetabularia: a useful giant cell, Sci. Amer. **215**:118-24, 1966.

Guberlet, M. L.: Seaweeds at ebb tide, Seattle, 1956, University of Washington Press.

Jackson, D. F., editor: Algae and man, New York, 1964, Plenum Press.

Lamb, I. M.: Lichens, Sci. Amer. **201**(4):144-156, 1959.

Lewin, R. A., editor: Physiology and biochemistry of algae, New York, 1962, Academic Press Inc.

Milner, H. W.: Algae as food, Sci. Amer. **189**:31-35, 1953.

Prescott, G. W.: How to know the fresh-water algae, Dubuque, Iowa, 1954, William C. Brown Company, Publishers.

Prescott, G. W.: The algae: a review, Boston, 1968, Houghton Mifflin Company.

Smith, G. M.: Cryptogamic botany, vol. 1, New York, 1955, McGraw-Hill Book Company.

Tiffany, L. H.: Algae, the grass of many waters, Springfield, Ill., 1958, Charles C Thomas, Publisher.

Weiss, F. J.: The useful algae, Sci. Amer. **187**:15-17, 1952.

14 Phylum Bryophyta—liverworts and mosses

SUBKINGDOM EMBRYOPHYTA

Embryophyta produce an embryo that is parasitic for some time within a multicellular female sex organ called an archegonium (Gr. *archegonos*, first of a race). The multicellular male sex organ is called an antheridium (Gr. *anthos*, flower). The sex organs are surrounded by a sterile, protective jacket layer of tissues.

Embryophytes reproduce by oogamy in which a small motile sperm cell fertilizes a larger nonmotile egg. In the reproductive cycle the embryophytes go through an alternation of generations, in which a spore-producing plant called a *sporophyte* alternates with a gamete-producing plant called a *gametophyte*.

Embryophytes are essentially terrestrial although some types may live in water. Chlorophyll is present in organized chloroplasts. The aerial parts of the plants may be protected by a waxlike cutin. Embryophyta include such plants as true mosses, ferns, cone-bearing plants, and flowering plants. Distinguishing characteristics are the presence or absence of chlorophyll; the absence or presence of true leaves, stems, and roots; the absence of presence of conducting (vascular) tissues such as phloem and xylem; the absence or presence of flowers; the ability to produce seeds; and the presence of exposed (naked) or enclosed seeds (protected by an ovary, or fruit).

Bryophyta

The members of the phylum Bryophyta (Gr. *bryon*, moss; *phyta*, plants) are terrestrial plants, although they require considerable moisture for growth and for the transmission of the sperm to the egg. Bryophytes possess chlorophyll-containing chloroplasts. In general the adult plant body is composed of blocks of rather thin-walled cells that form parenchymatous tissues in contrast to the simpler construction of the thallophytes. The mature plant is not filamentous, but the developmental protonema stage may be thread-like. None of the bryophytes grow to any great height.

The gamete-producing sex organs, the gametangia, are multicellular and possess a protective layer of sterile cells. With few exceptions the

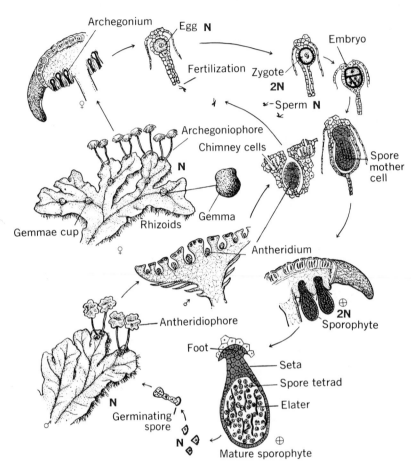

Fig. 14-1. *Marchantia,* a common liverwort. There is an alternation of generations between the gameto-phyte (male ♂ and female ♀ plants) and sporophyte ⊕ that develops from the fertilized egg of the female plant. A gemma, which is produced asexually, develops directly into a gametophyte plant.

gametangia of thallophytes (mentioned in previous chapter) are unicellular. All bryophytes show alternation of generations between a gamete-producing generation and a spore-producing generation. The sporophyte is more or less dependent upon the gametophyte.

Bryophytes are placed in the subkingdom Embryophyta because they develop a multicellular embryo from the zygote. The sporophyte develops from the embryo. Spores are produced from spore mother cells within the sporangium. Asexual reproduction may occur by fragmentation of the plant or through special bodies known as gemmae (L. *gemma,* bud). Bryophytes lack true conducting tissues, such as phloem and

xylem, which are found in higher plants. Rootlike rhizoids anchor the plant and absorb materials from the substratum. Liverworts and true mosses possess similar methods of reproduction and life cycles and are much alike structurally and functionally despite differences that casual observation may indicate.

CLASS HEPATICAE (LIVERWORTS)

The members of the class Hepaticae (L. *hepaticus,* liver) have bodies that are flattened dorsoventrally with distinct upper and lower surfaces (Fig. 14-1). The thalloid liverworts have flat, lobed bodies that seem to resemble the lobes of a liver of higher animals, whereas the "leafy" liverworts have bodies with a stemlike axis upon which leaflike structures grow.

Marchantia

Marchantia is the genus to which the flat, lobed thalloid liverworts belong. They are commonly found on moist rocks and soil along streams. The surface of the branched thallus body has rhomboidal areas, each of which has a pore in its center. Internally, the thallus has air chambers and columns of cells (chimney cells) containing chloroplasts. Rootlike rhizoids anchor the thallus and absorb materials from the substratum (see Fig. 14-1).

Marchantia is diecious, with separate male and female plants (Fig. 14-2), one thallus bearing antheridia and another thallus bearing archegonia. Stalklike antheridiophores, with lobed disks at the tip, grow on the male plant. The male antheridia are borne in cavities that open on the upper surface of the disks. Each antheridium is an enlarged, oval structure that produces coiled, biflagellated sperms (Fig. 14-3).

Stalklike archegoniophores, with small, terminal disks bearing fingerlike rays, grow on the female plant. The female archegonia are borne on the undersurface of the disks. Each archegonium has a hollow, tubular neck and an enlarged venter at the base of which is a single egg (Fig. 14-3).

The sperm swims through the water from the antheridium to the venter of the archegonium where the sperm and egg fuse. Through numerous cell divisions the fertilized egg forms a multicellular embryo from which the spore-producing sporophyte develops. The sporophyte consists of a *foot* embedded in the disk of the female gametophyte, a *seta,* and a *sporangium.* The sporangium produces numerous spores. Elongated, spiral-shaped, hygroscopic elaters (Gr. *elater,* driver) are moisture-sensitive and expel the spores from the sporangium. The spores germinate to form new male or female gametophytes.

There is an alternation of generations between the gamete-producing gametophyte and the spore-forming sporophyte. *Marchantia* may also reproduce asexually by the formation of special bodies, known as gemmae, in little gemma cups or by the process of fragmentation.

Porella

Porella is a common leafy liverwort (Fig. 14-4) that may form a green mass on moist soil, rocks, or rotten wood. Some species may grow on tree trunks in damp forests. *Porella* has three rows of

Marchantia

Gametophyte ⟶ Antheridia ⟶ Sperms
(Male) (N)   (N)   (N)
                                                    ⟶ Zygote ⟶ Young sporophyte ⟶ Mature sporophyte
                                                        (2N)   (2N)   (2N)
Gametophyte ⟶ Archegonia ⟶ Eggs
(Female) (N)   (N)   (N)

Gametophyte ⟵ Young gametophyte ⟵ Spores ⟵ Spore mother
(Male or female)   (Prothallus) (N)   (N)   cells (2N)
(N)

Fig. 14-2. Life cycles of *Marchantia,* showing the alternation of generations. In another thalloid liverwort *(Riccia)* the gametophyte contains both antheridia and archegonia on the same plant.

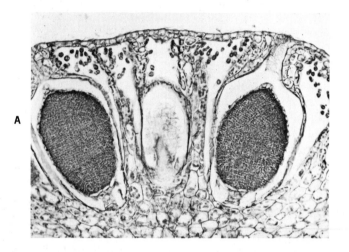

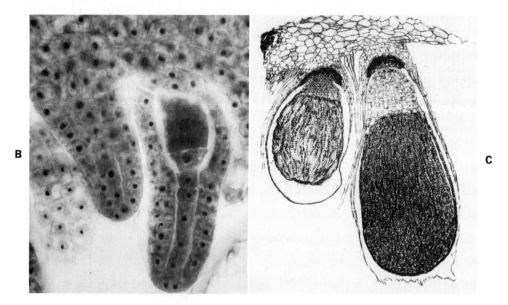

Fig. 14-3. *Marchantia* **A,** Top of the male disk showing saclike cavities with antheridia; **B,** bottom of the female disk with archegonia; **C,** bottom of the female disk with sporophytes, attached by means of a foot embedded in the female gametophyte.
(Courtesy General Biological Supply House, Inc., Chicago, Illinois.)

leaflike structures that grow on a stemlike axis, which may be branched. Rhizoids are attached to its lower surface. The leaflike structures are much simpler than the gametophyte of *Marchantia,* consisting of only one layer of cells without a midvein. The sporophyte of *Porella* is similar to that of *Marchantia,* consisting of foot, stalk, and sporangium. The sporangium bears spores and elaters, as in the thalloid liverworts. The life cycle is shown in Fig. 14-4.

Liverworts, mosses, and lichens, when growing on bare rocks, may mechanically and chemically convert the rocky surface into soil. Through their death they contribute valuable organic materials to the soil. Soon there may be sufficient soil formed to permit the growth of ferns and other simpler plants, and still later shrubs and trees may appear, thus creating what is called a plant succession.

CLASS MUSCI (TRUE MOSSES)

The members of the class Musci are small, green plants that usually grow upright. In some moist places they may form a mat of vegetation. Although true mosses may possess structures that superficially resemble leaves and stems, they are not true leaves and stems because they lack the vascular tissues (phloem and xylem) of such structures. Each individual plant consists of a stemlike axis with attached leaflike structures. The epidermal cells contain stomata (pores). Typical representatives of this class include the following.

Polytrichum

Polytrichum (po -lit' ri kum) (Gr. *poly-,* many; *thrix,* hair) is a common moss known as the hairy cap moss (Figs. 14-5 to 14-8).

Several antheridia are borne in a cluster at the tips of certain plants, and several archegonia are borne at the tips of other plants. *Polytrichum* has its sexes in separate plants and is therefore diecious. In species of mosses that are monecious (bisexual) the antheridia and archegonia are borne on the same plant.

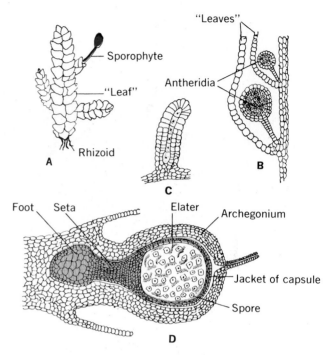

Fig. 14-4. *Porella,* a leafy liverwort. **A,** Female plant with an attached sporophyte; **B,** portion of the male plant bearing antheridia; **C,** young archegonium showing an egg at the base of the elongated neck; **D,** developing sporophyte attached to the parent female gametophyte plant.

In *Polytrichum* the antheridia are separated by multicellular, sterile hairs called paraphyses. Both antheridia and paraphyses are surrounded by a rosette of leaflike appendages that may be colored and resemble a "flower." Each antheridium consists of a short stalk and an enlarged vessel in which unicellular sperms are formed. The sperms are coiled, with two long, terminal flagella, and escape through the apex of the antheridium.

The archegonia are also separated by paraphyses. Each archegonium has a stalk supporting an enlarged venter that contains an egg. When mature, a canal opens through the long neck to the venter. During fertilization the motile sperm swims through water from the antheridium to the female plant. It travels down the canal to the

Fig. 14-5. Moss sporophytes. The sporangium is at the end of a long stalk that grows out of the gametophyte plant.
(Photograph by F. S. DeMartino, University of Dayton.)

Common moss

Gametophyte ⟶ Antheridia ⟶ Sperms
(Male) (N)   (N)   (N)
                                      ⟶ Zygote ⟶ Embryo ⟶ Sporophyte ⟶ Sporophyte
                                         (2N)   (2N)   (Young) (2N)   (Mature) (2N)
Gametophyte ⟶ Archegonia ⟶ Eggs
(Female) (N)   (N)   (N)

Gametophyte ⟵ Protonema ⟵ Spores ⟵ Spore mother
(Male or female)   (Young gametophyte)   (N)   cell (2N)
(N)   (N)

Fig. 14-6. Life cycles of mosses, showing the alternation of generations. Some species of peat moss possess male and female sex organs on the same plant.

Fig. 14-7. Hairy cap moss *(Polytrichum),* showing **(1)** archegonial, **(2)** sporophyte, and **(3)** antheridial plants.
(Courtesy Carolina Biological Supply Company, Burlington, North Carolina.)

venter where sperm and egg fuse. The fertilized egg is retained in the venter where it forms an embryo by numerous cell divisions. The embryo is parasitic on the female gametophyte plant from which it receives water, food, and protection. The embryo then grows to form a new plant, the sporophyte.

A sporophyte consists of a foot, which is attached to the female plant, and a stalklike seta, which bears a sporangium at its tip. The sporangium is covered with a hairy cap, or calyptra (Gr. *kalyptra,* veil). When the calyptra is removed, a lidlike operculum (L. *operculum,* lid) covering the sporangium can be seen. Beneath the operculum is a ring of hygroscopic teeth, the peristome (Gr. *peri,* around; *stoma,* opening). The teeth are affected by moisture, and their movements expel the spores from the sporangium. When immature, the sporangium contains spore mother cells, each of which undergoes reduction division to produce four spores. The four spores of each tetrad are of two kinds. One kind contains a small Y chromosome (sex chromosome) and produces a male plant; the other kind contains a large X chromosome and produces a female plant. This method of sex determination is similar to that in man, in whom there are also X and Y chromosomes.

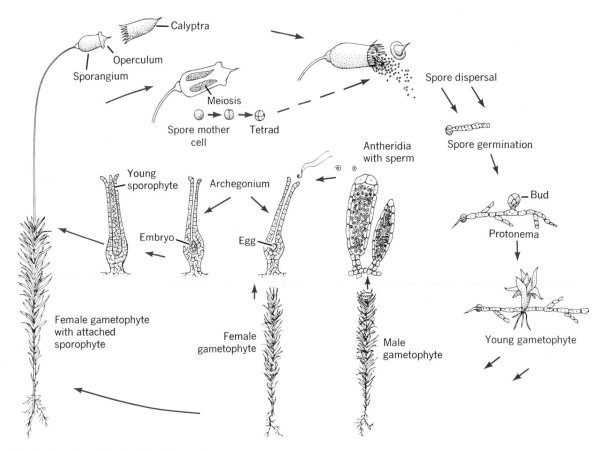

Fig. 14-8. Life cycle of *Polytrichum,* the hairy cap moss.

Each spore germinates to form a threadlike, branched protonema. The cells of the protonema bear chloroplasts. Rhizoids anchor the young plant and absorb materials from the soil. Buds appear on the protonema, and these produce a new male or female moss plant by cell division. There is an alternation of generations between the gametophyte plant and the sporophyte plant. Under certain conditions some mosses may reproduce asexually by a fragmentation of the plant or by the formation of gemmae.

Sphagnum

Sphagnum is the genus to which the peat or bog mosses belong. They commonly inhabit bogs, ponds, and other wet places. Their life cycles are similar to that of *Polytrichum.* The upright, branched axis may reach 1 foot in length; it bears leaflike appendages that contain two types of cells—one stores water and the other contains chloroplasts. The water storage cells are large and empty and have openings to the outside. *Sphagnum* can absorb water up to twenty times its weight.

In certain species antheridia and archegonia are present on the same plant (but on different branches). In others they are present on different plants. The fertilized egg develops into a sporophyte, which has a base embedded in the gametophyte, a short stalklike seta, and an enlarged sporangium. Because of the short seta, the gametophyte develops a structure known as the pseudopodium at the base of the foot in order to elevate the sporangium above the gametophyte. Alternation of generations occurs.

ECONOMIC IMPORTANCE OF SOME MOSSES

Sphagnum and other mosses grow on the edges of ponds and lakes where they may gradually fill in the entire body of water. During this process of filling in there may be masses of floating mosses.

Sphagnum is utilized in gardening to keep the soil porous and to increase its water-retaining capacity. Because of its water-holding quality, it is used by florists in packing cut flowers and in the development of seedlings. In the past, *Sphagnum* and other mosses have accumulated in bogs and swamps where they have slowly decomposed and become compacted and carbonized. This process produces peat, a valuable fuel, which can be used in place of coal. Peat moss is also used for the bedding of farm animals. Some mosses are used commercially as packing material in shipping. Mosses may prevent soil erosion because of their abundant growth. Also, their natural water-absorbing qualities enable them to assist in flood control by preventing rapid runoff of water.

Some contributors to the knowledge of
LIVERWORTS AND MOSSES

Marchant

A French botanist who studied the liverwort, *Marchantia* (1678), and the genus was named after him.

Johann Hedwig (1730-1799)

A German botanist who established the class Musci (1782) and is often called the "Father of Bryology."

Antoine Laurent de Jussieu (1748-1836)

He first established the class Hepaticae (1789).

Samuel F. Gray (1766-1828)

A British botanist who first recognized the bryophytes as a natural group (1821), although the name Bryophyta was not applied until nearly 50 years later.

Stephen L. Endlicher (1804-1849)

An Austrian botanist who classified all plants into two primary groups, the Thallophyta and Cormophyta (1836). The term Embryophyta is now used instead of his term Cormophyta (Gr. *kormos,* trunk; *phyta,* plants) and includes the Bryophyta (liverworts and true mosses).

William Starling Sullivant (1803-1873)

An American, who made significant contributions to the knowledge of liverworts and mosses, especially in their classification (1848).

John Merle Coulter, Sereno Watson, and Lucien Marcus Underwood

Americans, who contributed to the knowledge of classification of liverworts and mosses (1890).

Review questions and topics

1 List the characteristics of Embryophyta, Bryophyta, Hepaticae, and Musci.
2 What are the general habitats for bryophytes?
3 Discuss the economic importance of liverworts and true mosses.
4 Diagram the life cycle of a liverwort, showing the stages in correct sequence.
5 Diagram the life cycle of a true moss, showing the stages in correct sequence.
6 Explain what is meant by alternation of generations.
7 Define gametophyte and sporophyte. Why may the sporophyte be considered to be a parasite upon the gametophyte? Contrast gametophytes and sporophytes in every way possible.
8 Describe the method of sex determination in a moss.
9 Give specific reasons why most bryophytes are rather short plants.
10 Contrast antheridia and archegonia, giving examples of each.
11 Describe the structure and functions of elaters. What is the function of a so-called foot?
12 Contrast what is meant by diecious and monecious (bisexual), giving examples of each. What advantages and disadvantages can you list for each phenomenon?

Selected references

Bland, J. H.: Forests of Lilliput; the realm of mosses and lichens, Englewood Cliffs, N. J., 1971, Prentice-Hall, Inc.
Bodenberg, E. T.: Mosses, a new approach to the identification of common species, Minneapolis, 1954, Burgess Publishing Co.
Conard, H. S.: How to know the mosses, Dubuque, Iowa, 1956, William C. Brown Company, Publishers.
Grout, A. J.: Mosses with a hand lens, Newfane, Vermont, 1947, A. J. Grout.
Richards, P. W.: A book of mosses, Harmondsworth, England, 1950, Penguin Books, Ltd.
Smith, G. M.: Cryptogamic botany, vol. 2, New York, 1955, McGraw-Hill Book Company.
Steere, W. C.: Cenozoic and Mesozoic bryophytes of North America, Amer. Midl. Nat. **36:**298-324, 1946.
Steere, W. C.: Bryology, in a century of progress in the natural sciences, 1859-1953, San Francisco, 1955, California Academy of Sciences.
Thieret, J. W.: Bryophytes as economic plants, Econ. Bot. **10:**75-91, 1955.

15 Phylum Tracheophyta—club "mosses," horsetails, and ferns

Subphylum Lycopsida (club "mosses")
  Selaginella
Subphylum Sphenopsida (horsetails)
  Equisetum
Subphylum Pteropsida
  Class Filicinae (ferns)
  Polypodium

The phylum Tracheophyta (Gr. *tracheia*, vessel; *phyta*, plants) contains plants with vascular tissues of varying degrees of complexity. Tracheophytes have leaves, stems, and roots; supporting tissues for more or less upright growth; stomata or small openings for the exchange of gases; and a protective layer of waxlike cutin in certain parts of the plant.

Tracheophytes undergo alternation of generations. The gamete-producing gametophyte is small and inconspicuous, and the spore-producing sporophyte is relatively large and independent when mature. All tracheophytes form sex organs of two kinds, antheridia and archegonia. They all form embryos that develop from a fertilized egg. The sporangia are usually borne on leaves known as sporophylls. Sporangia may occur (1) singly on the upper surface of the sporophylls to form a clublike strobilus (Gr. *strobilos*, cone), as in club "mosses"; (2) in groups upon shield-shaped sporangiophores to

form conelike strobili, as in horsetails; or (3) in clusters known as sori (Gr. *soros*, heap), as in ferns.

In some species the sporophylls bear sporangia in which the spores are all alike (homosporous); in others there are two kinds of sporophylls, two kinds of sporangia, and two kinds of spores (heterosporous). Spores germinate to form different types of gametophytes: (1) colorless, rather bulky prothalli with antheridia and archegonia, as in club "mosses"; (2) thin, green, irregular gametophytes with antheridia and archegonia, as in horsetails; and (3) small, green, thin, heart-shaped prothalli with antheridia and archegonia, as in ferns. In general, club "mosses," horsetails, and ferns have somewhat similar methods of reproduction and life cycles. Although most Tracheophytes are land plants, some are aquatic. In accordance with the system of classification being used, this phylum contains the club "mosses," horsetails, ferns, cone-bearing plants, and flowering plants.

SUBPHYLUM LYCOPSIDA (CLUB "MOSSES")

The subphylum Lycopsida includes a variety of plants that are also commonly called club mosses or ground pines because they are prin-

193

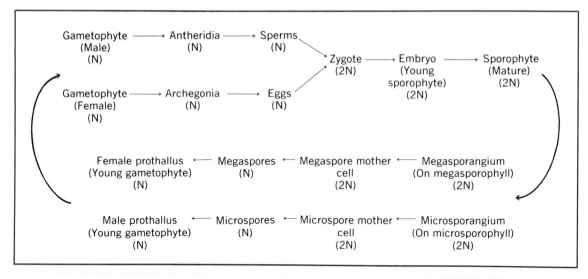

Fig. 15-1. Life cycle of a smaller club moss *(Selaginella),* showing the alternation of generations and two kinds of spores.

cipally perennial, creeping evergreens. The sporophyte plants of club mosses have rather simple vascular tissues. The leaves are usually spirally arranged, and the stems and roots are usually branched dichotomously (Gr. *dicha,* in two).

Sporangia occur singly on the upper surface of specialized leaves known as sporophylls (spore leaves). Usually the sporophylls and their sporangia are arranged at the tips of the stems to form strobili (cones). The spore-bearing organs are often foot- or club-shaped.

Selaginella

The genus *Selaginella* contains the perennial, smaller club mosses (Fig. 15-1) which are widely distributed, especially in the tropics. One species, known as the "resurrection plant," can withstand the dry conditions in the southwestern United States by rolling into a ball. Other specimens are grown as ornamental plants in greenhouses. Most species of *Selanginella* are creepers, although a few are erect.

The branched stems of the mature sporophyte

bear tiny, lanceolate, green leaves, usually in four rows, two of larger leaves and two of smaller ones. Roots anchor the plant and absorb materials. The vascular system is rather simple and varies with the species. The so-called vascular cylinder usually consists of a group of separate vascular bundles. Between these and the surrounding cortical tissue there is an air space when the stem is viewed in cross section.

Strobili at the tips of the branches are composed of spore-bearing sporophylls. In a cone the upper surface of each microsporophyll (male) has in its axil (upper angle) a small microsporangium that produces many small microspores. The megasporophylls (female) produce four large, thick-walled megaspores in each megasporangium located in the axil. Frequently, the microsporophylls are located toward the tip of the cone, and the megasporophylls are located lower on the same cone. Since *Selaginella* produces two kinds of spores (microspores and megaspores), it is heterosporous (Fig. 15-2).

A megaspore forms a megagametophyte within the megaspore while still within the megasporan-

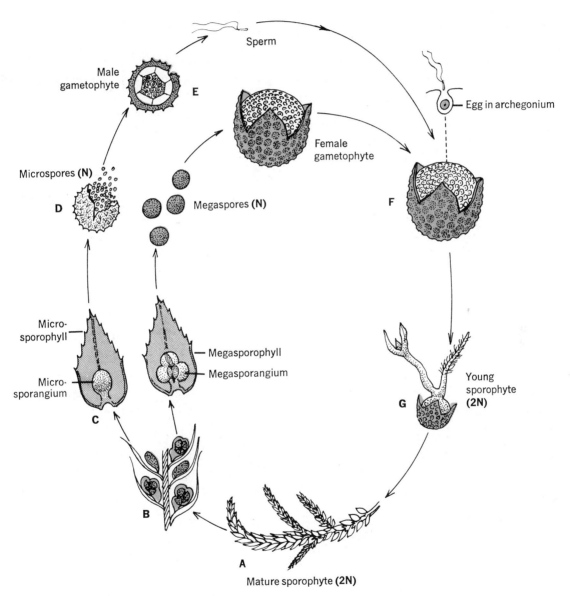

Fig. 15-2. *Selaginella* life cycle. **A,** Mature sporophyte. **B,** Section of a strobilus. **C,** Sporophylls with sporangia. **D,** Spores. **E,** Gametophytes. **F,** Fertilization. **G,** Young sporophyte.

gium. As the megagametophyte develops, it forms several egg-containing archegonia, stored foods, rhizoids, and chlorophyll.

A microspore develops within the microsporangium to form a small, parasitic microgametophyte, which is surrounded by the microspore wall, has no chlorophyll, and consists of one prothallial cell and one male antheridium. The antheridium produces a biflagellated sperm that swims to the female archegonium where the egg is fertilized to form a zygote. The zygote forms two cells by cell division, the upper one becoming a suspensor for pushing the developing embryo into the stored food of the megagametophyte. The other cell of the dividing zygote develops into the embryo, which forms a mass of cells, the foot, for absorbing food from the megagametophyte. Eventually, the embryo sporophyte forms stems, roots, and leaves. Later, the developing sporophyte becomes independent and photosynthesizes its own food. The formation of a suspensor by the zygote is somewhat similar to a phenomenon in higher, seed-forming plants. The megagametophyte, together with its embryo, resembles a seed of a higher plant in some ways, but it lacks some of the tissues found in a true seed.

SUBPHYLUM SPHENOPSIDA (HORSETAILS)

The sporophyte plants of the horsetails have a simple vascular system: small leaves (sometimes scalelike or wedgelike) arranged in whorls at the joints of the hollow stems; stems that are roughened by ribs and impregnated with silica, the ribs being opposite the vascular bundles and alternating with small, air-filled canals; and true roots. The roughened stems account for another name and use applied to them in pioneering America, that is, "scouring rushes."

In most species a horizontal, branched, underground stem known as a rhizome bears two types of aerial stems: (1) the sterile, green, branched, vegatative stems for food manufacture, and (2) the colorless, unbranched, fertile, reproductive stems with a single, terminal strobilus.

Sporangia are borne on a shield-shaped sporangiophore whose stalk attaches it to a central axis where, with many other sporangiophores, they form the strobilus. The spores are alike, and each has ribbonlike elaters that are affected by moisture changes to assist in the movement of spores. A germinating spore forms a small, green, ribbonlike young gametophyte with rhizoids and usually with antheridia and archegonia. The spiral, multiflagellated sperms produced by the antheridia swim to the egg-producing archegonium where the egg is fertilized. The egg develops into a multicellular embryo, which in turn develops into a sporophyte plant. Hence there is an alternation of generations.

Equisetum

In addition to the characteristics given, the following apply to the members of the genus *Equisetum* (L. *equus,* horse; *seta,* tail). The horsetails are widely distributed and vary from a few inches to over 3 feet in height, depending upon the species (Fig. 15-3). Photosynthesis occurs primarily in the stems. The vascular system consists of vascular bundles, each composed of phloem and xylem tissues. The rhizome has distinct nodes and internodes.

The spores, produced in tetrads, are all alike. Each spores has four elaters, or delicate bands with spoon-shaped tips. Moisture causes the elaters to uncoil. In so doing they assist in the dispersal of the spores. A spore germinates into a green, flat, ribbonlike gametophyte that consists of a thick mass of tissue from which grows thin, irregular, vertical lobes that may be branched. Rhizoids attach the gametophyte to the substratum. Near the bases of the vertical lobes are the archegonia, although only their necks may be visible. The antheridia are also present on the gametophyte and produce coiled, multiflagellated sperms. The egg is fertilized by a sperm to form a zygote from which the young embryo develops. In contrast to some species of club "mosses," however, the embryo is without a suspensor.

Horsetails have helped to form coal in early geologic times. In certain places they may help to

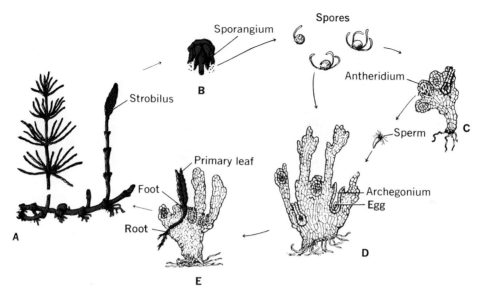

Fig. 15-3. Horsetail or scouring rush, *Equisetum*. **A,** Sterile, vegetative branch (left) and fertile, reproductive branch (right). **B,** One unit of the strobilus is much enlarged to show several sporangia. **C,** Prothallus with antheridia and an archegonium. **D,** Gametophyte with archegonia and antheridia; sperm from plant **C** swims to plant **D. E,** Embryo developing in the gametophyte. The embryo consists of two primary leaves, a stalklike foot, and a primary root. The prothallus eventually disappears and the embryo develops into a mature plant with both kinds of aerial branches, as shown in **A**. Sporophyte (2N) generation is shown in color.

prevent soil erosion. Their growth on the shores of ponds and lakes may alter the shore lines and may assist in filling in the pond completely.

SUBPHYLUM PTEROPSIDA

The subphylum Pteropsida contains most of those organisms one usually thinks of as plants. Their leaves have a more complex system of veins and they have a wider range of sizes. The Pteropsida includes the ferns—considered in this chapter—and the seed plants. There are over 250,000 species in this grouping.

Class Filicinae (ferns)

The sporophyte plants of ferns have a rather well developed vascular system: true leaves (fronds), which are usually large and conspicuous, underground stems called rhizomes (Fig.

15-4), and true roots. There is an alternation of generations between the conspicuous sporophyte and the small gametophyte. Multicellular sporangia are borne in clusters called sori on the lower surface of the leaves or on the margins of the leaves in certain species (Fig. 15-5). In some ferns the sori are covered with a membranous indusium (L. *indusium,* tunic). When sporangia are young, they contain spore mother cells that divide to form spores. In ferns the sporophyte at maturity is independent, free-living, and larger than the gametophyte, which is also independent but small.

The leaves of the true ferns have a characteristic way of unrolling as they grow, giving rise to the popular name of fiddleheads. Usually the leaves are compound and pinnate (L. *pinna,* feather). Fern leaves continue to grow at their tips for a long time, whereas in higher seed plants

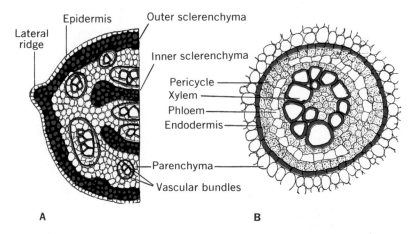

Fig. 15-4. Rhizome (underground stem) of fern. **A,** One half of a rhizome (cross section); **B,** vascular bundle (highly magnified).

Fig. 15-5. Underside of a fern leaf showing many sori. Each sorus contains many sporangia. *(Photograph by F. S. DeMartino, University of Dayton.)*

the growth period of leaves is short. Ferns are world wide in their distribution, but usually grow in moist, shady places. They require sufficient water to permit the transfer of sperm to the egg for fertilization purposes. They do not produce seeds.

A typical sporangium consists of a stalk, a thin-walled capsule, and an annulus (L. *annulus, ring*). The annulus is a band of cells, in which portions of the cell walls are thick and other parts thin. The annulus is affected by moisture, which causes it to bend and spring forward, thus dispersing the spores.

A germinating spore forms a small, thin, green, heart-shaped, haploid gametophyte (also called a prothallus or prothallium), which will eventually develop antheridia and archegonia on its lower surface (Figs. 15-6 and 15-7). The antheridia produce motile, flagellated haploid sperms. The sperms swim through water to fertilize the egg (N) in the archegonium, thus producing a diploid zygote (2N). By mitosis the zygote will develop eventually into a multicellular embryo (2N) from which a sporophyte will develop (2N).

Polypodium

The polypody ferns are common in the woods or clear areas of the temperate regions. The rhizome of *Polypodium* is long, slender, rarely branched, and grows a few inches below the surface of the soil. It grows at its anterior end and may die at the posterior end. The rhizome may even separate into parts, each existing as an individual plant. In structure the rhizome (Fig.

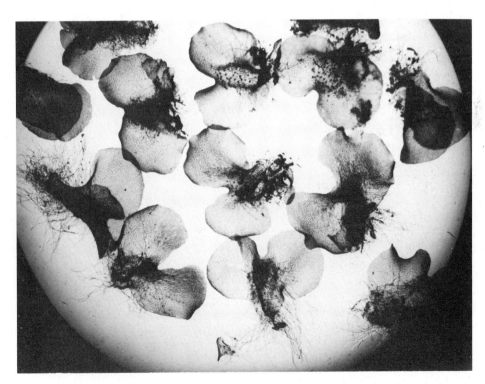

Fig. 15-6. Fern gametophytes. Each one is approximately ¼ inch in diameter. *(Courtesy General Biological Supply House, Inc., Chicago, Illinois.)*

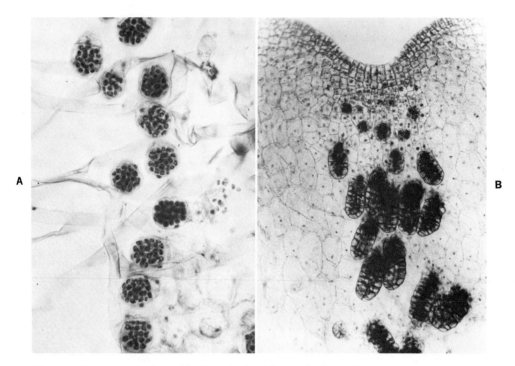

Fig. 15-7. Fern gametophyte detail. **A,** Antheridia. **B,** Archegonia.
(Courtesy General Biological Supply House, Inc., Chicago, Illinois.)

15-5) consists of an outer layer of thick-walled epidermal cells covering a thick layer of rigid mechanical tissue called the outer sclerenchyma, which gives rigidity. Much of the interior part of the stem consists of thin-walled parenchyma cells, which serve for storage. Two well-defined strands of mechanical tissue, called the inner sclerenchyma, are located near the center of the stem, and numerous vascular bundles (phloem and xylem) of various sizes are located in the center of the stem. As the rhizome elongates, it gives rise to many slender adventitious roots a short distance from the cap. Each root has a root cap and small root hairs. Roots that arise from stems or leaves are called adventitious. (L. *adventicius,* foreign). The rhizome resumes its growth each spring and summer.

Each leaf begins its development as a coiled structure attached to the rhizome underground.

Eventually it pushes into the air and uncoils, an action that is characteristic of nearly all ferns. A mature leaf consists of a slender petiole and a blade that is greatly divided. Internally, the structure of the leaf resembles that of higher seed plants. Each leaflet has a lower and an upper epidermis, a palisade layer of elongated cells, spongy tissue in which are numerous spaces, and veins for the conduction of liquids. Stomata are numerous in the lower epidermis.

All leaves of this fern are green sporophylls that bear sporangia (see Fig. 15-6). Within the sporangin are numerous spore mother cells, each of which forms four spores, as in the mosses. When spores are formed, the number of chromosomes is reduced from the diploid (2N) number to the haploid (N) number by reduction division. The cells of the annulus are thickened on all sides except the outer one and are sensitive to changes

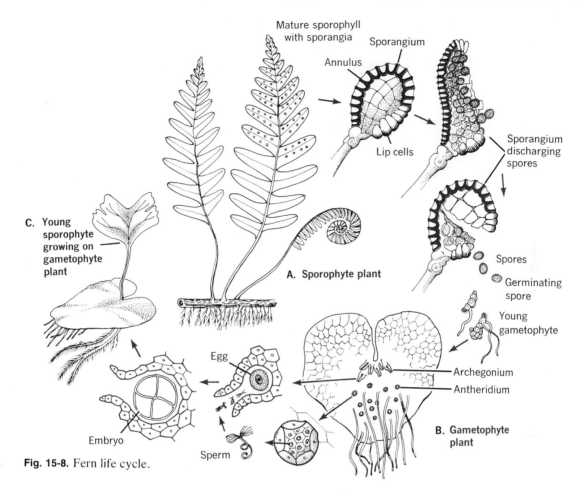

Mature sporophyll
with sporangia

Sporangium

Annulus

Lip cells

Sporangium
discharging
spores

Spores

Germinating
spore

Young
gametophyte

Archegonium

Antheridium

C. Young
sporophyte
growing on
gametophyte
plant

A. Sporophyte plant

Egg

Embryo

Sperm

B. Gametophyte
plant

Fig. 15-8. Fern life cycle.

in moisture. When dry, the annulus straightens, breaks the capsule, snaps forward, and disperses the spores.

The spores are shed in late summer. When a spore germinates, it forms a thin, green elongated plate (prothallus) with rhizoids (see Fig. 15-8). If moisture and light conditions are favorable, the prothallus develops typically into a flat, heart-shaped prothallus with a shallow notch at its anterior end. Prothalli usually are not more than ¼ inch in diameter and only one cell thick. The prothallus is the gametophyte of the fern, just as in the mosses, and produces antheridia and archegonia.

The small, dome-shaped antheridia are usually most numerous on the undersurface of the posterior (older) part of the prothallus where the rhizoids are most abundant. Numerous spiral-shaped, multiflagellated sperms (antherozoids) are produced within the antheridia. Archegonia are usually restricted to the undersurface of the prothallus just behind the apical notch and are constructed much like those of liverworts and mosses. The basal venter is embedded in the surrounding tissues and connects with a short neck to the outside. When mature, a passageway is developed to the egg. Although antheridia and archegonia are formed on the same gametophyte,

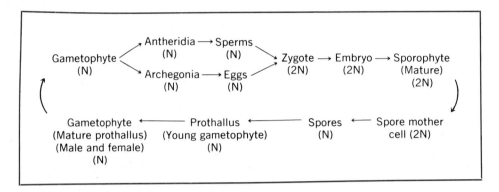

Fig. 15-9. Life cycle of a fern, showing the alternation of generations.

most antheridia mature and discharge their sperms before the archegonia on that plant mature. Thus sperm from one gametophyte usually unite with the egg of another gametophyte.

A sperm unites with an egg to form a zygote, which remains within the venter of the archegonium and develops into an embryo. The embryo becomes four lobed. One lobe develops into a foot that temporarily absorbs foods from the gametophyte; another lobe produces a primary root that grows into the soil; a third lobe produces a primary leaf that develops a simple, green blade; and the fourth lobe develops the stem. At first the young sporophyte is parasitic upon the gametophyte, but later, as secondary leaves and secondary or adventitious roots develop from the stem, the gametophyte dies. Thus the mature sporophyte develops from only one lobe of the embryo, namely that which developed the stem. The structures derived from the other three lobes of the embryo function only in the early stages of young sporophyte development. The life cycle of *Polypodium* fern, like that of other ferns, liverworts, and mosses, includes an alternation of generations between the gametophyte and sporophyte (Fig. 15-9).

Some contributors to the knowledge of
FERNS, CLUB "MOSSES," AND HORSETAILS

Wilhelm Hofmeister (1824-1877)

A German botanist, who described the life cycles of horsetails, certain club "mosses," and many ferns (1851) and made a great contribution in this field. He discovered alternation of sporophytic and gametophytic generations.

Karl E. von Goebel (1855-1932)

A German botanist who divided ferns into two groups (1880) on the basis of the structure and developmental history of the spore cases.

Charles E. Bessey (1845-1915)

An American botanist who first recognized the club "mosses" and horsetails as distinct groups (1907).

Edward C. Jeffrey (1866-1952)

An American botanist, who pointed out that on anatomic grounds the vascular plants could be divided into two fundamental groups. One group, including ferns and modern seed plants, tends to have relatively large leaves with a branching vein system and leaves with a stalk-like petiole and an expanded blade. The second group, including the club "mosses," tends to have relatively small, narrow leaves with one, unbranched midvein, and the leaves are not divided into petiole and blade.

Walter Zimmerman (1892-   )

A German botanist who is the author of the telome theory (Gr. *telos,* end), whereby he attempted to explain the origin of all leaves by modification of branch stems or branch-stem systems. He calls the ultimate branches of a dichotomously branching stem "telomes." Although certain parts of the theory are still controversial, the basic concepts are generally accepted.

Review questions and topics

1 List the characteristics of the phylum Tracheophyta.
2 List the characteristics of the subphyla Lycopsida, Sphenopsida, and Pteropsida.
3 Discuss the economic importance of club "mosses," horsetails, and ferns.
4 Explain how the sporophytes and gametophytes differ in club "mosses," horsetails, and ferns.
5 Diagram a fern life cycle, showing the stages in correct sequence.
6 Explain the significance of reduction division (meiosis).
7 Why are the leaves of ferns and their allies considered to be true leaves?
8 Why are ferns and their allies classed as embryophytes?
9 Contrast the sporophytes and gametophytes of ferns with those of mosses.
10 In what ways are the gametangia (sex organs) of tracheophytes similar to those of the bryophytes but different from those of the thallophytes?
11 Describe the structure and functions of stomata.
12 Describe the type of gametophyte formed from a germinating spore in club "mosses," horsetails, and ferns.
13 Describe how the multicellular sporangia are borne in club "mosses," horsetails, and ferns.
14 Why is water necessary for fertilization in ferns?
15 Describe the structure and function of annulus, suspensor, and foot.
16 Compare the alternation of generations in ferns with similar stages in the horsetails and club "mosses." Be specific in your comparisons.
17 Contrast the terms prothallus and protonema, with examples of each.

Selected references

Andrews, H. N.: Evolutionary trends in early vascular plants, Cold Spring Harbor Symposium, Quant. Biol. **24:**217-234, 1960.

Andrews, H. N.: Studies in paleobotany, New York, 1961, John Wiley & Sons, Inc.

Delevoryas, T.: Morphology and evolution of fossil plants, New York, 1962, Holt, Rinehart & Winston, Inc.

Foster, A. S., and Gifford, E. M.: Comparative morphology of vascular plants, San Francisco, 1959, W. H. Freeman and Co., Publishers.

Sporne, K. R.: The morphology of the pteridophytes, London, 1962, Hutchinson & Co. (Publishers), Ltd.

White, R. A.: Tracheary elements of the ferns, Part I, Amer. J. Bot. **50:**447-454, 1963.

16 Phylum Tracheophyta—gymnosperms

Order Coniferales (conifers)
   Pine trees
Order Cycadales (cycads)
   Zamia

The gymnosperms are placed in the subphylum Pteropsida and the class Gymnospermae (Gr. *gymnos,* exposed; *sperma,* seed). Seeds are formed on the exposed surface of modified leaves called megasporophylls and are not protected by an ovary wall, as in flowering plants. Gymnosperms are usually rather large and woody evergreen perennial plants, although some types are short and shrubby. They possess true leaves, stems, and roots. The leaves in certain cone-bearing evergreens are needlelike or scalelike (Fig. 16-1). Cone-bearing evergreens (conifers) and their relatives are included in the class Gymnospermae.

The sporophyte generation is large, complex, and independent, while the gametophyte generation is microscopic and parasitic upon the sporophyte. Two kinds of cones are usually present, in the conifers. The male cones, composed of microsporophylls, and the female cones, composed of megasporophylls, are present on the same plant or on different plants, depending upon the species (Fig. 16-2). Immature, undeveloped seeds called ovules are borne exposed on the megasporophylls. Two kinds of spores are formed, namely microspores, which develop into microgametophytes, and megaspores, which develop into megagametophytes. The two kinds of spores may be the same size, or the microspores may be larger than the megaspores.

Pollination is aided by wind dispersal; the pollen grains (mircrospores) land in the small opening or microphyle of the ovule. The pollen grains are drawn into the ovule by the drying and shrinking of a sticky material secreted by the female cone. A pollen tube is developed through which the sperms travel to the egg.

In gymnosperms one sperm is involved in fertilizing the egg. After fertilization the egg develops into the embryo, the rest of the female megagametophyte tissues form the endosperm (stored food), and the integument eventually forms the seed coat of the mature seed.

In pine trees the lapse of time between pollination and subsequent fertilization (actual union of sex gametes) is a distinct feature. For example, if pollination occurs in June, fertilization does not ordinarily occur until July of the next year. This time lapse varies with the species and locality, but usually a year elapses between pollination and fertilization. After fertilization the seed develops quickly, reaching maturity by the end of the year in which fertilization occurs.

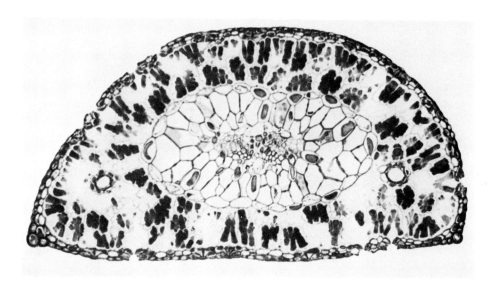

Fig. 16-1. Pine needle. Note the outer epidermis with stomata and air spaces beneath in the mesophyll tissue.
(Courtesy Carolina Biological Supply Company, Burlington, North Carolina.)

Fig. 16-2. Cluster of cones on the stem of pitch pine *(Pinus rigida).*
(Courtesy Jane Evelyn Richardson.)

Gymnosperms are considered to be higher plants than club "mosses," horsetails, and ferns because of (1) the transfer of the small microgametophyte directly through a pollen tube to the area near the megagametophyte; (2) the retention of the megaspore within the megasporangium where it germinates to produce the megagametophyte, or embryo sac in which the archegonia are formed; (3) the enclosure of the megasporangium and embryo sac by a covering integument; (4) the development of the young sporophyte upon and at the expense of the mother sporophyte; and (5) the release of dormant seeds that have a reserve supply of food and are protected by a seed coat.

The gymnosperms are of great economic importance. The cone-bearing trees produce much valuable timber, such as yellow pines, redwoods, pitch pines, firs, cedars, hemlocks, and white pines. Billions of board feet of lumber are used in the United States annually. Pine lumber is durable because of its composition and is easily worked and quite resistant to the attacks of insects, probably because of its resin content (Fig.

Fig. 16-3. Pine tree *(Pinus)*. **A,** One half of a needlelike leaf; **B,** stem (both somewhat diagrammatic cross sections). Note that there are two layers of xylem, one for each year of stem growth, and that the outer cells of each xylem layer represent fallwood and the inner layers represent springwood. Spring and fall growths of each xylem constitute an annual ring.

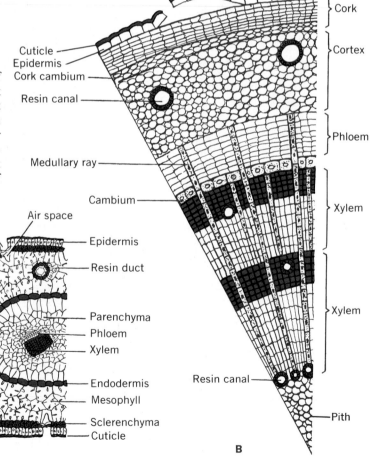

Fig. 16-4. General Sherman Tree, a giant sequoia, in winter. It is more than 270 feet high and is 26 feet in diameter. Its size can be compared with nearby trees, which are about 35 feet high. Its age is estimated to be over 2,500 years. It is located in Sequoia National Park, California.
(Courtesy National Park Service, U. S. Department of the Interior.)

16-3). Conifers are used extensively in the manufacture of wood pulp and paper. Red cedars are used in making pencils, cigar boxes, chests, trunks, and similar articles. Conifers yield large quantities of resins, oils, and amber products used in arts, industries, and medicine, such as spruce gum (for chewing gum), oil of juniper, and oil of savin. The seeds of certain pines may be used as food, for example, the edible or nut pine of western United States and the sugar pine of California. The barks of conifers, such as hemlock and spruce, furnish important materials for use in tanning animal skins for leather. Millions of youngsters and adults enjoy the decorated conifers at Christmas. The wood of certain conifers is particularly resonant, so that is is used in making certain musical instruments. The remains of conifers are often found as fossils and as fossil-resin amber in which other fossils have been embedded.

ORDER CONIFERALES (CONIFERS)

The conifers are placed in the order Coniferales (L. *conus,* cone; *ferre,* to bear) because most of them bear cones composed of sporangium-bearing leaves. There are over 500 species of conifers, including the various species of pine, spruce, fir, juniper, cedar, hemlock, larch, yew, cypress, and redwood. Some of the oldest and largest plants are conifers, for example, the giant sequoia trees of California (Fig. 16-4) may be more than 30 feet in diameter, 300 feet tall, and 4,000 years old.

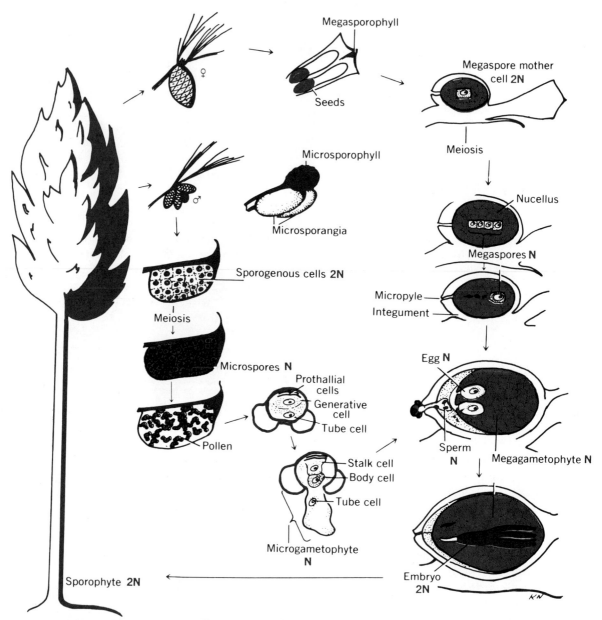

Fig. 16-5. Pine tree *(Pinus)* life cycle, showing female (♀) and male (♂) cones and detailed structures of each. In pollen development sporogenous (spore-forming) cells are 2N and after meiosis the microspores are N. Winged ''pollen grains'' have a tube cell and a generative cell. The latter forms a stalk cell and a body cell; the body cell forms two sperms (N), which fertilize eggs (N). In ovule development the megaspore mother cell (2N) produces megaspores (N). The megagametophyte contains an embryo with several cotyledons; the embryo develops into a mature sporophyte tree.

Pine trees

Pine trees *(Pinus)* have large, branched stems, and needlelike leaves borne in clusters. The number of leaves per cluster (two to five) and the length of the leaves vary with the species.

In the pines the male and female cones are borne on the same tree (Fig. 16-5). In other conifers the male and female cones are borne on separate plants. The staminate (male) cones are smaller than the female ones and are borne in a group. Each male cone is composed of micro-sporophylls that are spirally arranged and at-tached to a central axis. Each microsporophyll bears two microsporangia on the undersurface. Numerous microspore mother cells, each of which produces four microspores (pollen grains), are produced in the microsporangia. Each pollen grain develops into a microgametophyte by pro-ducing two prothallial cells and an antheridial cell. The latter divides to form a generative cell and a tube cell. During this process a pair of "wings" forms on the four-celled pollen grain to assist in its dissemination by the wind—in some cases for hundreds of miles.

The ovulate (female) cones are larger than the male ones and are usually borne singly. Each female cone is composed of scalelike mega-sporophylls attached to a central axis. Each megasporophyll bears two ovules on its upper surface. An ovule consists of an external, pro-tective integument that has a micropyle (small opening) for the entrance of a pollen grain and a central megasporangium. When young, the megasporangium contains one megaspore mother cell that produces four megaspores, three of which abort. The one remaining megaspore de-velops into the female megagametophyte, which produces two or three archegonia. Each arche-gonium contains an egg.

During pollination the wind carries the pollen grains to the female cones. The pollen enters through the micropyle and is pulled to the mega-sporangium by contraction of a sticky liquid. The pollen grains form pollen tubes through the mega-sporangium toward the archegonia. The genera-tive cell and the tube cell pass through the pollen tube, and the generative cell produces two nuclei that function as nonmotile sperms.

About 1 year after pollination the fertilization process occurs. It consists of the fusion of one sperm with the egg within the archegonium. The resulting zygote eventually produces an embryo and suspensor cells. The suspensor cells force the embryo into contact with the food endosperm (transformed female megagametophyte). Later, the embryo develops an epicotyl (ep i -kot' il) (Gr. *epi,* upon; *kotyle,* seed leaf) and a hypo--cotyl (hi po -kot' il) (Gr. *hypo,* under), which bear a number of primary, embryonic seed leaves known as cotyledons. The embryo is surrounded by the endosperm (food), which is covered by the seed coat (hardened integument). The seed thus formed was originally the ovule and contains a wing for wind dispersal. When a seed germinates, the embryo produces a young seedling that even-tually develops into a pine tree (sporophyte generation). Since gymnosperms produce two different kinds of spores (microspores and mega-spores), they are heterosporous.

Fig. 16-6. Cycad *Cycas* with a crown of young leaves unfolding.
(From Weatherwax, P.: Plant biology, ed. 3, Phila-delphia, 1959, W. B. Saunders Company.)

ORDER CYCADALES (CYCADS)

The cycads ("sago palms") (Fig. 16-6) belong to the class Gymnospermae and the order Cycadales (sik a da' lez). They are palmlike trees or low shrubs with unbranched stems, terminating in a tuft of thick, fernlike leaves that are often spiny edged. The trunk of *Cycas* contains starch, hence the common name of "sago palm" (Malay *sagu*, starch of sago palm), although the true sago palm is native to the East Indies.

Zamia

In *Zamia* (L. *zamia*, fir cone), which grows in Florida, the short, tuberous stem is not more than 4 feet tall and bears a crown of leathery, pinnate leaves. Sometimes much of the stem is underground. In general the cycads are inhabitants of the tropics or semitropics.

Zamia is diecious, since male and female cones are borne on separate plants. The carpellate cones (female) are composed of shield-shaped megasprophylls, each of which bears two ovules, with a micropyle in the enclosing integument. In the center of the ovule is the megasporangium. This contains one megaspore mother cell that produces four megaspores, three of which disintegrate. The nucleus of the remaining megaspore divides to form two nuclei, and further division results in numerous nuclei. Walls separate the nuclei, so this multicellular tissue becomes the female megagemetophyte, which produces two to six archegonia. Each archegonium consists of one large egg and a neck.

The staminate cones (male) are smaller than the female ones and consist of numerous microsporophylls, each of which bears thirty to forty microsporangia on the lower surface. Each microsporangium contains many microspore mother cells, which produce many microspores. These microspores are carried by the wind to the female cone where they pass through the micropyle of the ovule to the megasporangium. About 6 months after pollination a motile sperm (antherozoid) fertilizes an egg in the archegonium, forming a zygote that produces a multicellular embryo, two cotyledons (embryonic leaves), and a coiled suspensor to push the embryo in contact with the endosperm (stored food) of the megagametophyte.

Some contributors to the knowledge of
GYMNOSPERMOUS PLANTS

Robert Brown (1773-1858)

A Scottish botanist who first observed (1827) that pine trees and their kin have ovules (undeveloped seeds) and mature seeds that are "naked" (not protected by an ovary). He classified gymnosperms and separated them from true, flower-bearing angiosperms (1825).

Charles J. Chamberlain (1863-1943)

An American botanist who was an outstanding student of cycads, those gymnosperms that are somewhat fernlike with large, pinnately compound leaves and seeds that are usually aggregated into a simple, terminal cone.

Charles E. Bessey (1845-1915)

An American botanist who first recognized cone-bearing plants as a distinct group (1907) and treated the cycads as a separate group. His ideas have been accepted by many in succeeding years.

Rudolph Florin (1894-   )

A Swedish botanist who is an outstanding authority on cone-bearing plants, particularly fossil forms.

Review questions and topics

1 List the characteristics of the class Gymnospermae.
2 In what ways are gymnosperms considered to be higher plants than ferns, horsetails, and club "mosses"?
3 Discuss the differences and similarities between conifers and cycads.
4 Discuss the economic importance of gymnosperms.
5 Make a detailed diagram of the life cycle of the pine tree, including the various stages in correct sequence. Label each stage correctly.
6 Explain the phenomenon of alternation of generations in gymnosperms.
7 What is the sporophyte and what is the gametophyte in the pine? How do they differ concerning size and independence?
8 Explain what is meant by the term "evergreen."
9 Contrast pollination and fertilization, giving examples of each.
10 Explain the significance of heterospores.
11 Describe the structure and function of the micropyle.
12 What is the significance of the endosperm?
13 Discuss the structure and function of the pollen tube.
14 Contrast ovules and true seeds, giving an example of each.
15 What is the significance of the motile sperms of cycads and the nonmotile sperms of pine trees?

Selected references

Arnold, C. A.: An introduction to paleobotany, New York, 1947, McGraw-Hill Book Company.
Florin, R.: The systematics of the gymnosperms. In Kessel, E. L., editor: A century of progress in the natural sciences, San Francisco, 1955, California Academy of Sciences.
Sporne, K. R.: The morphology of gymnosperms; the structure and evolution of primitive seed plants, London, 1965, Hutchinson & Co. (Publishers) Ltd.
Zobel, Bruce J.: The genetic improvement of southern pines, Sci. Amer. **225**(5):94-104, 1971.

17 Phylum Tracheophyta—angiosperms

Flowering plants are placed in the subphylum Pteropsida and the class Angiospermae (Gr. *angio-,* enclosed; *sperma,* seed). The seeds are enclosed within an ovary that has probably been evolved by a folding together of the edges of the megasporophylls. At maturity the ovary constitutes the fruit.

Angiosperms constitute the dominant, the most economically important, and the largest class in the plant kingdom, comprising about 250,000 species. These species are widely distributed and are primarily terrestrial, although a few, such as water lilies, are aquatic.

Angiosperms possess true leaves, stems, and roots. Each of these structures contains conducting tissues composed of phloem and xylem. A unique characteristic of the angiosperms is the presence of flowers. The flowering plant is the conspicuous, independent, adult sporophyte part of the alternation of generations. The gametophyte is very small, without chlorophyll, and dependent upon the sporophyte. The sporophyte produces two kinds of spores, although, as in gymnosperms, the male microspores may be larger than the female megaspores. The cells of the sporophyte contain the diploid (2N) number of chromosomes, whereas the cells of the gametophyte contain the haploid (N) number.

In the evolution of plants from the lowest types to the angiosperms, there has been an increase in the size and independence of the sporophyte. Also, there has been a reduction in the size and independence of the gametophyte.

Angiospermae are divided into the subclasses Dicotyledoneae (Gr. *di-,* two; *kotyle,* seed leaf) and Monocotyledoneae (Gr. *mono-,* one), which may be differentiated as shown in Table 17-1.

ROOTS

Roots are an adaptation to life on land and are typically cylindric in form. They arise embryologically from the hypocotyl region of the embryo of the seed. The radicle (L. *radix,* root), or basal

Fig. 17-1. Close-up of a flowering plant.
(Photograph by F. S. DeMartino, University of Dayton.)

Table 17-1 Distinguishing characteristics of dicotyledonous and monocotyledonous plants

Dicotyledonous plants	Monocotyledonous plants
Two embryonic seed leaves (cotyledons) supply embryo with food during early development	One embryonic seed leaf (cotyledon) supplies embryo with food during early development
Flower parts usually in four's or five's or multiples of these	Flower parts typically in three's or multiples of three
Leaves net veined	Leaves parallel veined, usually long and narrow
Some have woody stems, others have soft herbaceous stems	Most have herbaceous stems (few exceptions)
Vascular bundles of stems usually arranged in circular cylinder	Vascular bundles scattered throughout stem in no definite pattern
Cambium (meristematic tissue) present between phloem and xylem of vascular bundle	Usually no cambium between phloem and xylem of vascular bundle
Seem more ancient because of resemblance, particularly in vegetative structure, to primitive gymnospermous condition	Seem more advanced than dicotyledonous plants
Examples: beans, sunflowers, roses, violets, clovers, snapdragons, buttercups, potatoes, elms, oaks, apples, maple, hickories, poplars, and lilacs	Examples: corn, wheat, bluegrass, lilies, irises, daffodils, and cattails

part of the hypocotyl, emerges from the germinating seed to form the first or primary root of the new plant. The primary root may form many branches called secondary roots. In general the primary root grows downward, whereas the secondary roots grow somewhat horizontally.

The root systems of plants are commonly called (1) taproot systems, in which the single, main primary root is distinctly larger than the secondary roots, for example, radishes, carrots, beets, and dandelions, or (2) diffuse root systems, comprised of numerous slender roots, many of which are of equal size, and usually without a main root that is larger than the others, for example, grasses, corn, and potatoes. Sometimes the larger fibrous roots become enlarged with stored foods, as in sweet potatoes and dahlias.

The principal functions of roots include (1) anchorage of the plant, thus binding the soil particles to reduce soil erosion by water or wind, (2) absorption of water and dissolved minerals and the conduction of substances through the roots, and (3) storage of foods, as in radishes, carrots, and turnips.

Some roots perform specialized functions, such as the following: (1) Adventitious prop roots arise from the aerial stem and grow downward to the soil, thus giving added support, as in corn plants. (2) Adventitious climbing roots (ivies and certain vines) anchor plants to walls and trees. (3) Specialized suckers called haustoria are formed by certain parasitic seed plants (mistletoe and dodder) for absorbing food from the host plant. (4) Spongy, water-absorbing aerial roots of epiphytic plants, such as epiphytic orchids, take water from the atmosphere and rain, anchor the plants to objects (poles, trees, or wires), and absorb materials from the debris that collects about the roots. (5) Root outgrowths called "knees" or pneumatophores are produced by swamp plants such as bald cypress trees and grow into the air above the water or swamp soil. (6) Reproductive functions are performed by roots of certain plants and by "suckers" that de-

velop into new plants. The roots of potatoes and dahlias are used for propagation purposes. (7) Spices and other aromatic substances are formed by roots of plants, such as horse-radish, sassafras, and sarsaparilla.

STEMS

Stems may be herbaceous or woody. Herbaceous stems are soft and usually small in diameter. They are usually green, have few hard tissues, and are chiefly annuals. Woody stems are tough, usually larger in diameter than herbaceous ones, are not green, have well-developed fibers and other strengthening cells, and are chiefly perennials. The stems of all gymnosperms are woody, as are the stems of many angiosperms. Some examples of dicotyledons with woody stems are oaks, elms, maples, hickories, poplars, apples, and lilacs. Woody stems in monocotyledons are uncommon, but are present in palms and lilies. The hardened stems of bamboo, although used as wood, are fibrous in nature and not true wood (xylem).

The principal functions of stems are (1) the production and support of leaves, flowers, or cones, (2) the conduction of substances, and (3) the storage of foods. Substances are conducted upward in stems by the xylem, downward by the phloem, and horizontally by the wood rays, which are found in woody stems. Any living cell may store foods, but the pith, the cortex, and the wood and phloem parenchyma, particularly in the rays, are the primary storage cells. Starch is the most commonly stored food, but fats and proteins may also be stored.

In addition to the general functions described above, certain stems are of special importance. For example, in dicotyledonous trees and shrubs the great part of the stem consists of wood tissue. Through the activity of the cambium a new concentric layer of woods cells is added annually to the outside of the woody cylinder, thus forming annual rings. Spring growth produces large, thin-walled cells, and summer growth produces smaller, more thickly walled cells. The dead annual rings in the centers of stems constitute the heart wood, whereas the functional, living part of the wood just under the bark is called the sapwood.

Certain parenchyma cells arranged in horizontal, ribbonlike bands among the woody cells constitute the wood rays that emanate from the central pith toward the periphery of the stem like the spokes of a wheel. These rays transfer materials horizontally within the stem and also store foods. (The stems of many conifers have specialized resin canals that secrete "pitch," an economically valuable product.) Tannin, various oils, and certain mucilaginous substances are secreted by special cells in the stems of numerous plants. The milky liquid, latex, is produced by various plants and is the source of natural rubber and other products.

Certain underground stems have lost their foliage-bearing abilities and have become shortened and thickened, constituting a tuber, as in the potato plant and Jerusalem artichoke. Such

tuberous underground stems produce buds, a characteristic of stems, and store large quantities of food.

One of the important foods derived from typical plant stems is extracted from the sap of sugarcane and crystallized into sugar. Other sources of sugar are the stems of sorghums and the sap of maple trees. Other modified stems that serve as foods include the bulbs of onions and the swollen stems of kohlrabi. Among chemicals and commerical products derived from wood are dyes, tannin, tars, oils, wood alcohol, cellophane, and rayon. The liquid cellulose called "viscose" is forced through minute pores to produce tiny streams that harden into fibers that may be spun into threads. If forced through a narrow slit, transparent films of cellophane are formed.

Among the commerical fibers produced by plant stems are those of flax plants, hemp plants, jute, and ramie. The bark of the Spanish oak tree produces a thick layer of cork tissue that can be harvested as commercial cork every few years. The barks of oak and hemlock yield tannin which

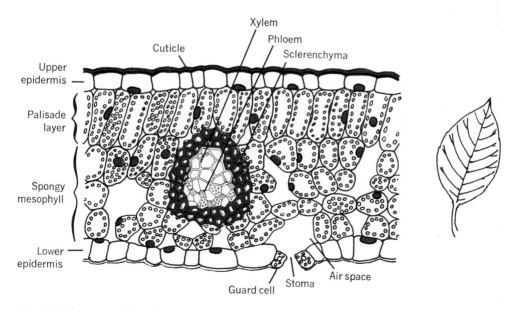

Fig. 17-2. Section of a leaf.

is used in tanning hides. The bark of the cinchona of South America and the East Indies yields quinine, an important medicine. Among the spices, cinnamon comes from the bark of the cinnamon tree of southeastern Asia.

LEAVES

Leaves arise from the terminal growing point of the stem. Unlike stems, leaves have no organized growth points and thus stop growing when mature. The blade of a leaf may be attached directly to the stem, but, most frequently it is attached by a stalklike petiole, which serves to transport materials and to expose the blade to light. Leaves may vary greatly in size, shape, texture, margin, and type of venation. They may vary from small scales (arborvitae) to needlelike leaves (pines), or to broad leaves (many deciduous trees). Leaves may vary from long and narrow (grasses) to circular forms (nasturtium).

A thin cross-section of a leaf blade (Fig. 17-2) shows (1) an external, protective cuticle, (2) an epidermis, whose outer cell walls contain a waxy cutin, (3) a palisade layer, which is composed of column-shaped cells closely arranged and containing many chloroplasts for photosynthesis, (4) a spongy mesophyll layer (Gr. *mesos,* middle; *phyllon,* leaf) beneath the palisade layer, which is composed of irregularly shaped, loosely packed storage cells (chloroplasts are also present in the spongy tissue cells), (5) veins, which are vascular bundles composed of xylem and phloem, as in the petiole and stem, and (6) a lower epidermis, which is similar to the upper epidermis except that it contains stomata for the exchange of gases. Each stoma is bordered by two guard cells that contain chloroplasts and also regulate the size of the stoma. The stomata open into intercellular spaces.

FLOWERS AND REPRODUCTION

Flowers develop from buds. Some buds produce only flowers (elm, poplar, and morning glory), whereas others produce both flowers and leaves (apple and buckeye). There are many variations in flowers in the different species of angiosperms. The number of flower parts, the sizes and colors of the petals, the various degrees of fusion of parts, and the relative position and arrangement of the various floral organs are among the chief variations.

Flower colors usually result from the presence of pigments called anthocyanins or carotenoids. The anthocyanins are blue, purple, or reddish, and the carotenoids are yellow, orange, or sometimes reddish pigments. Plant pigments are considered in more detail in Chapter 18.

Pollination of flowers may be accomplished by wind, insects, or birds. Many flowers that are pollinated by insects are brightly colored or scented or both. Their pollen is heavy, sticky, and not readily carried by the wind. Such flowers are often provided with nectaries (Gr. *nektar,* sweet) that secrete a fluid that has a rather high sugar content. Nectaries vary greatly in form and location, depending upon the species. Insect pollination is accomplished chiefly by bees, wasps, butterflies, and moths, although other insects may frequent flowers to gather pollen and nectar. Details of pollination, fertilization, and seed formation have been covered in Chapter 6.

The sense of smell of honeybees is sufficiently sensitive to enable them to detect flower odors (Fig. 17-3). Honeybees cannot distinguish red from dark gray or black. They can distinguish yellow, blue-green, and blue but not violet or purple. Butterflies frequently pollinate red flowers. Many types of cultivated plants depend upon insect pollination.

FRUITS AND SEEDS

The commonly accepted idea of what constitutes a fruit may differ from the botanical meaning of the term. To the average person a fruit probably means an apple, peach, or pear, or something similar at the food market. After fertilization occurs a ripened ovary of a flower, its seed, and sometimes the adjacent parts that may be associated with it develop into a fruit. Sometimes a group of ovaries may develop into a multiple or aggregate fruit.

After fertilization an ovary is composed of a thickened ovary wall called the pericarp (per′ i karp) (Gr. *peri*, around; *karpos*, fruit) and one or more ovules, which develop into mature seeds. The pericarp is derived from one or more carpels, which are modified sporangia, called megasporophylls. Fruits are important to plants for seed dissemination and they also serve as sources of food for animals and man. A fruit always develops from a flower and is always composed of a least one ripened ovary that may be fused with other floral parts or structures associated with the flower. Any edible part of a plant that does not conform to this definition of a fruit is classed as a vegetable.

Successful reproduction requires the formation of seeds and their dispersal. Seeds and fruits may be dispersed by wind (winglike structures of maples and ashes). In fleshy fruits, such as cherries, grapes, and tomatoes, the soft pulp is often colored and tasty, so that animals may disperse the seeds by eating them. In some instances spines and hooks (as in the burdock) permit dispersal by attachment to animals.

As an ovary matures or ripens into a fruit, its wall, the pericarp, may thicken and be differentiated into layers of tissues; (1) the outer exocarp, consisting of one or several layers of epidermal cells, (2) the middle mesocarp, consisting of one layer of cells, or a large mass of tissues, several inches thick, and (3) the inner endocarp, which varies greatly in thickness, structure, and texture in the fruits of different species.

Fruits may be classified on the basis of their structure and the number of ovaries from which they are derived. In some species flower parts

Fig. 17-3. Honeybees are important in plant pollination. *(Photograph by F. S. DeMartino, University of Dayton.)*

other than the ovaries may adhere to or enclose the mature ovaries to form so-called accessory fruits, which are considered later in this chapter. Fruits may be simple, aggregate, or multiple (compound), as described in the following.

Simple fruits

Simple fruits are derived from one ripened ovary (pistil) plus, in some species, such adherent parts as sepals and stamens. In some species the ovary is composed of more than one carpel (a member of a compound postil). The fruits of most angiosperms are simple fruits.

Dry fruits

When mature, the ovary wall of these fruits (pericarp) is dry and may become paperlike, leathery, or woody. The wall does not become fleshy.

DEHISCENT FRUITS

At maturity dehiscent fruits dehisce (open) along definite seams or at definite points and may contain several to many seeds.

Legume. Legumes (L. *legare,* to gather) develop from a single carpel and usually dehisce along two sides. The so-called pod is a single carpel and contains the seeds. However, in some legumes (peanut) the fruit does not open at maturity. Examples are string bean, lima bean, pea, locust tree, and peanut.

Follicle. Follicle (L. *folliculus,* small sac) dry fruits consist of a single carpel and dehisce along one seam only. Examples are mildweed, larkspur, peony, spiraea, and columbine.

Capsule. Capsule (L. *capsula,* little box) dry fruits develop from a compund pistil (two or more fused carpels). The capsule may dehisce in various ways. In the iris it may split along the back of the individual carpels. In the poppy it may open by forming a circular row of pores near the top of the fruit, seeds coming out through the pores. In the horse chestnut the capsule splits into three valves, thus releasing the seed. In the

Brazil "nut" the capsule is a thick shell, often 6 inches in diameter, and may contain 20 seeds, each with a hard seed coat. A lidlike cover at the tip of the fruit opens at maturity but does not permit the exit of seeds. Upon germination the seedlings grow out through the capsule openings. Examples are iris, lily, tulip, poppy, pansy, azalea, horse chestnut, snapdragon, violet, plantain, jimsonweed, Brazil "nut," and cotton.

INDEHISCENT FRUITS

Indehiscent fruits do not split at definite seams or points and usually contain one or two seeds.

Achene. Achenes are (Gr. *a-,* not; *chainein,* to open) rather small dry fruits that contain a single seed that nearly fills the cavity of the fruit. The seed coat does not adhere to the pericarp, except at the point of attachment of the seed. Examples are sunflower, buttercup, buckwheat, and dandelion.

Grain or caryopsis. Grains or caryopsis (Gr. *karyon,* nut; *opsis,* appearance) are one-seeded fruits and are similar to achenes, but the thin seed coat is fused to the inner surface of the pericarp and is thus not easily removed. These fruits serve as cereals. Examples are corn, wheat, oats, barley, rice, and most grasses.

Samara (key or wing fruit). Samara (L. *samara,* seed of elm) are dry, indehiscent winged fruits and are usually single seeded, the wings being formed from the pericarp for the purpose of wind dispersal. Examples: elms, ashes, basswood, and maples.

True nut. These one-seed fruits have a pericarp that is rather thick, hard, stony, or woody when mature. The edible inner part is the seed. Examples; filbert (hazelnut), beechnut, acorn (of oaks), chestnut, and hickory nut.

Fleshy fruits

When mature, the pericarp, or some part of it, is filled with sap and becomes fleshy, the pericarp has three parts: (1) an outer exocarp, which is usually thin, (2) a middle mesocarp, which is a fleshy pulp, and (3) an inner stony or fleshy endocarp. Fleshy fruits may be classified as true berry, false berry, drupe, and pome.

TRUE BERRY

The botanist's definition of a berry is somewhat different from that of a layman. In these fruits the entire pericarp ripens, becoming fleshy and often edible and juicy. In tomatoes the ovary becomes large and fleshy and bears many seeds. Citrus fruits are modified berries, although the peel, or rind (pericarp), is not usually edible. Sections of young citrus fruits develop multicellular out-growths from the surface of the carpel walls, and these outgrowths become large, juicy, and edible. Each wedgelike segment of oranges and grapefruits represents a carpel filled with the multicellular outgrowths. In citrus fruits the fleshy part with its seeds is endocarp, whereas the peel is exocarp and mesocarp. In grapes the thin skin is exocarp, and the fleshy material with its seeds is mesocarp and endocarp. Examples are tomato, grape, date, avocado, red pepper, citrus fruits (orange, lemon, and grapefruit), blueberry, and gooseberry.

FALSE BERRY

Some fruits may include not only a matured ovary but also other flower parts or closely related structures. When such structures constitute an important part of the fruit, it is said to be an accessory fruit. Most accessory fruits are simple fruits derived from both the pericarp and the floral tube (composed of basal parts of sepals, petals, and stamens). This tube is fused with the pericarp and becomes fleshy when ripe. Since the entire fruit is fleshy when ripe, as in true berries, such accessory fruits may be termed "false" berries. In contrast to true berries "false" berries have remnants of the flower persisting at the part of the fruit that is opposite the stem. The apical end of the cranberry has remnants of sepals, whereas the tip of the banana fruit has a large scar where flower parts have

fallen away. When the berry has a hard rind (watermelon, squash and cucumber), it is called a pepo. Examples are cucumber, pumpkin, watermelon, gourd, cantaloupe, cranberry, and banana.

DRUPE OR "STONE FRUIT"

In drupes (Gr. *dryppa,* berry) the pericarp is divided into an outer, thin skin, the exocarp; a fleshy, thick mesocarp, and a stony (or woody) endocarp. The endocarp ("pit" or "stone") encloses one seed (rarely two to three).

In olives and apricots the so-called pit is endocarp with the seed inside; the edible, fleshy part is mesocarp; and the skin is exocarp. In almonds the fleshy exocarp and mesocarp are leathery, inedible, and removed when the fruit is harvested. The stony shell (endocarp) contains the edible almond seed. The shell of the coconut is composed of fibrous exocarp and mesocarp, so that the coconut is often called a dry, or fibrous, drupe. The stiff fibers that compose the husk are used commercially in making coco mats, brushes, and cordage. The hard, woody endocarp that we purchase has germination pores and encloses a large seed. Examples are olive, apricot, coconut, peach, plum, almond, cherry, and walnut.

POME

Pomes (L. *pomum,* apple) are fleshy, indehiscent accessory fruits; the pericarp has parts that are somewhat like drupes. In the pome the true fruit (ripened ovary) constitutes the central, inedible "core." Thus in apples and pears the edible parts are not the matured ovary but the stem and floral tube tissues in which the true fruits (ovaries) are embedded. The outer fleshy part is exocarp and mesocarp, and the inner endocarp is leathery (papery). In apples and pears the inner leathery endocarp contains seeds. A group of vascular bundles, sometimes called the "core line," marks the outer boundary of the exocarp. Between the inner leathery endocarp and the exocarp is the mesocarp. The part of the apple extending from the "core line" to the outside of the fruit is derived from the floral tube. Apples and pears are accessory fruits because the remains of sepals and stamens are present at the tip of the mature fruit. Examples are apples, pears, quinces, and hawthornes (haws).

Aggregate fruits

An aggregate (L. *ad,* to; *gregare,* to collect together) fruit develops from a cluster of several to many ripened, simple ovaries (pistils) produced by one flower and borne on the same basal receptacle (axis of the flower stalk bearing floral organs).

In raspberries the simple fruits develop into tiny drupelets (drupes) that adhere to one another but separate as a unit from the dome-shaped receptacle. Blackberries are like raspberries, but the elongated receptacle is also fleshy and forms part of the fruit, thus making it an aggregate-accessory fruit. In strawberries the numerous simple ovaries develop into tiny achenes that are visible on the exterior of the fruit and are commonly called seeds. Each ovary contains one ovule that develops into a seed within the achene. The edible, accessary part of the fruit is the enlarged, fleshy receptacle (modified stem), thus making the strawberry an aggregate-accessory fruit. Examples are raspberry, boysenberry, blackberry, and strawberry.

Multiple (compound) fruits

Multiple fruits are formed from a cluster of several to many ripened ovaries produced by several flowers associated closely together on the same inflorescence, rather than from a single flower. The ovary of each flower forms a fruit, and at maturity all these fruits remain in a mass. As in aggregate fruits, the small fruitlets of a compound fruit may be berries, drupes, or nutlets.

In the pineapple from 100 to 200 sessile flowers are attached to the elongated, fleshy axis of the inflorescence, which is leafy at the top. The flowers are fused with each other, and together with their fleshy bracts (modified leaf asociated with a flower), they ripen at the same time. Each of the units visible on the surface of the mature fruit represents a flower together with its bract, thus making it a multiple-accessory fruit. The edible part is mostly the thickened, pulpy central stem in which the individual fruits are embedded. The edible part of a fig is the mature axis of the inflorescence, which is hollow and encloses many tiny achenes, each fromed from a single flower. Examples: pineapple, Osage orange ("hedge apple"), mulberry, and fig.

DICOTYLEDONOUS PLANTS

Legume family

The podlike fruits of these plants are called legumes. The family includes food plants, such as peas, beans, and peanuts, and forage crops, such as clovers and alfalfa, and ornamental plants, such as sweet pea, lupine, wisteria, Judas tree, and locust tree. Leguminous plants add nitrogenous materials to the soils through the fixation of free nitrogen by the actions of special types of bacteria that inhabit the enlarged nodules on the roots.

Garden bean

The common bean plant *Phaseolus* (L. *fabaceus,* from *faba,* bean) may be a short, bushlike, or vinelike annual, depending upon the variety. The stems may be long and slender, and because of unequal rates of growth on opposite sides, they have a tendency to twine spirally around objects with which they come in contact.

The flowers of the common bean are bilaterally symmetrical with petals that are various sizes and unequally spaced. They contain both stamens and pistils and all four sets of flower parts. The calyx is composed of four to five green sepals, which are more or less united. The corolla is butterfly-like and consists of four to five petals, some of which may be fused. In those species with definitely irregular flowers the large, recurved,

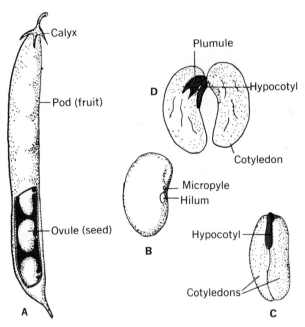

Fig. 17-4. Common bean *(Phaseolus).* **A,** Pod (legume) with part of the wall removed to show seeds; **B,** bean seed (side view); **C,** seed coat removed (face view); **D,** cotyledons spread to show structures between them.

somewhat contorted upper petal is called the standard, the two lateral petals are called wings, and the two lower petals are fused to form the keel, which may be spirally coiled.There are usually ten stamens, nine of which may be united into a thin sheath around the pistil. One stamen is free. Each stamen bears a pollen-producing anther at its tip.

The pistil is composed of an elongated ovary with one carpel that contains several ovules, a filamentous style, and a pollen-receiving stigma. Pollen tubes are formed through the style and extend from the stigma to the ovary. Fertilization occurs in the ovary, and the fertilized ovules develop into the seeds. There is no endosperm, but its place is taken by the two cotyledons. Each seed is attached to the pod by a stalklike funiculus (Fig. 17-4). The seeds, like those of other legumes, are developed in a bivalved, multi-seeded legume (pod). Since a ripened ovary is known as a fruit, the legume is a rather special type of fruit.

Each individual seed consists of (1) a small, prominent ridge above the hilum, where it was attached to the pod, (2) a prominent ridge above the hilum, the raphe, which is formed by the ovule beneath, (3) the micropyle, which is a small opening in the seed coat near the hilum for the entrance of pollen, (4) seed coats, which form the protective covering, (5) two cotyledons, which are the two fleshy halves of the bean for the storage of food, (6) the plumule (or epicotyl) with its true leaves folded over the growing tip, (7) the hypocotyl, and (8) the radicle, which is continuous with the hypocotyl and forms the embryonic root. All of these structures may best be seen in seeds that have been soaked to initiate germination.

Phloem with sieve tubes

Parenchyma cells with starch granules

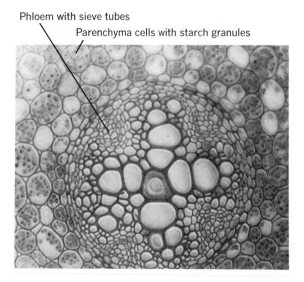

Fig. 17-5. Buttercup *(Ranunculus)* root showing details of the stele (cross section).

During germination the food of the cotyledons is digested and transferred to the plumule, hypocotyl, and radicle. The embryonic primary root is formed from the radicle, which bends downward under the influence of gravity. The hypocotyl elongates and carries the plumule and two cotyledons with it out of the soil. The two cotyledons spread to allow the foliage leaves of the plumule to develop. The cotyledons may develop chlorophyll and carry on photosynthesis for a time, but eventually they shrivel. The roots absorb water and nutrients from the soil and conduct them to the stem. (See also Fig. 6-3.)

Crowfoot family (Ranunculaceae)

Marsh marigold, hepaticas, anemones, columbine, and larkspurs belong to the Crowfoot family, along with the common buttercup, *Ranunculus.* The roots or leaves of certain species of the family are poisonous when eaten.

Buttercup

When a thin cross section of a buttercup root (Fig. 17-5) is studied, it will be noted that the xylem is shaped like a four-pointed cross and is composed of thick-walled cells for conducting materials upward toward the stem. Alternating with the xylem points are four areas of phloem, composed of thin-walled conducting cells. Surrounding these vascular bundles of xylem and phloem is a pericycle of thin-walled cells, and external to this is the ringlike endodermis, composed of thick-walled cells, although thin-walled cells may be present. The xylem, phloem, pericycle, and endodermis constitute the small, centrally located stele (Gr. *stele,* column). External to the endodermis is a wide cortex (L. *cortex,* bark) of thin-walled cells, which is surrounded by a hypodermis and an epidermis.

When a cross section of a buttercup stem is studied (Fig. 17-6) the following will be noted. An external, protective epidermis is present, beneath which is a cortex of thin-walled cells. Numerous vascular bundles are arranged is a circle toward the periphery of the stem. The outside of each bundle contains thin-walled phloem cells, separated from thick-walled xylem cells by a band of cambium between them. Internally, there is a large pith composed of thin-walled cells. (See also Fig.17-9.)

The leaves have typically lobed blades attached to a petiole and contain palmately netted veins that branch out from the petiole like the fingers of a hand. The veins connect with the vascular bundles of the petiole and stem. Some variations in the leaves will be found in different plants.

The flowers of the buttercup are similar in shape and bear both stamens and carpels. The numerous pollen-producing stamens and the many free, one-ovuled carpels are spirally arranged. There are five green sepals and five yellow petals, the latter having a nectar gland at the base of each. The ovule develops into an achene, a small, dry fruit that does not open at maturity and contains a single seed.

Composite family (Compositae)

Plants of this family have many closely compacted individual flowers (florets) that form a head that is commonly mistaken for a flower

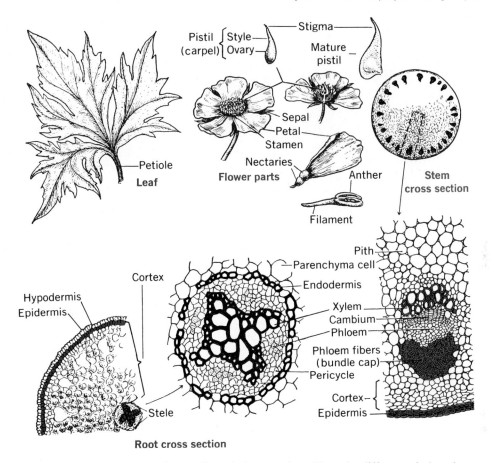

Fig. 17-6. Buttercup *(Ranunculus)*, a dicotyledonous plant. Note the difference in locations of xylem and phloem tissues in the stem and root and note the nectar-producing nectaries at the base of the petal.

(Fig. 17-7). They include dandelions, ragweeds, goldenrods, daisies, asters, zinnias, dahlias, marigolds, and sunflowers. Their enormous seed production and efficient dispersal devices have distributed them widely (Fig. 17-8).

Sunflowers

The stem of a sunflower plant contains nodes (joints) (L. *nodus,* knot), at which leaves are borne, and internodes, which are present between the successive nodes. Growth occurs by an elongation process at the internodes.

A cross section of a mature stem of a sunflower (Fig. 17-9), shows the following structures: (1) A central pith region is composed of thin-walled parenchyma cells. (2) Vascular bundles are arranged in a circle toward the periphery of the stem and are composed of (a) xylem, which is composed of thick-walled cells (single-celled tracheids and vessels), (b) phloem, which is composed of nonnucleated sieve tubes (with perforated sieve plates), and elongated, nucleated companion cells, and (c) a layer of cambium, which separates the xylem and phloem. (3) The pericycle is a cylinder of mechanical tissue that is external to the vascular bundles and is com-

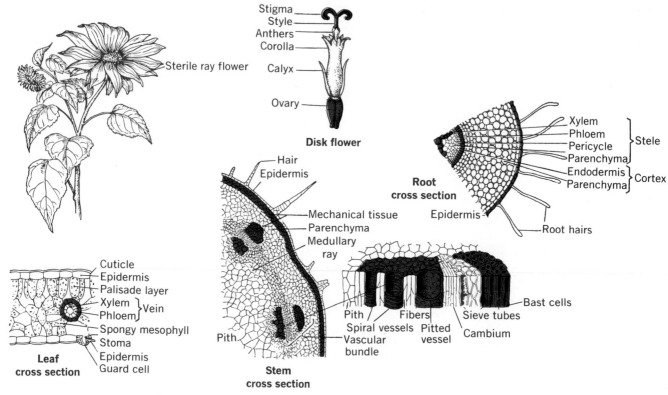

Fig. 17-7. Sunflower *(Helianthus)*, showing various parts in some detail and somewhat diagrammatically.

posed of thick- or thin-walled cells. Individual bundles may be separated by radial strands, medullary rays, which are composed of parenchyma cells to conduct materials across the stem. This entire central core of the stem constitutes the stele. (4) A layer of cortex is external to the stele and is composed of large, thin-walled parenchyma cells. (5) A layer of mechanical tissue is external to the cortex and is composed of thick-walled cells. (6) The external epidermis is composed of elongated cells whose outer walls contain a waxy cutin to make them impermeable to water. Certain epidermal cells may produce extensions known as hairs. A connection between the vascular bundle of a stem and a leaf is called a leaf trace.

A mature leaf from a sunflower consists of a broad blade, which is attached to the stem by a slender petiole. Net veins conduct materials throughout the leaf.

A composite flower, such as the sunflower, is composed of many individual flowers grouped together on a disklike head that resembles a single flower in a general way. At the edge of the head are two or more spirals of overlapping, flat, green bracts (L. *bractea,* thin plate).

There are two types of flowers on the head—ray flowers and disk flowers. Ray flowers form one or two circles at the edge, each in the axil of a small bract (modified leaf). These flowers consist of a strap-shaped corolla (petals), one side of which is modified into a broad, flat structure. The

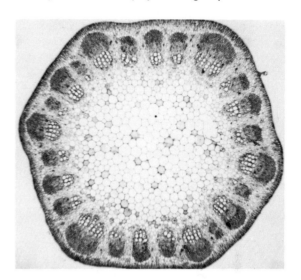

Fig. 17-9. Cross section of a sunflower stem.

Fig. 17-8. Close-up of dandelion head showing modification of seeds that aids in wind dispersal.
(Photograph by F. S. DeMartino, University of Dayton.)

male stamens and the female style may be abortive (poorly developed). These marginal ray flowers may be sterile, or they may contain only female pistils. Disk flowers, in the center of the head, consist of a basal ovary, containing one functional ovule.

MONOCOTYLEDONOUS PLANTS

Indian corn

Indian corn, *Zea mays*, (Gr. *zea*, corn) has an erect stem from which adventitious roots may

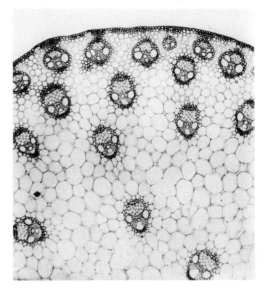

Fig. 17-10. Corn stem (cross section). Note the typical vascular bundles.
(Courtesy General Biological Supply House, Inc., Chicago, Illinois.)

form at the nodes. These roots assist the fibrous true roots in absorbing water and dissolved materials from the soil and in anchoring the plant. Such unusual adventitious roots are called "brace roots," or "prop roots."

Corn has a typical monocotyledonous stem, composed of parenchyma cells of various shapes and sizes, with numerous vascular bundles scattered throughout it. When a thin cross section of a corn stem is studied, it will be noted that the external, protective epidermis is composed of relatively small, thick-walled cells (Fig. 17-10). Beneath the epidermis is a narrow layer of mechanical sclerenchyma tissue whose cells are small and thick-walled. Each vascular bundle is surrounded by a sheath or layer of thick-walled sclerenchyma tissues. Internally, each bundle consists of phloem and xylem. There is no meristematic cambium separating the phloem and xylem, as in dicotyledonous stems, so there can be no secondary increase in size after the primary tissues are mature. Bundles lacking cambium are called "closed" bundles because of their inability to grow indefinitely.

The leaves of corn are characterized by numerous main veins running parallel to the long axis and all connected by a network of fine, inconspicuous branches (Fig. 17-11). The veins are actually vascular bundles that are connected with the vascular bundles of the stem. The broad portion of a leaf is called the blade.

The corn flowers (Fig. 17-12) are incomplete and on different parts of the same plant. The tassel at the tip of the stem consists of pollen-bearing male flowers.

The female flowers consist of a series of enlarged ovaries ("kernels") arranged on the corncob to form the corn "ear." A long style (the "silk" of corn) is attached to each ovary, and the tip of the style, the stigma, is sticky. Wind-disseminated pollen produces a pollen tube that grows through the style to the ovary. Fertilization takes place within the ovary.

A grain of corn is really a fruit because it consists of a ripened ovary. A mature grain of corn

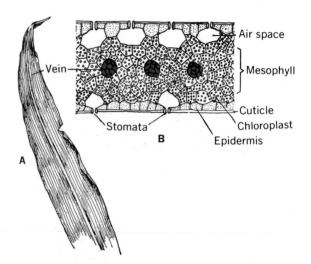

Fig. 17-11. Corn leaf. **A,** Entire. **B,** Detail.

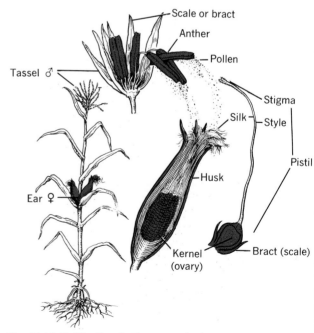

Fig. 17-12. Pollination in the corn plant.

consists of an outer pericarp firmly fused to the seed coat beneath. On the concave side of the grain, beneath the pericarp, is the embryo, which is embedded in the extensive endosperm (food). The endosperm is composed of three parts: (1) a single layer of cells next to the nucellus, called the aleurone layer, each cell in the layer being filled with grains of protein known as aleurone (a lu′ ron) (Gr. *aleuron,* flour), (2) an inner, starchy endosperm, and (3) an outer, horny endosperm containing proteins.

The embryo (Fig.17-13) consists of (1) one broad cotyledon for absorption of food from the endosperm, (2) a well-developed plumule consisting of a stem and one or more foliage leaves, (3) a radicle, which forms the primary root of the seedling, (4) a sheathlike coleoptile (ko le -op′ til) (Gr. *koleos,* sheath; *ptilon,* feather), which encloses the plumule, and (5) a sheathlike coleorhiza (ko le o -ri′ za) (Gr. *koleos,* sheath; *rhiza,* root), which encloses the radicle.

Upon germination the radicle breaks through the coleorhiza and forms a temporary primary root. The Plumule breaks through the protective coleoptile to form true leaves (Fig. 17-14).

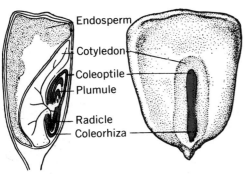

Endosperm
Cotyledon
Coleoptile
Plumule
Radicle
Coleorhiza

Fig. 17-13. Details of corn embryo and seed.

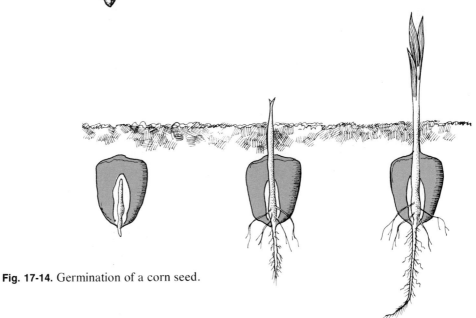

Fig. 17-14. Germination of a corn seed.

Review questions and topics

1 List the distinguishing characteristics of the class Angio-spermae, subclass Dicotyledoneae, and subclass Mono-cotyledoneae, including examples of each.
2 Describe the sporophyte and gametophyte generations in angiosperms, including their relative sizes and indepen-dence.
3 Compare the pollination process in angiosperms with that in gymnosperms, such as the pine tree.
4 Describe the structures and functions of the various parts of a complete flower.
5 Describe how seeds are produced and protected in an-giosperms.
6 List the chromosome number in gametes, sporophyte, and endosperm. How can you explain the number of chromo-somes in each?
7 Describe the formation and function of the pollen tube in angiosperms.
8 Contrast the phenomena of pollination and fertilization, giving examples.
9 Discuss the so-called double fertilization process and its significance. Contrast this with fertilization in gymno-sperms.
10 Define heterospory in angiosperms, including examples.
11 In a botanical sense contrast: ovary and fruit, ovule and mature seed, giving examples.
12 Discuss the embryologic origin of roots.
13 Discuss the various functions of roots.
14 Discuss the principal external differences between roots and stems.
15 Describe the structure of typical stems, including the func-tions of different tissues.
16 Contrast herbaceous dicotyledonous stems with herba-ceous monocotyledonous stems, including the significance of the differences.
17 Discuss the functions of various plant stems.
18 Define such terms as herbaceous, woody, vascular bun-dle, annual rings, wood rays, heartwood, sapwood, latex, and "viscose" (liquid cellulose).
19 Describe leaves, including the functions of the different tissues.
20 Describe the various kinds of leaves, including examples of each kind that you may be able to collect.
21 Define such terms as stomata, guard cells, palisade layer, spongy layer, veins, xylem, phloem, leaf trace, and chloroplast.
22 Describe the structure of a typical flower.

Selected references

Burns, G. W.: The plant kingdom, New York, 1974, The Macmillan Company.
Cronquist, A.: The evolution and classification of flowering plants, Boston, 1968, Houghton Mifflin.
Esau, K.: Plant anatomy, New York, 1965, John Wiley & Sons, Inc.
Evans, L. T., editor: Environmental control of plant growth, New York, 1963, Academic Press, Inc.
Foster, A. S., and Gifford, E. M.: Comparative morphology of vascular plants, San Francisco, 1959, W. H. Freeman & Co., Publishers.
Grant, V.: The fertilization of flowers, Sci. Amer. June, 1951.
Gundersen, A.: Families of dicotyledons, Waltham, Massa-chusetts, 1950, Chronica Botanica Co.
Johansen, A.: Plant embryology, Waltham, Massachusetts, 1950, Chromica Botanica Co.
Koch, W. J.: Plants in the laboratory, New York, 1973, The Macmillan Company.
Mangelsdory, P. C.: Wheat, Sci. Amer. July, 1953.
Swain, T., editor: Chemical plant taxonomy, New York, 1963, Academic Press, Inc.

18 Plant physiology

As in all living things, the various metabolic activities of plants are in a state of constant change. The rate and kind of activity vary with internal and external factors, such as nutrition, growth, age, light, and water supply. All these factors form the subject matter of that area of biology called physiology. Some of these activities are considered in this chapter.

PLANT COLOR

Plant color results from a number of pigments that perform a variety of functions. In general they may be grouped as (1) the chlorophylls, (2) the carotenoids, and (3) the anthocyanins. Some of the characteristics of these pigments are shown in Table 18-1.

Plastids are present in the cytoplasm of many plant cells and may be classified as (1) chloroplasts, which are bodies that contain chlorophyll, (2) chromoplasts, which are bodies of various shapes that contain nonchlorophyll pigments and are usually yellowish, and (3) leukoplasts (Gr. *leukos*, white), which are colorless, widely distributed, and associated with starch production.

Carotenoids are present in chromoplasts, either with chlorophyll or without it. The most abundant carotenoids are carotenes ($C_{40}H_{56}$) and xanthophylls (Gr. *xanthos*, yellow; *phyllon*, leaf) ($C_{40}H_{56}O_2$). Carotenes vary from creamy white to orange red and are present in carrots, sweet potatoes, pumpkins, and leafy vegetables. They are valuable because they are converted into vitamin A by the body. Xanthophylls produce yellow colors and are present in a variety of organisms.

The development of yellow color during the

Table 18-1 Pigments in higher plants

Pigment	Formula	Present in plastids	Color	Location
Chlorophyll a	$C_{55}H_{72}O_5N_4Mg$	Yes	Bluish green	Leaves and other plant parts
Chlorophyll b	$C_{55}H_{70}O_6N_4Mg$	Yes	Yellowish green	Leaves and other plant parts
Carotenes	$C_{40}H_{56}$	Yes	Creamy white to orange red	Associated with chlorophyll
Xanthophylls	$C_{40}H_{56}O_2$	Yes	Pale yellow	Associated with chlorophyll
Anthocyanins	All glucosides $C_3H_6O_5$	No (dissolved in cell sap)	Red, blue, purple, depending on type of anthocyanin	Flower petals, certain leaves (wandering Jew and purple cabbage), roots (red beets), fruits (grapes and plums)

ripening of fruits may involve the formation of carotenoid pigments and the disappearance of chlorophyll, as in the ripening of bananas. The anthocyanin pigments, such as red, blue, and purple, may be present in many parts of plants, for example, in red beets, red radishes, cherries, grapes, plums, asters, geraniums, tulips, and poinsettias. Both anthocyanins and carotenoids may be present in the same flower. In the nasturtium there are yellow carotenoids and red anthocyanins.

The chromoplasts and their pigments play important roles in the coloration of autumn leaves. One group of autumn colors, the purples, blues, and reds, are influenced by anthocyanins, whereas the yellows are produced by carotenoids and xanthophylls. As the quanity of chlorophyll diminishes in the autumn, the associated pigments are revealed. In many woody plants the anthocyanins are so abundant in the autumn that the carotenoids are concealed, forming shades of red and purple.

The production of anthocyanins is determined by the heredity of the particular plant and by such environmental factors as temperature and light. Many woody plants, such as lilacs and privets, develop no anthocyanins, hence their coloration is affected accordingly. Lower temperatures generally influence anthocyanin formation, although actual frosts are not necessary to produce autumn colors. Light stimulates anthocyanin formation, as revealed by the brighter shades on the outer leaves of trees in contrast with the yellow leaves within the same tree. Poison ivy colors may be bright if leaves have been subjected to strong light but yellow if the leaves have been exposed to less light.

The red and purple colors of the leaves of hard maples, oaks, sweet gums, and sumacs are caused by anthocyanins. This coloration is correlated with the presence of sugars. If the main vein of a leaf is experimentally cut in early autumn, the blade beyond the cut may become a deeper red because sugars cannot leave that part of the leaf, thus promoting anthocyanin formation.

RADIANT ENERGY

Radiant energy is the energy in the sun's rays (solar energy). When the electromagnetic waves of sunlight are passed through a prism, a spectrum (L. *spectrum*, image), or bands of different colors, is produced (Fig. 18-1). The visible part of the spectrum extends from the violet, whose wavelengths start at about 400 nm (1 nm, that is, 1 nanometer, is approximately 1 one-millionth of a millimeter), through the red, with wavelengths up to 750 nm.

The shorter, visible violet rays at one end of the spectrum range from the blue, green, yellow, and orange to the longer red rays at the opposite

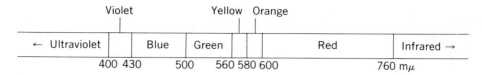

| ← Ultraviolet | | Blue | Green | | | Red | Infrared → |

Fig. 18-1. Light spectrum. Visible wavelengths of light extend from about 400 to 760 mμ.

end of the spectrum. At each end of the visible spectrum there are no visible wavelengths, or colors, but the rays continue to become shorter beyond the violet end and are called ultraviolet rays, whereas the rays beyond the red end become longer and are called infrared rays.

The red rays of the visible spectrum and those of the infrared are known as heat rays. The various rays of the spectrum produce different effects on living protoplasm. Of the various regions of the visible spectrum, the violet, blue, and red wavelengths are absorbed by chlorophyll, although lesser amounts of the orange and yellow may be absorbed as well.

Chlorophyll is prevalent particularly in the palisade layers of leaves, where maximum light is available. The green color of leaves is caused by the transmission of certain light rays to our eyes, whereas other rays are absorbed or transmitted through the leaves. The absorption of certain rays by chlorophyll is essential in manufacturing foods. Certain lower plants, such as the blue-green, the brown, and the red algae, as well as such higher plants as coleus and red cabbage, possess other pigments that mask the chlorophyll.

PHOTOSYNTHESIS

No one should be surprised to read that sunlight is energy and that man depends on it for both food and fuel. Most people are at least aware that when sunlight falls on a green plant, somehow carbon dioxide and water combine to form sug-

ars and oxygen is given off in the process. We are further given the impression that if we really knew how it all worked, most of the problems of food for mankind would be solved.

Although many details have been known for years, others are still somewhat hazy. Experiments have shown that plants grow neither in the absence of light nor in the absence of air containing carbon dioxide. When an aquatic plant was placed under a suitable container in the presence of light and carbon dioxide (CO_2), bubbles of oxygen O_2 were given off (Fig. 18-2). Chemical

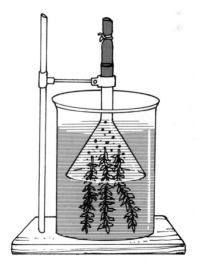

Fig. 18-2. Demonstration showing bubbles of oxygen given off during photosynthesis by *Elodea*. The clamp on the tubing can be opened to test for presence of oxygen gas.

analysis of the plant showed that sugars accumulated. All of this led to a statement of photosynthesis as follows:

$$6\ CO_2 + 6\ H_2O \xrightarrow[\text{green plants}]{\text{light}} C_6H_{12}O_6 + 6\ O_2$$

As analytic techniques were perfected, especially the use of radioactive isotopes to "label" chemical compounds, it was shown that in the sugar resulting from photosynthesis the carbon is from carbon dioxide, the hydrogen from water, and the oxygen also from water. What was not expected was that the oxygen given off also came from water.

This led to a more correct summary statement of photosynthesis as follows:

$$6\ CO_2 + 12\ H_2O \xrightarrow[\text{chloroplasts}]{\text{light}} C_6H_{12}O_6 + 6\ H_2O + 6\ O_2$$

Although this is still acceptable as a reasonable statement of photosynthesis, it is by no means complete or final. What follows is a closer look at the components of the process.

Conditions for photosynthesis
Chloroplasts

Except in the bacteria and blue-green algae, photosynthesis takes place exclusively in organelles called chloroplasts. These may occur in a variety of shapes and numbers. In some of the algae, such as *Chlorella*, there is only one relatively large chloroplast that seems to take up

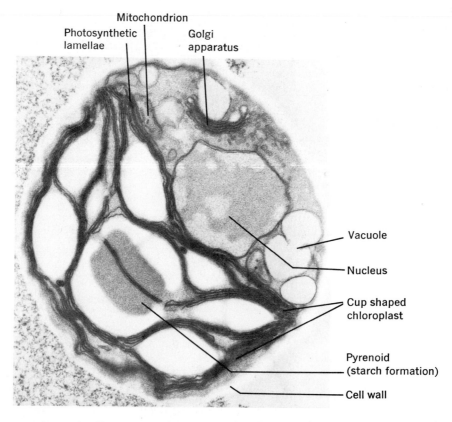

Fig. 18-3. *Chlorella,* a green alga, as seen by electron microscope. Note the very large chloroplast.
(Courtesy M. A. Hayat.)

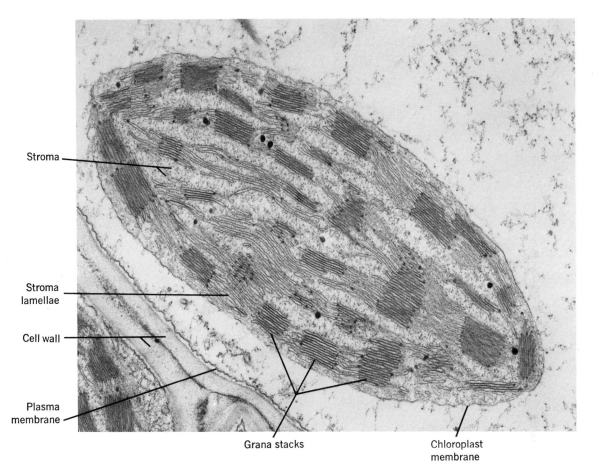

Stroma

Stroma
lamellae

Cell wall

Plasma
membrane

Grana stacks

Chloroplast
membrane

Fig. 18-4. Chloroplast of a crabgrass cell. The membranes of the grana and lamellae are the site of the light reaction of photosynthesis. The stroma contains various enzymes, ribosomes, and other elements necessary to the carbon-fixing cycle. *(Courtesy F. D. Schwelitz, University of Dayton.)*

most of the cell (Fig. 18-3). At the other extreme we find some of the higher plants that contain up to several hundred chloroplasts. At one time or another in a high school or college lab you have probably studied the water plant *Anacharis (Elodea)* and watched the numerous chloroplasts carried along in the cytoplasm.

Chloroplasts may be isolated from the intact cells and studied in some detail. Chemical analysis shows that they contain proteins, lipids, chlorophylls, carotenoids, nucleic acid (DNA

and RNA), inorganic matter, and water. Isolated chloroplasts can carry on cell activities of photosynthesis. However, if the intact chloroplast is disrupted, the chemical contents alone will not support photosynthesis.

Studies with the electron microscope show some details of the chloroplasts structure (Fig. 18-4). Typically there is an outer limiting membrane and an inner, highly complex membrane. This double membrane forms the boundaries of the chloroplast. The inner membrane is continuous with a complex system of elongate tubes connected to flattened disklike structures called thylakoids. These are arranged in layers and resemble stacks of coins. An individual stack is called a granum. For example, each chloroplast of a spinach cell contains about 50 grana. The grana contain the pigments and some of the enzymes necessary for photosynthesis. The space between the grana and associated lamella is called the stroma. It contains a watery mixture of chemicals, including ions, enzymes, ribosomes, various organic molecules, and the chloroplast DNA.

Light

With very few exceptions, the things that we see are visible only because they reflect part of the light that hits them from the sun or an electric light. It is what happens after the light source is involved that determines whether an object is red, green, or some other color.

Sunlight is a source of electromagnetic radiation with wavelengths of various sizes. When this light hits a surface, it is either reflected or is absorbed and its energy transformed into heat. Rarely is it reemitted as a fluorescence. The colors of nature, green leaves, red flowers yellow petals, and so forth, are based on a principle of preferential absorption. The plant cells contain many pigments that may absorb part of the light falling on them. The light that is reflected is missing the wavelengths that have been absorbed. When this light hits the eye, we see the reflected light and not the absorbed light. In this manner green plants do not really absorb green light, rather they absorb other colors and reflect the green that we see.

It is assumed that light energy comes in units called *photons,* with the shorter wavelengths containing the most energy. When light hits an atom of a substance, the atom absorbs a photon and becomes excited. After a short time the extra energy is emitted and the atom drops to its "ground" or less energetic state.

In photosynthesis, light energy is first absorbed by pigment molecules. The most significant of these are the chlorophylls. In higher green plants they appear in two similar forms called chlorophylls a and b. Other pigments in the chloroplast, primarily the yellow and orange carotenoids, absorb light and then pass the energy on to chlorophyll. It appears then, that the initial events of photosynthesis involve interaction of light energy and plant pigments.

Carbon dioxide and water

Carbon dioxide (CO_2) is a basic raw material that is incorporated into the carbohydrate end-products of photosynthesis. Its annual conversion into organic matter is estimated at over 100 billion tons. It provides the basic food for plants and man as well as the energy that drives most living activities. The wood, coal, and oil supplies of the world are the result of its conversion into parts of living organisms that died and accumulated. The famous White Cliffs of Dover, England, are almost pure calcium carbonate, which represents the skeletons of phytoplankton that settled to the bottom of ancient oceans.

The average concentration of carbon dioxide in the atmosphere is about 320 parts per million (0.03%). This may fluctuate somewhat during the day while photosynthesis is taking place and during the night when it is not. There are also seasonal variations.

Carbon dioxide involved in photosynthesis is sooner or later returned to the air as a result of decomposition or burning. Man has been burning fossil fuels at an increasing rate over the last 100 years. It is estimated that the return of carbon dioxide to the air has raised the atmospheric content from 290 to the present 320 parts per million. This is expected to increase to over 375 parts per million by the year 2000. Just how this will change photosynthesis and human activity is an interesting question.

The role of water in photosynthesis can be demonstrated by the use of radioactive isotopes.

It can be shown that the hydrogen of the product as well as the resulting oxygen originated from water. In 1941 a group of workers prepared a solution containing carbon dioxide and water and placed a suspension of a green microorganism, *Chlorella,* in it. The water had been prepared using the isotope ^{18}O instead of the normal ^{16}O, whereas in carbon dioxide the oxygen was all ^{16}O. When the solution was illuminated, bubbles of oxygen gas was given off. The ^{18}O content of the gas was the same as that in the water.

Although it can be shown that the oxygen comes from water, the details of the process are not at all clear. For the present we can assume that water dissociates in solution and that its components are available for photoactivities.

$$2 H_2O \rightleftharpoons 2 H^+ + 2 OH^-$$

$$2 OH \rightleftharpoons \frac{1}{2} O_2 + H_2O + 2e^-$$

The process of photosynthesis

It is important to remember that the components of the photosynthetic system are all present in the chloroplasts. These components may be in the form of ions, molecules, enzymes, pigments, carriers, and many other states. Since chloroplasts are parts of living cells, many different types of reactions are occurring at the same time. It is in the midst of all of this that photosynthesis takes place.

Recent schemes for expressing the details of photosynthesis use the terms "light reaction" and "dark reaction." These are more correctly stated as "light-dependent" and "light-independent" (light-independent reactions can occur in the dark). The light reactions result in the preparation of hydrogen for incorporation into a carbohydrate end product. Further, during this event a process called photophosphorylation takes place. The result of this is the formation of the energy-rich compound called ATP (*A*deno-sine *T*ri *P*hosphate).

In the dark, or light-independent stage, energy is used to drive a series of reactions that combine carbon dioxide and the hydrogen product from the light reactions. This involves many compounds of the chloroplast and cell solution and results in carbohydrate formation.

Light-dependent reactions

Light reactions depend on light energy. The energy may be that of the sun or some artificial source such as those used in greenhouses and growth chambers. Recent evidence suggests that the light reaction takes place in specific parts of the chloroplasts called reaction centers. There appear to be two separate reaction centers in the granum. Light energy hits these reaction centers at the same time, and several simultaneous events take place. If we consider them in sequence, the process may be similar to the following:

1. Light rays hit chlorophyll A in a reaction center activating it so that electrons are raised to a "high energy" state.
2. These electrons are now "excited" and are removed from chlorophyll.
3. If not trapped, the high energy is dissipated as heat. In photosynthesis, the high energy electrons are trapped by an electron acceptor called Q (probably plastoqionone).
4. The electrons are now transferred to a series of carriers similar to those formed in mitochondria.
5. During the transfer, energy is captured by inorganic phosphates to form ATP; this occurs when electrons are transferred from one carrier to another. Some of the carriers are cytochromes. ATP is made from ADP (adenosine diphosphate) and inorganic phosphate using the energy obtained through the transfer of electrons from one carrier to another.
6. The electrons are now at a low energy state and are captured by molecules of chlorophyll A, which previously had given off electrons in photosystem I. (This was a simultaneous reaction to 1.)

7. Chlorophyll A molecules that had lost "excited" electrons now react with electrons (from ionized water) to restore the original state of chlorophyll B. The result of this activity of photosystem 2 is two molecules of ATP.

The simultaneous reactions in photosystem I have also produced excited chlorophyll as follows:

A. Light rays striking molecules of chlorophyll A produce molecules of "excited" chlorophyll A, which give off electrons at a high energy level.

B. Excited electrons are now raised to a higher energy level and are trapped by a molecule called Z (iron-containing molecule).

C. These electrons then go through a transfer system involving an iron compound called ferredoxin.

D. From the transfer system they are trapped by NADP (nicotinamide adenine dinucleotide phosphate) from the chloroplast pool.

E. Hydrogen ions (from the ionization of water) now combine with the electron-rich $NADP^+$, forming the association $NADPH^+ + H^+$. For simplicity this is sometimes designated $NADPH_2$.

F. $NADPH_2$ is now ready to move into the carbon dioxide cycle in the dark and light-independent stages.

G. Chlorophyll A now captures electrons from photosystem II to restore its normal state as in step 6.

The dark reactions

These reactions, which take place in the light as well as in the absence of light, involve the conversion of carbon dioxide into a carbohydrate end-product. Although we will assume that this carbohydrate is glucose, this is only one of many possible products.

Both the energy necessary for the reaction and the hydrogen component of the end product come from the light reaction. Further, many of the reactants are normal components of the plant cell and if used up must be replaced. One of the most important of these is a phosphorylated 5-carbon sugar, ribulose diphosphate (RuDP). Melvin Calvin received the Nobel Prize in 1961 for work done during the 1940s, which showed the details of the dark reaction. Although the total scheme involved about fifteen reactions and as many enzymes, only a brief outline is given here.

Remember that the result of the light reaction is NADPH, H^+, ATP and water. Carbon dioxide is present in the cell solution along with ribulose diphosphate, enzymes, and other chemicals. Carbon dioxide combines with the 5-C ribulose diphosphate to form an unstable 6C molecule that immediately splits to form two molecules of phosphoglyceric acid (so named because it has the structure of an organic acid, a carbon skeleton similar to glycerol, and a phosphate group).

$$CO_2 + RuDP \xrightarrow{H_2O} 6\,C \rightarrow 2\,PGA$$

Phosphogylceric acid now combines with ATP to form a diphosphoglyceric acid and to return ADP to cell.

$$2\,PGA + 2\,ATP \rightarrow 2\,DPGA + 2\,ADP$$

The diphosphoglyceric acid now reacts with $NADPH_2$ to form phosphoglyceraldehyde (acid structure changed to aldehyde structure).

$$2\,DPGA + 2\,NADPH + 2H^+ \rightarrow$$
$$2\,PGAL + 2\,NADP^+ + \textcircled{P} \text{ (phosphate group)}$$

The 2 PGAL may now be used in a variety of reactions in cell metabolism. In order to form glucose, two PGAL combine to form a 6-carbon fructose diphosphate,

$$2\,PGAL \rightarrow P - fructose - \textcircled{P}\ (FDP)$$

which loses phosphate, is rearranged to glucose phosphate, which now loses the final phosphate, and is the 6-carbon sugar glucose.

$$FDP \rightarrow FMP + \textcircled{P} \rightarrow GMP \rightarrow Glucose + \textcircled{P}$$

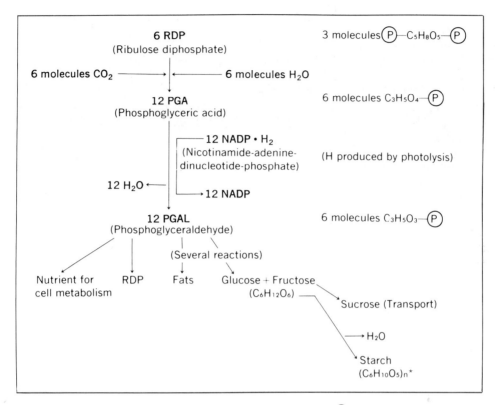

Fig. 18-5. Summary of carbon dioxide–fixing cycle (simplified). P represents the phosphate group, PO_3H_2. Ribulose diphosphate is a 5-carbon sugar to which two phosphates, P, have been added. Asterisk (*) refers to a number (N) of glucose units per molecule; twenty-four glucose units may be joined into a single molecule of starch ($C_{144}H_{240}O_{120}$), with the production of 24 H_2O.

Once all of this happens, the glucose is usually given as the end product, but something has been lost in the process. You will remember that the sequence started with the 5-carbon sugar ribulose diphosphate. If this is used up in the formation of glucose, how will the next carbon dioxide reaction take place?

The answer to this problem was suggested by Calvin in a more detailed scheme that involved starting with 6 molecules of carbon dioxide and 6 of RuDP (and multiples of 6 for the other reactants). In this way it was shown that

$$6\ CO_2 + 6\ RuDP \rightarrow 12\ PGA \rightarrow 12\ PGAL$$

Photophosphorylation

$$ADP + \text{P} \xrightarrow[\text{Nucleotides, cytochromes}]{\text{Energy}} ATP$$

Photolysis

$$H_2O + NADP \xrightarrow[\text{Chlorophyll}]{\text{Light energy}} NADP \cdot H_2 + O \uparrow$$

CO_2 fixation

$$RDP + CO_2 \xrightarrow[H_2O]{} PGAL$$

Fig. 18-6. Summary of photosynthetic activity.

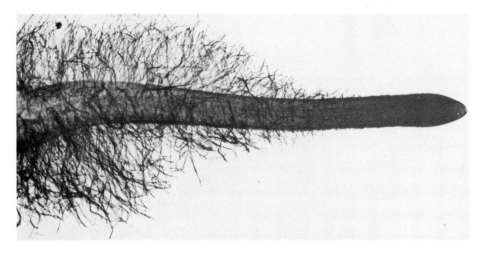

Fig. 18-7. Tip of a root, showing the root cap and root hairs.
(Courtesy General Biological Supply House, Inc., Chicago, Illinois.)

and two 2 PGAL were involved to form glucose. The remaining 10 PGAL (10 3-carbon sugars) were then rearranged in a series of reactions to restore 6 RuDP. The net reaction then was:

$$6 \ CO_2 + 18 \ ATP + 12 \ NADPH + 12H^+ \rightarrow$$
$$Glucose + 18 \ ADP + 12 \ NADP^+ + 18 \ \textcircled{P}i$$

As mentioned above, PGAL itself may be used for further metabolic activities of the plant cell. It may be used up as an energy yielding food. It may be converted, as we have seen, to glucose and other sugars. It may be converted to the more complex sugars such as sucrose, or leave the cell and be used by other parts of the plant, or be converted to starch (Fig. 18-5), lipids, or other organic materials and used for cell metabolism.

ABSORPTION AND TRANSLOCATION

Many materials necessary for the various activities of plants must be absorbed by the roots and particularly by the root hairs (Fig. 18-7). The absorption of water and of substances dissolved

in water involves several complex phenomena, including imbibition, diffusion, osmosis, and active transport. Imbibition is the process by which solid particles, chiefly colloidal particles, absorb liquids and increase in volume. Plant cell walls absorb water by this process, much like a dry sponge soaks up water and swells. Colloidal materials of plants, such as cellulose, the proteins of the protoplasm, pectic substances, and other organic compounds, imbibe large quantities of liquids.

Diffusion is the spreading out of the molecules of a substance from places of greater molecular concentration to places of lower concentration. The molecular actions of materials assist in their diffusion in plant roots.

Osmosis is the diffusion of water through a differentially permeable membrane from a region of high water concentration to a region of lower water concentration. Ordinarily, the osmotic concentration of the cell liquids of root epidermal cells is higher than that of the soil solution. Thus osmosis seems to permit continued absorption by osmotic action, for water usually diffuses from a region of low osmotic concentration of solutes (or high concentration of water) toward a region of high concentration of solutes (or low concentration of water).

Active transport is the concentration of a substance on one side of a membrane. Unlike osmosis and diffusion, this process is not merely a physical phenomenon. Cell metabolism provides the energy to drive the substances across the barrier. Also in contrast to osmosis and diffusion, active transport is selective; not all substances are able to pass through the cell membrane.

Ion exchange occurs between soil particles and the adhering root hairs without entering into solution in the soil. Root cells absorb solute ions from the soil solution and also from the surfaces of soil particles as a result of the exchange of cations (electropositive ions).

Both osmosis and imbibition assist in the transfer of water across the cortex of the root, either from cell to cell or between the cells in the walls (Fig. 17-5). Solutes enter the root by diffusion and by active transport. Entry into the central stele is restricted by the presence of suberin bands between the cells of the endodermis. Water and solutes appear to pass through the endodermis and probably provide a control of the composition of the xylem tissues of the root, stem, and veins of the leaves.

Materials in the leaves, including manufactured foods, are conducted downward through the phloem tissues of the leaves, stems, and roots. In woody stems wood rays conduct materials horizontally (radially) through bands of parenchymatous tissue, hence, they are often called vascular rays.

Water is important to plants because (1) the colloidal constituents of protoplasm are dispersed in water; (2) water and carbon dioxide are used to manufacture food by photosynthesis; (3) solid materials are usually dissolved in water before they can enter or leave a cell or be moved from one part of a plant to another; (4) it provides internal pressure to maintain form and give support to the plant; and (5) it is the chief medium in which most of the chemical reactions of living protoplasm occur.

Water and organic solutes, such as sugars and amino acids, move through the vascular tissues of the phloem and xylem. The transport of these organic compounds is called translocation.

In general the direction of solute movement is governed by the metabolic requirements of the plant rather than by the strict morphology of the conducting system. Prior to the full expansion of a growing leaf, sugars and amino acids move into its tissues. After it has reached 50% to 60% of its final size (Fig. 18-8), the leaf is an exporting organ. Most of its photosynthetic product is translocated in the form of sucrose to such regions as young leaves, shoot tips, root tips, and young fruits. This movement is primarily through the sieve tubes of the phloem (Fig. 18-9). At any one time different phloem elements of the same vascular bundle may be transporting solutes in different directions.

Prior to translocation, sugar enters the phloem of the smallest veins of the leaf (nearly 70 cm. in length in a square centimeter of sugar beet leaf). After translocation the sugar may move out of the sieve tubes into parenchyma or cortex cells where it is stored as starch or sugar. Analysis of the sieve tube contents reveals sucrose concentrations of 35% as well as smaller quantities of other compounds.

During translocation sucrose may move across the sieve tubes into the parenchyma or cortex cells, where it is stored as starch. Analysis of the sieve tube contents may reveal a sucrose content of up to 25% as well as much smaller quantities of sugar alcohols and amino acids.

Xylem sap contains significant amounts of nitrogenous compounds, especially amino acids. The main movement of xylem sap is from the roots to stems, leaves, and fruits.

The translocation (conduction) of liquids and organic solutes through the stem is a complex phenomenon that is incompletely understood. Water and nitrogenous compounds are conducted upward largely by the xylem tissues.

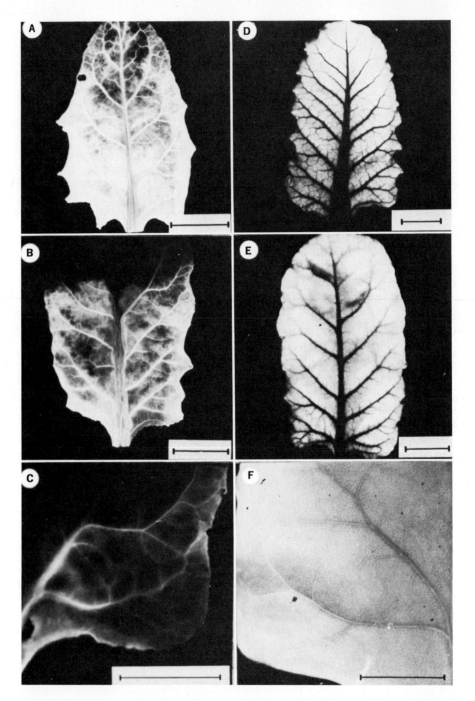

Fig. 18-8. Leaves that are 25% (**A** and **D**), 35% (**B** and **E**), 45% (**C**), and 50% (**F**) of their final size. The leaves on the left show import of radioactive sugar while those on the right show loading of sugars into the phloem in the leaf veins. Beginning of export (right) and the cessation of import moves across leaf from tip to base as it matures.
(Courtesy R. J. Fellows and D. R. Geiger, University of Dayton.)

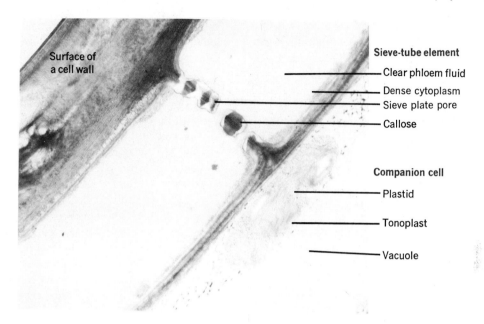

Surface of
a cell wall

Sieve-tube element

Clear phloem fluid

Dense cytoplasm

Sieve plate pore

Callose

Companion cell

Plastid

Tonoplast

Vacuole

Fig. 18-9. Detail of a sieve tube showing two sieve-tube elements and the sieve between them. The pores are lined with a whitish carbohydrate called callose. The clear phloem fluid is an approximately 35% sucrose sugar solution with K^+ ions, amino acids, and other materials. The fluid is under a pressure of approximately 350 pounds per square inch.
(Data and photograph courtesy R. Giaquinta and D. R. Geiger, University of Dayton.)

Proof of this can be seen by experimentally cutting away the phloem and other tissues external to the cambium without appreciably decreasing the upward flow of the water and solutes. Likewise, when the xylem of the stem is experimentally severed, the leaves wilt, and the aerial parts of the stem usually die. Great quantities of water evaporate from the aerial parts of plants. In addition water is used in food manufacture and growth so that a supply must be conducted upward almost constantly. The rate of ascent is influenced by many external environmental factors, including humidity, air temperature, and light intensity. Water absorbed by roots may ascend in the wood of a stem sometimes as fast as 15 feet per hour.

Among the various theories that have been proposed to account for the ascent of liquids in stems are the following examples. It can be observed that liquids will exude from the stumps of

some plants and from the cut branches or tapped sapwood of such trees as sugar maples. The force that causes this so-called bleeding is root pressure. Root pressure may assist in the ascent, but it is present in great amounts only during the spring and thus is lacking or is only a minor factor when the ascent is greatest. Likewise, cut stems, when placed in water, are able to translocate water upward even though no roots are present.

Another theory to explain the ascent of liquids in plant stems involves the force of atmospheric pressure. Atmospheric pressure can force up and maintain a column of water approximately 33 feet high (at sea level). It can do this only if the column of water is unbroken, if the atmospheric pressure is exerted directly on the surface of the water at the base of the column, and if there is a vacuum at the tip of the column. Evidence against this explanation includes the absence of an actual vacuum in the leaves and the fact that the water column is actually broken by cross walls in the conducting channels of stems. Since liquids rise for hundreds of feet in some trees, atmospheric pressure is at best only a minor factor in the ascent.

When glass tubes of very small diameter are placed vertically with their lower ends in water or similar liquids, the liquid will rise in the tubes to a level above that in the container in which they are standing. The smaller the diameter of the tube, the higher the level to which the liquid rises. This rise is caused by capillarity, or surface attraction, between the molecules of the liquid and those of the tube substance. This force of capillarity has been thought to be involved in the ascent of liquids in stems because the hollow tracheids and xylem vessels are somewhat comparable to the tubes just described, and function in a similar way. Possibly, capillary forces may assist in the ascent of liquids, but they could cause a rise of only a few feet at the most.

It is generally accepted that the major cause for the ascent of the xylem sap is the pull generated by capillarity in the minute spaces of the cell walls of leaves and other structures' water loss by evaporation. Because water coheres strongly, the tension or pull is transmitted to the root tips where it assists the entry of water. Thus a tremendous tension or "negative pressure" in the water columns is started by water evaporation in the leaves, resulting in definite decreases in the diameter of stems. A strong upward osmotic pull in the leaf cells will raise a tall column of water, such as that found in the conducting xylem of wood, because of the high cohesive power of a thin column of water. Within a plant the movement of water can be directed into cells that have a high solute content or that have a great deal of cytoplasm. If water is lost by these cells, they attract water by osmotic and by imbibitional forces.

TRANSPIRATION AND GUTTATION

Transpiration (L. *trans,* beyond; *spirare,* to breathe) is the loss of water vapor from aerial parts of plants, particulary through the stomata of leaves. Guttation (L. *gutta,* a drop) is the exudation of water from plants, usually in the form of drops. The early morning dew is often water from guttation. Of the amount of water absorbed from the soil, a plant uses only a fraction for its various activities, but much escapes from the leaves and stems as water vapor. A very high percentage of transpiration occurs through the leaves.

The structures of leaves are directly or indirectly involved in transpiration. Often leaves are broad, thus exposing large surfaces to the air. The presence of stomata in the lower and upper surfaces permits the escape of water vapor. The number of stomata per unit area of leaf varies with the species and is influenced by the age of the leaf and certain environmental factors. Usually, the lower surface has many more stomata than the upper. In fact, some plants, such as apple, oak, and orange, have no stomata in the upper surface but have thousands in the lower. The water lily, on the other hand, has stomata only in the upper surface. The bean leaf has 4,000 stomata per square centimeter in the upper surface and approximately six times that number in the lower. Corn has a ratio of 6,000 to 10,000, and the sunflower has a ratio of about 8,000 to about 16,000.

The mesophyll tissues inside the leaf are surrounded by many intercellular spaces that connect with large spaces just beneath the stomata. Hence large areas are exposed and therefore can pass much water into the air. Water lost from the cells is replaced by the movement of water from adjacent cells and eventually from the leaf veins.

Since much of the transpiration occurs through the stomata and since the rate of water loss even in the same plant varies from time to time, the relative sizes of the stomata are of significance, even though these openings occupy less than 1% of the leaf's surface. In general the stomata are closed at night and open in the day, although under drought conditions or wilting they may close during the day.

The most influential environmental factors that

Table 18-2 Comparison of photosynthesis and respiration

Photosynthesis	Respiration
1. Occurs only in chlorophyll-bearing cells of plants	1. Occurs in all living cells of plants and animals
2. Occurs only in presence of energy-supplying light	2. Occurs throughout life of cell; light is not required
3. Releases O_2 as by-product	3. Releases CO_2 and H_2O as end products
4. Uses CO_2 and H_2O	4. Uses O_2 and food
5. Solar (radiant) energy converted into chemical energy and stored	5. Energy released by oxidation, chemical energy being converted into heat and useful energy for various plant activities
6. Results in weight gain	6. Results in weight loss
7. Foods manufactured	7. Foods broken down

affect the transpiration rate include the humidity and temperature of the air, air currents, the intensity of the light, and the amount of water in the soil. If the humidity of the surrounding air is high, the amount of water evaporation will be reduced. Air movements may also remove the water vapor from around the leaf, thus affecting the rate of transpiration. Because of a lack of water in the soil, if there is a minimum of water circulating in a plant, the amount of transpiration might be affected.

Although most water is passed from plants as a vapor, it may leave in liquid form by the process of guttation, in which the drops of water are confined to the margins and tips of the leaves. Guttation usually occurs when water loss by transpiration is less than the amount being absorbed rapidly by the roots. Such is the case when warm days are followed by cool nights. Probably osmosis plays an important role in guttation. This phenomenon can be demonstrated experimentally in such plants as corn, potato, tomato, cabbage, nasturtium, and grass, especially if environmental condition are right and there is sufficient absorption by the roots.

RESPIRATION AND DIGESTION

Respiration

Respiration is a chemical oxidative process whereby living protoplasm breaks down organic substances with the release of energy necessary for various activities. Through respiration the common sugar, glucose, is oxidized into simpler substances with the release of energy as summarized by the following:

$$C_6H_{12}O_6 + 6 O_2 \rightarrow 6 CO_2 + 6 H_2O + Energy$$

From the standpoint of energy loss or gain, the reaction in the process of respiration is the opposite of photosynthesis. The process of respiration involves a series of complex stages, the use of a number of enzymes, and the formation of several intermediate compounds. Only the basic principles are summarized here.

The fundamental differences between photosynthesis and respiration are shown in Table 18-2.

Energy is the capacity to do work. The energy liberated in respiration is used in part to form heat and in part for such plant functions as the flowing of living protoplasm; the movement of cell parts; the process of assimilation, whereby food is converted into living protoplasm or its products; the synthesis of fats and proteins from sugars; the accumulation of solutes from the soil; and the synthesis of pigments, vitamins, enzymes, oils, resins; organic acids, and tannins.

Perhaps all chemical reactions within living cells are controlled by organic catalysts called enzymes. In respiration the action of certain enzymes results in the oxidation of glucose with the liberation of energy and the production of carbon dioxide and water. During the day photosynthesis usually proceeds at a rate several times that of respiration, so that foods and energy are stored. At night only respiration occurs, and the plant uses food and oxygen and liberates carbon dioxide.

The type of respiration considered so far occurs when a sufficient supply of free oxygen is available and is termed aerobic respiration. Under certain conditions respiration may occur in the absence of free oxygen; this is called anaerobic respiration (Gr. *an*, without; *aer*, air; *bios*, life). Both types of respiration are controlled by specific enzymes and release energy. Anaerobic respiration does not release as much energy or form the usual end products of carbon dioxide and water, but it produces a number of compounds, including alcohol, carbon dioxide, and organic acids. In fact the commercial productions of certain products are obtained from anaerobic respiration. An example of anaerobic respiration is shown by the following equation, in which several stages are omitted:

$$C_6H_{12}O_6 + \text{Specific enzymes} \rightarrow$$
Sugar   (in the absence of free oxygen)

$$2\ C_2H_5OH + 2\ CO_2 + \text{Energy}$$
Alcohol

Digestion

Insoluble stored foods, such as starch or protein, can neither be used in the cells in which they are stored nor translocated to other parts of plants until they are changed into a soluble, diffusible form through enzymatic actions. This process is known as digestion.

One of the most common digestive activities in plants is the change of starch into sugar through the catalytic activity of a specific enzyme called amylase (L. *amylum*, starch) with the production of maltose. Nearly all digestive enzymes change a complex compound into one or more simpler compounds through a reaction with water, which is known as hydrolysis. Digestion of starch may be shown as follows, with several stages and intermediate carbohydrates being omitted:

$$2(C_6H_{10}O_5)_n + n(H_2O) + \text{Amylase} \rightarrow n(C_{12}H_{22}O_{11})$$
Starch   **Maltose**

In plant tissues maltose is hydrolyzed by another enzyme, maltase, to form glucose, as shown by the following equation:

$$C_{12}H_{22}O_{11} + H_2O + \text{Maltase} \rightarrow 2\ C_6H_{12}O_6$$
Maltose   **Glucose**

Many other enzymes digest various types of carbohydrates in living plants.

Oils and fats in plants are digested by the enzyme, lipase, to form glycerin (glycerol) and fatty acids, such as palmitic, oleic, and linoleic.

Proteins are hydrolyzed into amino acids by enzymes known collectively as proteases. The process occurs in several stages, and intermediate products are formed. Proteases also catalyze the conversion of amino acids into proteins. There are many plant proteases, including bromelin in fresh pineapple, which liquefies (digests) the gelatin in desserts, and papain in papaya, a tropical fruit. Papain digests proteins and is used in tenderizing meats, in surgery, and in the tanning and brewing industries.

PLANT GROWTH FACTORS

There are many internal factors in plants that either independently or in conjunction with each other influence their growth and development. These are so numerous and complex that only a general outline can be given here.

Inherited traits

The genetic traits of an organism basically determine its structures and activities. Changes in the number and construction of the chromosomes have important effects. Plants with tetraploid chromosomes (four sets) often have larger organs with larger cells and wider leaves and fruits than diploid plants of the same species. Sometimes the differences in size and shape of a fruit (such as disk and spherical squashes) may result from a single gene difference, whereas in other plant phenomena the interaction of several genes may be responsible. Such internal structures as cell shape and size, the character of the cell wall, and the plane of cell division are influenced by heredity. Traits that result from inheritance may influence the development and activities of that plant.

Hormones

Growth substances may be formed in one part of an organism and in some cases transported to other parts where they produce their effects. There are a number of such substances that have been shown to be naturally occurring growth regulators. The present list includes auxins, gibberellins, kinins, ethylene, and abscisin, which are of general occurrence. In addition there are specialized substances and possibly a flowering hormone florigen. This latter messenger, whatever its nature, is often induced by specific photoperiods and appears to move to the shoot's tip in the phloem. In all cases the hormones are effective in exceedingly small quantities.

Auxins

Auxins (Gr. *auxein,* to increase) are growth substances of plants that have been studied extensively. They are formed chiefly in young, physiologically active parts of plants, for example young leaves, shoots, and root apices.

Auxin is concerned with phototropic (light) as well as geotropic (gravity) changes. In young oat seedlings or seedlings of other grains, the young shoot is surrounded by a light-sensitive, sheathlike coleoptile. This coleoptile will bend toward light because the cells on the darker side elongate more rapidly than those on the lighter side. If about 1 mm. of the tip of the coleoptile is experimentally removed, no bending toward light results (Fig. 18-10). When the cut tip is replaced on the stump, light sensitivity is restored. This

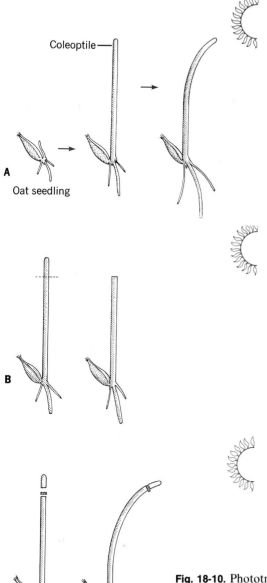

Coleoptile

Oat seedling

A

B

C

Fig. 18-10. Phototropic response of oat seedling. Early experiment demonstrating role of coleoptile tip. **A,** The coleoptile of a growing oat seedling grows toward the light. **B,** If the coleoptile tip is cut off, the seedling does not elongate. **C,** If, after cutting, the tip is replaced, even when separated by an agar block, the seedling grows toward the light, as if intact.

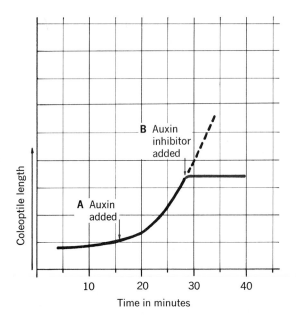

Fig. 18-11. Auxin promotes growth by "cell wall loosening," enabling cells to take up water and stretch the walls. When auxin is added at point **A**, coleoptile length increases to end of dotted line. In another study, if auxin inhibitor is added at **B**, coleoptile elongation stops.
(Data from M. J. Evans.)

growth substance can be extracted from the tip and applied to the stump with similar results. The auxin passes in only one direction from the tip toward the base of the coleoptile, showing what is called physiologic polarity.

There is evidence that the action of auxins in promoting growth consists of "wall-loosing" which allows turgor to stretch the walls and permits water uptake resulting in cell growth (Fig. 18-11). New wall synthesis then makes the growth permanent.

Several types of auxins are organic acids of the indoleacetic acid group and are produced by living protoplasm. Certain artifically synthesized substances, such as indolebutyric acid and naphthaleneacetic acid, are able to produce effects similar to those of the natural growth substances.

Some types of plants produce fruits even though pollination and fertilization do not occur, a phenomenon known as parthenocarpy (Gr. *parthenos,* virgin; *karpos,* fruit). Such a fruit is usually seedless, and the fruit seems to maintain an auxin content that is sufficiently high to continue growth. Parthenocarpy is found in seedless grapes, in some kinds of citrus fruits, and in cultivated varieties of banana and pineapple.

Seedless fruits of tomato, squash, cucumber, and watermelon may be produced artificially by using synthetic auxin-like growth substances. The use of plant hormones to increase fruit yield when plants are not easily pollinated may be of practical value.

Root formation is also stimulated by auxin. More roots will form on a cutting if leaves or buds are present because of the production of growth substances. If the basal end of a cutting is treated with the proper growth substances for a few hours before planting, more abundant roots will form.

In some plants the development of shoots is caused by an auxin deficiency rather than by any shoot-forming substance. The phenomenon of dominance and inhibition is associated with growth substances. When a terminal shoot or bud is present, the buds below may often fail to develop, but they will do so after the inhibiting terminal tip is removed. Evidently the tip secretes a substance that moves downward to inhibit bud development below.

Auxin is also influential in certain geotropic changes (Fig. 18-12). Roots placed in a horizontal position tend to bend downward. Auxin checks the elongation of root cells, so that the upper surface of such roots with less auxin grows faster. On the other hand, shoots placed in a horizontal position tend to bend upward because gravity causes auxin to accumulate on their lower side, where it stimulates elongation. The differences between root and shoot reactions to auxin probably result from roots being stimulated by weak auxin concentrations but being inhibited by the stronger ones that stimulate the growth of stem cells.

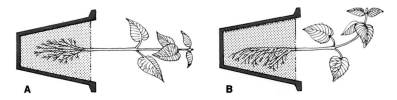

Fig. 18-12. Geotropism in plants. **A,** Potted plant on its side. **B,** Roots respond to auxin by growing down, stems by growing up.

Gibberellins

Other plant growth hormones, accidentally discovered in the last decade of the nineteenth century, are the gibberellins. Japanese rice farmers observed some greatly elongated seedlings, which never lived to maturity and only rarely produced flowers. The condition was named "foolish seedling" disease. It was discovered in 1926 that these seedlings were all infected with *Gibberella,* a fungus that could be transferred to healthy plants, thereby producing the disease. If grown in a nutrient medium in a flask, the fungus produces a substance that causes typical overgrowth symptoms when applied to a receptor plant. When this giberellic acid is applied to plants, enormously elongated stems result, and in some instances the leaf area is reduced.

Gibberellin is most effective in dwarf mutant plants. Chemical regulators such as Phosphon and CCC are used commercially to produce the opposite effect of gibberellin. These substances produce short mums and poinsettias and may inhibit the action of gibberellin. Gibberellin also triggers digestion of starch in seeds of grains such as barley and wheat. The fungus also produces prompt stimulation of flowering in those plants known as "long day plants." In the grape industry applications of gibberellins to seedless grape clusters result in the retention and development of greater number of large-sized grapes. Applied to celery, gibberellins produce larger, more succulent plants in a shorter time.

Other hormones

Many other substances, most of which are not completely understood, have been discovered and are being investigated. A high auxin concentration causes production of ethylene, which causes a number of effects including ripening of fruit and inhibition of cell growth. The kinins promote cell division and in combination with auxin appear to determine whether or not undifferentiated cells grow into roots or shoots. A high kinin/auxin ratio promotes shoot differentiation. Abscisin causes dormancy in shoots, promotes abscission of leaves, and lowers water loss from leaves.

ABSCISSION

Abscission (L. *abscissus,* cut off) is the separation of plant parts, such as leaves, stems, flowers, and fruits, usually by the natural dissolution of certain cell walls of the so-called abscission layer. The leaves of deciduous plants fall at the end of their growing season. The pectins in the cell walls of the abscission layer are dissolved through the action of enzymes, partially detaching the petiole from the stem and leaving only the vascular bundles to hold the leaf until it eventually falls. A layer of cork tissue develops at the petiole base for the protection of the exposed tissues. Because of tough vascular bundles and lack of an abscission layer, leaves may remain for some time, as in the case of red oak leaves. Pectic substances are gelatinous and viscous and tend to hold cells together. In evergreens the leaves remain for several seasons (4 years or more), but new leaves are produced to replace those that are shed periodically.

Abscission may be demonstrated experimentally by cutting away the blade portion of a leaf. In a few days the remaining part of the leaf will fall. However, if a plant hormone, such as indoleacetic acid, is applied to the cut surface of the petiole, the abscission layer will not form, and the petiole will remain on the plant. The abscission of flowers and fruits is also controlled in part by auxins. If certain immature fruits are sprayed by growth substances, their fall is often delayed. It is likely that auxin has its effect by regulating the production or action of abscission.

LIGHT AND FLOWERING

Light affects different plants in various ways. Nonchlorophyll-bearing plants do not require light; some are even adversely affected by it. On the other hand, chlorophyll-bearing plants require certain quantities and qualities of light for photosynthetic purposes. Some lower plants, such as algae and certain types of bacteria, have pigments (other than chlorophyll) that are affected in different ways by light.

In addition to its role in photosynthesis, light influences the development and functions of many parts of plants. The direction and intensity of light influence the amount and direction of the growth of stems and leaves. In darkness plants tend to have long, weak stems, poorly developed leaves, and tissues without chlorophyll, a condition called etiolation (Fr. *étioler,* blanch). Leaves growing in relatively bright light tend to have thicker epidermis and palisade layers and smaller intercellular spaces. Very intense light, particularly the ultraviolet rays, affects the protoplasm so that the structure and position of plant parts may be modified to subject as little of the plant as possible to the harmful rays.

A number of processes such as stem elongation, leaf expansion, seed germination, and inter-ruption of flower induction by light appear to be mediated by a protein pigment, phytochrome. Phytochrome is converted to one form by far red (longer wave red) light an back to the other by red (shorter wave red) light. Other pigments mediate phototropism and other light quality effects on plant form and function.

The young leaves of many plants serve as receptors of day length. Sometimes a single night of the proper duration is sufficient to induce production of the flowering signal. Cocklebur and Japanese morning glory are examples of short day plants that can be induced to bloom with one long night. Chrysanthemums can be made to bloom in about 2½ months if part of the plant is covered for a few hours at the beginning or end of the day during the summer. The entire plant will come into flower and will continue to flower through the autumn when the control plants normally do so. The black covering must be in place for the number of hours that would reduce the length of the longer summer day to the same length as the shorter autumn day when the plant ordinarily begins to flower.

On the basis of the effect of light on flower production, plants may be divided into (1) "long day plants," such as red clover and spinach, which form flower buds as soon as the day length exceeds a certain critical length; (2) "short day plants," such as poinsettia, aster, dahlia, cosmos, and chrysanthemum, which form flower buds only when the night is longer than a certain critical period; and (3) "indifferent plants," including many types, which appear to be unaffected by day length.

If light is withheld for part of each day, "short day plants" can be induced to flower entirely out of season. If artificial light is provided for an extra period at the end of the day, they can be prevented from flowering, even though they are mature. "Long day plants" respond in the opposite way to artificially controlled periods of illumination. If poinsettias are desired at Christmas, they are placed in the dark until noon each day, starting in November, and then returned to the light all afternoon to allow for sufficient photosynthesis. Such responses of plants to periodic light and dark conditions are phenomena of photoperiodism. Flower induction is regarded as an interaction between a built-in "clock" in the plant and the timing of the dawn and dusk to which the plant is exposed.

TEMPERATURE

Plants are constantly influenced by variations in the temperature of the air and soil. Most of them grow best with an optimum temperature of 70° to 90° F. Plant activities tend to decrease as temperatures approach freezing or 100° F. The maximum, optimum, and minimum temperatures for various plants and their activities vary with species, age, water content, and inherited abilities.

Plants fit generally into two classes: (1) chill sensitive plants, generally of tropical origin, which are injured at temperatures below 45° to 50°, and (2) chill tolerant. These latter, of more temperate origin, can withstand temperatures down to or below freezing. The lower freezing point of lipids in the cell membranes of this latter group seem to make the difference.

In general, seeds with low water content tend to resist extremes of temperature better than actively growing plants. Certain seeds and spores have existed for months in a temperature as low as −226° F and for shorter periods at temperatures as high as 220° F (boiling point of water is 212° F). Certain bacteria and blue-green algae live in the water of hot springs just below the boiling point.

If germinating seeds of winter wheat are exposed to low temperatures for a time, the developing plants will flower earlier than normal. If an embryo is exposed to a low temperature during early seedling development, the early stages of development seem to be accelerated and the development time is shortened.When such a young seedling emerges, it is in a more advanced stage of development than if it had developed at a higher temperature. The explanation and use of this phenomenon may have far-reaching effects in future plant developments.

GROWTH MOVEMENTS

Unequal growth rates in different parts of a plant organ may occur. These include the following. (1) Nutations (L. *nutare,* to nod) are nodding movements of the growing apices of certain organs. Different growth rates are internally induced on different sides of the growing apex, as shown in the movements of the epicotyls. (2) Twining movements are discernible, spiral movements of the growing apex of stems, for example, the twining of morning glories, hops, and beans.

Tropisms (Gr. *trope,* a turning) are rather slow bending movements of cylindrical organs, such as stems, leaf petioles, flower stalks, and roots, in response to external stimuli. Growth occurs more rapidly on one side than on the opposite, which results in a bending that usually tends to place the structure in a more favorable position as far as that stimulus is concerned. Tropisms are common in fungi, bryophytes, ferns, and seed plants.

Among the common tropisms caused by an unequal distribution of auxin because of a specific stimulus are (1) phototropism (Gr. *phos,* light), which is a light-induced growth movement that is usually positive for aerial parts, such as leaves and stems, and places them in advantageous positions; (2) chemotropism (Gr. *chemeia,* chemical), which is a growth reaction to chemical substances; (3) hydrotropism (Gr. *hydro,* water), which is a growth reaction to water whereby roots obtain necessary supplies; (4) thigmotropism (Gr. *thigma,* touch), which is a growth reaction to contact stimuli; and (5) geotropism (Gr. *ge,* earth), which is a growth response to gravity that is positive when roots grow toward soil and negative when stems bend away from a horizontal position (Fig. 18-12).

TURGOR MOVEMENTS

Turgor movements are caused by a change in the water pressure of certain tissues of an organ and do not result from irreversible growth of cells. Generally they result from the uptake or release of potassium ions from cells on the upper and lower sides of leaflets and other parts that exhibit turgor movements. Some, like the thig-

motropic (touch) response of the sensitive plant *Mimosa,* occur very rapidly.

Some of the more common types of turgor movements are *nastic,* contact, and rapid movements. Nastic movements are those of such flattened organs as petals, leaves, and bud scales in which inflation and deflation of cells resulting from changes in osmotic pressure cause closing or opening of an organ. The so-called sleep movements are nastic movements that are induced in many plant leaves because of changes in light intensity. Such changes in leaf positions are shown by white clover leaflets and by garden beans, which assume a somewhat horizontal position in bright light and a vertical position in diminished light.

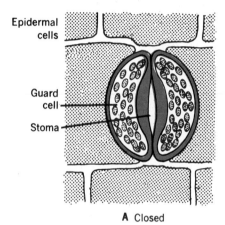

Epidermal cells

Guard cell

Stoma

A Closed

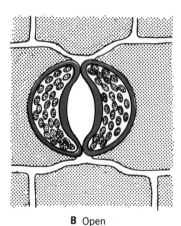

B Open

Fig. 18-13. When light hits the cells of **A,** photosynthesis begins and CO_2 is used. Lowered CO_2 increases potassium ion K^+ uptake and osmotic pressure is raised. **B,** Swellings at sides of guard cells cause stoma to open.

Review questions and topics

1 Explain how materials are absorbed and translocated by plants, including the roles of imbibition, osmosis, diffusion, root pressure, atmospheric pressure, capillarity, and cohesion.
2 Which theory, or theories, seems to explain the ascent of plant liquids in the most logical manner? Give specific reasons why you believe this.
3 Explain the phenomena of transpiration and guttation, including the functions of both.
4 Explain how chlorophyll-bearing plants photosynthesize foods. What foods are made? Might we consider photosynthesis as a chemicophysical process? Explain the origin of fats, proteins, and such special products as cellulose and aleurone grains.
5 Discuss the raw materials, source of energy, end products and role of chlorophyll in photosynthesis. Give reasons for dividing photosynthesis into such stages as photolysis and carbon dioxide fixation.
6 List all the characteristics of chlorophyll that you can. How many kinds of chlorophyll are there? How many carotenoids function in photosynthesis?

Contact movements are induced by contact and pressure stimuli, such as displayed by leaves and flower parts. The contact turgor movements of the two parts of a Venus's flytrap leaf result in a closing and possible trapping of an insect. The rapid turgor movements of certain flower parts are beneficial in efficient insect pollination, in some cases covering the stimulating insect with pollen.

Another rapid turgor movement, displayed by the reactions of the leaves of the sensitive plant *Mimosa,* may be initiated by such stimuli as contact and temperature change. Other adjacent leaves react successively as the excitation chemicals (possibly auxin) are transported from the original point of stimulation. The leaves return to normal after several minutes (see Fig. 2-10).

The closing, opening, and change in size of the stomata of plants are caused by changes in the surrounding paired guard cells. The internal water pressure in cells is referred to as turgidity (L. *turgere,* to swell).

Potassium uptake raises the osmotic pressure in the guard cells, water enters, and the guard cells increase in turgor. Uneven thickening of the guard cell walls causes these cells to bend, thus opening the stomata (Fig. 18-13). When light strikes the leaf, carbon dioxide concentration inside is lowered, potassium enters the guard cells, and the stomata open. Wilting of the leaf may cause loss of turgor, closing stomata and conserving of water.

7 Explain why all living plants and animals must respire. Contrast aerobic and anaerobic respiration.
8 Contrast photosynthesis and respiration in as many ways as possible.
9 Explain the process of digestion in plants.
10 Explain how leaves and other plant parts may fall and why this is desirable in certain regions of the world.
11 List ways in which plants, or plant products, may be of economic importance, giving specific examples to prove your points.

Selected references

Arnon, D. T.: The role of light in photosynthesis, Sci. Amer. **203**:105-18, 1960.
Bold, H. C.: The plant kingdom, ed. 3, Englewood Cliffs, New Jersey, 1970, Prentice-Hall, Inc.
Clevenger, S.: Flower pigments, Sci. Amer. June, 1964.
Crofts, A. S.: Translocation in plants, New York, 1961, Holt, Rinehart & Winston, Inc.
Denison, W. C.: Life in tall trees, Sci. Amer. **228**(6):74-82, 1973.
Epstein, E.: Roots, Sci. Amer. **228**(5):48-57, 1973.
Galston, A. W.: The life of the green plant, Englewood Cliffs, New Jersey, 1964, Prentice-Hall, Inc.
Jensen, W. A., and Kavaljian, L. G., editors: Plant biology today, Belmont, Calif., 1963, Wadsworth Publishing Co., Inc.
Meyer, B. S., Anderson, D. B., and Bohning, R. H.: Introduction to plant physiology, ed. 2, Princeton, New Jersey, 1973, D. Van Nostrand Co., Inc.
Rabinowitch, E. J., and Govindjee: The role of chlorophyll in photosynthesis, Sci. Amer. 213:74-83, 1965.
Salisbury, F. B.: Plant growth substances, Sci. Amer. **196:** 125-134. 1957.
Salisbury, F. B.: The flowering process, New York, 1963, The Macmillan Company.
Steeves, T. A., and Sussex, I. M.: Patterns in plant development, Englewood Cliffs, New Jersey, 1972, Prentice-Hall, Inc.
Whittingham, C. P.: The chemistry of plant processes, New York, 1964, Philosophical Library, Inc.

Some contributors to the knowledge of the
BIOLOGY OF CERTAIN HIGHER PLANTS

Theophrastus (370-285 B.C.)

A student of Aristotle and "Father of Botany" who observed the association of certain fungi and the roots of oak trees. Such an association of filamentous fungi and roots is known as mycorhiza (Gr. *mykes*, fungus; *rhiza*, root), in which, from a food standpoint, both types of plants may benefit.

John Ray (1628-1705)

An English Botanist who studied plant physiology and described experiments explaining the rise of sap. He recognized differences between dicotyledonous and monocotyledonous plants.

Stephen Hales (1677-1761)

An English physiologist who conducted critical experiments on the manufacture of foods by plants and on the transportation of materials within plant bodies. He is the "Father of Modern Plant Physiology."

Robert Brown (1773-1858)

An Englishman who discovered the minute "areola" (spaces) in pollen grains, which he regarded as points for the production of pollen tubes.

N. T. de Saussure (1767-1845)

A Swiss who published the first quantitative treatment of photosynthesis expressed in somewhat modern terms (1804).

René Dutrochet (1776-1847)

A French physiologist who recognized that chlorophyll was necessary for photosynthesis (1837).

Julius Sachs (1832-1897)

A German who proposed experimental methods for studying photosynthesis, respiration, and transpiration in plants.

Wilhelm Pfeffer (1845-1920)

A German botanist who observed the role of fungi in the mycorhiza of nonchlorophyll-bearing plants (1877). Such plants, including the Indian pipe *(Monotropa),* depend on the fungi for food, water, and minerals.

W. J. Beal

An American botanist who experimentally showed the longevity of seeds of common plants (1879), some of which retain their ability to germinate after 90 years.

H. H. Dixon (1869-1953)

A British botanist who proposed the theory of cohesion, which is generally accepted as a principal explanation of the rise of liquids in the xylem tissues of plants. Root pressures are also of importance, especially in shorter plants.

Went

Woodward

W. W. Garner (1875-1956)

An American botanist who discovered photoperiodism, or the effect of natural day length on flowers (1920), and noted that natural day length could be extended by artificial light or shortened by placing plants in the dark, or by at least covering them.

Richard Willstäter (1872-1942)

He and co-workers suggested the chemical nature of chlorophyll (1913), although many investigators since that time have made additional contributions.

Peter Boysen Jensen (1883-1959)

A Danish botanist who performed experiments that suggested that stem growth resulted from a stimulus transmitted by a chemical that was formed in the tip and migrated down into the region of bending.

Frits W. Went (1903-   )

An American botanist who proved the existence of plant hormones (1926) by growing decapitated coleoptile tips of oats on blocks of agar. *(Wide World Photos, Inc.)*

Robert Woodward (1917-   )

Woodward, of Harvard University, won a Nobel Prize (1965) for his synthesis of chlorophyll, cortisone, cholesterol, and quinine. The synthesis of chlorophyll required fifty-five separate and very complicated steps. *(Wide World Photos, Inc.)*

PART FOUR

Animal biology

19 Sponges and coelenterates

METAZOANS

All multicellular animals are collectively known as the Metazoa. This classification implies that these higher animals have both a specialization of cells and a multicellular structure. Bodies of metazoan animals are usually composed of distinct tissues, organs, and often several organ systems. Metazoan cells contain centrioles, and the cells do not have cell walls or cuticles. Metazoans possess multicellular reproductive organs, and development passes through distinct embryonic and, typically, larval stages.

It is thought that metazoans evolved from some ancestral, unicellular type of life and then became multicellular, retained and improved their ability to locomote, but lost any photosynthetic abilities that may have been present originally. In general, metazoan animal bodies possess a structural architecture that reflects a way of life based on alimentation and locomotion, while plant body architecture reflects a way of life based on photosynthesis and attachment, or sessilism.

The two groups to be considered in this chapter, the sponges and the coelenterates, are Metazoan. There are some qualifications, however, that must be made in describing them.

The sponges are a rather unusual group of organisms in many ways. Although they have different types of cells, the cells are not clearly arranged into tissues. They have a body cavity, but it is more of a water filtering system than a digestive site.

Sponges were once thought to be plants because they lack movement and have an indefinite structure. Even now they are not considered to be on the direct line of animal evolution and, therefore, are placed in the branch Parazoa (Gr. *para,* beside) as an offshoot of the Metazoa.

The coelenterates, by contrast, appear in the

evolutionary line as the lowest animals with tissues. They have a digestive cavity (an enteron) in which food is broken down by enzymes. Several cellular aggregates are present and function as tissues. A nerve network is also present.

CLASSIFICATION OF SPONGES (PHYLUM PORIFERA)

Class Calcarea (L. *calcarius*, lime)
  Calcareous spicules, which may be straight or branched; marine types. Examples: *Sycon* and *Leucosolenia*.
Class Hexactinellida (Gr. *hex*, six; *aktin*, rays)
  Six-rayed siliceous spicules arranged in three planes, which in some types are fused into a skeleton resembling spun glass; marine types. Example: Venus's flower-basket (*Euplectella*).
Class Demospongiae (Gr. *demos*, people; *spongos*, sponge)
  Network of fibers of protein-like spongin alone or associated with siliceous spicules; complex system of canals; mostly marine types. Example: Commercial sponges.

Canal systems

There are three kinds of canal systems that are found in the various types of sponges (Fig. 19-1).

Ascon type

The ascon type of canal passes directly from the external ostia (L. *ostia*, pores) to the spongocoel (Gr. *spongos*, sponge; *koilos*, cavity). This type of canal is lined with collar cells. *Leucosolenia*, a freshwater sponge, is a sponge with this type of canal system.

Sycon type

The sycon type of canal system has an incurrent system that ends blindly near the central spongocoel and an excurrent system of radial canals that ends blindly near the body surface. Both canals are connected by small openings. Only the radial canals are lined with collar cells. *Sycon* is sponge with this type of canal system.

Leucon type

The leucon type of canal system is complex with many branches. These multibranched canals contain numerous chambers that are lined with collar cells. Commercial (bath) sponges have this type of canal system.

Class Calcarea
Sycon

Sycon is a vase-shaped, sessile, radially symmetrical, marine sponge with calcareous (limy) skeletal spicules. It may be up to 1 inch long and often forms a colony by budding (Fig. 19-2). The surface has a fuzzy appearance and has numerous incurrent pores, or ostia, which open into the incurrent canals just beneath. These canals end blindly near the central cavity, the spongocoel (Fig. 19-1). Openings in the spongocoel wall are called apopyles (Gr. *apo*, away from; *pyle*, gate) and are surrounded by perforated cells known as porocytes (Gr. *poros*, pore *kytos*, cell). The apopyles lead into radial (excurrent) canals that run toward the outer surface and end blindly. The incurrent canals and radial canals are connected by small openings called prosopyles (Gr. *proso*, forward; *pyle*, gate). The incurrent canals are lined with dermal epithelium. The spongocoel and radial canals are lined with gastral epithelium. Radial canals are well supplied with flagellated collar cells, the choanocytes (Gr. *choane*, funnel; *kytos*, cell), to propel water and food. Water enters the spongocoel and eventually goes out through an opening, the osculum (L. *osculum*, small "mouth"), located at the free end of the sponge. Cells called myocytes (Gr. *myo-*, muscle; *kytos*, cell) around the ostia contract and expand to regulate the size of the opening and thus the flow of water.

The body wall of the *Sycon* is composed of an outer dermal epithelium and an inner gastral epithelium. A noncellular jellylike layer, the mesenchyme (Gr. *mesos*, middle; *en*, in; *chein*, to pour), is present between them (Figs. 19-3 and 19-4). This middle layer contains wandering ameboid cells, the amebocytes, for ingesting

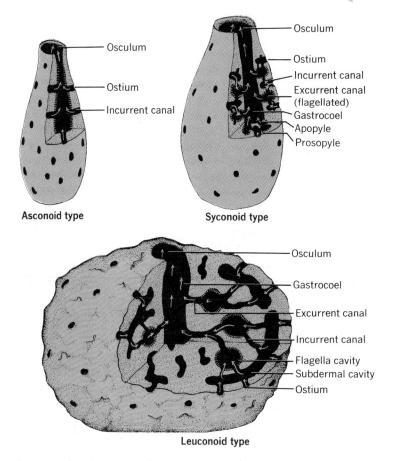

Fig. 19-1. Sponges, showing types of canal sytems. Simple asconoid type, such as *Leucosolenia;* syconoid, such as *Sycon;* leuconoid, such as commerical (bath) sponge. Central cavity, or gastrocoel, is also called spongocoel. Flagellated cells are known as choanocytes. Observe the increase in complexity, particularly of the number of chambers lined with flagellated choanocytes.

foods, forming reproductive cells, and producing skeletal materials.

The body wall contains many calcareous spicules (calcium carbonate), which are interlaced to give protection and support. *Sycon* has four types, namely, short monaxon (straight), long monaxon, triradiate (three rays in one plane), and polyaxon (many rays). All spicules originate from specialized ameboid cells called scleroblasts (Gr. *skleros,* hard; *blastos,* bud or origin).

Sycon ingests microscopic organisms and small pieces of organic materials, drawing them in through the incurrent canals by the beating action of the flagella of the collar cells. Digestion is intracellular (within the collar cells). The collar cells engulf the foods by means of pseudopodia and form food vaculoes in which the foods are digested by digestive enzymes. Digested foods are circulated from cell to cell by diffusion and by the wandering amebocytes of the mesenchyme.

Respiration occurs by absorbing oxygen from

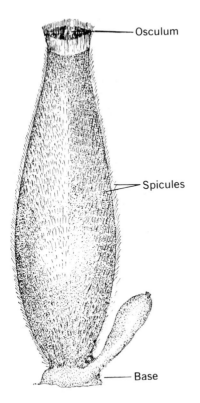

Osculum

Spicules

Base

Fig. 19-2. Simple sponge with a young bud. *(From Braungart, D. C., and Buddeke, R.: An introduction to animal biology, ed. 7, St. Louis, 1968, The C. V. Mosby Co.)*

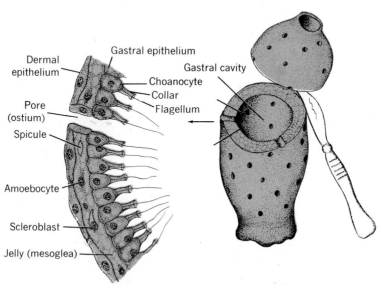

Dermal epithelium

Gastral epithelium

Gastral cavity

Choanocyte

Collar

Flagellum

Pore (ostium)

Spicule

Amoebocyte

Scleroblast

Jelly (mesoglea)

Fig. 19-3. *Leucosolenia,* showing some of the structures in detail (cross section). The jellylike mesoglea is also called mesenchyme.

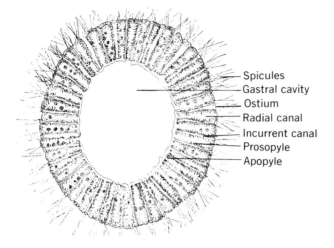

Spicules
Gastral cavity
Ostium
Radial canal
Incurrent canal
Prosopyle
Apopyle

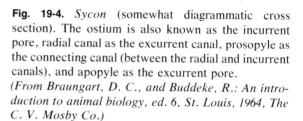

Fig. 19-4. *Sycon* (somewhat diagrammatic cross section). The ostium is also known as the incurrent pore, radial canal as the excurrent canal, prosopyle as the connecting canal (between the radial and incurrent canals), and apopyle as the excurrent pore. *(From Braungart, D. C., and Buddeke, R.: An introduction to animal biology, ed. 6, St. Louis, 1964, The C. V. Mosby Co.)*

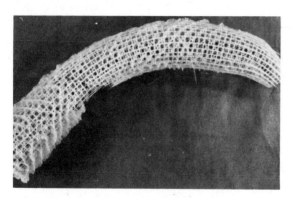

Fig. 19-5. Skeleton of Venus's flower-basket *(Euplectella)* from the Philippine Islands.

the water and giving off carbon dioxide to the water as it circulates. Wastes are carried with the water and out through the osculum; hence, the osculum functions more as an anus than as a mouth.

Sycon reproduces asexually by budding, and many may be produced to form a cluster, or colony, of individuals. Sexual reproduction occurs by sperms and eggs, both of which are formed by special ameboid cells and are present in the same animal (monecious). The fertilized egg (zygote) forms a embryo that becomes a

flagellated larval stage, called the amphiblastula. One half of the amphiblastula bears flagella, whereas the other does not. The motile larva escapes from the parent, swims freely, and finally attaches and grows to become an adult.

Class Hexactinellida
Venus's flower-basket

Venus's flower-basket, a glass sponge, belongs to the genus *Euplectella* and is composed of siliceous (silicon dioxide), six-rayed spicules interwoven into a beautiful cylindrical network that resembles spun glass (Fig. 19-5). The spicules surround the numerous pores and canals and form a complex architectural pattern. Many of the spicules are star shaped. The chambers on the canals are lined with collar cells. *Euplectella* is found in the deeper regions of the ocean near the Phillippines where it is attached by a mass of long glasslike threads.

Class Demospongiae
Commercial (bath) sponges

Commercial bath sponges have a skeleton of spongin fibers, which are composed of a protein-like material (Figs. 19-6 to 19-7). The fibers are branched and form a network that supports and protects other parts of the sponge. Bath sponges, when alive, are frequently black, but

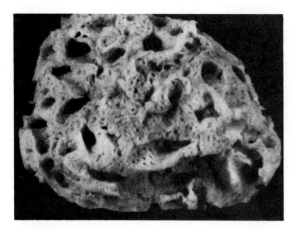

Fig. 19-6. Commercial sheep's wool sponge from Nassau.

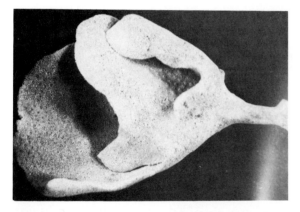

Fig. 19-7. Paper sponge *(Phyllospongia)* from Africa.

some types are yellow, brown, or grayish. The canals are complex, very branched, and have numerous chambers lined with collar cells.

Commercial sponges are collected by diving, hooks, and dredging. They are killed by being placed in shallow water in order to rot the tissues, thus leaving the skeleton. They are then beaten, washed, and bleached in the sun. They may be farmed artificially by fastening small pieces of suitable species in the proper type of water and waiting several several years for them to reach a usable size.

Economic importance of sponges

The skeletons of sponges have a great number of uses for bathing, washing, absorbing fluids during operations, packing materials, and soundproofing. Even though artificial sponges made of cellulose and other materials are widely used today, natural sponges are still in great demand. The best commercial sponges are found in the warm waters of the Gulf of Mexico, the West Indies, and the Mediterranean Sea. Over one million dollars' worth of sponges have been collected off the Florida coast in a single year.

Sponges are usually not used as food by other animals because of the skeletal materials, their bad odors or tastes, and their relatively small supply of nutritive materials. Some smaller animals may seek protection in or among sponges. Certain boring sponges attach themselves to the shells of clams and oysters and, by means of a secretion, bore the limy shells so full of holes that they destroy the animal inside. Siliceous sponges and protozoans of the order Radiolaria may initiate the process of flint formation. It is believed that beds of flint may be made from a mass of sponge skeletons in a few years.

Freshwater sponges may attach themselves to water pipes, water filtration equipment, and reservoirs and, together with other living organisms, form a feltlike mass that inteferes with the water system.

Sponges may starve oysters and other shelled mollusks by attaching themselves to the shells and taking food from the mollusks. Fossil sponges that are similar to present-day forms have been found in chalk and flint formations that are estimated to be one-half billion years old.

CLASSIFICATION OF COELENTERATES (PHYLUM COELENTERATA)

Class Hydrozoa (Gr. *hydor,* water; *zoon,* animal)
Marine and freshwater types; solitary or colonial;

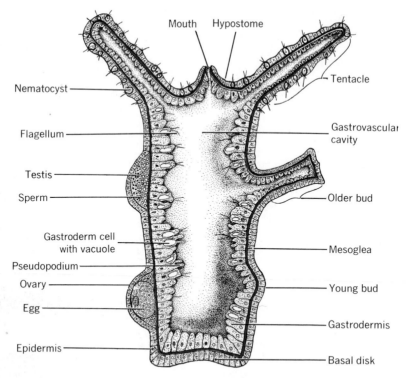

Fig. 19-8. *Hydra*, a freshwater coelenterate, in section (much enlarged and somewhat diagrammatic).

gastrovascular cavity is not held in position with membranous mesenteries; medusae (when present) with a velum (membrane on the undersurface); usually with alternation of generations of asexual polyps and sexual, umbrella-shaped medusae, although medusae, although one or the other may be absent. Examples: *Hydra, Obelia, Gonionemus,* and *Physalia.*

Class Scyphozoa (Gr. *skyphos,* cup; *zoon,* animal) Marine types; solitary; gastrovascular cavity is held in position with mesenteries; polyp stage is absent or inconspicuous; umbrella-shaped medusa stage is large, without a velum, and typically with eight notches on the margin and supplied with sense organs. Examples: large jellyfishes.

Class Anthozoa (Gr. *anthos,* flower; *zoon,* animal) Marine types; solitary or colonial; polyp stage but no medusa stage radially arranged mesenteries (eight or more) extend from the body wall to the central gullet. Examples: sea anemone *(Metridium)* and corals.

Class Hydrozoa
Hydra

Hydra (Figs. 19-8 to 19-10) is a common, cylindrical, radially symmetrical, solitary, freshwater animal, which may be about one fourth inch long when extended.

Two common species of *Hydra* are the green *Hydra (Chlorohydra viridissima),* which is green because algae live within it body symbiotically, and the brown *Hydra (Pelmatohydra oligactis).*

The free end (oral end) contains a mouth in the center of a hypostome (Gr. *hypo,* under; *stoma,* mouth), which is surrounded by a circle of about six slender, extendable tentacles. The mouth opens into a gastrovascular cavity (enteron) that connects with the hollow tentacles.

The body wall consists of an outer epidermis (ectoderm) and an inner gastrodermis (endo-

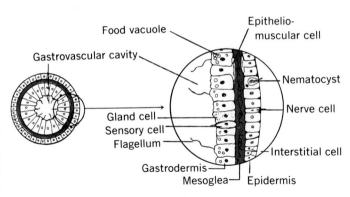

Fig. 19-9. *Hydra* (cross section and somewhat diagammatic). The central cavity is the gastrovascular cavity (enteron).

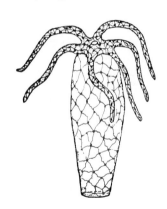

Fig. 19-10. *Hydra*, showing the nerve net (diagrammatic). Special staining techniques show the netlike arrangement of the nervous system and its concentration in the mouth region.

derm). A thin, noncellular mesoglea is present between them (Fig. 19-9).

The epidermis cells are typically cube-shaped and include such types as (1) epitheliomuscular cells, which possess contractile fibers; (2) cnidoblasts, which contain the saclike nematocysts; (3) interstitial cells, which give origin to nematocysts, buds, and reproductive gonads; (4) gland cells, which secrete mucus on the basal disk for attachment; and (5) nerve and sensory cells, which form part of the so-called nerve net.

The gastrodermis cells are typically column-shaped and line the gastrovascular cavity. Among such gastrodermal cells are (1) digestive-muscular cells, which digest foods (intracellular) and possess flagella for moving water and fibers for contraction; (2) gland cells, which secrete mucus and albumin; (3) interstitial cells, which are thought to give rise to other types of cells; and (4) sensory and nerve cells.

The mesoglea consists of a jellylike substance and contractile fibers from the two cellular layers.

A characteristic of *Hydra* and of coelenterates in general is the presence of stinging cells, or nematocysts (Gr. *nema*, thread; *kystis*, sac), which are saclike, filled with fluid, and contain a

coiled thread. A triggerlike cnidocil (Gr. *knide*, nettle; L. *cilium*, hair) projects from the cell and, when stimulated (by acetic acid), cause the nematocyst to explode, thus uncoiling the thread. probably increased fluid pressure in the sac is responsible for this phenomenon. There are several kinds of nematocysts in *Hydra*, which perform such functions as paralyzing organisms by a poisonous secretion, producing an adhesive secretion for attachment, and coiling around their prey. When a nematocyst is discharged, the surrounding cnidoblast cell is destroyed and must be replaced. Most nematocysts do not harm man, but those of large jellyfishes and the Portuguese man-of-war may be dangerous.

Hydra feeds on small animals an plants, overpowering its prey with its tentacles and nematocysts. Food is forced into the gastrovascular cavity, where certain gastrodermal cells secrete digestive enzymes and the food is digested. Some food particles are engulfed by the pseudopodia of the endodermal epitheliomuscular cells, where digestion occurs within food vacuoles.

Indigestible materials are forced out through the mouth. Digested materials may diffuse from cell to cell. The flagella and body movements may assist in circulating the contents of the gastrovascular cavity. Respiration and excretion are performed by individual cells, since there are no special organs. Necessary energy is obtained by oxidizing foods within cells. Wastes such as urea and carbon dioxide pass from the cells into the water.

Hydra may attach itself by its basal disk, or it may move in different ways, such as by: (1) gliding on the basal disk; (2) a type of "somersaulting," whereby the basal disk is suddenly detached and the animal turns over completely and reattaches itself; and (3) a "measuring-worm" type of movement, in which *Hydra* bends over and attaches its tentacles, then detaches the basal disk and slides it along toward the tentacles, which are then released, and the *Hydra* assumes its upright position.

The nervous and sensory equipments of coelenterates consist of a plexus of nerve cells, considered to be a nerve net, which is connected with nerve fibers within the epidermis (Fig. 19-10). The nerve net is connected to sensory cells, in order to receive stimuli, and to epitheliomuscular cells, in order to perform movements. The nervous system functions especially well in the hypostome and basal disk regions, and sensory cells are most numerous in these areas. Sometimes the sensory-nerve net is referred to as the neuromuscular system.

Hydra responds to external and internal stimuli. The body and tentacles contract and expand, the movements being slower, as a rule, when *Hydra* is well fed. The movements caused by the contractile fibers in the body wall result from the proper transmission of impulses through the nerve net. The degree of response depends partly on the kind and intensity of the stimulus. Usually, if the stimulus is strong, *Hydra* will respond negatively. Mild stimuli may cause only localized responses. *Hydra* tends to seek moderately intense light and avoids very strong light. It reacts negatively to strong chemicals. Cooler water seems to be most acceptable, for *Hydra* has a tendency to move away when temperatures approach 20° C. When subjected to electric current, the mouth end tends to orient toward the anode (positive charge) and the basal end toward the cathode (negative charge). By trial-and-error, *Hydra* seems to seek conditions best fitted for its existence.

Under natural conditions, particularly in the autumn when there is a reduction of water temperature, *Hydra* may produce several testes, which are rounded outgrowths near the mouth end. A single, larger, somewhat conical ovary is produced nearer the basal end. Numerous sperms, which are set free in the water, swim (each has one long flagellum) to the ovary where one will fertilize the egg. In the egg the normal number of chromosomes has been reduced from twelve to six. A similar reduction has occurred in the sperm, so that the fertilized egg (zygote) again has the normal number of twelve. Within the ovary the zygote undergoes divisions to form a hollow blastula, which is followed by a solid gastrula. The embryo passes from the parent and is surrounded by a protective, shell-like cyst. When favorable conditions are encountered, the shell breaks and the young *Hydra* emerges. Both male and female gonads may be present on the same individual, in which case self-fertilization may occur. When gonads are formed at different times, cross-fertilization must occur.

Asexual buds may project from the sides as outpocketings of the entire body wall, even with an internal extension of the parent gastrovascular cavity. The bud develops a mouth and tentacles, eventually constricts at its base, and detaches to live separately.

Hydra has a great ability to restore lost parts through regeneration. When cut into pieces, each part may regenerate a new individual. If the hypostome region is cut off, a new individual is formed. If the mouth region is split, an individual with two "heads" is produced. Parts of individuals may be grafted together experimentally, even between different species. The ability of animals to regenerate lost parts was first discovered in 1740 in the *Hydra*. Even if a *Hydra* is turned inside out, the cells of the epidermis and gas-

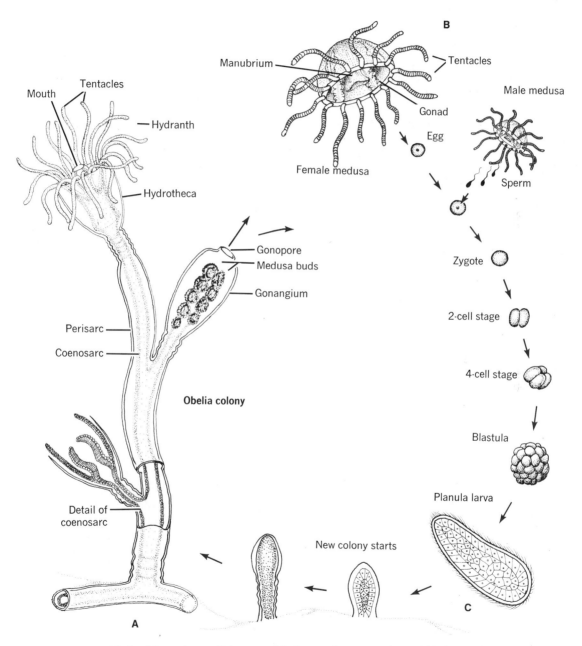

Fig. 19-11. *Obelia* life cycle. **A,** Colony. **B,** Medusae of separate sexes. **C,** Planula larvae swims, settles, and changes to form a new colony. Colony is highly branched; size is about ½ to 1 inch. Other stages are microscopic.

trodermis may migrate past each other through the mesoglea to their original positions. Regeneration occurs to a greater or lesser extent in all animals; it is correlated with the increase in complexity of animal types and influenced by many external and internal factors.

Obelia and Gonionemus

Obelia is a marine, colonial animal that illustrates the principle of alternation of generations, in which the budding, asexual generation is represented by a conspicuous, polyp-type, hydroid stage, and the sexual generation is represented by a small, inconspicuous, jellyfish-type, medusoid stage. *Gonionemus,* another marine hydrozoan, also represents alternation of generations because its asexual generation is represented by a tiny, inconspicuous, hydroid stage (polyp) and its sexual generation by a conspicuous, jellyfish-type, medusoid stage. The two most conspicuous generations of these different animals will be studied, and the two inconspicuous stages will not be considered.

The colonial, hydroid stage of *Obelia* (Fig. 19-11) is attached to a substrate such as rock and possesses many branches that contain numerous polyps of two types: (1) the nutritive hydranths (Gr. *hydor,* water; *anthos,* flower) and (2) the reproductive gonangia (Gr. *gone,* offspring; *angios,* case). The hydranths of *Obelia* somewhat resemble *Hydra,* for they possess a hypostome and mouth, which are surrounded by many tentacles with nematocysts for capturing prey.

The vase-shaped gonangia contain an internal blastostyle on which are borne asexual buds that develop into umbrella-shaped medusae with tentacles around the margin. The medusae give rise to sperms and eggs, and sperms being produced in testes in one individual and the eggs being produced in ovaries in another individual (diecious).

The entire animal has an outer, horny, protective layer, the perisarc (Gr. *peri,* around; *sarx,* flesh). Inside the perisarc is the hollow, tubular coenosarc (Gr. *koinos,* common), consisting of

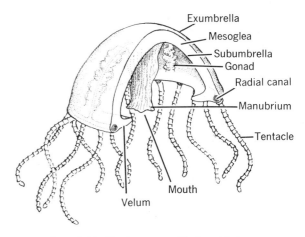

Fig. 19-12. *Gonionemus.* Medusoid stage.

epidermis, mesoglea, and gastrodermis. The internal cavity, called the gastrovascular cavity, is continuous throughout the colony.

The medusoid stage of *Gonionemus* is umbrella-shaped and about ½ inch in diameter (Fig. 19-12). The numerous tentacles, which bear nematocysts, fringe the margin. Each tentacle bends sharply near its tip and bears an adhesive pad for attachment. The convex, or aboral, side is called the exumbrella, whereas the concave, or oral, side is called the subumbrella. A shelflike membrane, the velum (L. *velum,* covering), partly covers the concave side. There is a circular opening in the velum through which water is forced by body contractions propelling the animal in a type of jet propulsion.

Hanging downward from the subumbrella is a tubular manubrium (L. *manubrium,* handle) with a mouth at its tip. The mouth leads into the gastrovascular cavity, and four radial canals lead to a ring canal, near the margin. Attached to the radial canals are testes in some individuals and ovaries in others. Hence, the medusoid stage of *Gonionemus* is diecious.

The egg fertilized by the motile sperm forms a zygote, which by repeated cell divisions produces a multicellular blastula. This becomes a free-swimming ciliated planula, which eventually

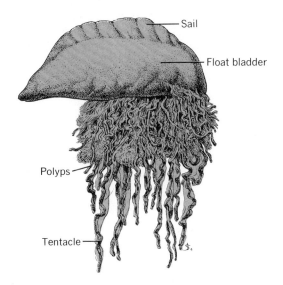

Sail

Float bladder

Polyps

Tentacle

Fig. 19-13. Portuguese man-of-war *(Physalia)*. This colonial coelenterate floats on the surface of the sea. The male and female zooids, vegetative polyps, and long tentacles with nematocysts are suspended from the float bladder.

settles down, attaches itself, and starts a new colony of the hydroid type again.

Physalia (Portuguese man-of-war)

The Portuguese man-of-war *(Physalia)* is a marine colony composed of several different types of polyps suspended in the water from a gas-filled, colered flat (pneumatophore), which may be as much as 1 foot long (Fig. 19-13). The chief polyps include (1) nutritive zooids that capture prey, which is digested in a digestive sac, thus nourishing the colony, (2) long defensive zooids with powerful, poisonous, stinging nematocysts, and (3) reproductive zooids. A sail-like crest on the float assists in carrying the animal along on the surface of tropical seas. The float and crest are beautiful, having different iridescent colors. The defensive zooids may be more than 50 feet long, and the nematocysts may be strong enough to injure or kill a man. Certain kinds of fish are captured for food, but other species of fish are immune to attack and even live

in association with the tentacles, thereby obtaining protection.

Class Anthozoa
Sea anemones

The sea anemone *(Metridium)* is a marine coelenterate with a cylindrical body, soft tough skin, and circles of hollow tentacles arranged around a mouth (Fig. 19-15). A pedal disk attaches it to rocks and wood. Sea anemones may move by gliding along on the pedal disk. Foods consist of smaller animals that they capture with their tentacles and nematocysts. Respiration and excretion occur by diffusion through the cells and body wall. Sea anemones are conspicuous, beautiful, and when expanded, resemble a garden with flowerlike crowns of various colors. In nature they are not so flowerlike since they are death-traps for many smaller animals.

The mouth opens into a gullet (stomodeum), on either side of which is a ciliated groove called the siphonoglyphe (L. *siphon,* tube; *glyphein,* to carve). The gullet leads to the gastrovascular cavity, which is divided into six radially arranged chambers by six pairs of primary mesenteries (septa), which extend from the body wall to the gullet. Water passes from chamber to chamber through ostia in the mesenteries, which are open below the gullet. Numerous smaller mesenteries extend inwardly from the body wall, partially subdividing the larger chambers, but they do not reach the gullet. The free edges of the mesenteries below the gullet are thickened into digestive filaments, which secrete digestive enzymes. Near the base these filaments bear long thread-like acontia (Gr. *akontion,* dart), which are armed with nematocysts and gland cells. The acontia capture prey by being extended through the mouth and from pores in the body wall.

Sex organs (gonads) are present on the margins of the mesenteries, the sexes being separate (diecious). The zygote develops into a ciliated larva that later forms an adult. No medusa stages have ever been discovered in any Anthozoa. Asexual reproduction may occur by budding.

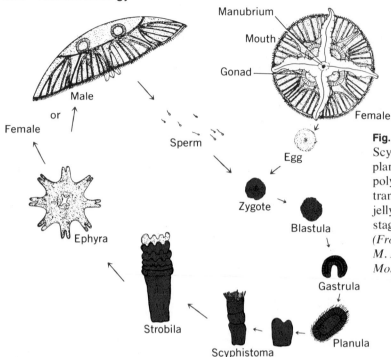

Manubrium

Mouth

Gonad

Female

Male

or

Female

Sperm

Egg

Zygote

Blastula

Gastrula

Planula

Ephyra

Strobila

Scyphistoma

Fig. 19-14. Jellyfish *(Aurelia)* of class Scyphozoa life cycle. The ciliated, larval planula settles down and forms a small polyp, the scyphistoma, which divides by transverse fission to form a series of small jellyfishes. Each free-swimming ephyra stage is produced by the strobila stage. *(From Goodnight, C. J., and Goodnight, M. L.: Zoology, ed. 1, St. Louis, The C. V. Mosby Co.)*

Fig. 19-15. Sea anemone *(Metridium)* showing numerous sensitive tentacles that surround the mouth. *(Courtesy Carolina Biological Supply Company, Burlington, North Carolina.)*

Corals

The coral (Fig. 19-16) is a polyp with short tentacles and lives in a stony cuplike structure. The polyp is constructed somewhat like the sea anemone, but it secretes a calcareous skeleton (coral) around itself and between its mesenteries. Thus the cup is divided radially by ridges. The living polyps can withdraw into these cups. A number of individuals usually live in a colony, although some are solitary. They may display many branches or be rounded masses. Varieties of colors are found in different species. Corals live in the shallower waters of the sea. The polyps are alive only on the surface of coral masses, and they are rarely found below 100 feet. Many characteristics and physiologic processes of coelenterates are found in corals. The great coral reefs of southern seas and smaller, circular atolls that enclose lagoons are of coral origin. The Great Barrier Reef off the Australian coast in more than 1,000 miles long.

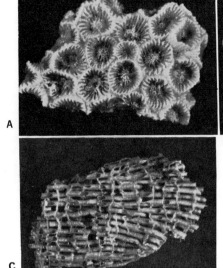

Fig. 19-16. Types of corals. **A,** Colony of northern coral *(Astrangia).* Limestone cups into which delicate polyps with their tentacles retract. **B,** Eyed coral *(Oculina)* from the Bahamas. **C,** Organ-pipe coral *(Tubipora)* from Singapore.

Economic importance of coelenterates

Coelenterates have limited economic importance. Some animals utilize certain types as food. Some flatworms and mollusks may eat hydroids and use the stinging nematocysts for their own defense. Sea anemones may be used for food in certain areas where better foods are unavailable. Corals are important because, through years of growth, they may build calcareous materials that can be used for building purposes, including roadway construction. Certain types of corals are used in making ornaments and jewelry.

Review questions and topics

1 What are the characteristics of sponges and coelenterates?
2 What advances in structures and functions do we find in sponges and coelenterates in each of the following: (1) methods of obtaining and digesting foods, (2) excretion of wastes, (3) coordination of body parts, and (4) sensory equipment and defense mechanisms?
3 What advancement is made in the middle layer in these animals?
4 What is the significance of the different types of symmetry in these animals?
5 Contrast intracellular and extracellular digestion, giving examples of each.
6 What advancement is noted in the gastrovascular cavity?
7 Of what significance is an alternation of generations as shown in certain coelenterates?
8 What is meant by a nerve net? Is such a mechanism, with some modification, found in man?
9 Is the organization of an animal into oral and aboral ends an improvement over lower animals?
10 Discuss the types of skeletal materials found in sponges and certain coelenterates.
11 What is the improvement in the nervous mechanism as we proceed from the sponges to the coelenterates?
12 Discuss the economic importance of sponges and coelenterates.

Selected references

Barnes, R. D.: Invertebrate zoology, ed. 3, Philadelphia, 1974, W. B. Saunders Company.
Burnet, A. L., editor: Biology of Hydra, New York, 1973, Academic Press.
DeLaubenfels, M. W.: A guide to the sponges of Eastern North America, Miami, 1953, University of Miami, Marine Laboratory Special Publication.

Fry, W. G., editor: The biology of the Porifera, New York, 1970, Academic Press, Inc.

Harrison, F. W., and Cowden, R. R., editors: Aspects of sponge biology, New York, 1976, Academic Press, Inc.

Hyman, L. H.: The invertebrates; vol I, Protozoa through Ctenophora, New York, 1940, McGraw-Hill Book Company.

Hyman, L. H.: Coelenterata. In Edmondson, W. T., editor: Fresh-water biology, New York, 1959, ed. 2, John Wiley & Sons, Inc.; 1st ed. by Ward, H. B., and Whipple, G. C., pub. 1918.

Jewell, M.: Porifera, In Edmondson, W. T., editor: Fresh-water biology, New York, 1959, ed. 2, John Wiley & Sons, Inc.; 1st ed. by Ward, H. B., and Whipple, G. C., pub. 1918.

Lenhoff, H. M., and Loomis, W. F., editors: The biology of Hydra and of some other coelenterates, Coral Gables, Florida, 1961, University of Miami Press.

Meglitsch, P. A.: Invertebrate zoology, ed. 2, New York, 1972, Oxford University Press, Inc.

Moore, H. F., and Goltsoff, P. S.: Sponges. In Tressler, D. K., and Lemon, J. M.: Marine products of commerce, New York, 1951, ed. 2, Reinhold Publishing Corp.

Runcorn, S. K.: Corals as paleontological clocks, Sci. Amer. Oct., 1966.

Russell-Hunter, W. D.: A biology of lower invertebrates, New York, 1968, The Macmillan Company.

Windsor, M. P.: Starfish, jellyfish and the order of life, New Haven, 1976, Yale University Press.

20 Helminths—flatworms and roundworms

Phylum Platyhelminthes (flatworms)
    Class Turbellaria (free-living flatworms)
    Class Trematoda (flukes)
    Class Cestoda (tapeworms)
    Economic importance of flatworms
Phylum Nematoda (roundworms)
    Class Nematoda
    Economic importance of roundworms

In early systems of classification all wormlike animals were known collectively as Helminthes (Gr. *worm*). These organisms have both bilateral symmetry (that is, two sides are mirror images) and at least a semblance of a head and tail.

The groups considered in this chapter, the Platyhelminthes and the Nematoda, are members of different and distinct phyla.

The Platyhelminthes have a distinct middle or third germ layer, called a mesoderm, in place of the mesoglea of the coelenterates. In addition to other organs, this tissue produces distinct muscle layers, which permit more varied movement. Because they have a compact mesoderm instead of an open body cavity, these organisms are often called acoelomates, that is, without a coelom, or lined body cavity. This coelom is not to be confused with the enteron, or digestive cavity.

The second group considered in this chapter, the roundworms or nematodes, have a body cavity in which the intestine and the reproductive system are supported more or less loosely. Neither these organs nor the body wall is lined, however, with mesodermal tissue. Therefore, roundworms are referred to as pseudocoelomates.

PHYLUM PLATYHELMINTHES (FLATWORMS)

Class Turbellaria (L. *turbella*, a stirring)
    Usually free living; flattened dorsoventrally; external epidermis is ciliated and contains secretory cells and rodlike rhabdites, mouth usually on the ventral surface; reproduction is usually monecious. Example: *Dugesia tigrina*.

Class Trematoda (Gr. *trematodes*, sucker)
    Parasitic; leaflike or cylindrical in shape; flattened dorsoventrally; thick external cuticle without cilia; ventral suckers, and sometimes hooks; usually monecious. Examples: flukes, such as *Fasciola hepatica* and *Clonorchis sinensis*.

Class Cestoda (Gr. *kestos*, ribbonlike)
    Parasitic; tapelike in shape; flattened; thick external cuticle without cilia; body divided into a series of proglottids; an anterior scolex with suckers or hooks, or both, for attachment; no digestive or sen-

sory systems; usually monecious and self-fertilizing. Example: tapeworms, such as *Taenia saginata*.

Class Turbellaria (free-living flatworms)

The free-living flatworms of this class are commonly called planarians. They and related forms occur in freshwater ponds and streams throughout the world. Planarians avoid light, and sometimes hundreds of them may be found on the underside of a submerged leaf or rock.

Dugesia

Dugesia tigrina, a common type of planarian, will be discussed here to illustrate the characteristic structure and activities of the free-living flatworms. The animal is a small, flat, wormlike form, which is usually about 10 to 20 mm. (about ½ inch) in length. The broadly triangular anterior end has two lateral projections called auricles. A pair of dorsal, light-sensitive eyespots is located between the auricles. The structure of the eyespots is such that they give the animal a "crosseyed" appearance. The body surface of *Dugesia* often has a brownish, mottled appearance. The ventral surface has two openings, the mouth and the genital pore, which are not easily observed.

An epidermis, which is formed by a single layer of cells resting on a basement membrane, covers the body. Most of these cells are alike and have numerous embedded rhabdites whose function is not known. Numerous mucous-secreting gland cells are also present, especially on the lower surface. Many cilia are also located on the ventral surface, which aid in locomotion over a mucous surface formed by gland cells.

The interior of *Dugesia* is filled with an unorganized parenchyma, which is a spongy mass with many undifferentiated spaces. Many individual ameboid cells are also present. There is no distinct body cavity, since the various organs are surrounded by the parenchyma.

Circular, longitudinal, and dorsoventral muscles control the contractions that are necessary for crawling and wriggling movements. In addition to helping the organism perform obvious

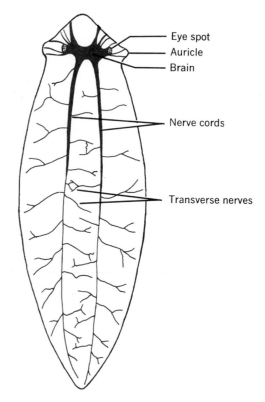

Fig. 20-1. *Dugesia* nervous system. Note the nerves leading from the brain to the auricles.

movement, muscles are important in the protusion of the pharynx to grasp or to tear food. Other muscles aid in the movement of food through the digestive system and throughout the parenchyma.

Dugesia's nervous system (Fig. 20-1) consists of a "brain," which is composed of two thickened cerebral ganglia and two longitudinal nerve cords, which extend toward the posterior end of the organism. Transverse connectives run between the nerve cords, and many peripheral nerves connect the surface and the main nerves.

Sense organs and several kinds of receptors are present. The most obvious of these are the eyes. The eye consists of an inverse pigment cup and light-sensitive neurosensory cells. In *Dugesia* these neurosensory cells are stimulated only by

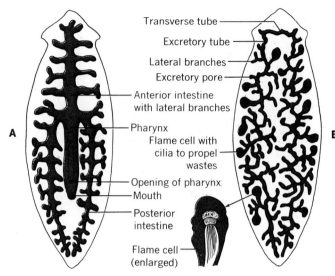

Transverse tube

Excretory tube

Lateral branches

Excretory pore

Anterior intestine
with lateral branches

Pharynx

Flame cell with
cilia to propel
wastes

Opening of pharynx

Mouth

Posterior
intestine

Flame cell
(enlarged)

A

B

Fig. 20-2. *Dugesia*. **A,** Digestive system. **B,** Excretory system.

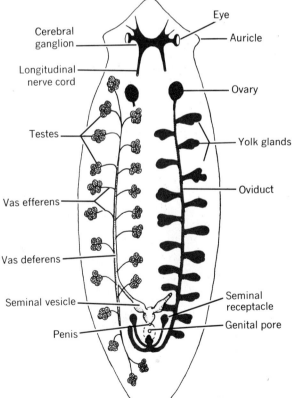

Cerebral
ganglion

Eye

Auricle

Longitudinal
nerve cord

Ovary

Testes

Yolk glands

Vas efferens

Oviduct

Vas deferens

Seminal vesicle

Seminal
receptacle

Penis

Genital pore

parallel light rays. This sensitivity to light rays provides a means of orientation for the organism.

Ciliated pits at the base of the auricles serve as chemoreceptors. *Dugesia,* like all planarians, gives immediate and highly positive responses to such substances as meat juice and blood.

Touch receptors, consisting of single neurosensory cells with bristles, are common on the external surface. Other cells may also serve to orient the animal to food, water currents, and the oxygen concentration of the water.

The excretory system consists of two longitudinal, multibranched canals with numerous terminal, hollow flame cells, which contain a tuft of vibrating cilia to propel wastes. The two main canals open through minute dorsolateral excretory pores (Fig. 20-2). Respiration occurs through the body surface.

Dugesia reproduce asexually by dividing transversely behind the pharyngeal region, each resulting part reorganizing into a complete animal. Sexual reproduction also occurs, since each organism has both testes and ovaries, hence it is monecious (Fig. 20-3). Each of the many spher-

Fig. 20-3. *Dugesia* reproductive system. Organs of each sex are paired on both sides.

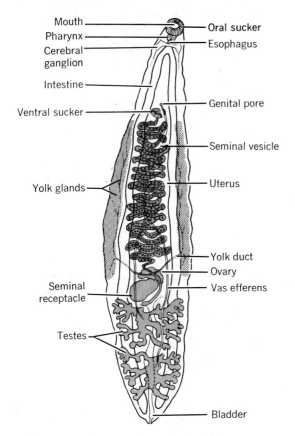

Mouth — Oral sucker
Pharynx — Esophagus
Cerebral ganglion
Intestine
Ventral sucker — Genital pore
— Seminal vesicle
Yolk glands — Uterus
— Yolk duct
— Ovary
Seminal receptacle — Vas efferens
Testes
— Bladder

Fig. 20-4. Anatomy of Chinese liver fluke, *Clonorchis sinensis.*

ical testes has a tubular vas efferens. All the vasa efferentia on one side of the body join to form a larger tubular vas deferens. The right and left vasa deferentia enter a seminal vesicle where sperms are stored until discharged through a muscular penis.

Two ovaries, near the anterior end, are connected to two oviducts that carry eggs to the one vagina. Several yolk glands along the oviducts supply yolk (food) to the eggs. Connected to the vagina is a rounded uterus (seminal receptacle) that stores sperms from the mating partner. Although *Dugesia* is monecious, cross-fertilization is practiced.

During copulation each organism inserts its penis into the other, and there is a mutual exchange of sperms from the seminal vesicles of each animal. The sperms then pass through the oviducts to fertilize the eggs, forming a zygote. As the zygotes move down the oviducts, yolk (food) is added, and finally shells are formed around the eggs. A number of eggs, encased in a single cocoon, are passed into the water. After development, immature planaria emerge from the cocoon.

Dugesia has the remarkable ability to regenerate lost parts. When cut in two, the anterior part will regenerate a new tail, and the posterior part will regenerate a new head. The organism illustrates the theory of the axial gradient. Metabolic activity is measured by the amount of oxygen consumed and the amount of carbon dioxide given off. The metabolic rate is greatest at the anterior end of the animal and gradually and progressively decreases toward the posterior end. The metabolic rate of any part will depend on its position with respect to this axial gradient. In any fragment a head will develop where metabolic activity is greatest and a tail where it is lowest. *Dugesia* may be grafted in many ways to produce animals with two heads and two tails.

Class Trematoda (flukes)

Flukes are the flatworm parasites of a wide variety of vertebrates. Many of them are large enough to be seen without a microscope. The fact that they occur in parts of animals used for food has been known for hundreds of years. In spite of this it was not until 1883 that the complete life cycle of a fluke was described.

As will be shown, flukes produce a fantastic number of offspring during a rather complex life cycle. Since some stages of this cycle occur in man, flukes are of great medical importance in some areas of the world.

Oriental liver fluke

The average size of *Clonorchis (Opisthorchis) sinensis* is about 3 by 15 mm. It is covered with a rough cuticle. It has an anterior (oral) sucker

Table 20-1 Some common flukes

Fluke	Adult stages in primary (definitive) host	Immature stages in intermediate host	Where found
Oriental liver fluke (*Clonorchis [Opisthorchis] sinensis*)	Man, cat, dog, fish-eating mammals	Snails (different species) and freshwater fish	Very common in China, Thailand, Japan, Southeast Asia, Korea, Philippines
Sheep liver fluke (*Fasciola hepatica*)	Sheep, cow, pig	Snail	Europe, United States, Cuba, Australia
Blood fluke (*Schistosoma mansoni*)	Man (blood vessels of mesenteries), also monkey	Snail (*Biomphalaria*)	Africa, Near East, Puerto Rico, South America
Blood fluke (*Schistosoma japonicum*)	Man (blood vessels), also cat, dog, pig, etc.	Snail (*Oncomelania*)	Japan, China, Philippines
Lung fluke (*Paragonimus westermani*)	Man, cat, dog, pig, tiger	Snails, crabs (freshwater), crayfish	China, Japan, Korea, Philippines, America
Skin fluke or "swimmer's itch" fluke (*Trichobilharzia*)	Aquatic birds	Snails (many species)	Widespread in northern lakes and coastal regions

around the mouth and a ventral sucker below the oral one (Fig. 20-4).

The digestive system consists of a muscular pharynx attached to two long, unbranched intestines. There is no anus; the excretory system consists of two tubes whose branches terminate in flame cells to propel wastes. One posterior excretory pore exists. The nervous system, as in planarians, consists of two cerebral ganglia connected to two ventral, longitudinal nerve cords that are transversely connected.

Like the planarians, the fluke is monecious, and cross-fertilization is normal. The reproductive system generally resembles that of *Dugesia*.

LIFE CYCLE

The adult stage of the oriental liver fluke is normally spent in the bile ducts of man. The eggs, each containing a ciliated larva called a *miracidium*, are shed with the feces into water. When the eggs are ingested by certain species of snails, the miracidium is transformed into an elongated *sporocyst*, which produces one generation of *rediae* (after Redi, an Italian scientist). The rediae pass into the snail's liver where they produce tadpole-like *cercariae* that escape into

water where they bore into the muscles of certain fishes. In the fish the cercariae lose their tails and encyst to form *metacercariae* (Gr. *meta,* after). When incompletely cooked fish is eaten, the protective cyst is dissolved, and the metacercariae move up the bile ducts to become adults. They may live in the bile ducts for many years, and serious infections may cause a severe liver disease and death. Additional information about flukes is given in Table 20-1.

Sheep liver fluke

Fasciola hepatica (L. *fasciola,* small bandage; Gr. *hepar,* liver) is parasitic in the liver of sheep, cows, and pigs, where it may cause great damage. The general characteristics and life cycle are shown in Figs. 20-5 to 20-7. There may be as many as 200 adults in a sheep liver, and if all live, they could produce a total of 100 million eggs. The larval stage develops after the eggs are passed into water. The ciliated larvae (miracidia) bore into the body of a certain freshwater snail where a complex series of developmental stages occurs. Finally, they pass from the snail and become encysted on grass. When this grass in eaten by a sheep, the cysts are dissolved to liberate the

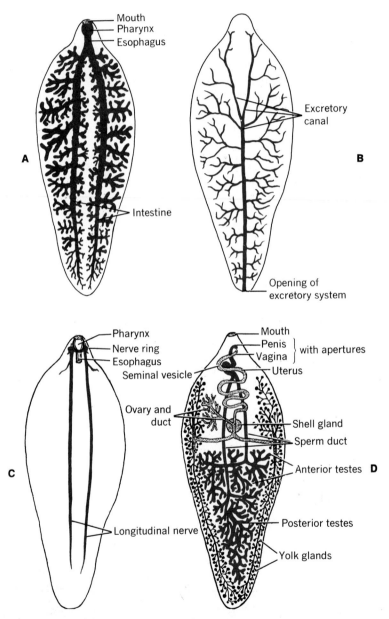

Fig. 20-5. Sheep liver fluke *(Fasciola hepatica)*. **A,** Digestive system; **B,** excretory system; **C,** nervous system; **D,** reproductive system. (All much enlarged and somewhat diagrammatic.)

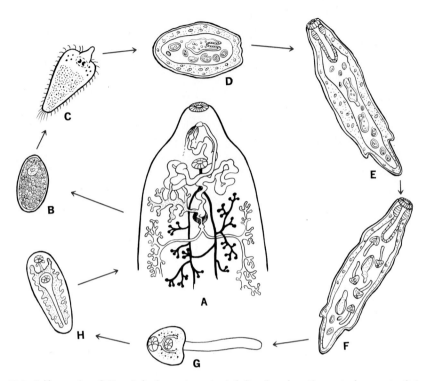

Fig. 20-6. Life cycle of *Fasciola hepatica*. **A,** Adult, showing the anterior part of the reproductive system, oral sucker, and ventral sucker (found in sheep); **B,** egg and yolk cells within a shell (found in feces); **C,** miracidium, showing cilia and awl-like proboscis for boring (found in water); **D,** saclike sporocyst with germ cells that develop into rediae I (in freshwater snail); **E,** redia, showing the gut, birth pore, and germ cells that develop into rediae II (in snail); **F,** redia, showing the gut, birth pore, germ cells, and cercariae (in snail); **G,** free-swimming cercaria with oral and ventral suckers, branched gut, and tail (in water); **H,** metacercaria (encysted cercaria) with suckers, branched gut, and cyst (on grass). Sheep (or man) eats grass, and adult fluke develops.

young fluke, which finally enters the liver to form an adult, thus completing the complex life cycle. Occasional cases of human liver infections are reported.

Blood flukes

Blood flukes are the cause of very serious diseases in many parts of Africa, Asia, Puerto Rico, and other areas of the world. For example, it is reported that in many areas of Egypt, up to 90% of the population may suffer from the effects of these flukes.

An event of significant historical interest occurred in the 1950s concerning blood flukes. The armies of mainland China were about to launch an attack on the Nationalist China forces on Taiwan. The mainland troops were massed, and supplies were gathered. Yet, nothing happened. It was later revealed that thousands of cases of schistosomiasis had occurred among the main-

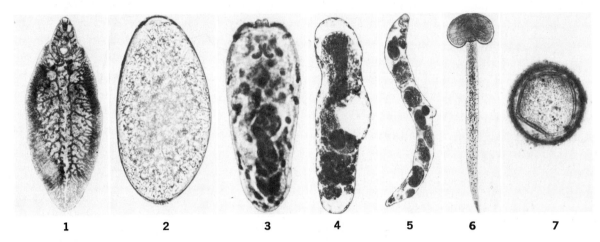

1 2 3 4 5 6 7

Fig. 20-7. Stages in the life cycle of *Fasciola hepatica*. **1,** Adult. **2,** Egg. **3,** Ciliated mira-cidium. **4,** Sporocyst. **5,** Redia. **6,** Cercaria with tail. **7,** Metacercaria in cyst. Adult is about 1½ inches; others are microscopic in size.
(Courtesy Carolina Biological Supply Company, Burlington, North Carolina.)

land soldiers. As we know, a sick man cannot fight.*

Blood flukes are placed in the genus *Schistosoma* (Gr. *schistos,* divided; *soma,* body) because the male has a ventral, splitlike fold in his body in which he carries the smaller female. Unlike most flukes, the sexes are in separate individuals (diecious) in blood flukes and the two branches of the digestive tube are united into one at the posterior part of the body. The life cycles of the various species of blood flukes are similar and generally resemble those of other flukes. However, the cercariae bore directly into the skin.

Class Cestoda (tapeworms)
Beef tapeworm

The adult *Taenia saginata* (Fig. 20-8) lives in the human alimentary canal, where it may become 10 to 30 feet long. Larval stages are found in the muscles of cattle. An enlarged anterior scolex has four suckers. The body surface is covered with a thick, protective cuticle. The animal is composed of a linear series of proglottids that may number over, 1,000. New proglottids are produced in the neck region by a process of transverse budding, so that the most immature proglottids are nearest the scolex, and the broader, mature ones are toward the posterior end.

There is no digestive system, and foods are absorbed from the surroundings. Two lateral excretory canals run throughout all the proglottids and possess flame cells to propel wastes. A transverse excretory canal connects the lateral canals in each proglottid. Two lateral nerve cords run throughout the proglottids and terminate in a nerve ring in the scolex. Each mature proglottid has a complete set of male and female sex organs. Fertilization of eggs may occur by sperms of the same proglottid (self-fertilization), from adjacent proglottids, or from another tapeworm (cross-fertilization). As the posterior, mature proglottids break off, they pass with the feces and by muscular contraction expel eggs and embryos on

---

*Many cases of this disease have been reported in the United States among visitors or immigrants from Puerto Rico, Yemen, and various African and Mid-eastern countries.

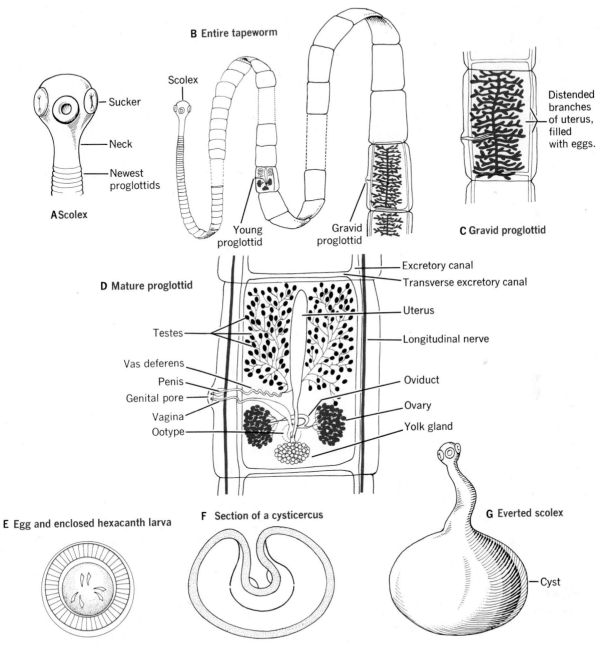

B Entire tapeworm

Scolex

Sucker

Neck

Newest
proglottids

A Scolex

Young
proglottid

Gravid
proglottid

Distended
branches
of uterus,
filled
with eggs.

C Gravid proglottid

D Mature proglottid

Excretory canal
Transverse excretory canal

Uterus

Longitudinal nerve

Testes

Vas deferens
Penis
Genital pore
Vagina
Ootype

Oviduct

Ovary

Yolk gland

E Egg and enclosed hexacanth larva

F Section of a cysticercus

G Everted scolex

Cyst

Fig. 20-8. Life cycle of the beef tapeworm *Taenia saginata*. **A,** Scolex, showing suckers. **B,**
Entire tapeworm. **C,** Detail of a gravid proglottid. **D,** Detail of a mature proglottid. **E,** A
mature egg with hexacanth larva. **F,** Section of a cysticercus. **G,** Everted scolex.

Table 20-2 Numbers of beef carcasses infected with *Cysticercus bovis*, United States, 1967*

State	Cases
Arizona	303
California	10,455
Colorado	334
Texas	777
All other states	2,538
TOTAL United States	14,407

*Source: Liverstock Slaughter Inspection Division, U.S.D.A. (from Morbidity and Mortality Weekly Reports, United States Public Health Service, April, 1968).

grass and soil, where they may be ingested by cattle.

The male reproductive system consists of many spherical testes, all of which are connected to a sperm duct (vas deferens) leading to the genital pore. The female reproductive system consists of a multibranched ovary; an oviduct; a yolk gland; a shell gland; a saclike, egg-containing uterus; and a tubular vagina leading to the genital pore. As the uterus matures with its developing eggs, it may become branched and almost fill the proglottid.

LIFE CYCLE

When cattle swallow a proglottid or an egg, the shell is dissolved. The six-hooked embryo that develops from the egg burrows through the in-testinal wall to enter the blood and lymph vessels. It finally encysts in skeletal muscles and becomes a saclike bladder worm, called a cysticercus. In 10 to 20 weeks each of the cysticerci forms as invaginated (turned in) scolex with suckers.

The larval stages associated with the muscles of cattle are called *Cysticercus bovis*. The greatest incidence of bovine cysticercosis in the United States occurs on the West Coast (Table 20-2).

Man becomes infested with *Taenia saginata* if he eats inadequately cooked beef that is infected with cysticercosis. The infection can be spread to cattle by indiscriminate defecation by humans in feed pens or pastures or by allowing sewage and septic tank effluent to pollute pastures where cattle graze. A related species, *Taenia solium,* is transmitted to humans when they eat uncooked or rare pork. There are a few differences in its morphology; for example, the scolex has numerous hooks. The life cycle is the same except that it occurs in pigs instead of cattle.

The case described on the following pages illustrates the complexity of the problems involved in dealing with *Taenia saginata.*

Economic importance of flatworms

The free-living types of flatworms are of little direct importance to man, but many of the flukes are dangerous parasites, especially in the Far

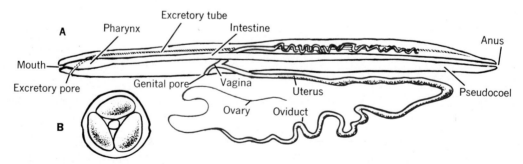

Fig. 20-9. *Ascaris.* **A,** Internal anatomy of a female. **B,** Front view showing three lips around the mouth.

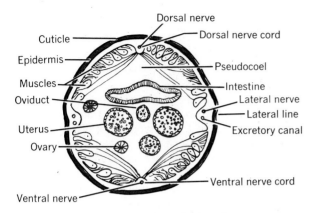

Fig. 20-10. Cross section of a female *Ascaris*.

Labels on figure:
Cuticle
Epidermis
Muscles
Oviduct
Uterus
Ovary
Ventral nerve
Dorsal nerve
Dorsal nerve cord
Pseudocoel
Intestine
Lateral nerve
Lateral line
Excretory canal
Ventral nerve cord

PHYLUM NEMATODA (ROUNDWORMS)

In contrast to flatworms, roundworms have a pseudocoel, which is a body cavity that is not lined by a special layer of epithelial cells, the mesenteries, as it is in higher animals. The interior organs lie free in this cavity. Because their body structure consists of an intestine in a cavity, which is surrounded by a body wall, roundworms are often referred to as having a "tube within a tube" structure.

A great many species of roundworms are known. While many of them are free-living, many others are predatory or parasitic. The latter are of much interest because they cause damage to crops and diseases in man and other animals.

Roundworms are often placed in the phylum Aschelminthes, along with many other groups of worms. Since these other worms are not considered in this text book, we are using the phylum Nematoda, as well as the class Nematoda, for classification purposes.

Class Nematoda

Nematoda are slender elongate animals with a complex cuticle lacking cilia. They have a complete digestive system. The sexes usually separate, and the males are generally smaller than the females. Examples of nematodes are the human roundworm *(Ascaris lumbricoides)*, the hookworm *(Necator americanus)*, and the vinegar eelworm *(Turbatrix aceti)*.

East, and the tapeworms cause damage in all parts of the world. The various types of flukes may affect such organs and tissues as liver, intestine, blood, lung, and skin. Flukes spend one part of their life cycle in one animal host and the other part of the cycle in another animal host, thereby affecting two different animals. Tapeworms also alternate between two hosts in their life cycle. The particular tissues or organs involved depend on the species of tapeworm. In the dog tapeworm *(Echinococcus granulosus)* a larval stage, known as a hydatid cyst (Gr. *hydatis,* watery sac), may occur in the liver and other organs of man, cattle, horses, and sheep, where the fluid-filled sac may assume large proportions.

Table 20-3 Some common tapeworms

Tapeworm	Site of adult stages (primary host)	Site of immature larval stages (secondary host)
Pork tapeworm *(Taenia solium)*	Man (intestine)	Pig (muscle)
Beef tapeworm *(Taenia saginata)*	Man (intestine)	Cattle (muscle)
Broad tape worm *(Diphyllobothrium latum)*	Man, dog, cat, fox (intestine)	Freshwater fish, certain crustaceans (cyclops)
Dog tapeworm *(Dipylidium caninum)*	Dog, cat, sometimes children	Dog lice, fleas
Hydatid tapeworm, or dog tapeworm *(Echinococcus granulosus)*	Dog, wolf (intestine)	Many mammals, including man, monkey, cattle, sheep, cat (in liver, lungs, brain, etc.)
Sheep tapeworm *(Moniezia expansa)*	Sheep, goats	A small mite *(Galumna)*

Cases of tapeworm disease

1. Bovine cysticercosis—Texas*

An epizootic of bovine cysticercosis has been reported from northern Texas. During the period from March 15 to April 1, 1968, 771 cattle from two large commercial feedlots were slaughtered in packing plants under the United States Department of Agriculture Inspection Program; 346 cattle (45 percent) were found to be infected with *Cysticercus bovis*. Infected cattle are known to have been shipped [to] plants in Oklahoma, Nebraska, Colorado, Missouri, Kansas, Iowa, Texas, and Florida.

An investigation in now in progress to determine the mode of spread of this zoonosis and the extent of the movement of infected cattle into retail channels.

2. Taeniasis—Rhode Island*

On May 9, 1968, a 40-year-old female X-ray technician from Rhode Island recognized tapeworm proglottids in her stool. For approximately 2 months the patient had experienced mild abdominal cramps, borborygmi [rumbling], and a change in her bowel habits from relative constipation to bowel movements on arising each morning. Because she believed that she was infected with pinworms, she had mistakenly been looking at her stools each day until May 9 when she first sighted the tapeworm segments.

The patient gave no history of recent travel. She ate rare beef, often sampling raw hamburger during its preparation, but she rarely ate pork. She purchased all her meat in a single Rhode Island supermarket.

The patient's stool was examined and found to contain *Taenia* eggs. She was treated with oral Atabrine, but this was not successful. She was then treated with Niclosamide and she stated that within a day her bowel habits returned to the previous normal pattern. Her stools will be examined

*Three separate articles reprinted from Morbidity and Mortality Weekly Reports, United States Public Health Service. March, May, June, 1968.

periodically to see if the entire worm was removed.

The commerical sources of beef for the single supermarket from which the [patient] purchased her meat were traced. It was found that two of the sources were slaughter houses in Nebraska and Iowa that had processed Texas cattle infected with *Cysticercus bovis* during the epizootic that first appeared in mid-March (MMWR, Vol. 17, No. 16). An investigation is now underway to determine if there are more cases of taeniasis in this region.

At the same time that this autochthonous [domestic] case was found, two imported cases of taeniasis, one in an Ethiopian and the other in a Lebanese, were also reported in Rhode Island.

3. Follow-up bovine cysticercosis—Texas

Intensive investigation of the epizootic of bovine cysticercosis which occurred in March 1968 in cattle from feedlots in northern Texas (MMWR, Vol. 17, Nos. 16 and 23) has recently been completed. The following information summarizes the epidemiologic findings at the feedlots near Gruver and Hereford, Texas, where the infected cattle originated.

INVESTIGATION NEAR GRUVER TEXAS, AT FEEDLOT A

Feedlot A is a commercial feedlot with a capacity for 8,000 cattle. The cattle are shipped to the feedlot from many sources for intensive feeding. On arrival, cattle are put in one of 34 pens at the feedlot. While at the feedlot, one pen of animals is never mixed with another pen. Cattle are fed on consignment and careful records are maintained of the amount and type of feed given to cattle in each pen.

From January 1 to March 15, of 1,398 cattle from Feedlot A that were sent to slaughter, only one was found infected with *Cysticercus bovis*. However, from March 15 to June 12, of 5,870 cattle originating in Feedlot A, 743 were infected (overall infection rate 12.7 percent). Investigation into the sources for all cattle slaughtered since

---

Cases of tapeworm disease—cont'd

January 1, 1968, revealed that the cattle were assembled from a variety of sources by several commercial buyers.

In the infected animals, cysts were found in the masseter muscles, esophagus, liver, heart, and diaphragm, and the majority of cysts were degenerated and caseous, indicating that the infections were probably several months old.

The personnel at the feedlot were investigated for *Taenia saginata*. In addition to the manager, seven employees worked at Feedlot A at the time of the inquiry. Two former employees, who worked at the feedlot between the time the cattle entered the feedlot and were slaughtered, were also questioned. None of the present employees gave a history suggestive of taeniasis *(T. saginata)*, but a former employee stated that 1½ years ago he had passed motile, flat worms, approximately 2 cm. long, in his stool. He had not been treated for his condition. The likelihood that he was the source of infection is also increased by temporal considerations. Since the cattle were not infected in pasture and since the majority of cysts were degenerated and caseous, the most likely time of infection would have been mid-October when the cattle first entered the feedlot. The former employee had started working in the feedlot at that time.

Human roundworm

Ascaris lumbricoides is one of the largest parasites of man. Females may reach a length of 12 to 14 inches with a diameter of ¼ inches. Males are much smaller and can be recognized by their strongly curved posterior. Females taper smoothly at both ends (Fig. 20-9).

The body wall is covered with a smooth cuticle, which is secreted by epidermal cells. The anterior mouth opening is modified by one dorsal and two ventral lips.

The body wall itself is relatively simple in structure. From the outside inward it consists of the cuticle, an epidermis that is largely syncytial (that is, nuclei are not separated by definite cell boundaries), and a muscle layer of large longitudinal cells.

The epidermis projects inward in four definite ridges on the top, bottom, and both sides (Fig. 20-10). The lateral ridges surround longitudinal excretory canals that run the length of the body. The dorsal and ventral ridges contain longitudinal nerve cords.

The muscles contain contractile fibers and are innervated by extensions that pass directly to the nerve cords. The muscles are separated into four groups by the projection of the epidermal ridges. Movement occurs by the alternate contraction of four muscle groups.

The sensory system is apparently poorly defined. There is a nerve ring around the pharynx, and from it nerves pass anteriorly to the lips and posteriorly to other areas. In addition to the dorsal and ventral cord there are longitudinal nerves along the length of the lateral ridges.

A pseudocoel is present, which is lined by mesoderm only on the surface toward the body wall. No mesoderm covers the digestive tract. The digestive system consists of a mouth; a short, muscular, sucking pharynx; a long, nonmuscular intestine for absorption; a short rectum; and an anus at the posterior end. There are no respiratory or circulatory systems. An excretory tube without flame cells is present in each of the two lateral lines. The two tubes unite and empty wastes through a ventral excretory pore just posterior to the mouth.

In the male one coiled, threadlike testis produces sperms that are conducted by a vas deferens to an enlarged seminal vesicle where they are stored. A muscular ejaculatory duct empties into the cloaca. Two bristlelike, chitinous spicules attach the male during copulation.

In the female two coiled, tubular ovaries produce eggs. Two coiled oviducts carry the eggs to two larger, tubular uteri that unite and empty into a short vagina that opens through the genital pore about one third of the way behind the anterior end. Fertilization occurs in the uteri, and the egg is enclosed by a thick, rough shell and passed out through the genital pore.

LIFE CYCLE

A female may produce eggs at the rate of 200,000 daily, and one worm may contain over 25 million eggs at one time. The eggs pass with the feces, are deposited in soil, and develop into immature embryo worms inside their shells. Eggs usually develop embryos in 3 weeks, and when taken into the intestine in this state, they will produce small larvae (0.2 mm.) that burrow through the intestinal wall into the veins and lymph vessels. From here they may pass through the heart to the lungs, break into the air passages, travel into the trachea (windpipe) and down the esophagus, and finally go to the intestine where they mature in 2 to 3 months. Great damage may be done by the migrating embryos, especially in the lungs. When very numerous, they may migrate to the bile ducts, appendix, sinuses, and nose.

In spite of the efforts of State and Federal Public Health Services and the availability of safe drugs, sizable numbers of cases of ascariasis still exist. A number of studies in several southeastern states in the last ten years showed infection rates up from 20% to 60% of 5- to 9-year-old school children with *Ascaris.*

Hookworm (Necator)

The American hookworm, *Necator americanus* (L. *necare,* to kill), was once common in the southeastern United States.* Adult worms

---

*Occasional cases are still found but the disease is no longer a major problem except in isolated rural areas.

are about 12 mm. long, males being slightly shorter than females. They are called hookworms because of a bend in the anterior of the male. Cutting, platelike teeth in the mouth cut holes in the human intestine. They suck blood and other fluids by means of a sucking pharynx and digest them in the intestine. An anticoagulant in the mouth prevents blood clotting in the host.

The methods of reproduction and the life cycles are somewhat similar to those of *Ascaris.* Larvae (0.5 mm.) may live for weeks or months. When infested soil contacts the skin, the larvae burrow into the blood, thus causing an irritation known as "ground itch." Although the larvae usually enter human beings through the skin of the feet, they may also be taken orally. Their journey within the human body is similar to that described for *Ascaris,* finally reaching the intestine. They may remain in a host for years, where their presence may cause anemia, loss of energy, and retarded physical and mental growth. Hookworm disease is still common in many parts of the world, especially in rural areas. For example, recent surveys in Iran showed an incidence of 100% in some villages. Preventing the entrance of larvae through the skin and proper disposal of feces may reduce possible infestations. Proper medication will eliminate them from the human body.

Pork roundworm (trichina)

The human parasite *Trichinella spiralis* is about 2 to 4 mm. long and may cause a serious disease called trichinosis. Man becomes infested by eating improperly cooked pork. The adults live in the small intestine, where the female burrows into the intestinal wall and produces larvae (0.1 mm. long), which burrow into veins or lymph vessels and are carried to the skeletal muscles, especially those of the neck, eyes, tongue, and diaphragm. In the muscles they coil and are surrounded with a limy cyst. In the encysted stage they may live for years. When cysts are ingested with infested meat, the larvae are liberated in the intestine and reach maturity in a few days.

In addition to man, mammals such as pigs,

Trichinosis from bear meat—Alaska*

In Alaska, in July and August 1968, three persons developed clinical trichinosis and five others had serologic evidence of infection after ingestion of bear meat. Symptoms for the clinically ill persons included fever, myalgia, vomiting, diarrhea, periorbital edema, and muscle ache. In addition, one patient had a generalized maculopapular rash. All three recovered after therapy with thiabendazole.

The incriminated meat was from two black bears killed on June 6. The meat from Bear 1 was frozen in two home freezers that could maintain a minimum temperature of 0° F. Then it was served at meals on June 28, July 7, and July 11 (see figure). The meat served on June 28 was "well-cooked" on a charcoal grill while that served on July 11 was cubed and sauteed in butter and was described as "pink in the center." All five guests of G. K., Sr., who ate the meat on June 28 had clinical and/or serologic evidence of *Trichinella spiralis* infection. Three of these persons ate meat from Bear 1 again on July 11. One other person at the July 11 meal who had not eaten the June 28 meal also became ill. No persons at the July 7 meal became ill.

*From Morbidity and Mortality Weekly Reports, United States Public Health Service, Sept. 1968.

The meat from Bear 2 was cut into stew meat and chops, aged for 5-6 days, flash-frozen in a commercial locker at −25° F., and then stored at 0° F. It was served at five meals. Although no one at these meals developed clinical illness, two had serologic evidence of *T. spiralis* infection. The charcoal card flocculation test for trichinosis was used to identify the symptomatic and asymptomatic infections.

Microscopic examination of meat samples from both bears demonstrated heavy *T. spiralis* infections (see table). Pepsin hydrochloric acid digestion in the laboratory of the meat 81 days after the death of each animal showed definite sluggish movement of larvae in the meat from Bear 1 and no larval movement in the meat from Bear 2. The larvae in Bear 2 showed no signs of calcification.

All 30 persons who had consumed the bear meat were contacted, symptomatic persons were treated, and the other persons maintained under surveillance for illness for 30 days. All the remaining meat was destroyed under the supervision of the USDA in Anchorage.

Table 1 Larvae per gram of bear meat

Animal	Origin of sample	Larvae per gram
Bear 1	Hind quarter	110
Bear 2	Chops	80
Bear 2	Stew meat	190

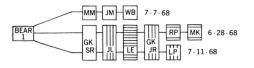

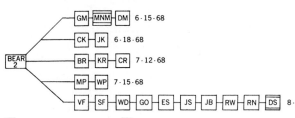

☐ PERSONS EATING BEAR MEAT   ▥ CLINICAL AND SEROLOGIC EVIDENCE OF INFECTION
▤ SEROLOGIC EVIDENCE ONLY OF INFECTION

Incidence of trichinosis in persons eating infected bear meat by date of meal.

bears, dogs, cats, and rats, may become infested by eating meat containing trichina cysts. Many of the people of the United States have trichina infestations, but most cases are mild and are not reported. Some human symptoms inculde intestinal disturbances, fever, muscular pains, and edema. Precautionary measures must include thorough cooking of pork. See p. 285 on cases of trichinosis from bear meat.

Vinegar eelworm

The free-living worm *Turbatrix aceti* (L. *turbatio,* disturbance; *acetum,* vinegar) lives in cider vinegar. The sexes are separate (diecious), the female being about 2 mm. long and the male about 1.5 mm. long. This difference illustrates sexual dimorphism. There is a false body cavity (pseudocoel) and a cuticle in the eelworm. In general the various systems are rather typical of nematodes.

The eggs are fertilized within the female worm where they develop into active larvae. Sometimes eggs develop into embryos within the mother's body but are not nourished by her bloodstream. Such a condition is called ovoviviparous (o vo vi -vip′ a rus) (L. *ovum,* egg; *vivus,* alive; *parere,* to bear).

Economic importance of roundworms

In general, the free-living roundworms do not seem to be of great direct importance to man. The parasitic nematodes are very numerous and affect large numbers of animals and plants. Some species of these nematodes live in plants where they may be of great economic importance because they cause crop loss. It is thought that there are thousands of unclassified nematodes, and when some of them are studied, their economic importance may be ascertained. Filarial worms *(Wuchereria bancrofti)* live in the lymphatic system of man, where they obstruct the flow of lymph and may cause a severe condition known as elephantiasis. Enormous swelling and growth of connective tissue, especially in the arms and legs, occur, thus giving the appearance of the limb of an elephant.

Adult pinworms may live in the cecum, appendix, and large intestine of children. Females with eggs may migrate to the anal regions at night to deposit eggs, thus causing an irritation. Scratching may transfer them to the mouth if sanitary precautions are not taken. No intermediate host is necessary, and eggs taken into the intestine hatch into adults. Thus a reinfestation occurs. Each generation lasts 3 to 4 weeks and will die out unless reinfestation occurs. In 1969, pinworm eggs were found in fossilized human feces estimated to be 10,000 years old.

Damage produced by ascaris worms, hookworms, and trichina worms has been considered earlier. Guinea worms *(Dracunculus)* are found in Africa, India, South America, and places in the Orient. The females are long, some 1/25 inch in diameter and up to 4 feet long. They are present in dogs and minks in the United States. The female lies beneath the skin where it may appear like an enlarged vein. The anterior end of the worm protrudes through a sore, and immature stages may be discharged. If the larvae are taken into the body of a certain crustacean *(Cyclops),* they will develop there. If water containing *Cyclops* is taken by man, the worms pass to the intestine and then migrate to areas beneath the skin to become mature in about a year. The World Health Organization (1975) estimated that there are over 3 billion human roundworm infections.

Some contributors to the knowledge of
FLATWORMS AND ROUNDWORMS

Manson

Avicenna (981-1037 A.D.)

A Persian physician who described four kinds of worms, apparently including several types of tapeworms and roundworms. *(Historical Pictures Services, Chicago.)*

Jehan de Brie

He discovered (in 1379) the first fluke, *Fasciola hepatica,* the cause of sheep liver rot. It was described more accurately by **Gabucinus** in 1547.

Avicenna

Francesco Redi (1626-1697)

An Italian scientist who recognized the larval, cysticercus stage of the tapeworm *Taenia* as an animal. Much later, **E. Küchenmeister** proved by feeding experiments that these bladder worms represented the immature stages in the life cycle of the tapeworm, and that as a rule, they required a different host from that of the adult worm (1851). Küchenmeister also worked out the life cycle of the pork tapeworm, *Taenia solium* (1885).

J. F. McConnell

He discovered the Oriental liver fluke, *Clonorchis,* in 1874.

Carolus Linnaeus (1707-1778)

A Swedish biologist who recognized the genus *Ascaris* in 1758.

Carl T. E. von Siebold (1808-1884)

A German parasitologist who extended the idea of worms alternating between hosts. He proved that intestinal parasites did not arise spontaneously.

R. Leuckart and A. P. Thomas

They discovered (in 1882 and 1883, respectively) the complete life cycle of the sheep liver fluke, involving a snail as a required host.

Patrick Manson (1844-1922)

An English pathologist who proved that the blood-sucking mosquito served as the larval host for Brancroft's filaria (1878) and that the periodicity of these microfilaria in the human peripheral blood seemed to be related to the life cycle. *(Historical Pictures Services, Chicago.)*

C. J. Davaine

He first discovered that ascaris larvae hatched from eggs in the small intestine of experimental rats (1863).

Eduard van Beneden (1846-1910)

A Belgian cytologist who studied for the formation and maturation of sperms and eggs of the roundworm ascaris (1883).

A. Looss

He first demonstrated the life cycle of hookworms, including the infectiveness of larval stages that usually entered through the skin (1896-1911).

Theodor Boveri (1862-1915)

A German zoologist who did pioneer work on ascaris, especially with the chromosomes in the cell nucleus.

N. A. Cobb (1859-1933)

An American who made valuable contributions to nematode biology.

B. H. Ransom and E. B. Cram

They traced the course of migration of *Ascaris* larvae (1921).

Review questions and topics

1 What are the general characteristics that describe the flatworms as a group?
2 What are the characterisitcs that distinguish planarians, flukes, and tapeworms from each other?
3 Discuss the status and significance of the head (cephalization) and sensory organs in certain flatworms, including specific examples.
4 Discuss the importance of polarity and axiate organization in planarians.
5 What is the significance of bilateral symmetry?
6 Name the advantages and disadvantages of monecious reproduction?
7 What are the effects of prolonged parasitism on the host and upon certain structures and systems of the parasite itself? Give specific examples.
8 Of what importance are meat inspections and the proper cooking of meats?
9 Discuss the two types of hosts that certain flatworms commonly use in their complete life cycles, including specific examples.
10 What are the general characteristics that describe the roundworms as a group?
11 Discuss the significance of a complete digestive system that extends from a mouth to an anus, including the improvement over a gastrovascular cavity.
12 Discuss the status and significance of cephalization and sensory organs in certain roundworms, including specific examples.
13 Discuss parasitism as found in certain roundworms, including the number of hosts usually involved in the life cycle.
14 Discuss the role of proper inspections and cooking of meats to curtail the spread of roundworm parasites.
15 Explain how serious infestations by hookworms may affect human beings.

Selected references

Baer, J. G.: Animal parasites, New York, 1971, McGraw-Hill Book Company.
Cheng, T. C.: General parasitology, New York, 1973, Academic Press, Inc.
Faust, E. C., Beaver, P. C., and Jung, R. C.: Animal agents and vectors of human disease, Philadelphia, 1974, Lea & Febiger.
Faust, E. C., Russell, P. F., and Jung, R. C.: Clinical parasitology, ed. 8, Philadelphia, 1970, Lea & Febiger.
Hyman, L. H.: The invertebrates; vol. II, Platyhelminthes and Rhynchocoela, New York, 1951, McGraw-Hill Book Company.
Hyman, L. H., and Jones, E. R.: Turbellaria. In Edmondson, W. T.: Fresh-water biology, New York, 1959, John Willey & Sons, Inc.; 1st ed. by Ward, H. B., and Whipple, G. C., pub. 1918.
Meglitsch, P. A.: Invertebrate zoology, ed. 2, New York, 1972, Oxford University Press, Inc.
Noble, E. R., and Noble, G. A.: Parasitology; the biology of animal parasites, Philadelphia, 1971, Lea & Febiger.
Rogers, W. P.: The nature of parasitism, New York, 1962, Academic Press, Inc.
Sasser, J. N., and Jenkins, W. R., editors: Nematology, Chapel Hill, North Carolina, 1960, University of North Carolina Press.
Schmidt, G. D., and Roberts, L. S.: Foundation of parasitology, St. Louis, 1977, The C. V. Mosby Co.
Thorne, G.: Principles of nematology, New York, 1961, McGraw-Hill Book Company.
Wardle, R. A., and McLeod, J. A.: The zoology of tapeworms, Minneapolis, 1952, University of Minnesota Press.
Yamaguti, S.: Systema helminthum; vol. 3, The nematodes of vertebrates, New York, 1961, Interscience Publishers, Inc.

21 Phylum Annelida—segmented worms

Class Oligochaeta
   Earthworm
Class Hirudinea
   Leeches
Economic importance of annelids
Classification of segmented worms

    Segmented worms are often called "real" worms because, unlike all other wormlike organisms, their bodies are composed of distinct ringlike segments. Annelids and all animals considered in the following chapters have a distinct body cavity, the coelom, which is lined by epithelium.

CLASS OLIGOCHAETA

Earthworm

    Lumbricus terrestris is the common earthworm. Its body is elongated, soft, and segmented. It burrows in the soil, moving debris through its alimentary tract and passing this debris on the surface as "castings." A conspicuous, saddle-like clitellum is present on segments XXXI to XXXVII. A true body cavity, or coelom, is present. Membranous, internal septa separate adjacent segments (Fig. 21-1).

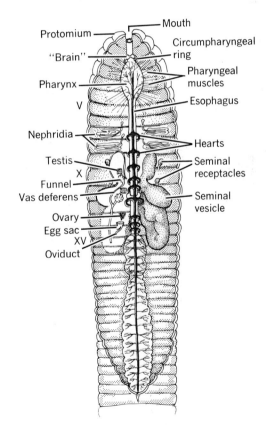

Fig. 21-1. Earthworm, internal anterior segments. Left seminal receptacle is cut away to show reproductive structures. Only a few nephridia are shown.

289

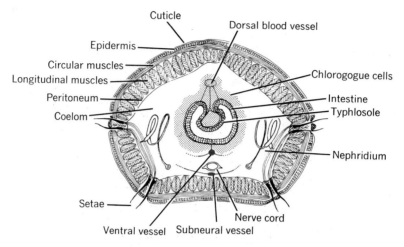

Fig. 21-2. Earthworm anatomy. Cross section.

INTEGUMENT

A thin, noncellular, transparent cuticle contains pores of glands located in the epidermis just beneath. Four pairs of chitinous, bristle-like setae are located in each segment (Fig. 21-2).

LOCOMOTION

An outer layer of circular muscles and an inner layer of longitudinal muscles in the body wall aid in locomotion. Setae are moved by protractor and retractor muscles at their bases. The setae are angled to provide friction or to reduce it as desired.

INGESTION AND DIGESTION

The foods of earthworms consist of dead plant and animal materials and soils rich in organic substances. Living materials are rarely taken. A fleshy prostomium (Gr. *pro,* before; *stoma,* mouth) projects over the mouth at the anterior end. The mouth leads into a buccal pouch that connects with a thick, muscular, sucking pharynx. Muscles attached to the outside of the pharynx contract and expand it to produce suction. A narrow esophagus connects the pharynx with the large, thin-walled crop, which is used for storage. Posterior to the crop is the thick, muscular gizzard used for grinding food with grains of sand and similar materials. The gizzard leads to the long intestine, which has a deep, dorsal fold, the typhlosole (Gr. *typhlos,* blind; *solen,* channel). The typhlosole increases the absorbing surface of the intestine. It is filled with chlorogogen cells, which probably aid in the digestion of foods and the elimination of wastes. The anus is at the posterior end the earthworm.

Three pairs of calciferous glands (L. *calx,* lime; *ferre,* to bear) secrete calcium carbonate into the nearby esophagus to neutralize acid foods. The calcium carbonate also lines the burrows, thus preventing their collapse. These plastered tunnels, through which the worm crawls, allow moisture and air to penetrate the soil. In this way the earthworms cultivate the soil.

Enzymes from the digestive juices act on foods in a manner similar to that of higher animals. Absorption in the earthworm occurs through the walls of the intestine, assisted by the ameboid action of some of the lining epithelial cells. Some absorbed foods are placed in the coelomic cavity where they are circulated by the coelomic fluid. Other absorbed foods are placed in the circulatory system to be taken to various parts of the body.

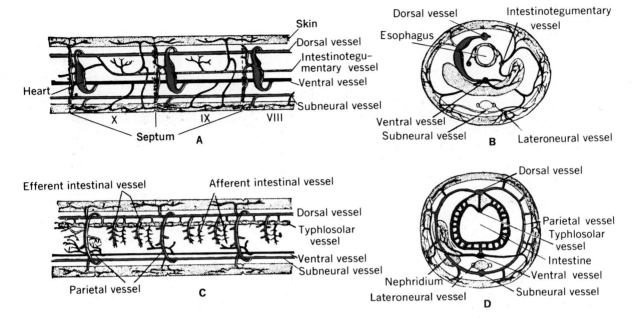

Fig. 21-3. Earthworm circulatory system. **A,** Longitudinal view in segments VIII, IX, and X; **B,** cross section of the same region; **C,** longitudinal view in the region of the intestine; **D,** cross section of the same region.
(From Hegner, R. W.: College zoology, New York, The Macmillian Company.)

CIRCULATION

The blood vessels form a so-called closed type of circulatory system (Fig. 21-3), in which the blood flows, more or less continuously, in the vessels and only a limited amount of the blood's constituents pass in and out through the vessel walls. The more important parts of the system are (1) a dorsal blood vessel above the digestive tract, (2) a ventral vessel below the digestive tract, (3) five pairs of pulsating, looplike heart arches in segments VII to XI, which connect the dorsal and ventral vessels, (4) a subneural vessel beneath the ventral nerve cord, (5) two lateral neural vessels on either side of the nerve cord, and (6) numerous branches from the vessels with their thin-walled capillaries to supply blood to all body parts. The blood is propelled through the vessels by the peristaltic contractions of the hearts and dorsal blood vessel, thus forcing it from the posterior part of the dorsal vessel toward the anterior end. Valves in the hearts and dorsal vessel prevent backward flow. Blood is returned from the body wall to the lateral neural vessels, in which it flows posteriorly, and eventually reenters the dorsal blood vessel.

The blood of the earthworm consists of (1) liquid plasma with hemoglobin, and oxygen-carrying red pigment, dissolved in it and (2) numerous colorless white blood cells, which resemble those of human blood. The blood carries absorbed foods from the digestive tract to all parts of the body, transports wastes rapidly from the tissue to the organs of elimination, and exchanges oxygen and carbon dioxide by flowing near the body surface.

RESPIRATION

There is no respiratory system, but oxygen is obtained and carbon dioxide eliminated through

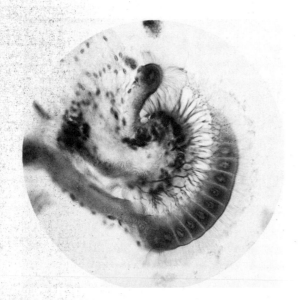

Fig. 21-4. Ciliated, funnel-shaped nephrostome of the earthworm from a cross section just posterior to the clitellum.
(Courtesy General Biological Supply House, Inc., Chicago, Illinois.)

the moist body surface. Many thin-walled capillaries just beneath the cuticle make the exchange of gases possible. Excess water around the animal interferes with respiration, which partly explains why earthworms are "rained out" after a rain.

EXCRETION AND EGESTION

A pair of nephridia (Gr. *nephros,* kidney) is present in each segment, except in the first three and the last one. The internal, free end of each is a ciliated, funnel-shaped nephrostome that filters the wastes from the coelomic fluid (Fig. 21-4). The nephridia collect the wastes and pass them through ciliated tubes to the exterior through openings called nephridiopores. These are located on the ventral side of the body just posterior to the segment in which their particular nephridia are located. Chlorogogen cells, covering the intestinal wall and filling the typhlosole, may act to eliminate wastes. Solids are eliminated through the anus.

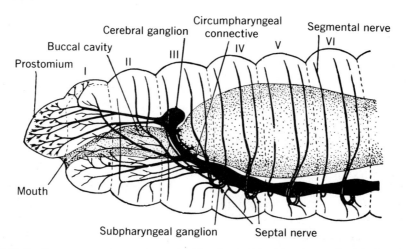

Fig. 21-5. Earthworm nervous system. A side view of the anterior end with the cerebral suprapharyngeal ganglion and larger nerves.
(After Hess, from Hegner, R. W.: College zoology, New York, The Macmillan Company.)

COORDINATION AND SENSORY EQUIPMENT

A bilobed brain (suprapharyngeal ganglion) is located above the pharynx near segment III (Fig. 21-5). The circumpharyngeal ring, or commissure, encircles the pharynx and connects the brain with the subpharyngeal ganglion below the pharynx. The ventral nerve cord extends posteriorly from the subpharyngeal ganglion and has an enlarged ganglion that gives origin to three pairs of nerves in each segment. These ganglia serve as subordinate "brains" where nerve impulses are received and redirected. Nerves connect the body segments to coordinate their various activities. The muscles of the setae are controlled to make them perform their functions properly.

There are epidermal sense organs in the peripheral tissues that, when stimulated, send impulses over nerves. Sensory hairs penetrate the cuticle and are connected with the nervous system. Earthworms react to light, contact, moisture, chemicals, and sound.

REPRODUCTION

Both male and female sex organs are present in the same earthworm (monecious) (Fig. 21-1). The female organs include one pair of small ovaries (segment XIII), which are not visible from the dorsal side; one pair of small oviducts, which are modified nephridia (segment XIII); one pair of egg sacs connected with the oviducts (segment XIV); one pair of oviduct openings on the ventral side (segment XIV), two pairs of seminal receptacles (spermatheca) in segments IX and X; and two pairs of seminal receptacle openings between segments IX and X and X and XI. The male organs include two pairs of testes (segments X and XI), which are covered by the seminal vesicles and not visible from the dorsal surface, one pair of vasa deferentia (sperm ducts) with ciliated funnels (segments X to XV); one pair of vasa deferentia openings on the ventral surface (segment XV); and three pairs of large, conspicuous seminal vesicles (segments IX to XII). The bases of these vesicles are attached in segments

Fig. 21-6. Copulation in the earthworm. The anterior ends point in opposite direction, and the ventral surfaces are held together by mucous bands secreted by the clitellum of segments XXXI to XXXVII. Male sperms pass out of the vasa deferentia of each worm, through the seminal channels on the ventral surface, to the openings of the seminal receptacles of opposite worms. After a mutual exchange of sperms the worms separate.
(Courtesy General Biological Supply House, Inc., Chicago, Illinois.)

IX to XII, although they may extend beyond them.

During copulation the ventral surfaces of two earthworms are in contact and the anterior ends point in opposite directions. A slimy band secreted by the clitellum encircles the two worms. A pair of temporary seminal channels is formed on the ventral surface of each worm, so that sperms expelled from the vasa deferentia of one worm travel to the openings of the seminal receptacles of the other, in which the sperms are stored. Copulation (Fig. 21-6) results in a mutual exchange of sperms. No discharge of eggs or fertilization occurs at this time. One earthworm cannot fertilize its own eggs, but there is a mutual cross-fertilization in the cocoon.

After copulation the worms pull away from each other, and half of the slimy band is slipped over the anterior end of each worm. In doing so the eggs from the oviducts are discharged into the slimy tubes, which also receive sperms from the

Fig. 21-7. Earthworm cocoons, in which young earthworms develop in the soil. Each cocoon is about ⅛ inch long.

seminal receptacles (segments IX and X). The elastic ends of the cocoons close to imprison sperms, eggs, and a liquid food for the developing embryos. A young worm eventually breaks from the cocoon and shifts for itself in the soil. After a few weeks the embryo becomes an adult.

CLASS HIRUDINEA

Leeches

Leeches, or "blood suckers," have anterior and posterior suckers, which are used to attach themselves to the host, and chitinous jaws to obtain blood from their victims (Fig. 21-8). They usually have 34 true segments, and each true segment is subdivided into several false grooves. Leeches have definite annelid characteristics, such as segmented bodies, a series of nephridia for excretion, a series of ganglia on the ventral nerve cord, and reproductive organs (gonads) in the coelom, although they do not have bristlelike setae.

The medicinal leech *(Hirudo medicinalis)* is 4 to 5 inches long, and its anterior sucker surrounds the mouth, which is supplied with three jaws of chitinous teeth. The salivary glands secrete an anticoagulant (hirudin) into the wound to prevent blood clotting in the attacked animal. Great amounts of the blood are sucked up by a muscular pharynx and are stored in a large crop that is supplied with many pairs of pouch-like ceca. Digestion occurs at intervals for months.

Fig. 21-8. Common leech of class Hirudinea, showing segments and suckers.

Respiration, circulation, and excretion occur in much the same way as in other annelids. The nervous system is annelid-like, and there are sense organs of taste, touch (tactile), and "sight" (light sensitive). The muscular system consists of circular, oblique, and longitudinal bands.

Leeches have both sexes in one animal (monecious), but there is cross-fertilization.

Because excess blood was believed to cause certain abnormal bodily conditions, the medicinal leech was used for years in medical practice for extracting blood from human beings. During those times pharmacies kept a supply of living leeches for this purpose.

ECONOMIC IMPORTANCE OF ANNELIDS

Earthworms are nature's great cultivators of the soil because they redistribute soil ingredients as they move along. By their migrations they help to aerate it and redistribute the chemicals that are present in the various layers. They bring such food elements as potassium and phosphorus from the subsoils toward the surface. They carry leaves and other organic materials into their burrows, thus bringing them closer to plant roots. Earthworms can ingest their own weight in soil every 24 hours. In his classic work *The Forma-tion of Vegetable Mould Through the Action of Worms,* Charles Darwin 1809-1882) showed the great value of worms in the mixing and aerating of soils. He estimated that up to 18 tons of earth per acre pass through their bodies annually and are thus mixed, which is of great importance in agriculture.

Leeches are widely distributed as parasites on a great many kinds of animals, including man.

CLASSIFICATION OF SEGMENTED WORMS

Class Oligochaeta (Gr. *oligos,* few; *chaite,* bristle)
Internal and external segments; relatively few bristlelike setae; no fleshy parapodia; brandlike clitellum around the body; both sexes in the same animal (monecious). Example: earthworm *(Lumbricus).*

Class Polychaeta (Gr. *poly-,* many; *chaite,* bristle)
Internal and external segments; fleshy parapodia with numerous bristlelike setae; distinct head with eyes, tenacles, and fingerlike sensory palps; no clitellum; sexes are usually separate (diecious). Example: sandworm *(Nereis).*

Class Hirundinea (L. *hirudo,* leech)
Body flattened; fixed number of segments (usually 34), each of which may be subdivided into several false segments; no setae or parpodia; anterior and posterior suckers; coelom is reduced by connective tissues and muscles; both sexes in the same animal (monecious). Example: leeches *(Hirudo).*

Review questions and topics

1 What general characteristics describe the segmented worms as a group?
2 How must a body cavity be constructed in order to qualify as a coelom?
3 Discuss the value of ganglia on the ventral nerve cord, especially if present in a series of segments.
4 What is meant by a so-called closed circulatory system? What are some advantages of such a construction?
5 What improvements are to be found in the digestive system of an earthworm? Explain the structure and functions of the typhlosole and calciferous glands.
6 Of what value is monecious reproduction when we consider the habitats of earthworms?
7 Because earthworms obtain their oxygen through their body surface, might this explain the erroneous statement that "it rains earthworms"? Give an explanation for the emergence of earthworms from their earthen tunnels during excessive rains.

Selected references

Barnes, R. D.: Invertebrate zoology, ed. 3, Philadelphia, 1974, W. B. Saunders Co.
Dales, R. P.: Annelids, London, 1963, Hutchinson & Co. (Publishers) Ltd.
Edmondson, W. T., editor: Fresh-water biology, New York, 1959, John Wiley & Sons, Inc.; 1st ed. by Ward, H. B., and Whipple, G. C., pub. 1918.
Hyman, L. H.: The invertebrates; vol. II, Platyhelminthes and Rhynchocoela, New York, 1951, McGraw-Hill Book Company.
Laverack, M. S.: The physiology of earthworms, New York, 1963, The Macmillan Company.
Mann, K. H.: Leeches (Hirudinea); their structure, physiology, ecology and embryology, New York, 1962, Pergamon Press, Inc.
Meglitsch, P. A.: Invertebrate zoology, ed, 2, New York, 1972, Oxford University Press, Inc.
Pennak, R. W.: Freshwater invertebrates of the United States, New York, 1953, The Ronald Press Company.
Prosser, C. L.: The nervous system of the earthworm, Quart. Rev. Biol. **9:**181-200, 1934.
Russell-Hunter, W. D.: A biology of higher invertebrates, New York, 1969, The Macmillan Company.

22 Phylum Mollusca

The name Mollusca (L. *soft*) refers to the fact that the animals placed in this phylum have a soft, unsegmented body although this is often not apparent in those forms that have distinct shells or leathery skins.

A number of forms are included in the phylum that do not seem to be related to the common types of mollusks until they are studied carefully. In general mollusks have an anterior head, a dorsal visceral mass, and a ventral foot. The presence of the foot is indicated by the "poda" in the names of several of the classes of mollusks. Table 22-1 gives details of classification and examples of the phylum.

CLASS PELECYPODA

Clams

A freshwater clam, such as *Lampsilis*, has a pair of limy shells held together by internal muscles and an elastic, ligamentous hinge. Each shell has concentric lines that indicate successive growth stages, several being formed each season (Fig. 22-1). The shell has an outer horny layer, a middle layer of calcium carbonate, and an inner, limy, iridescent layer of nacre or mother-of pearl. Layers of nacre may form natural pearls.

Internally (Fig. 22-2), the soft body consists of the visceral mass and a muscular foot. The foot may be extended forward between the valves into sand or mud. The thicker, distal end of the foot anchors the animal. One lobe of a mantle adheres to the inner surface of each valve, and together they form a mantle cavity that encloses the entire body mass. The mantle secretes the calcium carbonate used in forming the shell.

Posteriorly, the mantle is modified to form a short, tubular, dorsal excurrent (exhalant) siphon and a short, ventral, incurrent (inhalant) siphon, both of which regulate the output and intake of

Table 22-1 Characteristics of mollusks

Characteristic	Amphineura (Gr. *amphi,* both; *neuron,* nerve)	Pelecypoda (Gr. *pelekys,* hatchet; *pous* foot)	Gastropoda (Gr. *gaster,* belly)	Scaphopoda (Gr. *skaphe,* boat)	Cephalopoda (Gr. *kephale,* head)
Shell	Eight limy, dorsal plates	Two limy, lateral valves (bivalved)	Usually limy, coiled (univalved); may be flat, or absent (in some species	Limy, tubular, tusk shaped	May be present or absent (octopus); if present, may be external or internal (as in squid)
Mantle	Present	Sheetlike, bilobed; muscular siphons present	Present	Tubelike around body	Thick, muscular; muscular siphon present
Foot	Flat, ventral (absent in some)	Hatchet-shaped for burrowing	Flat, ventral for creeping	Conical, trilobed for boring	Modified into a "head-foot"
Head	Small, with rasping, tonguelike radula; no eyes; no tentacles	Absent; no radula; no eyes; no tentacles; labial palps present	Distinct, with radula, eyes, and tentacles	Rudimentary; mouth with tentacles	"Head-foot" with radula, paired eyes; eight to ten long tentacles with suckers
Respiration	Row of gills (between foot and mantle); all marine	Paired, platelike gills on either side of mantle cavity; aquatic or marine	Lungs (in terrestrial types); gills (marine and aquatic)	No gills; all marine	Paired gills (squids); two pairs of gills (nautili); all marine
Sexes	Diecious	Diecious (usually)	Diecious or monecious (depends on species)	Diecious	Diecious
Symmetry	Bilateral	Bilateral	Bilateral (head and foot); asymmetry (visceral hump)	Bilateral	Bilateral
Examples	Chitons	Clams (*Unio, Anodonta, Lampsilis*); oysters; scallops; shipworm (*Teredo*)	Snails (*Helix, Polygyra, Physa*) slugs (*Limax*); abalones	Tooth shells (*Dentalium*)	Squids (*Loligo*); octopuses; nautili; cuttlefish (*Sepia*)

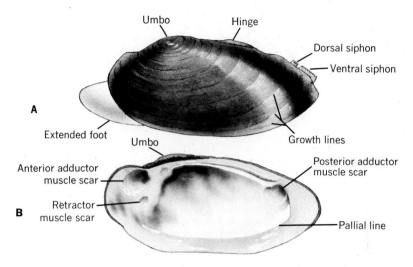

Fig. 22-1. Freshwater clam. **A,** External and **B,** internal shell features.

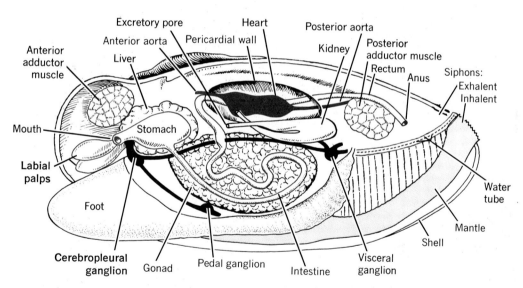

Fig. 22-2. Freshwater clam. Internal anatomy. Mantle, gills, and some tissues removed from one side to permit observation.

water. Food and oxygen in the water are drawn into the mantle cavity through the ventral, incurrent siphon by the action of cilia. Water and feces go out the excurrent siphon. In contaminated waters this siphon system may result in the buildup of harmful bacteria. (See Chapter 1 for an example.) The digestive system consists of (1) the mouth, located between two pairs of fleshy, ciliated, flaplike labial palps, which carry food to the mouth; (2) a short esophagus leading to (3) an

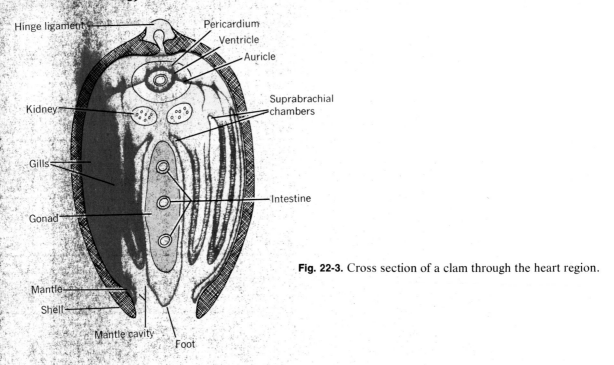

Hinge ligament
Pericardium
Ventricle
Auricle
Kidney
Suprabrachial chambers
Gills
Intestine
Gonad
Mantle
Shell
Mantle cavity
Foot

Fig. 22-3. Cross section of a clam through the heart region.

enlarged stomach, which receives digestive enzymes from the paired liver (digestive gland) surrounding the stomach; and (4) a coiled intestine followed by a rectum (surrounded by the pericardium), which leads to an anus. The anus opens into the mantle cavity near the dorsal excurrent siphon. Internally, the rectum has a longitudinal fold, the typhlosole, to increase the surface area.

Foods consist of small plants and animals carried by water inwardly through the incurrent siphon. Mucous secretions on the mantle and gills catch the food particles, and cilia move the masses of mucus and food toward the mouth. Foods are digested within cells of the digestive gland ("liver"), which surrounds the stomach.

The circulatory system is an "open" one, consisting of a heart within a pericardial cavity, arteries, sinuses (L. *sinus*, cavity), and veins. The heart consists of two auricles and one ventricle. Blood is pumped from the ventricle through an anterior aorta (artery) to the foot and viscera and backward through a posterior aorta

to the mantle and rectum. Some of the blood picks up oxygen in the mantle (becoming oxygenated) and returns directly to the auricles. Other blood circulates through sinuses, travels through a vein to the kidneys, to the gills to be oxygenated and back to the auricles. From the auricles it goes to the ventricle. Wastes and carbon dioxide are carried to the kidneys and gills for elimination.

There are two pairs of thin gills, one pair on either side of the foot (Fig. 22-3). Each gill is formed by two platelike lamellae (L. *lamella*, small plate). Each lamella consists of many vertical, filamentous gill bars, strengthened by chitinous rods. Cross partitions between the lamellae divide the gill into numerous vertical water tubes. The gill surface has many small water pores (ostia) through which water passes by the action of cilia.

The excretory system consists of two U shaped kidneys located just below the pericardium. Each kidney empties into a ciliated tube leading to a

Fig. 22-4. Clam glochidium.

bladder that empties into the suprabranchial chamber.

The nervous system consists of three pairs of ganglia (cerebropleural, pedal, and visceral) that are connected by nerve connections known as commissures. Sensory organs include tactile organs along the mantle margins; light-sensitive organs on the siphon margins; a pair of hollow statocysts, each containing a limy statolith and located in the foot for equilibrium purposes; and a yellowish osphradium (Gr. *osphradion,* strong scent) located over each visceral ganglion, which the clam may use to test the water chemically.

Reproduction is diecious. Gonads (testis or ovary) are located among the intestines within the foot. The sperm ducts (vasa deferentia) in the male and the oviduct in the female open into the suprabranchial chamber near the openings from the kidneys.

Sperms pass to the outside through the excurrent siphon where water carries them to a female, which they enter through the incurrent siphon.

Eggs are fertilized internally. The resulting zygote undergoes cleavage and forms a hollow mass of cells, the blastula. The blastula forms a gastrula, which in turn produces a bivalved glochidium (glo -kid' i um) (Fig. 22-4).This is discharged from the excurrent siphon of the female and attaches itself to a fish by closing the two valves. The fish tissues grow around it forming a ''blackhead'' that lives parasitically for 2 to 3 months. The miniature clam then breaks out, sinks to the bottom, and continues development. Information on clams and other mollusks is given in Table 22-1.

CLASS GASTROPODA

Snails

There are many kinds of land snails (Figs. 22-5 and 22-6). The imported European snail *(Helix)* is a typical one. The well-developed head has two pairs of retractile, sensory tentacles and a pair of light-sensitive eyes on the hollow, longer tentacles. A muscular foot is attached to the head and is supplied with mucous glands to assist in locomotion. The internal organs (visceral mass) form a hump surrounded by a mantle, which secretes the external, coiled shell. A muscle can pull the body within the shell.

The digestive system consists of a mouth; a pharynx with a radula of chitinous teeth for rasping green vegetation; an esophagus; a thin-walled crop; a stomach; a long, coiled intestine; and an anus, located in the margin of the mantle at the edge of the shell. A pair of salivary glands pours their secretions into the pharynx to assist in the digestive process. A large liver, high up in the spiral shell, pours its secretions into the stomach. Since *Helix* is a land snail, it has a membranous lung, well supplied with blood vessels (veins) to aerate the blood. Air is drawn into and forced out of the lungs through the respiratory pore, which is located in the margin of the mantle at the edge of the shell. The circulatory system consists of one heart (one auricle and one ventricle), arteries

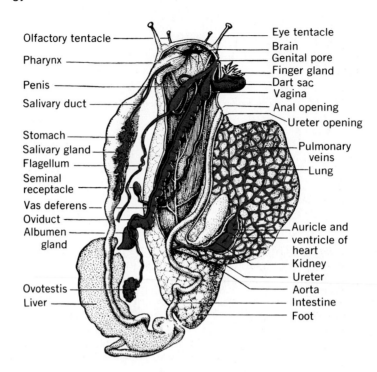

Olfactory tentacle
Pharynx
Penis
Salivary duct
Stomach
Salivary gland
Flagellum
Seminal receptacle
Vas deferens
Oviduct
Albumen gland
Ovotestis
Liver

Eye tentacle
Brain
Genital pore
Finger gland
Dart sac
Vagina
Anal opening
Ureter opening
Pulmonary veins
Lung
Auricle and ventricle of heart
Kidney
Ureter
Aorta
Intestine
Foot

Fig. 22-5. Snail *(Helix pomatia)*. The roof of the pulmonary sac is cut and turned to the right, the pericardium and visceral sac are opened, and the internal organs (viscera) are somewhat separated. The finger gland is also called the mucous gland.

to carry blood to various organs, and veins to return blood. The excretory system consists of a kidney (nephridium) to secrete wastes, which are carried by a tubular ureter to the mantle cavity.

The nervous system consists of a concentration of ganglia (cerebral, buccal, pedal, and visceral) that encircle the pharyngeal region and nerves leading to various organs. Sensory organs include light-sensitive eyes, olfactory organs on the tentacles, chemical and tactile (touch) organs on the head and foot, and a pair of statocysts (Gr. *statos,* stationary; *kystis,* sac) near the pedal ganglia for equilibrium.

Helix is monecious, but cross-fertilization is common. The gonad is called an ovotestis (producing both eggs and sperms) and is located high in the visceral hump and surrounded by the liver.

A hermaphroditic duct (Gr. *hermaphroditos,* containing both sexes) leads from the ovotestis and connects with the albumin gland. A tubular vas deferens connects with the hermaphroditic duct and carries sperms to the penis at the genital pore on the right side near the mouth. Another tube, the oviduct, also connects with the hermaphroditic duct and leads to the enlarged vagina, which also empties through the genital pore. To the vagina are connected the duct from the seminal receptacle, the dart sac, and the fingerlike oviductal glands. The penis has a slender, tubular flagellum in which sperms are formed into spermatophores (bundle of sperms).

During copulation each snail inserts its penis into the vagina of its partner and transfers a spermatophore. After separating each snail deposits its fertilized eggs into shallow burrows in moist soil. Young, small snails develop soon after this.

When dry weather exists, snails may form a temporary covering, or epiphragm, of limy mucous secretions, which can close the opening in the shell.

CLASS CEPHALOPODA

Squids

The common Atlantic coast squid (Fig. 22-7), known as *Loligo pealei,* has a torpedo-shaped body, which may be up to 1 foot long. The skin contains pigment cells of blue, purple, yellow,

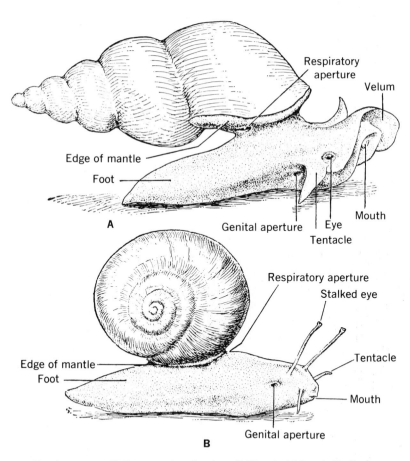

Fig. 22-6. A, Freshwater snail *(Lymnaea);* **B,** land snail *(Humboldtiana).* Bodies are expanded from the shell.
(From Potter, G. E.: Textbook of zoology, ed. 2, St. Louis, The C. V. Mosby Co.)

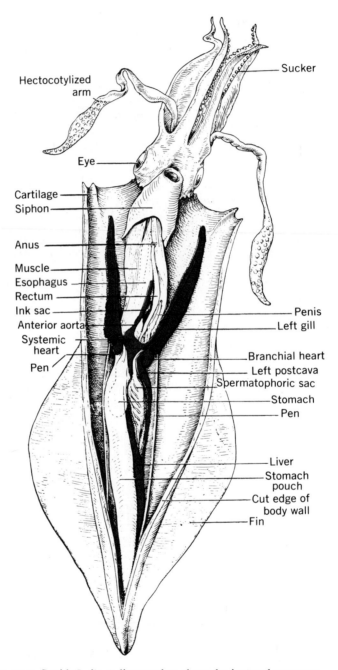

Fig. 22-7. Squid *(Loligo)* dissected to show the internal anatomy.

Fig. 22-8. Octopus or devilfish of class Cephalopoda.

and red, so that color changes for protective concealment can be produced quickly by changing the amount of the various pigments to blend with the surroundings. The head bears a mouth surrounded by ten sucker-bearing arms for capturing prey. One pair of arms (retractile tentacles) is longer than the others. A pair of well-developed eyes is present. A mantle encloses the internal organs within a mantle cavity. The mantle ends just in back of the head, and its free margin is called a collar. Beneath the collar is a cone-shaped siphon, which propels water outward by a contraction of the mantle. By directing the siphon backward and forcing water through it the animal is jet-propelled forward; by directing the siphon forward the animal is propelled backward. A pair of posterior, triangular fins assists in locomotion. A feather-shaped plate (shell), called the pen, is just beneath the skin of the back near the anterior end and gives some support.

Squids capture fishes, other mollusks, and crustaceans for food.

Octopuses

The octopus (Gr. *okto,* eight; *pous,* foot), or devilfish (Fig. 22-8), has a body with the foot divided into eight sucker-bearing tentacles. Two large eyes are also evident. There is no internal shell. In general the characteristics of octopuses somewhat resemble those of other cephalopods. Some types are rather small and harmless, but the giant octopus of the Pacific may have a length of 30 feet and become unsociable and dangerous. Octopus eggs are protected by a tough case from which the young, immature octopus emerges.

ECONOMIC IMPORTANCE OF MOLLUSKS

Great quantities of clams and oysters are used for human food. More than 120 million pounds of oysters are collected in the waters of the United States annually. Extensive oyster beds have been cultivated along our Atlantic and Pacific coasts, and those in Chesapeake Bay and Long Island Sound are noteworthy. It takes about 4 years for

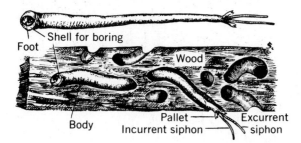

Fig. 22-9. Shipworm *(Teredo)* of class Pelecypoda. The shipworm (below) is in its burrow in wood (somewhat diagrammatic).

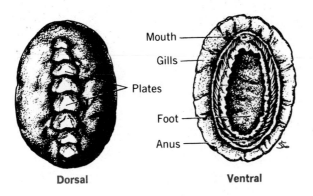

Fig. 22-10. Chiton *(Katharina)* of class Amphineura.

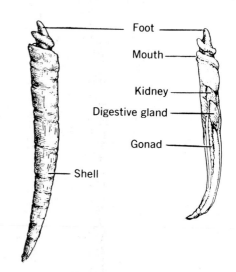

Fig. 22-11. Tooth shell *(Dentalium)* of class Scaphopoda. At right the shell is removed to show the internal organs.

an oyster to reach commercial size, and since oysters have natural enemies, their cultivation has become a great commercial endeavor. Many clams are used for chowder and other dishes. Valuable pearls are formed within oysters and clams when concentric layers of nacre are built up until a usable size is reached. Many types of irritating materials can initiate pearl formation. The Japanese have artificially cultured pearls by introducing foreign particles within the mantle, placing the oyster in an enclosure in the sea, and removing the pearl a few years later.

The shipworm *(Teredo)* (Fig. 22-9) burrows into the wood of wharves and ships by means of two movable valves on its anterior end. These burrows weaken the wood and permit it to de-

compose more rapidly. Snails are widely used as food in Europe, particularly in France. Garden snails and slugs (without shells) are great destroyers of plants. Certain species of snails may serve as intermediate hosts for such parasites as sheep liver flukes, blood flukes, and many others. Certain parts of squids may be used as food. A gastropod, the abalone is also used as food. A type of squid, the cuttlefish *(Sepia),* produces sepia-colored ink, which is used commercially. This cuttlefish has a shell ("cuttlebone") that is used as a source of calcium for caged birds.

Important findings in neurophysiology and animal behavior have been made in recent years by using squids and octopuses as experimental animals.

Review questions and topics

1 What general characteristics describe the mollusks as a group?
2 What are the characteristics that distinguish members of the different classes of Mollusca?
3 Describe the structure and functions of the mantle.
4 Is it unusual for such high types of animals as mollusks not to be segmented?
5 Discuss the construction and function of the foot in different types of mollusks.
6 Explain the significance of the head and the "head-foot" in certain kinds of mollusks.
7 Describe the structure and function of the radula.
8 Explain the structures and functions of gills and lungs, giving examples of each.
9 Discuss the status of the various sensory organs in different mollusks.
10 Describe a glochidium, including its function.
11 Discuss the circulatory systems found in different mollusks, including the significance of the so-called open type, giving example.

Selected references

Abbott, R. T.: American sea shells, Princeton, New Jersey, 1954, D. Van Nostrand Co., Inc.

Boycott, B. B.: Learning in the octopus, Sci. Amer. **212:** 56-66, 1956.

Fretter, V., and Peake, J., editors: Pulmonates: functional anatomy and physiology, New York, 1975, Academic Press.

Keynes, R. D.: The nerve impulse and the squid, Sci. Amer. December, 1958.

Mead, A. R.: The giant African snail; a problem in economic malacology, Chicago, 1961, University of Chicago Press.

Meglitsch, P. A.: Invertebrate zoology, ed. 2, New York, 1972, Oxford University Press, Inc.

Morris, P. A.: A field guide to the shells of our Atlantic coast, Boston, 1947, Houghton Mifflin Company.

Morris, P. A.: A field guide to shells of the Pacific coast and Hawaii, Boston, 1952, Houghton Mifflin Company.

Morton, J. E.: Molluscs, London, 1958, Hutchinson & Co. (Publishers) Ltd.

Purchon, R. D.: The biology of the Mollusca, New York, 1968, Pergamon Press, Inc.

Raven, C. P.: Morphogenesis; the analysis of molluscan development, New York, 1958, Pergamon Press, Inc.

Tressler, D. K., and Lemon, J. M.: Marine products of commerce, ed. 2, New York, 1951, Reinhold Publishing Corp.

Yonge, C. M.: Oysters, London, 1960, William Collins Sons & Co., Ltd.

23 Phylum Arthropoda

Arthropods are members of the largest phylum of animals. The number of different species is well over 1 million. Arthropods occupy every type of habitat, consume every type of food, exhibit every type of relationship, and show every type of motility. An entire course in biology could be structured using only arthropods as examples. Table 23-1 gives some of the types and characteristics of the animals included in the phylum Arthropoda.

CLASS CRUSTACEA

Crayfishes

The two most common genera of freshwater crayfishes in North America are *Cambarus*, found east of the Rocky Mountains, and *Astacus*, found principally west of the Rocky Mountains. In general the various species are structurally similar (Figs. 23-1 and 23-2) and resemble the large lobster, *Homarus*.

Members of the genus *Cambarus* are about 4 inches long and possess a calcareous exoskeleton. The body is divided into a cephalothorax (fused head-thorax) and an abdomen. The cephalothorax consists of twelve somites (segments); the abdomen consists of six. Each somite bears a pair of jointed appendages. The paired anterior antennules are not considered to be true, serially metameric appendages, because they are developmentally different from the eighteen pairs posterior to them. The antennules are actually paired, prostomial sense organs, arising from a structure that seems to be a homologue of the prostomium of annelids. The stalked, compound eyes are also considered to be prostomial sense organs. The cephalothorax is covered dorsally

Table 23-1 Some characteristics of arthropods

Class	Antennae	Legs	Wings	Respiration	Habitat	Miscellaneous	Examples
Crustacea (L. *crusta*, shell)	2 pairs	Numerous	None	Gill breathing	Aquatic for most species	Body composed of head and thorax, which may be fused into cephalothorax, and abdomen	Crayfish Crab Barnacle Sow bug
Onychophora (Gr. *onyx*, claw) *phorein*, to bear)	1 pair	Numerous (with 2 claws on each)	None	Tracheal (air breathing)	Terrestrial (moist places)	Primitive worm-like arthropods; numerous paired nephridia	*Peripatus*
Chilopoda (Gr. *cheilos*, lip; *pous*, foot)	1 pair	Numerous (1 pair on most segments)	None	Tracheal (air breathing)	Terrestrial (moist places)	Long, slender bodies, flat-tended dorso-ventrally; swift moving	Centipedes
Diplopoda (Gr. *diploos*, double; *pous*, foot)	1 pair	Numerous (2 pairs on most segments)	None	Tracheal (air breathing)	Terrestrial (moist places)	Long, slender bodies, sub-cylindrical; slow moving	Millipedes
Insecta (L. *in-sectus*, cut into)	1 pair	3 pairs (on thorax)	2 pairs, 1 pair, or none (depends on species)	Tracheal (air breathing)	Terrestrial or aquatic	Body composed of head, thorax, and abdomen; wings, if present, attached to thorax	Bees Butterflies True bugs Grasshoppers Flies Beetles
Arachnoidea (Gr. *arachne*, spider; *eidos*, like)	None	4 pairs	None	Tracheal and book lungs (air breathing)	Terrestrial	Head and thorax fused into cephalothorax; abdomen present	Spiders Scorpions Horseshoe crab Ticks

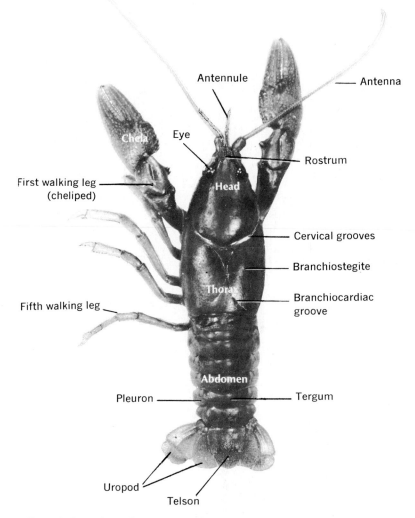

Fig. 23-1. Dorsal view of crayfish.
(From Hickman, Cleveland P., Sr., Hickman, Cleveland P., Jr., and Hickman, Frances M.: Integrated principles of zoology, ed. 5, St. Louis, 1974, The C. V. Mosby Co.)

and laterally by a shell-like carapace (Sp. *carapacho,* covering).

The eighteen pairs of appendages show considerable variations, depending on the functions they perform, but they all have fundamentally the same plan of structure, as can be noted from

Table 23-2. With a few exceptions each appendage consists of a basal protopodite that bears a median endopodite and a lateral exopodite.

Any appendage composed of two branches is called biramous. All the metameric appendages of the crayfish arise embryologically from bi-

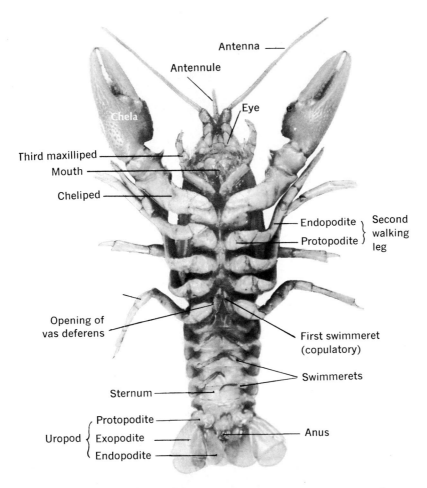

Fig. 23-2. Ventral view of male crayfish.
(From Hickman, Cleveland P., Sr., Hickman Cleveland P., Jr., and Hickman, Frances M.:
Integrated principles of zoology, ed. 5, St. Louis, 1974, The C. V. Mosby Co.)

ramous structures. When structures are constructed along basically similar plans and have similar embryologic origins, they are said to be homologous structures, even though their functions may differ. When such structures appear in a serial sequence on a body, they illustrate serial homology (Fig. 23-3).

The crayfish coelom is reduced in size and divided into several separate compartments, such as those enclosing the gonads and those enclosing the excretory green glands. The hemocoels contain blood and surround the digestive tract (Fig. 23-4 and 23-5).

The digestive system consists of a mouth (be-

Table 23-2 Segmental appendages of the crayfish

Appendage	Protopodite	Endopodite	Exopodite	Function
Prostomium (No segment)				
Antennule				
(first antenna)	(Not considered to be segmental appendage)			Sensory
Eye (stalked)	(Not considered to be segmental appendage)			Sensory
Cephalothorax (12 segments)				
I. Antenna	2 segments; excretory pore in basal segment	Long, many-jointed "feeler"	Thin, broad, dagger-like	Touch; taste; balance
II. Mandible	2 segments; heavy jaw	Small; 2 distal segments of palp	None	Chew food
III. First maxilla	2 thin, medial platelike lamellae	1 small, plate-like lamella	None	Handle food
IV. Second maxilla	2 bilobed, platelike lamellae; broad plate, epipodite	1 small, pointed segment	Dorsal plate, "bailer" (scaphognathite)	Handle food; create water currents in gill chamber
V. First maxilliped	2 thin segments extending inwardly; epipodite extending outwardly	2 small segments (reduced)	Long, basal segment with jointed filament	Handle food; touch; taste
VI. Second maxilliped	2 segments; gills	5 segments	Similar to V	Similar to V
VII. Third maxilliped	Similar to VI	5 larger segments	Similar to V	Similar to V
VIII. First walking leg (pincer, chela, or cheliped)	2 segments; gills	5 segments; distal 2 form large pincer	None	Offense; defense; walking; touch
IX. Second walking leg	Similar to VIII	Similar to VIII, but pincer smaller	None	Walking; grasping
X. Third walking leg	Similar to VIII; female bears genital pore	Similar to IX	None	Similar to IX
XI. Fourth walking leg	Similar to VIII	Similar to IX, but no pincer	None	Walking
XII. Fifth walking leg	Similar to VIII; male bears genital pore	Similar to XI	None	Walking; cleaning abdomen and eggs
Abdomen (6 segments)				
XIII. First abdominal (first swimmeret, or pleopod)	Fused with endopodite to form tube (in male); reduced or absent (in female)	Reduced or absent (in female)	None (in male); reduced or absent (in female)	Transfers sperms to female
XIV. Second abdominal (swimmeret, or pleopod)	Fused with endopodite (in male); 2 segments (in female)	Filament (in female)	None (in male); filament (in female)	As in XIII (in male); as in XV (in female)

XV. Third abdominal	2 segments	Filament	Filament	Creates water currents; used for attaching eggs and young (in female)
XVI. Fourth abdominal	2 segments	Similar to XV	Similar to XV	Similar to XV
XVII. Fifth abdominal	2 segments	Similar to XV	Similar to XV	Similar to XV
XVIII. Sixth abdominal (uropod)	1 short, broad segment	Flat, oval plate	Flat, oval plate divided crosswise into 2 parts	Swimming; egg protection (in female)
Telson (No segment)	(Not considered to be segmental appendage)			Swimming; egg protection

hind the mandibles); a short esophagus; a large stomach, composed of an anterior cardiac portion, which contains the gastric mill, and a posterior pyloric portion; this portion connects with a short midgut followed by a long intestine that exits through an anus beneath the telson. In the gastric mill chitinous teeth grind the food. A strainer of hairlike setae permits only fine particles to pass into the pyloric end. A pair of digestive glands ("liver") discharge secretions with enzymes into the midgut. All of the digestive system except the midgut is lined with chitin. Digested foods are absorbed by certain parts of the intestine and the digestive glands. Nearly anything that is edible seems to serve as food for a crayfish.

Respiration occurs in a series of delicate, featherlike gills attached to certain thoracic appendages. By moving a chitinous plate ("bailer"), located on the second maxilla, water currents are created over the gills.

Circulation is accomplished by an "open" system, consisting of a dorsal, muscular, rhythmically contracting heart, arteries, capillaries, and blood sinuses. The heart is located within a saclike pericardial sinus in the thorax. Blood from the heart enters seven main arteries (supplied with valves) and passes into fine capillaries that carry it to blood sinuses, which are actually spaces between tissues. These blood sinuses constitute the hemocoel ("blood cavity"). The blood sinuses return the blood to the pericardial sinus. Blood also passes from the sinuses to the gills, where oxygen and carbon dioxide are exchanged, and then returns to the pericardial sinus. Arthropod blood is commonly colorless and contains a number of ameboid cells and a respiratory, oxygen-carrying pigment called hemocyanin. The blood also has the ability to clot, which prevents some loss when an injury occurs.

Excretion is accomplished by green glands opening by a duct at the base of the antenna.

The well developed central nervous system consists of a series of paired ganglia with a supra-

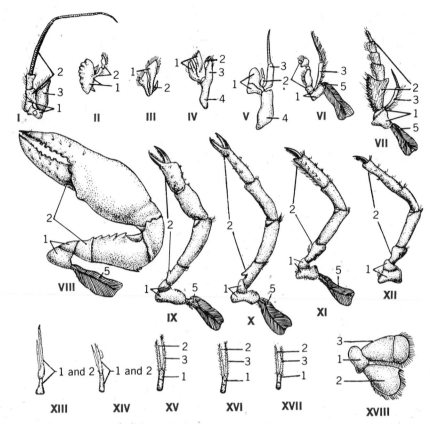

Fig. 23-3. Segmental appendages of the male crayfish *(Cambarus)* of class Crustacea, showing homology. The segmental appendages are removed from the left side, numbered from **I** to **XVIII**, and drawn somewhat to scale. **1**, Protopodite; **2**, endopodite; **3**, exopodite; **4**, epipodite; **5**, gills. Appendage **I** is the antenna; **II**, the mandible; **III** and **IV**, first and second maxillae; **V** to **VII**, first, second, and third maxillipeds; **VIII** to **XII**, walking legs; **XIII** to **XVII**, swimmerets (pleopods); **XVIII**, uropod (sixth swimmeret). Antennules (first antennae), stalked eyes, and telson are not considered to be segmental appendages.

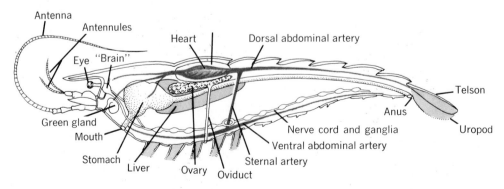

Fig. 23-4. Internal anatomy of a female crayfish. Muscles and segmental appendages omitted.

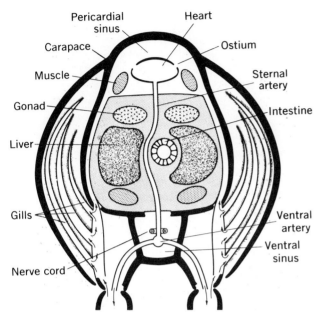

Pericardial sinus

Heart

Carapace

Ostium

Muscle

Sternal artery

Gonad

Intestine

Liver

Gills

Ventral artery

Ventral sinus

Nerve cord

Fig. 23-5. Crayfish (cross section).

esophageal "brain." A circumesophageal connective connects the brain with the subesophageal ganglia. A double ventral nerve cord extends posteriorly from the subesophgeal ganglia. Nerves pass from the various ganglia to different body parts.

The sinus gland at the base of the eye stalk produces endocrine hormones that control the spread of pigment in the epidermis and the compound eyes. They also regulate molting and the placing of limy salts in the exoskeleton.

The sensory organs are well developed. Two very efficient compound eyes at the ends of movable stalks are each composed of approximately 2,500 single eyes, called ommatidia (Gr. *ommation*, little eye). Each tubelike ommatidium does not take a complete picture but only a dark or light spot. The arrangement of these dark or light spots at the base of the eye gives the crayfish a flat, mosaic picture of the object. Moving objects can thus be detected.

The sense of equilibrium is regulated by two

small, chitin-lined sacs (statocysts) in the base of the antennules. Grains of sand rolling within these sacs stimulate sensory hairs, thus helping the crayfish to ascertain its position. The crayfish is without a sense of hearing. The entire external body surface is sensitive to touch (by means of tactile hairs) and taste, but the pincers are most sensitive to touch and the outer (expodite) antennules are most sensitive to taste.

Reproduction is diecious. Paired, hollow testes (or ovaries) are found in the center of the thorax. Sperms pass out of each testis through a vas deferens (sperm duct), which opens as a genital pore on the protopodite of the last (fifth) pair of walking legs. Eggs pass out of each ovary through an oviduct, which opens as a genital pore on the protopodite of the third pair of walking legs. As many as 300 eggs may be discharged in a stringlike manner at one time.

When copulation occurs, usually in the autumn, sperms are transferred to the seminal receptacle of the female (the cavity between the fourth and fifth pairs of walking legs) by using the first and second pairs of abdominal appendages of the male as a tube. Seminal receptacles are absent in certain kinds of crayfishes. The eggs are attached to the female swimmerets until the young larvae hatch in 5 to 6 weeks.

The larva, which looks like a miniature crayfish, sheds its external skeleton soon after hatching by molting. This occurs at intervals during the life of the animal, which averages about 4 years. Growth occurs in the short periods between moltings, and a new exoskeleton is formed by the underlying tissues.

In general the eyes or any appendage may be regenerated when lost. Complete regeneration may require several moltings. The regenerated structure is not always the same as the lost one. If only part of an eye stalk is cut off, a normal eye will be regenerated; if an entire eye stalk is removed, an antennalike structure may be produced.

A crayfish may automatically break off walking legs at a specific predetermined joint. The power

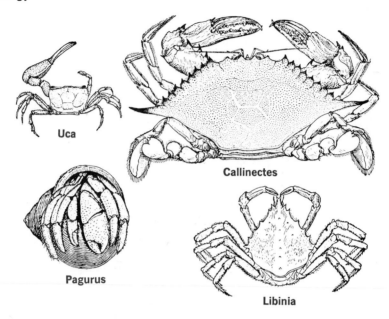

Fig. 23-6. Representative crabs of class Crustacea. Fiddler crab *(uca);* blue edible crab *(Callinectes);* hermit crab *(Pagurus);* spider crab *(Libinia).* *(Courtesy General Biological Supply House, Inc., Chicago, Illinois.)*

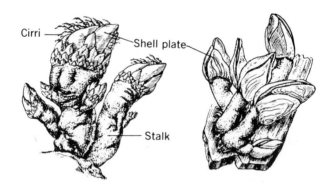

Fig. 23-7. Barnacles of class Crustacea. These animals live within their calcareous shell plates and are attached to various objects in the sea; they possess jointed legs.

of self-mutilation (autotomy) is accomplished by a special muscle at the "breaking point." Regeneration occurs in the regular manner. In case such an appendage is caught, it can be severed, thus giving the animal a means of escape.

CLASS CHILOPODA (CENTIPEDES)

In addition to the information given in Table 23-1, the following may be added. All body segments bear one pair of jointed legs (Fig. 23-8),

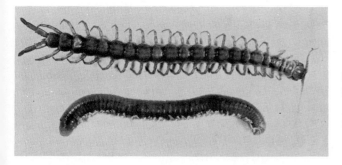

Fig. 23-8. Centipede of class Chilopoda (above); and millipede of class Diplopoda (below).

except the last two and the one just posterior to the head. The latter bears a pair of poison claws, maxillipeds, with which small animals may be killed. Centipedes live under stones and the bark of trees. Some tropical centipedes, which may become 1 foot long, are poisonous and dangerous to man. The common house centipede, with fifteen pairs of long legs, lives in damp places and destroys insects but is harmless to human beings.

CLASS DIPLOPODA (MILLIPEDES)

In addition to the information in Table 23-1 the following may be added. The two pairs of segmented appendages present on almost every body segment have probably come into being by a fusion of two segments. The mouth contains one pair of mandibles and one pair of maxillae. A series of scent glands may secrete an objectionable, odoriferous fluid for defensive purposes. Some millipedes may roll themselves into a ball. They live primarily on dead plant materials but may attack living plants. Common types may be found in gardens and grassy places.

CLASS INSECTA (INSECTS)

Insects are air-breathing arthropods with a head, thorax, and abdomen. The head bears one pair of antennae. The mouthparts vary with different types of insects and include: chewing, as in grasshoppers; sucking, or siphoning (with a tube for sucking foods), as in butterflies; or modifications of the two. Sometimes the sucking types are

modified for piercing-sucking, as in the mosquitoes, flies, and true bugs.

The insect thorax has three segments, each with a pair of jointed legs, and it may have one or two pairs of wings or be wingless, depending upon the species, age, and sex. Sense organs may include simple eyes (ocelli), compound eyes, receptors for touch on various body parts, and receptors for sounds.

Grasshopper

The head of the grasshopper (Fig. 23-9) bears one pair of antennae, and the thorax bears three pairs of jointed legs. Although different species of grasshoppers vary in certain respects, the following descriptions apply to most species that are available for study. Grasshoppers have enlarged hind legs, sound-producing and sound-receiving structures, leathery forewings, and membranous hind wings. One either side of the head is a compound eye. One top of the head are simple eyes called ocelli. Grasshoppers have chewing mandibles and belong to the order Orthoptera (Gr. *orthos,* straight; *ptera,* wings).

INTEGUMENT AND SKELETON

A flexible, noncellular, chitinous cuticle also serves as an exoskeleton to which organs, muscles, and tissues are attached on the inside. The cuticle is secreted by a cellular hypodermis beneath it. Beneath the hypodermis is a basement membrane.

Internally, the cavity in the body is not a true coelom but a hemocoel (Gr. *haima,* blood; *koilos,*

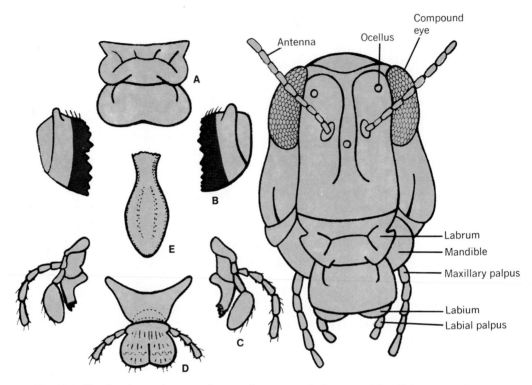

Fig. 23-9. Head and mouth parts of a grasshopper. **A,** Labrum. **B,** Mandible. **C,** Maxilla. **D,** Labium. **E,** Hypopharynx.

hollow). It is filled with organs and a colorless blood. When the animal grows, it sheds its chitin at intervals by the process of molting, or ecdysis (ek′ di sis) (Gr. *ekdyein,* to shed). The liquid secreted by the hypodermis hardens into new chitin.

MOTION AND LOCOMOTION

Grasshoppers may walk or jump by means of the three pairs of jointed legs or fly by means of two pairs of wings (Fig. 23-10). Each segmented leg consists of (1) a coxa (L. *coxa,* hip) attached to the thorax, (2) a trochanter (Gr. *trochanter,* runner), (3) a femur (L. *femur,* thigh), (4) a tibia (L. *tibia,* shin), and (5) a tarsus (Gr. *tarsos,* sole of foot). The tarsus is segmented, the proximal segment bearing three pads and the distal one a pair of claws. Between the claws is a fleshly pul-

villus. The forewings are leathery and unfolded and cover the folded membranous hind wings. Chitinized, tubular veins in the wings give strength. The fine, strong, striated muscles attached to the inside of the chitinous skeleton help to move the wings, legs, and mouthparts.

INGESTION AND DIGESTION

The food of grasshoppers consists of vegetation. The principal parts of the digestive system (Fig. 23-11, *A*) include (1) a mouth, with a pair of salivary glands that secrete digestive juices, and the various mouthparts; (2) a tubular esophagus; (3) and enlarged crop for storage; (4) a gizzard (proventriculus) for grinding; (5) a stomach, with glandular, cone-shaped gastric ceca that secrete digestive juices; and (6) an intestine, which expands into a rectum opening through the anus.

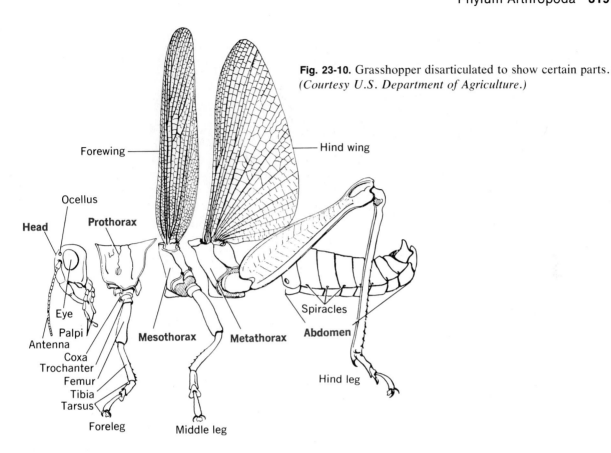

Fig. 23-10. Grasshopper disarticulated to show certain parts. *(Courtesy U.S. Department of Agriculture.)*

CIRCULATION

A single, tubular heart in the dorsal side of the abdomen is divided by valves into a series of chambers, each with a pair of ostia for the entrance of blood from the surrounding pericardial sinus (Fig. 23-11, *B*). Valves close the ostia when the heart contracts. A tubular aorta (Gr. *aorta*, the great artery) extends anteriorly from the heart and opens into the hemocoel in the head region. The hemocoel contains the internal organs and circulates the blood. This so-called open system of circulation causes the blood to flow in vessels only part of the time. Most of the time it flows in tissue spaces or sinuses in the body and appendages. Eventually, the blood returns to the pericardial sinus. The liquid plasma of the blood contains colorless cells.

RESPIRATION

Several pairs of external openings or spiracles (L. *spiraculum,* air hole) open into the tracheal (respiratory) system on either side of the thorax and either side of the abdomen. The spiracles permit the entrance of oxygen and the exit of carbon dioxide. The tubular tracheae ramify to all parts of the body and may have enlargements called air sacs (Figs. 23-11, *C* and 23-12). The blood does not play an important role in respiration.

EXCRETION AND EGESTION

The coiled malpighian tubules in the hemocoel collect wastes and empty them into the large intestine. Solid materials are eliminated through the anus.

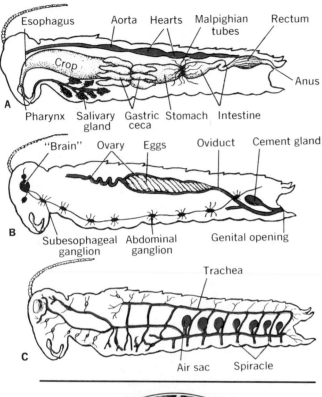

Esophagus   Aorta   Hearts   Malpighian tubes   Rectum

Crop

Anus

A

Pharynx   Salivary gland   Gastric ceca   Stomach   Intestine

"Brain"   Ovary   Eggs   Oviduct   Cement gland

B

Subesophageal ganglion   Abdominal ganglion   Genital opening

Trachea

C

Air sac   Spiracle

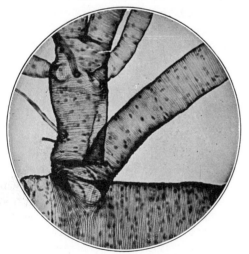

Fig. 23-12. Portion of the grasshopper trachea with its branching tubules, both large and small (photomicrograph). The tracheal rings and nuclei are quite distinct. (*Courtesy General Biological Supply House, Inc., Chicago, Illinois.*)

Fig. 23-11. Internal anatomy of the grasshopper. **A,** Digestive and circulatory systems; **B,** reproductive and nervous systems; **C,** respiratory system.

COORDINATION AND SENSORY EQUIPMENT

A dorsal brain consisting of three pairs of ganglia is connected by a pair of circumesophageal connectives with a subesophageal ganglion. The ventral nerve cord continues posteriorly, with a pair of large ganglia in each thoracic segment and five pairs of ganglia in the abdomen. A sympathetic nervous system supplies the spiracles and muscles of the digestive system.

The compound eyes are covered with a cuticular cornea divided into numerous hexagonal facets. Each facet is the external surface of a unit called an ommatidium, composed of a long visual rod, and the various units are separated from each other by a layer of dark pigment cells. Such an arrangement gives mosaic vision, in which each receptor receives a portion of the image.

Each simple eye, or ocellus, consists of a group of cells, the retinulae, a central optic rod, or rhabdom, and a transparent, cuticular "lens." The ocelli probably function as light-perception organs.

The pair of jointed, threadlike antennae bears sensory bristles, probably for olfactory purposes. Organs of taste are located on the mouthparts. Hairlike organs of touch are present on various body parts but particularly on the antennae. The pair of sound-receiving auditory organs located on the side of the first abdominal segment consists of a membranous tympanum (Gr. *tympanon*, drum), which covers an auditory sac.

REPRODUCTION

The sexes are separate. The female possesses a conspicuous ovipositor (L. *ovum*, egg; *ponere*, to place) at the tip of the abdomen for depositing eggs. In the female a pair of ovaries produces eggs that are discharged into a pair of oviducts. The oviducts unite to form a vagina, connected with the genital pore between the parts of the

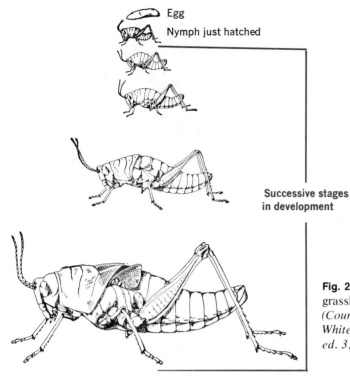

Egg

Nymph just hatched

Successive stages
in development

Fig. 23-13. Incomplete metamorphosis of the
grasshopper.
*(Courtesy Eleanor Sloan Hough; from
White, E. G.: A textbook of general biology,
ed. 3, St. Louis, The C. V. Mosby Co.)*

ovipositor. A seminal receptacle (spermatheca),
connected with the vagina, receives sperms from
the male during copulation and releases them to
fertilize eggs. A secretion of the cement gland
may stick eggs together as they are deposited.

In the male a pair of testes discharges sperms
into a pair of vasa deferentia (sperm ducts),
which unite to form the ejaculatory duct that
opens through a penis. Accessory glands secrete
a fluid into the ejaculatory duct to aid in the transfer of sperms to the female. Eggs are fertilized by
sperms when they are deposited. A young grasshopper that hatches from an egg is called a nymph
and resembles an adult without wings. As the
grasshopper grows it must shed its chitinous
exoskeleton at certain intervals by the process of
ecdysis (molting). Adult wings are eventually
formed from wing buds (Fig. 23-13).

Honeybee

The honeybee, *Apis mellifica* (L. *apis,* bee;
L. *mellificus,* honey), is placed in the order
Hymenoptera (Gr. *hymen,* membrane; *ptera,*
wings) because of its membranous wings.
Honeybees are more highly specialized in life
habits and structure than grasshoppers. Colonies
of honeybees consist of (1) workers, which are
females with undeveloped reproductive organs,
(2) male drones, and (3) female queens. A typical
colony may contain 50,000 workers, a few
hundred drones, and one adult queen (Fig. 23-
14). The drones and queen are for reproductive
purposes. Hymenoptera are characterized by
having their mouthparts modified for both sucking and biting (chewing). The body is divided into
head, thorax, and abdomen. The thorax is divided into an anterior prothorax, a middle

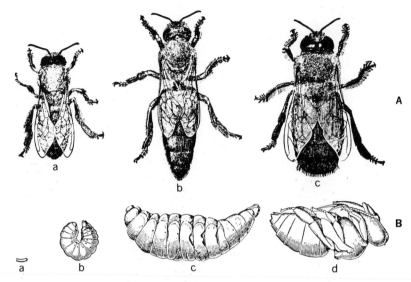

Fig. 23-14. Castes and development of the honeybee *(Apis)* of order Hymenoptera. In **A**, adults: **a**, worker; **b**, queen; **c**, drone. In **B**, immature stages: **a**, egg; **b**, young larva; **c**, old larva; **d**, pupa.
(Courtesy U.S. Department of Agriculture.)

mesothorax, and a posterior metathorax. The abdomen is also segmented.

INTEGUMENT AND SKELETON

A tough, flexible cuticle covers the body and serves as an exoskeleton to which muscles and organs are attached internally. The cavity is a hemocoel, which carries blood.

MOTION AND LOCOMOTION

Locomotion is accomplished by two pairs of wings and three pairs of jointed legs. The wings consist of a double layer of transparent membranes between which is a network of veins to strengthen them. When at rest, the wings are folded. During flight they are extended, and the forewings and hind wings are locked together by a row of tiny hooks on the front margin of the hind wings. Wings may vibrate over 400 times per second during flight.

The three pairs of legs are much more specialized than those of the grasshopper. Because of their complexity and differences, each leg will be considered separately (Fig. 23-15).

1. Prothoracic leg (first). The segments consists of (1) an oblong coxa, next to the thorax; (2) a short trochanter; (3) a long femur with branched, pollen-carrying hairs; (4) a tibia with pollen-carrying hairs and a flat, movable, spinelike velum; and (5) a segmented tarsus, the proximal segment of which may be called the metatarsus and which bears a semicircular antenna comb. This comb, together with the velum, constitutes the antenna cleaner, through which the antenna may be drawn to remove materials. On the opposite margin of the tibia from the velum is the pollen brush, composed of curved bristles. The last (distal) tarsal segment has claws with a padlike pulvillus between them. The pulvillus secretes a sticky substance for adherence.

2. Mesothoracic leg (second). The segments are the same as on the first pair of legs. A long pollen spur on the distal end of the tibia is used to

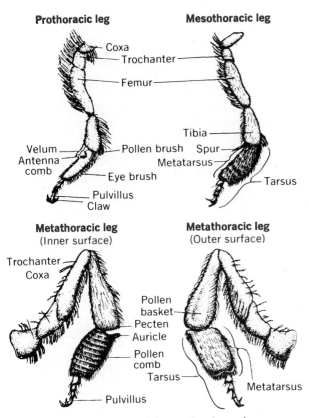

Prothoracic leg **Mesothoracic leg**

Coxa
Trochanter
Femur
Tibia
Velum
Antenna
comb — Pollen brush   Spur
Eye brush   Metatarsus
Pulvillus   Tarsus
Claw

Metathoracic leg **Metathoracic leg**
(Inner surface) (Outer surface)

Trochanter
Coxa
Pollen
basket
Pecten
Auricle
Pollen
comb
Tarsus
Pulvillus   Metatarsus

Fig. 23-15. Legs of the worker honeybee.

remove pollen from the pollen basket and to clean wings.

3. Metathoracic leg (third). The pollen basket is on the outer, concave surface of the tibia and the long hairs curve over its depression to cover it somewhat. A pincerlike structure between the tibia and metatarsus is composed of rows of spines, the pacten (L. *pecten,* comb), and a liplike auricle (L. *auricula,* small ear). The pecten and auricle convey pollen to and pack pollen into the pollen basket. On the inner surface of the metatarsus are numerous transverse rows of stiff, bristlelike pollen combs to comb out pollen from various body parts and to handle wax. The wax is secreted in flat scales by a glandular area on the underside of the abdomen. The wax is masticated

by the mandibles before it is used in building the "cells" of the honeycomb.

INGESTION AND DIGESTION

The mouthparts may be studied from Fig. 23-16. One pair of smooth mandibles lies beneath the upper lip (labrum) and is used in masticating wax. The sucking mouthparts assist the suction of the pharynx to convey fluids into the digestive tract. A long esophagus extends from the pharynx to the large honey sac (crop) in the abdomen. A large cylindrical stomach leads into the intestine, which joins the rectum, ending in the anus.

The nectar of flowers is sucked up and stored in the honey sac where it chemically changes into honey (Fig. 23-17, *A*). The honey is regurgitated into the "cells" of the honeycomb. Here, the honey is still further dehydrated by currents of air, caused by the rapid vibrations of the wings. An average colony of bees in an average season collects about 40 pounds of honey.

Pollen is rich in the protein that honey lacks, so pollen ("bee bread") is essential in the diet of bees. The pollen collected in the pollen basket is placed in certain "cells" of the honeycomb. "Bee-glue" or propolis (Gr. *pro,* for; *polis,* city) is a resin collected from plants, and it is used in filling cracks and cementing loose parts. Various types of cells constitute the honeycomb.

CIRCULATION

A long, delicate, tubular, muscular heart (Fig. 23-17, *B*) in the middorsal region of the body discharges the colorless blood toward the head region. Blood enters the heart through five pairs of ostia, each pair leading into a chamber of the heart. Valves prevent the backflow into the body during contraction. Blood discharges at the head region and passes into the hemocoel, from which it reenters the heart chambers. The blood plasma contains white blood corpuscles.

RESPIRATION

Respiration (Fig. 23-17, *C*) occurs through pairs of very small spiracles located along the

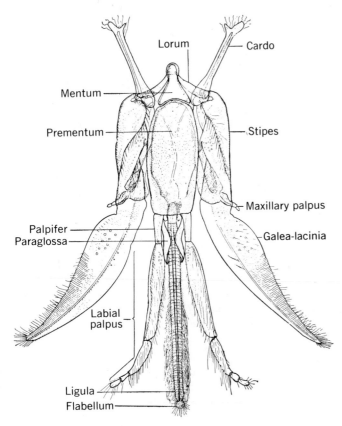

Lorum — Cardo

Mentum —

Prementum —

Stipes

Maxillary palpus

Palpifer —
Paraglossa —

Galea-lacinia

Labial
palpus

Ligula —
Flabellum —

Fig. 23-16. Mouthparts of the worker honeybee (much enlarged). Galea-lacinia is also known as the maxilla; the ligula is known as the glossa, or tongue; the flabellum is also known as the bouton, or labellum. Labial palpus and ligula together constitute the lower lip, or labium.
(From Braungart, D. C., and Buddeke, R.: An introduction to animal biology, ed. 6, St. Louis, 1964, The C. V. Mosby Co.)

sides of the thorax and abdomen and leading into a branched system of tracheae to convey air to all body parts. Certain tracheae may possess enlarged air sacs.

EXCRETION

Numerous, hollow, glandular, threadlike malpighian tubules excrete wastes into the intestine, much in the same manner as in grasshoppers.

COORDINATION AND SENSORY EQUIPMENT

A large "brain" (supraesophageal ganglion) in the dorsal part of the head supplies nerves to the eyes and antennae. The brain is connected by a ring (nerves) to the subesophageal ganglion, which supplies nerves to the mouthparts. A ventral nerve chain extends posteriorly from the subesophageal ganglion along the midventral side of the body. The chain is a double nerve strand and connects with two thoracic ganglia and five abdominal ganglia (Fig. 23-17, *D*).

The hairlike end organs of the sense of touch are present on various body parts but are particularly numerous on the tip of the antennae. The pair of jointed, hairy antennae has numerous, sound-sensitive pits that are thought to be for auditory purposes. Other pits on the antennae are thought to be for olfactory purposes. The so-called tongue bears numerous, bristlelike taste setae. Bees can be trained to estimate time intervals because some have been trained to come to a source of food at regular intervals. The pair of large compound eyes on the top and side of the head is constructed and functions similarly to that of the grasshopper. The color sense of bees is better adjusted to the shorter wave-lengths of the light spectrum, that is, toward the blue end of the spectrum. Three small simple eyes (ocelli) are present on the dorsal side of the head.

The sting is a modified ovipositor that is used for protection (Fig. 23-18). It is composed of two straight, grooved lancets (darts) with barbs at the tips and with muscles for their operation. A large storage poison sac is connected with the base of the sting. Two acid glands and an alkaline gland mix their secretions to form the poisonous material that is injected when the bee stings. After stinging, the worker bee leaves the sting, poison sac, and glands and dies. Males do not have a sting.

REPRODUCTION

The worker honeybee contains only vestigial (L. *vestigium*, trace) reproductive organs, since it is an undeveloped female. The reproductive

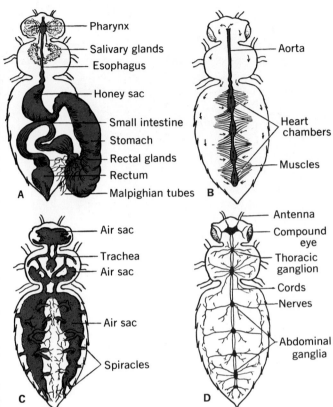

A

- Pharynx
- Salivary glands
- Esophagus
- Honey sac
- Small intestine
- Stomach
- Rectal glands
- Rectum
- Malpighian tubes

B

- Aorta
- Heart chambers
- Muscles

C

- Air sac
- Trachea
- Air sac
- Air sac
- Spiracles

D

- Antenna
- Compound eye
- Thoracic ganglion
- Cords
- Nerves
- Abdominal ganglia

Fig. 23-17. Internal anatomy of the worker honeybee. **A,** Digestive system; **B,** circulatory system; **C,** respiratory system; **D,** nervous system.

organs of the male (drone) include a pair of bean-shaped testes, which produce sperms that are carried away by a pair of slender vasa deferentia. These expand to form the seminal vesicles for sperm storage. The two seminal vesicles combine to form one ejaculatory duct, which leads to the copulatory mechanism. One pair of large accessory glands secretes and empties nourishment into the ejaculatory duct.

In the female (queen) a pair of large ovaries produces eggs, which are carried by a pair of oviducts. These unite to form a tubular vagina leading to the exterior. A spermatheca attached to the vagina stores sperm received from the male during copulation. The queen is fertilized once in a lifetime, during a nuptial flight when swarming, and the sperms remain alive for years in the spermatheca. As an egg passes down the ovary toward the oviduct, it receives a shell with a small

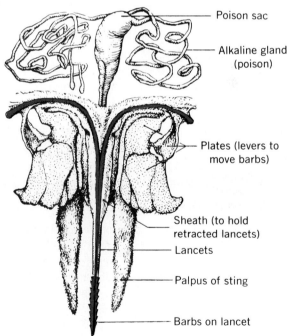

- Poison sac
- Alkaline gland (poison)
- Plates (levers to move barbs)
- Sheath (to hold retracted lancets)
- Lancets
- Palpus of sting
- Barbs on lancet

Fig. 23-18. Sting of the worker honeybee (parts somewhat separated and enlarged).

opening, the micropyle, through which a sperm may enter. A queen may lay an unfertilized egg to develop a drone or fertilized eggs to develop females, either queens or workers. A queen may lay 1,500 eggs per day for weeks at a time, and she may live several years. The eggs are small, oblong, and bluish white. Fertilized eggs are placed in worker or queen cells of the honeycomb, and unfertilized eggs, in the drone cells. A wormlike, whitish larva ("grub") hatches from the egg in 4 days. All larvae are fed on a specially prepared and predigested mixture of honey and pollen ("royal jelly") for a few days, after which the drone and worker larve are fed on plain honey and pollen, whereas the queen larva is kept on the "royal jelly" diet. This special food causes the larva to develop into a queen instead of a worker. After 6 days a larva develops into a pupa (L. *pupa,* baby) enclosed in a silken coccoon. A worker pupa changes into an adult bee in about 13 days, a queen in about 7 days, and a drone in about 15 days (Fig. 23-14).

Insect development (metamorphosis)

Many insects undergo marvelous changes as they develop from the egg to the adult stage. These changes are called metamorphosis.

Ametamorphosis

In ametamorphosis (or no metamorphosis) the egg develops into a young form that resembles the adult in all respects excepts size. After several molts the adult stage is reached.

Incomplete metamorphosis

In incomplete metamorphosis the egg develops into a larva that may display some resemblances to the adult and gradually transforms into an adult after a series of molts. Such larval forms are usually called nymphs, and the changes in body form occur gradually and are not very great between successive stages (Fig. 23-13).

If such larvae are aquatic, they are called naiads and often differ from the adult in having a respiratory system developed for aquatic

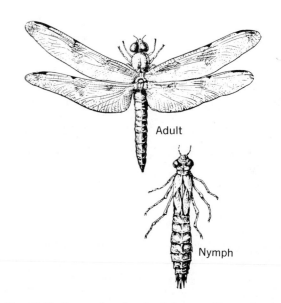

Fig. 23-19. Dragonfly of order Odonata, illustrating incomplete metamorphosis. The nymph has large eyes and developing wings and is aquatic.

breathing. The changes in body form are quite extensive (Fig. 23-19). In some classifications those possessing nymph stages are said to undergo gradual metamorphosis, whereas those with naiad stages are said to undergo incomplete metamorphosis.

Complete metamorphosis

In complete metamorphosis the egg develops into a segmented larval stage, to be followed by a quiescent pupal stage, which in turn develops into an adult stage (Fig. 23-20). In the larval stage the process of feeding and growth are outstanding; in the pupal stage the animal is enclosed in a case such as a fibrous cocoon or a membranous chrysalis, and the pupa is completely transformed into an emerging adult; in the adult stage the propagation of the species is a chief function, although feeding and other activities may take place.

Hormones

Experiments have demonstrated that crustaceans and insects produce chemical substances called hormones which regulate such activities as metamorphosis, molting, growth, and pigmentation. Work in the field of genetics also has shown that minute quantities of gene-controlled chemical substances are released by various tissues and transported to different parts of the animal to influence metabolic activities in various regions. The production and release of these hormones are controlled by a number of internal and external environmental factors, possibly even several tissues may be involved in hormone secretion. Much information concerning insect hormones has been obtained from studies of the embryonic development of certain forms.

V. B. Wigglesworth in England experimented with *Rhodnius,* which has an elongated head and neck that can be cut at various levels or entirely removed. The larval nymphs suck blood and pass through a series of five molts before becoming adults. The experiments showed that two hormones are involved in the metamorphosis: a juvenile hormone, which causes the retention of nymphal characteristics, and a growth-and-differentiation (GD) hormone, which stimulates growth and differentiation of tissues and organs to the adult status. The juvenile hormone is secreted by a special cerebral gland, the corpus allatum (L. *corpus,* body; *allatus,* brought to, changed), which lies behind the brain. If this gland is properly removed from the larva of this insect, molting is prevented. It was shown that the brain cells secrete the GD hormone. In this insect it is possible either to cut the head in such a way that the brain may be removed, leaving the corpus allatum intact, or to remove both areas. When the brain alone is removed, the nymph remains juvenile and undergoes extra molts. If both brain and corpus allatum are removed, the nymph changes to a miniature adult. When nymphs that were secreting large quantities of the juvenile hormone were grafted to headless adults, the latter molted and even developed certain nymphlike traits.

Experiments on the giant silkworm moth *(Cecropia)* by C. M. Williams of Harvard University showed that the juvenile hormone and the GD hormone were both present. These moths undergo complete metamorphosis, which makes it possible to analyze the actions of the hormones in more detail. The juvenile hormone is also secreted by a corpus allatum in moths, and the removal of the corpus allatum is quickly followed by transformation into the pupa stage. If a corpus allatum is surgically transplanted into a catepillar that is ready to form a pupa, this stage of metamorphosis will be prevented. If extra glands that produce juvenile hormones are transplanted into mature larvae, the larvae can be induced to undergo additional larval molts and grow to giant size before pupating and eventually forming giant adults.

The pupal stage of the giant silkworm moth undergoes a period of dormancy of about 5

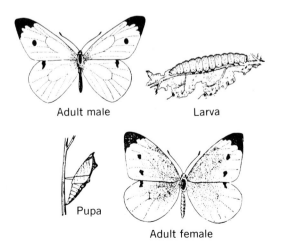

Adult male   Larva

Pupa

Adult female

Fig. 23-20. Complete metamorphosis of the cabbage butterfly *(Pieris).* The egg develops through successive stages of the larva into the pupa; the adult emerges from the pupa.
(Courtesy General Biological Supply House, Inc., Chicago, Illinois.)

Table 23-3 Orders of the class Insecta (insects)

Name of order	Examples	Wings	Mouthparts	Meta-morphosis	Miscellaneous
Thysanura (Gr. thysanos, tassel; oura, tail)	Bristletails, silverfish	Wingless	Chewing	None (larva resembles adult)	Primitive insects; 2 to 3 long, caudal appendages; common in dark, moist places
Collembola (Gr. kolla, glue; ballo, to put)	Springtails, snow fleas	Wingless	Chewing or sucking	None	Primitive insects; springing organ on ventral side of abdomen; sticky tubelike projection on abdomen; common under decaying leaves or bark
Orthoptera (Gr. orthos, straight ptera, wings)	Grasshoppers, katydids, cockroaches, crickets, walkingsticks, praying mantises	2 pairs (usually); forewings straight, leathery, hind wings folded	Chewing	Incomplete	Most species live on vegetation; mantises feed on other insects
Isoptera (Gr. isos, equal; ptera, wings)	Termites	2 pairs; long, narrow, held flat on back; or wingless	Chewing	Incomplete	Social insects with several castes; build earthen tubes for passageways
Neuroptera (Gr. neuron, nerve; ptera, wings)	Dobson flies (hell-grammite), aphis lions (lacewing flies)	2 pairs; similar, usually with many cross veins	Chewing	Complete	Some species predacious; larvae may suck blood from their prey
Ephemerida (Gr. ephemeros, living but a day) or Ephemeroptera	Mayflies	2 pairs; membranous; forewings large; hind wings small, or absent	Vestigial (in adults); chewing (in naiads)	Incomplete	Adults take no food and live for 1 day; serve as fish food; naiads have lateral, tracheal gills
Odonata (Gr. odon, tooth)	Dragonflies, damselflies	2 pairs; membranous; hind wings as large or larger than forewings	Chewing	Incomplete	Adults and naiads both predacious; body long, tusklike; large head; large, compound eyes
Plecoptera (Gr. plekos, folded; ptera, wings)	Stoneflies	2 pairs; membranous; hind wings folded under forewings when at rest	Chewing	Incomplete	Larval naiads live in water under rocks; often with tufts of tracheal gills
Corrodentia (L. corrodens, gnawing)	Book lice, bark lice	Book lice wingless; bark lice have 2 pairs membranous wings that have few distinct veins	Chewing	Incomplete	Book lice minute; soft bodied; found in old papers; bark lice have oval bodies and feed on plants

Order	Examples	Wings	Mouthparts	Metamorphosis	Characteristics
Mallophaga (Gr. *plekos*, folded; *phagein*, to eat)	Biting lice, such as chicken lice, cattle lice	Wingless	Chewing	Incomplete	Flat, broad body and head; sharp claws; on skin, hair, feathers of mammals and birds
Thysanoptera (Gr. *thysanos*, fringe; *ptera*, wings)	Thrips, such as wheat thrips, fruit thrips, onion thrips	2 pairs fringed with long hairs	Piercing-sucking	Incomplete	Minute, slender body; suck plant juices
Anoplura (Gr. *anoplos*, unarmed; *oura*, tail)	Sucking lice, such as human lice, body lice	Wingless	Piercing-sucking	None	Flat broad body; claws for clinging to hair; parasites on mammals
Hemiptera (Gr. *hemi*, half; *ptera*, wings)	True bugs, such as squash bugs, assassin bugs	2 pairs; forewings thickened at base, which overlap on back; membranous hind wings	Piercing-sucking (jointed beak)	Incomplete	Beak arises from front of head; many species on vegetation
Homoptera (Gr. *homos*, same; *ptera*, wings)	Cicadas, leafhoppers, aphids	2 pairs (usually) held rooflike over back	Piercing-sucking	Incomplete	Beak arises from hind part of lower side of head; many species on vegetation
Dermaptera (Gr. *derma*, skin; *ptera* wings)	Earwigs	2 pairs; forewings leathery, hind wings folded under forewings	Chewing	Incomplete	One pair pincerlike appendages at tip of abdomen; elongated body; not common in United States
Coleoptera (Gr. *koleos*, sheath; *ptera*, wings)	Beetles, such as tiger beetles, ladybird beetles, June beetles, ground beetles	2 pairs; forewings (elytra) hard, sheathlike; hind wings folded under elytra	Chewing	Complete	Chitinous covering usually heavy; sizes vary from small to very large; common in many places
Mecoptera (Gr. *mekos*, long; *ptera*, wings)	Scorpion flies	2 pairs; long, narrow, membranous, some species wingless	Chewing	Complete	Clasping organ at tip of male abdomen resembles sting of scorpion
Trichoptera (Gr. *trichos* hair; *ptera*, wings)	Caddis flies	2 pairs; membranous, covered with hairs	Vestigial (in adults)	Complete	Soft, mothlike body; larva usually lives in case formed of pebbles and sand
Lepidoptera (Gr. *lepis*, scale; *ptera*, wings)	Butterflies, skippers, moths	2 pairs; membranous and covered with overlapping scales	Sucking (coiled under head)	Complete	Moth antennae usually featherlike; butterfly larvae called catepillars; some form cocoons; pupa often called chrysalis

Continued.

Table 23-3 Orders of the class Insecta (insects)—cont'd

Name of order	Examples	Wings	Mouthparts	Meta-morphosis	Miscellaneous
Diptera (Gr. *dis*, two; *ptera* wings)	True flies, such as houseflies, fruitflies, crane flies, mosquitoes, gnats	1 pair only; hind wings represented by pair of knobbed threads, called halteres	Piercing-sucking (proboscis)	Complete	Head attached to thorax by slender neck; compound eyes usually large; larvae called maggots
Siphonaptera (Gr. *siphon*, tube; *a-*, without; *ptera*, wings)	Fleas, such as human fleas, dog and cat fleas, rat fleas	Wingless	Piercing-sucking	Complete	Body compressed sideways, legs for leaping; parasitic on mammals and birds
Hymenoptera (Gr. *hymen*, membrane; *ptera*, wings)	Honeybees, wasps, ants, gall flies	2 pairs; wings on each side held together by hooks	Sucking or chewing	Complete	Abdomen of female usually has sting; many species live in colonies; some parasitic on other insects

months at room temperature, which is called the diapause (Gr. *diapauvein,* to make to cease). If pupae are placed in a temperature of 3° to 5° C. for about 6 weeks, then at room temperature the pupae will become adults in about 1 month.

When brains from chilled pupae are surgically implanted into other pupae during their diapause, the latter metamorphose into adults. Even chilled pupae that are surgically grafted to other pupae in their diapause stage cause the latter to metamorphose into adults. If a pupa is cut into a head-thorax section and an abdominal portion and each of these sections is fastened to a plastic cover glass, the head-thorax piece transforms into an adult anterior end, but the abdominal section does not develop. The introduction of glandular tissue from the anterior end, however, causes the abdominal section to develop into an adult posterior end.

It was observed that chilled brains alone did not cause the change but were effective only if administered together with tissue from the so-called prothoracic gland. Williams proved that chilling stimulates certain gland cells in the insect brain to secrete a hormone that stimulates the prothoracic glands to release the GD hormone. The absence of either the brain hormone or the prothoracic hormone prevents metamorphosis into an adult.

Coloration, color patterns, and protective resemblance

Many insects possess colors of various kinds, especially the butterflies, moths, and beetles. Colors may be classed as chemical colors and physical colors. Chemical colors are caused by pigments that in insects are mostly black or brown, although yellow, orange, red, and white may be present. Physical colors are caused by the structure of the surface that may diffract (scatter light into its different colored rays) or refract (deflect rays of light from a straight path to the oblique, passing from one medium to another). Common physical colors in insects include violet, blue-green, silver, and gold, as represented in the metallic colors of certain beetles, the iridescence

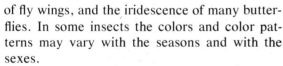

Fig. 23-21. Dead leaf butterfly.

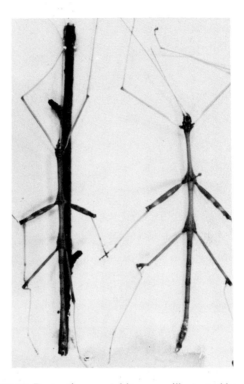

Fig. 23-22. Protective resemblance, as illustrated by the so-called walking stick; because of its resemblance to a dried twig, it may be afforded a certain degree of protection.

of fly wings, and the iridescence of many butter-flies. In some insects the colors and color patterns may vary with the seasons and with the sexes.

Certain insects have colors that blend in with their natural surroundings, thus giving them some protection. This may be a result of color, of the peculiar structures they possess, or of a combination of the two. Green katydids living among green plants are less conspicuous. The underwing moth has a mottled gray color resembling the bark of trees. The color of other insects resembles that of flowers that they customarily visit. The dead-leaf butterfly *(Kallima)* of India has the upper side of its wings brightly colored. However, the underside is colored like dead leaves,

and when at rest with wings held together above the back, the butterfly is difficult to detect (Fig. 23-21). It must rest on a twig in a particular way or it does not resemble a dead leaf, hence it may not be protected. The common walking-sticks are difficult to detect because their peculiar structure and coloration resemble a stick (Fig. 23-22).

A brief study of typical representatives of some of the important orders (Fig. 23-23) is given in Table 23-3.

Harmful insects

Space will not permit a detailed consideration of all the ways in which insects harm or benefit man, but there are probably more harmful than

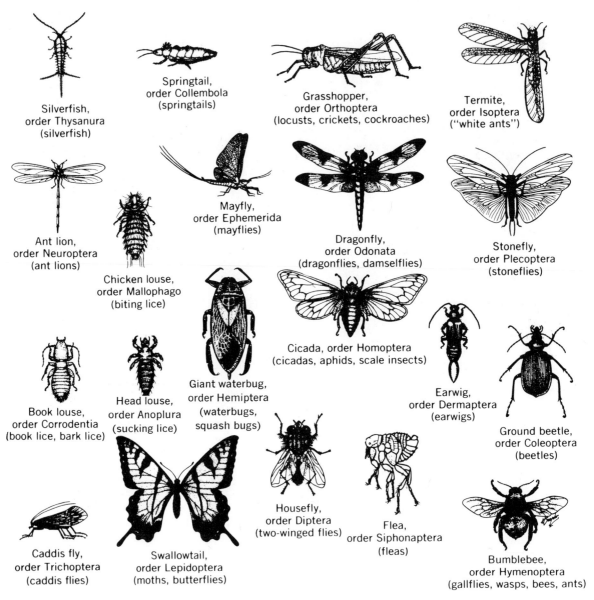

Fig. 23-23. Representatives of various orders of insects (not drawn to scale). *(Courtesy General Biological Supply House, Inc., Chicago, Illinois.)*

beneficial types. Harmful types (Fig. 23-24) may be grouped as follows.

1. Many insects destroy or damage plants, lumber, fruit trees, plant products, stored grains, fruits, and vegetables. The great numbers of in-sects involved are of many varieties. In recent estimates by the United States Department of Agriculture, insects cause $4,000,000,000 damage annually to farm crops, stored foodstuffs, forests, and domestic animals. Common exam-

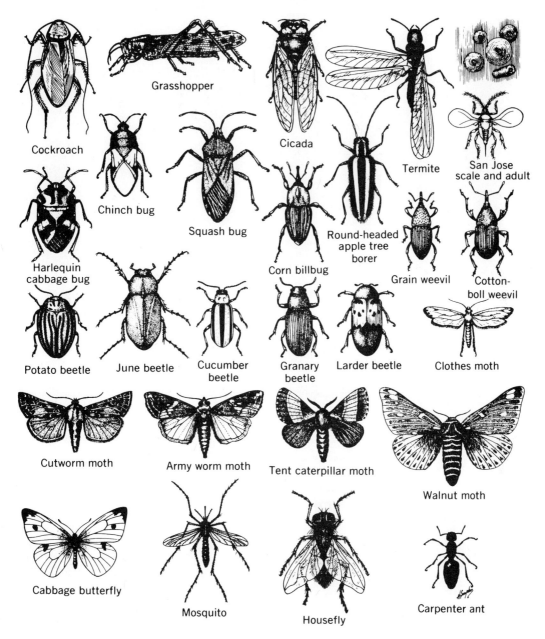

Fig. 23-24. Harmful or detrimental insects from various orders (not drawn to scale). *(Courtesy General Biological Supply House, Inc., Chicago, Illinois.)*

ples include chinch bugs, Hessian flies, corn borers, cotton-boll weevils, wireworms, grain weevils, army worms, grasshoppers, potato beetles, plant aphids (plant lice), and scale insects.

2. Insects that affect domestic animals and animal products include fleas, biting lice, sucking lice, mosquitoes, and various types of flies, such as horse botflies, hornflies of cattle, and warble flies, whose larvae burrow into the skin of cattle.

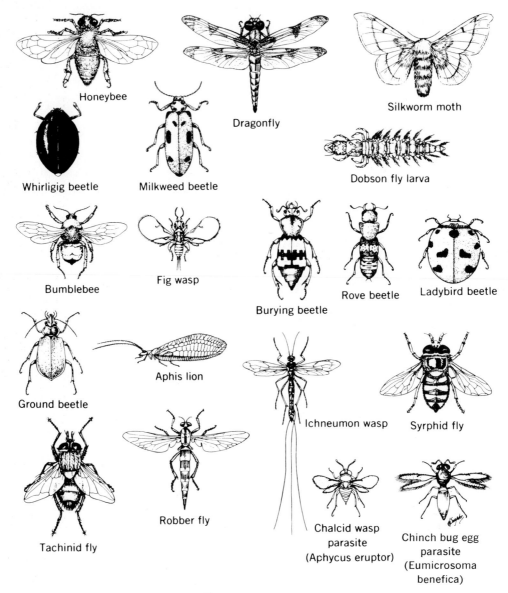

Fig. 23-25. Beneficial insects from various orders (not drawn to scale). *(Courtesy General Biological Supply House, Inc., Chicago, Illinois.)*

3. Insects are very destructive in houses. Termites may destroy wooden buildings and wooden contents. Furs, feathers, clothing, carpets, and fabrics may be damaged by clothes moths and carpet beetles. Papers may be destroyed by the simple silverfish, common in dark, damp places. Foods may be destroyed or damaged by ants, cockroaches, and weevils. Household pests also include mosquitoes, flies, and bedbugs.

4. Various diseases may be transmitted by

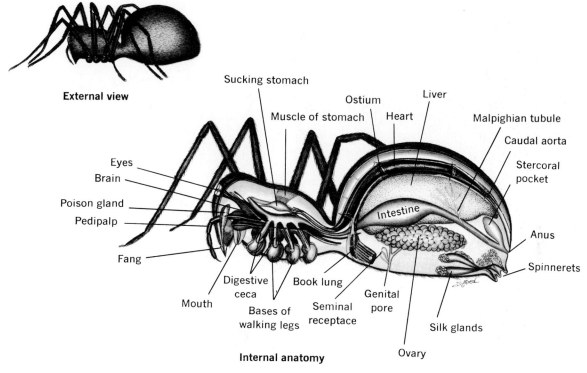

External view

Sucking stomach

Ostium

Liver

Muscle of stomach

Heart

Malpighian tubule

Caudal aorta

Eyes

Brain

Stercoral pocket

Poison gland

Pedipalp

Intestine

Anus

Fang

Spinnerets

Digestive ceca

Book lung

Mouth

Genital pore

Bases of walking legs

Seminal receptace

Silk glands

Ovary

Internal anatomy

Fig. 23-26. Anatomy of the spider.
(From Hickman, Cleveland P., Sr., Hickman, Cleveland P., Jr., and Hickman, Frances M.: Integrated principles of zoology, ed. 5, St. Louis, 1974, The C. V. Mosby Co.)

many types of insects. Examples include malaria and yellow fevers, carried by mosquitoes; typhoid, dysentery, and diarrhea, carried by flies; African sleeping sickness, carried by tsetse flies; bubonic plaque, carried by fleas; and typhus fever, carried by body lice.

Beneficial insects

It might seem that few insects are beneficial. Those activities of insects that are of benefit to man, however, (Fig. 23-25) might be grouped as follows.

1. Several insects make products of importance to man, such as honey, beeswax, shellac (lac) wax, silk, and cochineal. Honeybees produce 500,000,000 pounds of honey in the United States annually, and about 10,000,000 pounds of the beeswax is used annually. Lac insects secrete a wax from which shellac is made. Each cocoon of the silkworm moth contains about 1,000 feet of spun thread; about 25,000 such cocoons are required to produce 1 pound of silk. About 50,000,000 pounds of silk are produced in the world annually. The dried bodies of a scale insect that lives on cactus may be used in making the dye cochineal, but aniline dyes have largely supplanted this source.

2. Several types of insects are necessary in the cross-fertilization of flowers of many of our important and useful plants. Bees and certain other insects pollenize the flowers of clovers, berries, apples, pears, and other fruits and crops. The

Fig. 23-27. Scorpion of class Arachnoidea, showing the segmented appendages and sting at the tip of the abdomen.

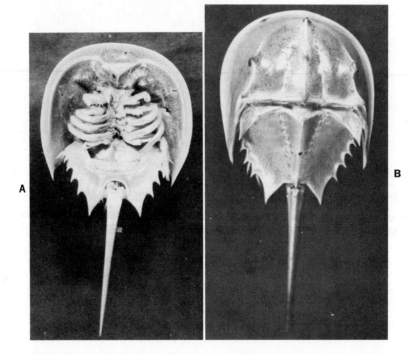

Fig. 23-28. Horseshoe crab *(Limulus)* of class Arachnoidea. **A,** Ventral view; **B,** dorsal view.

Smyrna fig could not be grown in California until a small fig wasp *(Blastophaga)* was imported to pollenize the flowers.

3. Predacious insects are beneficial because they destroy great numbers of other insects, many of which are injurious. Common examples include ground beetles, ladybird beetles, aphis lions, praying mantises, and certain types of wasps. An Australian ladybird beetle was imported to California to destroy scale insects that parasitized orange and lemon trees. Many insects lay their eggs in the larvae of harmful types, and the young hatched from these eggs slowly destroy the parasitized host.

4. Several types of insects act as scavengers, the larvae eating great quantities of dead plant

Fig. 23-29. *Peripatus* of class Onychophora, sometimes considered as the connecting link between annelids and arthropods because it possesses certain traits of both.
(Courtesy Carolina Biological Supply Company, Burlington, North Carolina.)

and animal materials, thus preventing their immediate decay and undesirable odors. Included in this group are flesh-eating flies, water-scavenger beetles, burying beetles, and tumble bugs. Insects also supply foods for animals, such as birds and fishes.

5. The maggots of certain common blowflies have been used in the healing of certain types of wounds. The maggots eat the dead tissues and secrete a substance called allantoin, which stimulates healing. It is an oxidation product of uric acid and is also present in the allantoic fluid of cows and in beet juice.

CLASS ARACHNOIDEA

Spiders, scorpions, ticks, mites, horseshoe crabs, and daddy longlegs are grouped together in the class Arachnoidea (Gr. *arachne,* spider; *eidos,* like). Many arachnids are predacious, sucking foods from their prey. Many types actually destroy injurious insects. The bite of the black widow spider *(Latrodectus)* may cause death in man. This is a black spider with an orange-red ''hourglass'' on the underside of the abdomen. The term widow is erroneously applied, because it was thought that she devoured her male immediately after mating. The sting of scorpions may be painful. Certain mites may

transmit diseases, damage plants by sucking their juices, or cause annoying skin irritations.

The web-making behavior of spiders is now being used to advantage by pharmaceutical firms in the testing of drugs. The spiders make various types of abnormal webs under the influence of drugs.

Scorpions have elongated, segmented bodies and a poisonous sting at the tip of the abdomen (Fig. 23-27). They are primarily inhabitants of the tropics and subtropics, where some forms may harm human beings. In courtship scorpions seize each other's claws and perform a mating dance.

Ticks and mites are arachnids with no external segments; even the cephalothorax and abdomen are fused. The mites are quite small and are found in many places. The female of the itch mite burrows into the skin to lay her eggs. Another species causes mange in dogs and other domestic animals. Chiggers (red ''bugs'') are the larvae of red mites and have only three pairs of legs. They are quite commonly found in vegetation at times. When a chigger attaches itself to the skin, it eats out a depression with the aid of a digestive secretion. This irritating fluid causes itching and red blotches. Ticks are usually larger than mites. They suck blood from vertebrate animals, and one large feeding may last for some time. The larvae are present in bushes and attach themselves to a host as it passes. Ticks may transmit such diseases as Texas cattle fever, relapsing fever, and Rocky Mountain spotted fever.

Horseshoe (king) crabs are common in shallow water along the Atlantic coast. These "living fossils" represent a very ancient group of animals. An unsegmented, horseshoe-shaped, dorsal cover and a broad, hexagonal abdomen with a long tail-like telson characterize them (Fig. 23-28). Ventrally, the cephalothorax bears five pairs of walking legs and a pair of chelicerae, whereas the abdomen bears six pairs of thin, broad appendages fused in the median line. On certain abdominal appendages are exposed book gills for respiratory purposes. One pair of compound eyes and a pair of simple eyes are present.

Important studies on the physiology of vision are being made using the horseshoe crab as the subject.

Some contributors to the knowledge of
ARTHROPODS

Swammerdam

Jan Swammerdam (1637-1680)

A Dutchman who studied the anatomy and life cycles of insects. (*Historical Pictures Service, Chicago.*)

René de Reaumur (1683-1757)

A French natural philosopher and inventor who studied insects and described their characteristics. He also invented a thermometer.

Charles Bonnet (1720-1793)

A Swiss naturalist who proved that certain insects (plant aphids) reproduced by parthenogenesis (without fertilization).

Jean Marie Léon Dufour (1780-1865)

A French naturalist who was referred to as the "Wizard of Landes," where he practiced as a doctor.

Francis Comte de Castelnau de la Prote (1812-1880)

A naturalist who published an excellent book on insects that influenced the work of Henri Fabre.

Thomas H. Huxley (1825-1895)

An English biologist who made important contributions in the studies of arthopods.

Fabre

J. Henri Fabre (1823-1915)

A French naturalist and teacher who was a great student of insects. The Belgian writer Maeterlinck called Fabre "the insect's Homer." Darwin called him "a savant who thinks like a philosopher and writes like a poet." Fabre wrote *The Mason Bees* (1914); *Insect Life; The Life of the Spider,* which described his observations of their interesting behaviors. (*The Bettmann Archive, Inc.*)

Karl von Frisch

An Austrian zoologist who about 40 years ago made extensive studies of bees and discovered a method of communication among them, particularly by means of so-called dances. (*Historical Pictures Service, Chicago.*)

Frisch

Review questions and topics

1 What general characteristics describe the arthropods as a group?
2 What are the characteristics that distinguish members of the different classes of arthropods, including examples of each class?
3 Discuss the significance of jointed appendages and an exoskeleton of chitin.
4 Discuss the significance of the division of the body into head, thorax, and abdomen.
5 Discuss the importance of wings as found in certain insects.
6 Discuss the process of respiration by means of tracheae, book lungs, and gills, including examples of each type.
7 Describe the structure and functions of the "open" type of circulatory system and the presence of a hemocoel, including examples.
8 Explain the necessity of molting during the growth of arthropods.
9 Describe the structure and functions of typical arthopod nervous systems.
10 Discuss the status of sensory organ development in arthropods, including specific examples of each type of organ.
11 Describe the different types of metamorphosis, with examples of each type.
12 Describe the anatomy and functions of each part of the various systems in (1) the crayfish, (2) the grasshopper, and (3) the honeybee.
13 Discuss the biology of centipedes, millipedes, and spiders.
14 Discuss coloration, color patterns, and protective resemblance, give examples of each.
15 Discuss hormones and insect activities, giving specific examples.
16 Discuss harmful and beneficial insects, giving examples of each.
17 Discuss the complex societies of bees.
18 List the distinguishing characteristics of each order of insects, giving examples of each order.

Selected references

Baker, E. W., and Wharton, G. W.: An introduction to acarology, New York, 1952, The Macmillan Company.
Borrer, D. J., DeLong, D. W., and Triplehorn, C. A.: An introduction to the study of insects, ed. 4, New York, 1976, Holt, Rinehart and Winston, Inc.
Campbell, F. L., editor: Physiology of insect development, Chicago, 1959, University of Chicago Press.
Carlisle, D. B., and Knowles, F.: Endocrine control in crustaceans, New York, 1969, Cambridge University Press.
Esch, H.: The evolution of bee language, Sci. Amer. Apr., 1967.
Fox, R. M., and Fox, J. W.: Introduction to comparative entomology, New York, 1964, Reinhold Publishing Corp.
Gertsch, W. J.: American spiders, New York, 1949, D. Van Nostrand Co., Inc.
Green, J.: A biology of Crustacea, London, 1961, H. F. & G. Witherby, Ltd.
James, M. T., and Harwood, R. F.: Herm's medical entomology, New York, 1969, The Macmillan Company.
Jaques, H. E.: How to know the insects, ed. 2, Dubuque, Iowa, 1947, William C. Brown Company, Publishers.
Jones, J. C.: The sexual life of a mosquito, Sci. Amer. **218:** 108-16, 1968.
Meglitsch, P. A.: Invertebrate zoology, ed. 2, New York, 1972, Oxford University Press, Inc.
Metcalf, C. L., Flint, W. P., and Metcalf, R. L.: Destructive and useful insects, ed. 4, New York, 1962, McGraw-Hill Book Company.
Michener, C. D.: The social behavior of the bees, Cambridge, Mass., 1974, Harvard University Press.
Patton, R. L.: Introductory insect physiology, Philadelphia, 1963, W. B. Saunders Company.
Petrunkevitch, A.: The spider and the wasp, Sci. Amer. Aug., 1952.
Richards, A. G.: The integument of arthropods, Minneapolis, 1951, University of Minnesota Press.
Roeder, K. D.: Moths and ultrasound, Sci. Amer. **212:**94-102, 1965.
Romoser, W. S.: The science of entomology, New York, 1973, The Macmillan Company.
Snodgrass, R. E.: A textbook of arthropod anatomy, Ithaca, New York, 1952, Comstock Publishing Associates.
Waterman, T. H., editor: The physiology of Crustacea, 2 vol., New York, 1960, Academic Press Inc.
Williams, C. M.: The metamorphosis of insects, Sci. Amer. Apr., 1950.
Wilson, D. M.: The flight-control system of the locust, Sci. Amer. **218:**83-90, 1968.
Wilson, E. O.: The insect societies, Cambridge, 1971, Harvard University Press.

24 Phylum Echinodermata

Class Asteroidea (starfishes)
Class Echinoidea
   Sea urchins
   Sand dollars
Class Holothuroida (sea cucumbers)
Economic importance of echinoderms

All echinoderms are marine animals. In contrast to most of the higher animals the "spiny-skinned" animals have radial symmetry. They share an endoskeleton and some embryologic developments with the chordates. A unique feature of the echinoderms is the presence of a water-vascular system. Table 24-1 shows many of the characteristics of the animals included in the phylum Echinodermata.

CLASS ASTEROIDEA (STARFISHES)

The starfish, *Asterias* (Fig. 24-1), has on its upper (aboral) surface numerous short spines attached to the calcareous skeletal plates, a circular, sievelike madreporite plate, which is the entrance to its water-vascular system, and an anus. The calcareous plates are held together by muscles and connective tissues. Surrounding the spines are minute, pincerlike pedicellariae (Fig. 24-2), supplied with jaws that are operated by

Fig. 24-1. The starfish, *Asterias*.

muscles. The pedicellariae are used to capture food, clean the body surface, and protect the soft, saclike dermal branchiae ("skin gills"), which carry on respiration.

On the under (oral) surface are a centrally located mouth and five ambulacral grooves (L. *ambulacrum,* covered), one in each arm, from which rows of muscular tube feet extend (Fig. 24-3). Other species of starfish may have more

340

Table 24-1 Characteristics of echinoderms (phylum Echinodermata)

Class	Body	Ambulacral grooves	Tube feet	Pedicellaria (pincerlike organs)	Anus and madreporite plate	Examples
Asteroidea (Gr. *aster*, star; *eidos*, like)	Typically, 5 arms indistinctly marked off from central disk	Open on oral (under) side of animal	With suckers	Present	Anus aboral; madreporite aboral	Starfishes (*Asterias*)
Ophiuroidea (Gr. *ophis*, snake; *oura*, tail; *eidos*, like)	Typically, 5 long, slender, jointed arms sharply marked off from central disk	Absent or covered with ossicles (plates)	Without suckers	Absent	Anus absent; madreporite oral	Brittle stars (*Ophiura*)
Echinoidea (Gr. *echinos*, hedgehog; *eidos*, like)	Globe shaped (sea urchins), or disk shaped (sand dollars); compact skeleton (test); movable spines	Covered with ossicles	With suckers	Present	Anus aboral (sea urchins); anus marginal (sand dollars); madreporite aboral	Sea urchins (*Arbacia*); sand dollars
Holothuroidea (Gr. *holothurion*, sea cucumber; *eidos*, like)	Soft and cucumber shaped; no arms; no spines; 10 to 30 circumoral tentacles (modified tube feet); scattered ossicles (dermal plates) confined to body wall; cloaca usually with respiratory tree	Concealed	With suckers	Absent	Anus aboral; madreporite internal	Sea cucumbers (*Thyone*)
Crinoidea (Gr. *krinon*, lily; *eidos*, like)	Attached during part or all of life by aboral stalk of ossicles; some have 5 branched arms each bearing feather-like pinnules; no spines	Open, ciliated, and on oral surface	Tentacle-like in ambulacral grooves; without suckers	Absent	Anus oral; no madreporite	Sea lilies; feather stars (*Antedon*)

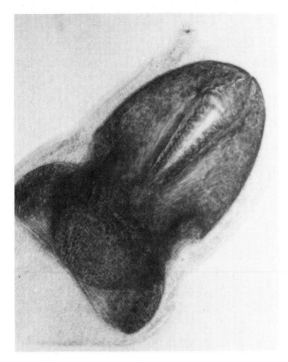

Fig. 24-2. Pedicellaria of a starfish.

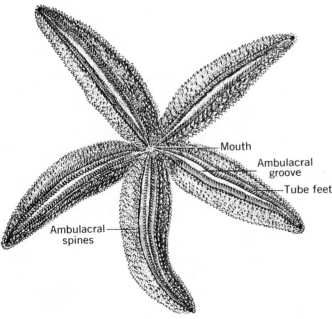

Mouth

Ambulacral groove

Tube feet

Ambulacral spines

than 40 arms. Starfish arms are moved by the action of muscles in the walls.

The water-vascular system is unique with the echinoderms (Fig. 24-4). In the starfish a porous, aboral madreporite leads to a short stone canal, which enters a circular ring canal around the mouth. From the ring canal five radial canals, one per arm, pass just above the ambulacral grooves and toward the tip of each arm. The radial canals give off connecting canals to which are attached the tube feet, with a bulb-shaped ampulla (L. *ampulla,* flask) at the inner end and a sucker at the outer end. The ampullae contain circular muscles, and the tube feet contain longitudinal muscles. When ampullae contract, valves in the connecting canals prevent the backflow of water into the radial canal, thus forcing water into the tube feet and causing the sucker to attach. The contraction of longitudinal muscles in the tube feet draws the animal forward. By regulating the internal water pressure, the tube feet function in locomotion, in capturing food, and in adhering to rocks. Nine small, spherical Tiedemann's bodies, located on the inner wall of the ring canal, are thought to produce the amebocytes of the water-vascular system.

Internally, a large coelom lined with cilia contains a fluid in which circulate the amebocytes, which function in excretion, respiration, and circulation. The digestive system contains a mouth and a short esophagus, which leads into a large thin-walled stomach (Fig. 24-5). The stomach consists of a large, oral cardiac portion and a small aboral pyloric portion. From the pyloric portion a tube passes into each arm and divides

Fig. 24-3. Starfish from the oral or underside, showing rows of tube feet extending from the ambulacral grooves. Long, movable ambulacral spines protect the tube feet when they are retracted within the groove; similar oral spines surround the mouth for protection. *(From Braungart, D. C., and Buddeke, R.: An introduction to animal biology, ed. 6, St. Louis, 1964, The C. V. Mosby Co.)*

into two branches called digestive glands, or hepatic (pyloric) ceca. These are greenish in the living animal and together with the pyloric part of the stomach secrete digestive enzymes. Above the stomach is a slender intestine that exits at the anus. A pair of brownish, branched, pouchlike rectal ceca arise from the intestine and probably have an excretory function.

Foods of the starfish consist of oysters, clams, fishes, snails, barnacles, and worms. When a starfish opens an oyster or clam, it arches its body and attaches its tube feet to the two shells of the mollusk. A steady pull eventually draws the shells apart. After some time, because of fatigue, the mollusk relaxes, and the starfish everts its stomach through its mouth and between the shells. Digestive juices begin digestion of the mollusk within its shells. The partially digested food is placed into the stomach, which is drawn back through the mouth.

The body fluid in the coelom is circulated by cilia, thus carrying absorbed foods to various body parts. The reduced circulatory system consists of a circumoral vessel (around the mouth) that connects with one radial vessel in each arm located beneath the radial canal.

Excretion is performed by amebocytes of the coelomic fluid, which collect wastes and pass them to the outside through the ciliated walls of the dermal branchiae.

Respiration is accomplished by soft, saclike

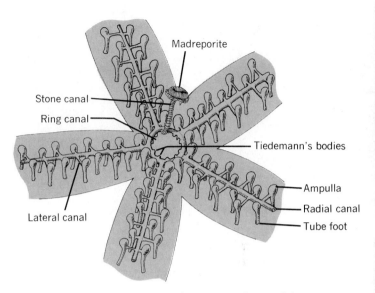

Fig. 24-4. Water-vascular system of a starfish.

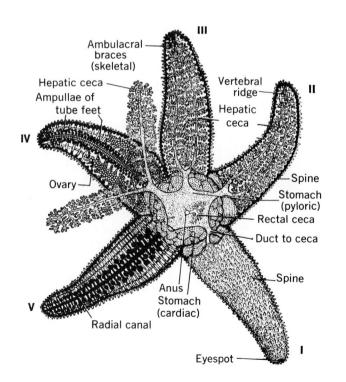

Fig. 24-5. Common starfish (*Asterias*) dissected from the aboral side to show the digestive, locomotor, reproductive, and skeletal systems. **I,** Arm (ray), showing the spines and eyespot; **II** and **III,** arms with the aboral surface removed; **IV,** arm with the aboral surface removed and the hepatic ceca moved to show the bulblike ampullae of the tube feet, etc.; **V,** arm with the internal organs and vertebral ridge removed to show the rows of ampullae of the tube feet, the connecting canals, and the radial canal (all parts of the water-vascular system).

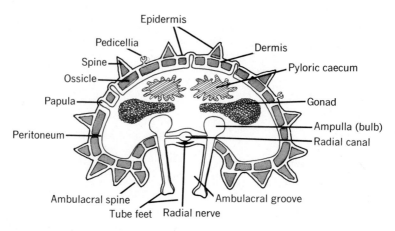

Fig. 24-6. Cross section of a starfish arm.

dermal branchiae, thin protrusions of the coelom lining that pass through small openings in the skeleton (Fig. 24-6). Oxygen and carbon dioxide are exchanged through the walls of these branchiae.

The nervous system consists of a single oral nerve ring around the mouth, which is connected with a radial nerve cord running beneath the radial canal of the water-vascular system in each arm. A double nerve ring above the other ring also passes nerves to various tissues; the tube feet are particularly sensitive. A light-sensitive eyespot and a small tactile tentacle are present at the end of each arm.

Reproduction is diecious. Five pairs of saclike, branched gonads (testes or ovaries) are present between the bases of the arms. Fertilization occurs in the water, and the zygote develops regularly through the blastula and gastrula stages. The larva formed is called a bipinnaria (L. *bis,* twice; *pinna,* feather), which is bilaterally symmetrical and swims by means of two bands of cilia. One female may release over 200 million eggs per season, and a male produces many times that number of sperms.

An arm may be automatically broken off near its base by a process known as autotomy, and any arm with part of its central disk may regenerate an entire body. A missing arm may be re-generated to form a complete animal again (Fig. 24-7).

CLASS ECHINOIDEA

Sea urchins

Typical sea urchins are somewhat globular and have a shell (test) composed of ten pairs of columns of calcareous plates, bearing long, sharp,

Fig. 24-7. Starfish, showing the regeneration of one arm.

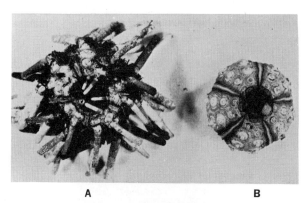

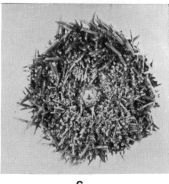

Fig. 24-8. Sea urchins of class Echinoidea. **A,** Oral view: Note the whitish teeth (Aristotle's lantern) in center; **B,** skeleton ("test") with the spines removed and showing the plates of the ambulacral and interambulacral zones; tubercles for the attachment of spines and pores for the tube feet are visible on the skeleton; **C,** oral view of another species.

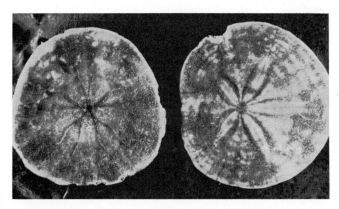

movable spines (Fig. 24-8). Five pairs of these columns are homologous to the five arms of a starfish and bear pores for the long tube feet. Spines are moved freely on knoblike elevations by muscle action. The three-jawed pedicellariae capture food and keep the body clean.

The water-vascular system consists of a madreporite; stone canal; ring canal; five radial canals; five interradial, pouchlike polian vesicles; and tube feet ampullae.

Plant and animal materials on the sea bottom are ingested by a complex apparatus called Aristotle's lantern, in which pairs of teeth are attached. Respiration occurs in ten branched, pouchlike gills located around the mouth. Many other structures resemble comparable ones in the starfish. Some sea urchins possess very long poisonous spines. Sea urchins may be green, purple, black, gray, white, and other colors.

Sand dollars

Sand dollars have flat, disk-shaped bodies with small calcareous spines (Fig. 24-9). There are no free arms; the spaces between them are more or less filled in. The aboral surface shows a starlike arrangement of areas, or "petals." The oral (under) surface shows five furrow-like areas radiating from the central mouth region. Many other structures resemble comparable ones in other echinoderms. The anus is located in the marginal edge of the animal.

CLASS HOLOTHUROIDEA (SEA CUCUMBERS)

Sea cucumbers (Fig. 24-10) have leathery bodies elongated along an oral-aboral axis and a mouth surrounded with ten to thirty retractile tentacles at the oral end and an anus at the aboral end. The tentacles are modified tube feet that

Fig. 24-9. Sand dollar of class Echinoidea. Furrowlike ambulacral areas shown radiating from the central mouth in the oral view at left; aboral areas or "petals" shown at right.

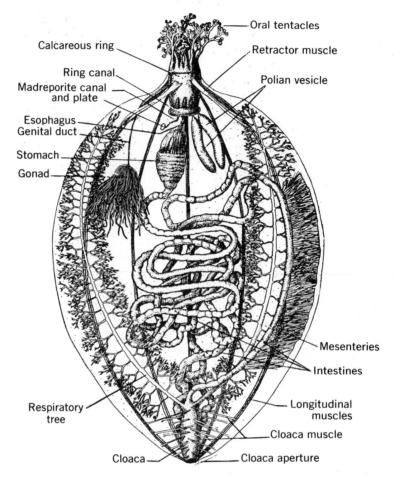

Fig. 24-10. Sea cucumber *(Thyone)* of class Holothuroidea (in longitudinal section and somewhat diagrammatic).

capture small marine organisms and transport them to the mouth. Small calcareous plates are embedded in the body. The dorsal (upper) side bears two longitudinal zones of tube feet, and the ventral side has three similar zones. Sea cucumbers move slowly by means of the tube feet and the action of body wall muscles. The coelom is extensive. The water-vascular system consists of a small, internal madreporite, a ring canal with two saclike polian vesicles, and five radial canals that are connected to the rows of tube feet.

The digestive system consists of a mouth; a short esophagus; an oval, muscular stomach; a long, looped intestine; an enlarged, muscular cloaca; and an anus. Two long, branched respiratory trees attached to the cloaca pump water in and out of the tree for respiration and excretion purposes. Respiration is also accomplished by the cloaca, tube feet tentacles, and body wall.

The nervous system includes a nerve ring with five radial nerves. Sensory organs include those affected by touch and light.

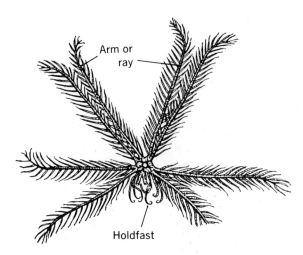

Fig. 24-11. Feather star *(Antedon)* of class Crinoidea, showing only four of the five branched arms, each with many small pinnules (oral view). Holdfasts attach it to the ground in the sea.

Reproduction is diecious, with the gonads opening to the outside behind the tentacles. When strongly stimulated, these animals cast out much of their viscera by contracting the body wall. This phenomenon is known as autotomy. They also have great powers of regeneration of lost parts. Sea cucumbers have a special kind of larval stage, known as auricularia (L. *auricula*, small ear), which bears short processes with diffuse cilia. The adult bodies may have such colors as brown, yellow, red, black, pink, and purple.

ECONOMIC IMPORTANCE OF ECHINODERMS

Starfishes destroy large numbers of mollusks, particularly oysters and clams. One adult may eat nearly a dozen oysters or clams daily. Control measures in shellfish beds include catching them with nets or spreading quicklime, which destroys them. Echinoderms may be used as agricultural fertilizers because of their lime and nitrogen content. Because of their nature, adults are not good sources of food, but eggs may be used. Echinoderm eggs are used in biologic experiments, especially those of embryologic development, including artificial parthenogenesis. The latter phenomenon was first discovered in 1900 by Jacques Loeb, when it was found that eggs would develop without the use of sperms when the chemical composition of the sea water was changed. Dried sea cucumbers (trepang) are used for preparing soup in the Orient.

Some contributors to the knowledge of
ECHINODERMS

Alexander Agassiz (1835-1910)

An American who made important studies of the structure and relationships of echinoids (1872-1874) and of North American starfishes (1877). *(The Bettmann Archive, Inc.)*

Oskar Hertwig (1849-1922)

He studied the fertilization of sea urchin eggs by sperms (1875). *(Historical Pictures Service, Chicago.)*

H. Ludwig

He wrote a classic monograph on the holothurians (1892).

Agassiz

Hertwig

H. L. Clark

He made systematic descriptions of most species of echinoderms in New England (1904).

E. W. MacBride

He contributed a most important account of echinoderms (1906).

Wesley R. Coe

Coe, of Yale University, made detailed studies of various echinoderms, especially those of Connecticut (1912).

Review questions and topics

1 What general characteristics describe the echinoderms as a group?
2 What are the characteristics that distinguish members of the different classes of echinoderms?
3 Discuss the significance of calcareous skeletons and spines, as found in echinoderms, including examples of the various types.
4 Describe the structure and function of the various parts of the unique water-vascular system.
5 Discuss the phenomena of bilateral symmetry in larval stages and radial symmetry in adults. When the positions of certain structures in adults are carefully considered, might one apply the term biradial?
6 Explain the structure and function of the pincerlike pedicellaria.
7 Explain the structure and function of the dermal branchiae ("skin gills").
8 Explain the structure and function of the respiratory tree as found in sea cucumbers.
9 Is it unusual for echinoderms not to have complex organs of excretion?
10 Discuss such phenomena as parthenogenesis, autotomy, and regeneration as displayed by certain echinoderms.

Selected references

Barnes, R. D.: Invertebrate zoology, ed. 3, Philadelphia, 1974, W. B. Saunders Company.

Binyon, J.: Physiology of echinoderms, Oxford, N.Y., 1972, Pergamon Press, Inc.

Feder, H. M.: On the methods used by the starfish *Pisaster ochraceus* in opening three types of bivalved mollusks, Ecology **36**:764-767, 1955.

Galtsoff, P. S., and Loansanoff, V. L.: Natural history and methods of controlling the starfish (*Asterias forbesi* Desor.), U. S. Bureau of Fisheries Bulletin **49**:75-132, 1950.

Harvey, E. B.: American Arbacia and other sea urchins, Princeton, New Jersey, 1956, Princeton University Press.

Hyman, L. H.: The invertebrates; volume IV, Echinodermata, New York, 1955, McGraw-Hill Book Company.

Meglitsch, P. A.: Invertebrate zoology, ed. 2, New York, 1972, Oxford University Press, Inc.

Millott, N., editor: Echinoderm biology, New York, 1967, Academic Press Inc.

Nichols, D.: Echinoderms, London, 1962, Hutchinson & Co. (Publishers) Ltd.

Russell-Hunter, W. D.: A biology of higher invertebrates, New York, 1969, The Macmillan Company.

25 Phylum Chordata

Distinguishing characteristics of the Chordata (Gr. *chorde*, cord) include (1) a semirigid skeletal axis, the dorsal notochord, which is present at some time during the life cycle; (2) a hollow dorsal nervous system, which is dorsal to the digestive system, and (3) paired pharyngeal clefts (gill slits), which connect the pharynx with the exterior at some stage in the life cycle.

All lower chordates up to and including the fishes are marine animals and respire by means of gills throughout life. In higher vertebrates gills or gill slits or traces of them are usually present only in the larval or embryonic stages. In mammals the gill slits never open but are modified for other purposes.

The phylum Chordata includes the subphyla Urochordata, Cephalochordata, and Vertebrata. Most chordates have a segmented vertebral column, which places them in the latter phylum. Since the following chapters deal with chordates (vertebrates) in detail, only the general characteristics of the subphyla will be considered in this chapter.

SUBPHYLUM UROCHORDATA

Urochordata are widely distributed marine animals; examples include ascidians, sea squirts, and tunicates. The body of the adult Urochordata is usually barrel shaped or saclike. In some species the body is brightly colored. The body is enclosed by a covering called a tunic (L. *tunica*, mantle). The animals are generally attached to the marine substratum.

A notochord is present in the tadpole-like, free-swimming larva, but is absent in the adult animal. In the larval stage there is a dorsal neural tube in the tail, which enlarges in the trunk and ends in the vesicle (''brain''). In the adult there is a single nerve ganglion with nerves. Respiration takes place through numerous pharyngeal gill slits, which are ciliated.

349

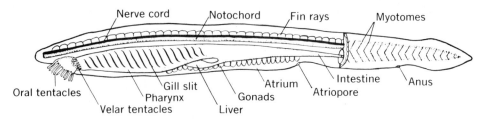

Fig. 25-1. *Amphioxus,* a protochordate, showing the internal structures (somewhat diagrammatic). Note especially the three characteristics of the chordate animal, namely, the dorsal notochord, dorsal nerve cord, and gill slits (pharyngeal clefts).
(From Hickman, Cleveland P., Sr., Hickman, Cleveland, P., Jr., and Hickman, Frances M.: Integrated principles of zoology, ed. 5, St. Louis, 1974, The C. V. Mosby Co.)

SUBPHYLUM CEPHALOCHORDATA

Cephalochordata are marine, sand-dwelling animals that swim with swift body movements. Amphioxus *(Branchiostoma)* is an example (Fig. 25-1). The body of *Amphioxus* is transparent and fishlike and is pointed at both ends. The body is compressed laterally, and there is no distinct head present. There are no paired lateral appendages, although there is a long dorsal fin along the entire length of the body. There are numerous V shaped muscles (myotomes) along the body.

A notochord is present along the entire length of the body throughout all stages of the life cycle. The nerve cord is tubular and extends along the entire dorsal side of the body. Like the notochord, it is present throughout the life of the animal. A small anterior cerebral vesicle is also present. Respiration occurs through numerous pairs of pharyngeal gill slits.

SUBPHYLUM VERTEBRATA

Vertebrates include cyclostomes (jawless fishes), sharks (cartilaginous fishes), bony fishes, amphibians, reptiles, birds, and mammals.

All vertebrates have an axial dorsal rod (notochord) present at some time during the life cycle. This notochord is present throughout the life in some of the lower vertebrates but disappears before adulthood in others. In the higher vertebrates the notochord becomes modified into a segmented vertebral column. The vertebrae may be either cartilaginous or bony. A dorsal nerve tube, which is connected anteriorly with the brain, is also present.

Pharyngeal clefts (gill slits) are present during some stage of the life cycle. These slits are present only during the larval or embryonic stage in the higher vertebrates.

The vertebrate body is divided into a head, neck (generally), and trunk. In some species a tail is present. Vertebrates usually have two pairs of lateral appendages and possess an internal jointed skeleton.

The heart is situated ventrally and has from two to four chambers. Red blood corpuscles are present. These animals are either cold or warm blooded, depending upon the class.

Class Cyclostomata (lampreys, hagfishes)

The notochord of these animals is present throughout life. A brain and dorsal nerve cord are also present, and there are from eight to ten pairs of cranial nerves. The skeleton is cartilaginous and fibrous and the body is round and slender. There are no scales present and the skin is soft. Mucous glands are present. These animals possess unpaired fins with rays of cartilage (Fig. 25-2).

The respiratory system consists of from six to 14 pairs of gills in gill pouches. The digestive

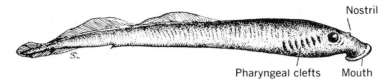

Fig. 25-2. Lamprey of class Cyclostomata. Note the circular sucking mouth, median un-paired nostril, and seven pharyngeal clefts (gill slits). Lampreys frequently attack and kill fishes.

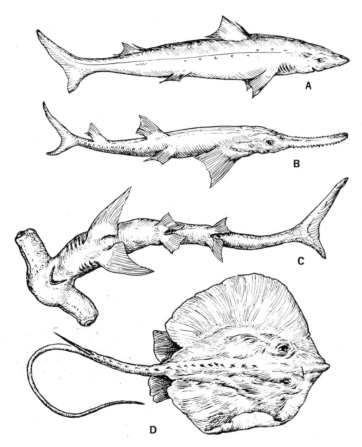

Fig. 25-3. Representatives of class Elasmobranchii (not drawn to scale). **A,** Spiny dogfish shark *(Squalus);* **B,** sawfish *(Pristis);* **C,** hammerhead shark *(Sphyrna);* **D,** southern stingray *(Dasyatis)* that is common in the Gulf of Mexico.

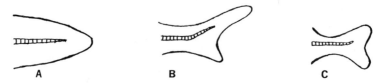

Fig. 25-4. Symmetries as shown by the tails of fishes. **A,** Diphycercal (Gr. *diphyes,* of double nature; *kerkos,* tail), as in the lungfish, in which the vertebral column extends to the posterior part of the body, and the tail develops symmetrically above and below the column; **B,** heterocercal (Gr. *heteros,* different), as in sharks, in which the vertebral column extends into the larger dorsal lobe; **C,** homocercal, as in most modern fish, in which the vertebral column extends slightly into the dorsal lobe but the tail is symmetrical.

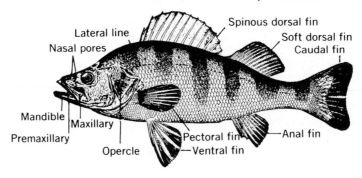

Fig. 25-5. External features of the perch.
(From Hegner, R. W.: College zoology, New York, The Macmillan Company.)

system consists of a sucking mouth with no jaws and a rasping tongue. An intestine with a fold (typhlosole) is present, but there is no stomach. The heart consists of one auricle and one ventricle; there are aortic arches in the gill region. The blood has both red and white corpuscles; these animals are cold blooded.

The excretory system consists of a pair of kidneys (pronephros) and paired pronephric ducts that empty into the cloaca. Lampreys are diecious; hagfishes are monecious. Fertilization in all animals of this class is external.

These animals possess the senses of taste, sight, smell, and hearing. There is one nasal sac present for smell; the auditory organs consist of two semicircular canals.

Class Elasmobranchii (Chondrichthyes)—sharks, rays

Animals of this class possess a notochord throughout life. A brain and dorsal nerve cord are

present and there are ten pairs of cranial nerves.

The endoskeleton is composed entriely of cartilage, and separate vertebrae are present. The body is spindle shaped and there are platelike (placoid) scales on the skin. There are mucous glands present on the skin also. These animals have a caudal (tail) fin that is asymmetrical and paired pectoral and pelvic fins. Two unpaired dorsal and median fins and fin rays are also present (Fig. 25-3).

Class Osteichthyes (Pisces)—bony fishes

Bony fishes possess a notochord that may persist in part throughout life. A brain and a dorsal nerve cord are present, and there are ten pairs of cranial nerves. The skeleton is more or less bony and replaces the original cartilaginous skeleton of earlier classes.

Bony fishes have various body shapes (Fig. 25-4). The caudal fin is usually symmetrical (homocercal), and there are paired and unpaired fins with fin rays of bone or cartilage. The skin may possess different types of dermal scales, although some fishes are scaleless. Mucous glands are present in the skin. Bony fishes have a terminal mouth with jaws and teeth (Fig. 25-5).

The respiratory system consists of paired gills that are supported by bony arches. There may be a gill cover (operculum) and an air bladder in some species.

The circulatory system consists of a heart with one auricle and one ventricle (Fig. 25-6). There are four pairs of aortic arches, and both arterial and venous systems are present. The red blood cells are nucleated. These animals are cold-blooded. The excretory system consists of a pair of kidneys (mesonephros) and paired ureters that empty into a cloaca. Fishes are diecious and have paired gonads. Fertilization is external; the eggs are laid in the water by the female.

Class Amphibia (frogs, toads, salamanders)

The notochord in amphibians does not persist throughout life. A brain and dorsal nerve cord are present, and there are ten pairs of cranial nerves. The skull has two occipital condyles. The endoskeleton is made up mostly of bone. Ribs may be present or absent, depending on the species. There are usually two pairs of appendages and the feet are often webbed. Some amphibians, however, have small limbs or are legless. Amphibians may have various body shapes. A head and trunk are present, and a tail may be present, depending upon the species. The smooth and moist skin is usually scaleless. Many glands are present and pigment cells are common.

A terminal mouth is present with small teeth in the jaws.

The respiratory system consists of paired gills or lungs. The external gills of larval stages persist throughout the life of some species. The skin may assist in respiration.

The circulatory system is made up of a heart with two auricles and one ventricle. There is double circulation through the heart. Numerous blood vessels are present in the skin. Amphibians are cold blooded.

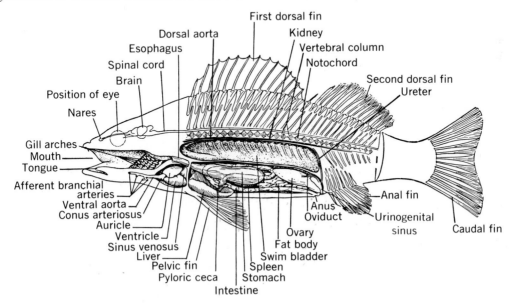

Fig. 25-6. Yellow perch *(Perca)* of class Pisces (median longitudinal section; somewhat diagrammatic).
(From Braungart, D. C., and Buddeke, R.: An introduction to animal biology, ed. 7, St. Louis, 1968, The C. V. Mosby Co.)

Fig. 25-7. Tiger salamander *(Ambystoma).*
(Courtesy General Biological Supply House, Inc., Chicago, Illinois.)

Fig. 25-8. Representatives of class Amphibia (not drawn to scale). **A,** ''Hellbender'' salamander *(Cryptobranchus);* **B,** mud puppy *(Necturus).*
(Courtesy General Biological Supply House, Inc., Chicago, Illinois.)

Fig. 25-9. A toad (*Bufo*). The skin bears numerous poison glands. The toad poison may be fatal to small animals.
(*Photograph by F. S. DeMartino, University of Dayton.*)

The excretory system consists of a pair of kidneys (mesonephros) and paired ureters, which empty into a cloaca.

Amphibians are diecious and possess paired gonads. Fertilization may be external or internal. Eggs with a jellylike cover are laid in water.

Amphibians possess the senses of taste, sight, hearing, smell, and touch. There are paired nostrils connected with the mouth.

The toad has a "warty" skin with many poison glands (Fig. 25-9). This poison can be fatal to some small animals. The frog, a typical amphibian, is discussed in detail in Chapter 27.

Class Reptilia

Reptilia include turtles, lizards, chameleons, snakes, crocodiles, and alligators. These animals possess a brain and a dorsal nerve cord. There are twelve pairs of cranial nerves present. The skull has one occipital condyle.

The endoskeleton of reptiles is bony, and ribs and a sternum are present. They may have various body shapes. The skin is dry and rough, and usually has horny scales or bony plates. There are few glands in the skin. There are usually two pairs of limbs present, although snakes have no appendages. Most reptiles have five toes on each appendage to aid in running, climbing, or paddling.

Paired lungs are present throughout life. The heart consists of two auricles and one ventricle, except in crocodiles, which have two auricles and two ventricles. There is usually one pair of aortic arches. All reptiles are cold-blooded. The excretory system consists of a pair of kidneys (mesonephros) and paired ureters, which empty into the cloaca.

Reptiles are diecious and possess copulatory organs. Fertilization is internal and the eggs have a leathery, limy shell. The embryo develops within the amnion.

Class Aves (birds)

Birds may have various body shapes, but they are frequently spindle shaped. Feathers cover the body, except the legs which are covered by scales. There is an oil gland at the base of the tail for preening. Paired wings are present for flying, although some birds possess only vestigial wings. There are paired hindlimbs for walking, perching, and swimming. Birds typically have four toes.

The endoskeleton of birds is bony and some cartilage is present. Birds have small ribs, and a sternum with a keel is present. The skull has one occipital condyle. The nervous system is made up of a brain, a dorsal nerve cord, and twelve pairs of cranial nerves. A well developed system of endocrine glands assists in coordination.

Paired lungs are present, and there are air sacs in the skeleton and among the viscera. A voice box (syrinx) is present at the junction of the trachea and the bronchi.

Birds possess a terminal mouth with paired jaws and a horny beak; there are no teeth present. Many birds have a crop for storage and a muscular gizzard for grinding.

The circulatory system consists of a heart and an aortic arch on the right side. The heart has two auricles and two ventricles. There are separate systemic and pulmonary systems. The red blood cells are nucleated. Birds are warm blooded (homoiothermic).

Fig. 25-10. Water snake *(Natrix).*
(Photograph by F. S. DeMartino, University of Dayton.)

The excretory system consists of a pair of kidneys (mesonephros) and paired ureters, which empty into a cloaca. There is no urinary bladder and urine and feces are eliminated together.

Birds are diecious. The male possess paired testes; the female has one ovary (left). There are no copulatory organs in most birds. Fertilization is internal. The eggs have a limy shell and much yolk is present within them. The embryos develop within the amnion.

Class Mammalia

Mammals include man, monkeys, horses, dogs, cats, pigs, whales, bats, and so on. The mammalian endoskeleton is bony with some cartilage present. Elongated tails may or may not be present. There are ribs and a sternum present. The skull has two occipital condyles.

Mammals have various body shapes. There is a unique muscular diaphragm that separates the thorax from the abdomen. The skin possesses hair at some time in life. In some types of mammals there are scales present on the tail. Glands include sweat, tear, mammary, scent, and sebaceous (hair) glands.

Mammals possess two pairs of appendages, although they are reduced in some animals. The appendages are adapted for walking, digging, climbing, swimming, or flying, depending upon the species. There are typically five digits on each appendage. All mammals (except certain whales and a few others) possess a mouth with teeth on both jaws.

The brain of mammals is well developed, as is the dorsal nerve cord. There are twelve pairs of cranial nerves. There is also a well developed endocrine system for coordination and other purposes. The respiratory system consists of paired lungs, a voice box (larynx), and an epiglottis.

The circulatory system consists of a heart with two auricles and two ventricles, and one aortic arch on the left side. There are separate systemic and pulmonary systems. The red blood cells are nonnucleated. Mammals are warm blooded.

The excretory system consists of a pair of kidneys (metanephros). Paired ureters empty into a urinary bladder and out through the urethra.

Mammals are diecious and possess paired testes or ovaries, and copulatory organs are present. The unique mammary glands provide nourishment for the young. The young develop in the amnion.

Mammals possess the senses of sight, smell, taste, touch, and hearing. There are external ears and paired movable eyelids.

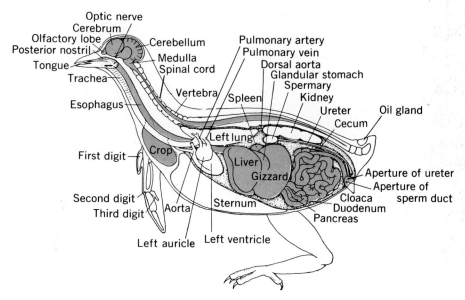

Fig. 25-11. Internal structures (diagrammatic) of a bird (class Aves). *(From Metcalf, R. L.: Economic zoology, Philadelphia, Lea & Febiger.)*

ECONOMIC IMPORTANCE OF VERTEBRATES

The economic importance of the various members of such a large and varied group is so extensive that limited space will not permit a complete and detailed consideration.

Hagfishes live in the mud of the sea and destroy fishes, especially if they have been caught in nets or on lines. Lampreys are serious enemies of game and food fishes. The Atlantic sea lamprey *(Petromyzon)* has reduced the commercial catch of fishes in the Great Lakes. Sea lampreys suck blood and injure fishes in other ways.

Dogfish sharks destroy large numbers of food fishes, lobsters, and crabs, besides destroying the nets of fishermen. In some places the meat of sharks is eaten by man. Sharkskin leather is used in manufacturing bags, shoes, bookbindings, and jewel cases. Shark livers are good sources of vitamin A.

Many fishes serve as sources of food and provide human recreation. Over 2 billion pounds of codfishes are caught annually, and cod-liver oil is a good source of vitamins A and D. The salmon of the Pacific coast are caught and canned in great quantities. The eggs of the sturgeon and other fishes are made into caviar. Certain types of fishes may help to keep waters clean by destroying mosquitoes and other insect larvae.

Frogs and toads are great enemies of insect pests. Frog skins may be used in making glue and bookbindings. Frog legs and the meat of mud puppies are eaten by man. Frogs are widely used in laboratories for dissection and experimental purposes. They are used in a test for human pregnancy.

Reptiles kill many destructive rodents and harmful insects. Turtles and tortoises are widely used as human food. Canned rattlesnake meat is eaten by man. The skins of crocodiles and alligators are used in making leather goods. The tortoise shell obtained from the horny covering of the hawkbill turtle is used in making combs and ornaments.

Birds are of great importance because they eat large quantities of harmful insects and the seeds of weeds. However, they may also eat great quantities of fruits and seeds of beneficial plants. Some birds supply food and feathers for man. In some regions the excrement, called guano, of certain sea birds accumulates in such quantities as to be a valuable source of fertilizer. Certain game birds afford much recreation and some food for hunters.

The economic importance of mammals is great and varied. Domestic mammals are of enormous value to man, serving as foods, beasts of burden, pets, sources of clothing, and sources of recreation. In the latter category might be listed polo, horse racing, and dog racing. Ivory is obtained from the tusks of elephants and walruses; oil from the fat of whales; ambergris, used in manufacturing perfumes, from the intestine of whales; furs are obtained from the muskrat, mink, fox, otter, weasel, marten, wolverine, badger, beaver, skunk, raccoon, bear, and rabbit. The worst mammalian pest is the rat, which causes damage amounting to hundreds of millions of dollars annually. Rats also carry the flea that harbors the cause of the dreaded bubonic plague ("black death"). When certain mammals are first imported into a country, they may be beneficial but later may become serious pests. After rabbits were introduced into Australia they became such pests that large numbers had to be destroyed. The mongoose destroys lizards, snakes, and rats in India. After it was imported into Jamaica it was very beneficial, but later it became a serious pest by destroying fruits, birds, poultry, and domestic animals.

Some contributors to the knowledge of certain
VERTEBRATES

Francis Willughby (1635-1672)

A British naturalist who laid the foundation for modern ornithology by writing *Ornithology,* which was completed (1676) by John Ray (162?-1705), Willughby's teacher.

Georges Cuvier (1769-1832)

A Frenchman who was largely responsible for establishing the science of comparative anatomy. He emphasized the relation between structure and function. He studied not only living animals but fossil remains as well.

John J. Audubon (1785-1851)

An American who published one of the great works on birds, *The Birds of America* (1828-1838), with unsurpassed artistic plates. The Audubon Society was named after him. *(Historical Pictures Service, Chicago.)*

Alphonse Toussenel (1803-1885)

He was the author of a number of valuable works on birds.

Richard Owen (1804-1892)

An English anatomist who made comparative studies of fishes, birds, apes, etc. He proposed the concepts of homology and analogy. *(The Bettman Archive, Inc.)*

Louis Agassiz (1807-1873)

An American, was a great investigator and professor of zoology and geology at Harvard University, where he founded the Museum of Comparative Zoology.

Ernest Haeckel (1834-1919)

He gave us a theory of embryonic germ layers and a principle of recapitulation, or "biogenetic law" (embryos in their development repeat in an abbreviated manner the stages of the ancestral history of the race). Both theories have required modification in recent times. *(Historical Pictures Service, Chicago.)*

Thomas H. Huxley (1825-1895)

An English biologist, made important contributions in the fields of comparative anatomy and paleontology.

John Burroughs (1837-1921)

An American naturalist who wrote many books on the lives and habits of living organisms.

Edward D. Cope (1840-1897)

An American naturalist and comparative anatomist who studied many vertebrates, including fossil reptiles and mammals.

David Starr Jordan (1851-1931)

An American biologist who made a lifelong study of fishes, and his publications and teachings assisted in the establishment of this phase of biology in the United States.

Henry Fairfield Osborn (1857-1935)

An American paleontologist who studied many types of living and fossil vertebrates.

Review questions and topics

1 Discuss the characteristics that apply to the vertebrates as a group.
2 List the characteristics by which the classes of the subphylum Vertebrata may be distinguished from each other, giving examples of each class.
3 Which systems seem to be developed best in the vertebrates? Which the least?
4 In the series of vertebrates from the lowest to the highest, discuss the status of (1) notochord and skeleton, (2) nervous system, (3) gills and lungs, (4) body and integument, (5) appendages, (6) digestive system, (7) circulatory system, (8) excretory system, (9) sensory organs, and (10) reproductive system.
5 Discuss the relative importance of two-, three-, and four-chambered hearts, giving examples of each.
6 Discuss the importance of systemic and pulmonary circulatory systems. What important consequence results from such an arrangement?
7 Discuss the importance of the division of the coelom into such cavities as pericardial, visceral, and thoracic, giving examples of each.
8 List the characteristics that distinguish mammals, giving examples.
9 Describe the structure and functions of a muscular diaphragm in mammals.
10 Discuss how the paired jointed appendages may be modified and adapted in various vertebrates to perform certain functions.
11 Discuss what happens to the pharyngeal (gill) slits, especially in higher vertebrates.
12 Describe the development of the nervous system, especially in higher mammals.
13 Discuss the well developed endocrine systems of higher vertebrates, including some functions performed.
14 Are most vertebrates diecious? Does fertilization occur externally or internally? Include examples of each.
15 What characteristics do all protochordates as a group have in common?
16 Discuss the characteristics that distinguish the members of the three subphyla from each other. Include examples of each subphylum.
17 Discuss each characteristic animals must have in order to qualify as chordates.
18 Which systems in the chordates seem to be developed to the greatest extent? Which the least?
19 What is the status of the muscular system in such a type as *Amphioxus?*
20 What is the status of the development of the head and sensory equipment in protochordates?
21 What might be some disadvantages to an animal of being attached (sessile)?
22 Is there any significance to the fact that all protochordates are marine?
23 System by system, do the chordates as a group show greater complexity of structure and function when compared with some of the higher nonchordates?

Selected references

Applegate, V. C., and Moffett, J. W.: The sea lamprey, Sci. Amer. **192:**36-41, 1955.
Barbour, R. W.: Amphibians and reptiles of Kentucky, Lexington, Ky., 1971, University of Ky. Press.
Barrington, E. J. W.: The biology of Hemichordata and Protochordata, London, 1965, Oliver & Boyd, Ltd.
Berrill, N. J.: The origin of vertebrates, New York, 1955, Oxford University Press, Inc.
Blair, W. F., and others: Vertebrates of the United States, New York, 1957, McGraw-Hill Book Company.
Braun, M. E., editor: The physiology of fishes, New York, 1957, Academic Press Inc.
Gans, C.: Biomechanics: an approach to vertebrate biology, Philadelphia, 1974, J. B. Lippincott Co.
McCauley, W. J.: Vertebrate physiology, Philadelphia, 1971, W. B. Saunders Company.
McCoy, C. J.: Diversity of life: vertebrates, New York, 1968, Reinhold Publishing Corp.
Orr, R. T.: Vertebrate biology, ed. 3, Philadelphia, 1971, W. B. Saunders Company.
Parker, T. J., and Haswell, W. A.: A textbook of zoology, ed. 7, New York, 1962, The Macmillan Company.
Pooley, A. C., and Gans, C.: The Nile crocodile, Sci. Amer. **234:**114-124, April, 1976.
Ruibal, R., editor: The adaptations of organisms, Belmont, Calif., 1967, Dickerson Pub. Co., Inc.
Torrey, T. W.: Morphogenesis of the vertebrates, New York, 1962, John Wiley & Sons, Inc.
Welty, J. C.: The life of birds, ed. 2, Philadelphia, 1975, W. B. Saunders Co.
Wessells, N. K., editor: Vertebrate adaptations; readings from Scientific American, San Francisco, 1968, W. H. Freeman and Company Publishers.
Young, J. Z.: The life of vertebrates, ed. 2, New York, 1962, Oxford University Press, Inc.

26 The frog

The common leopard or grass frog is known as *Rana pipiens* (L. *rana*, frog; *pipiens,* piping) (Fig. 26-1). Its body is smooth and covered with mucus, which is secreted by glands in the skin. The frog has the ability to change color because of changes in the black and yellow pigment cells in the skin. Because of its coloration, the frog is afforded a certain degree of protection from enemies. When in water, it need keep only the tip of its nose above the surface because of the location of the nostrils or external nares. Two large eyes are located on the top of the head. The tympanum or eardrum is external and just posterior to each eye. The body may be divided into head and trunk, and the latter bears two pairs of appendages.

Fig. 26-1. Grass frog *(Rana pipiens).* *(Courtesy Carolina Biological Supply Company, Burlington, North Carolina.)*

INTEGUMENT AND SKELETON

The loose-fitting skin is composed of (1) a rather thin outer layer, called the epidermis and (2) a thicker, inner layer, the dermis (Fig. 26-2). The epidermis consists of several layers of cells. The outer cells, composing the stratum corneum (L. *stratum,* layer; *corneus,* horny), are flat, compact, and horny (they are shed several times during the active season when the frog molts) and the inner ones located next to the dermis and

361

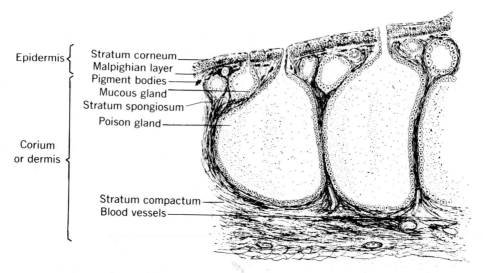

Epidermis {
Stratum corneum
Malpighian layer

Corium
or dermis {
Pigment bodies
Mucous gland
Stratum spongiosum
Poison gland

Stratum compactum
Blood vessels

Fig. 26-2. Skin of the frog (cross section and somewhat diagrammatic).
(From Braungart, D. C., and Buddeke, R.: An introduction to animal biology, ed. 6, St. Louis, 1964, The C. V. Mosby Co.)

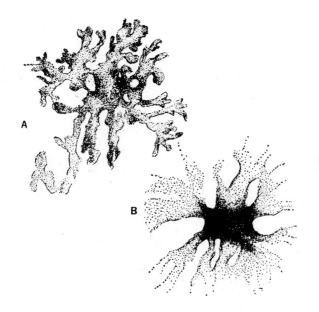

A

B

Fig. 26-3. Pigment melanophore from the frog *(Rana temporia).* **A,** Pigment distributed in response to light; **B,** pigment contracted.
(Redrawn and modified from Noble, G. K.: Amphibia of North America, New York, McGraw-Hill Book Company; from Potter, G. E.: Textbook of zoology, ed. 2, St. Louis, The C. V. Mosby Co.)

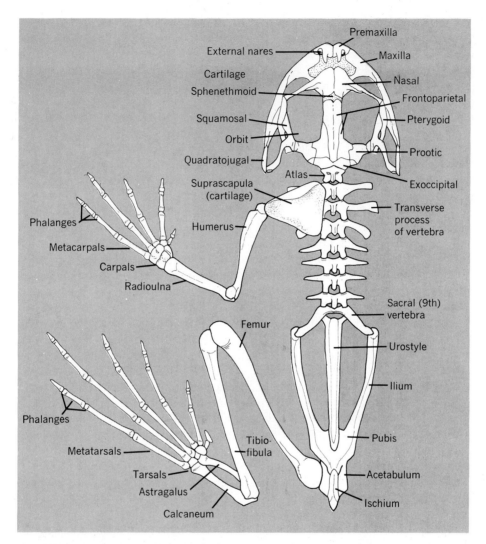

Fig. 26-4. Dorsal view of a frog skeleton. Note that humerus does not attach to the suprascapula but to bones on the ventral surface that are not shown.

composing the malpighian layer are columnar and give origin to the outer layer. The dermis consists of connective tissues that contain glands, blood vessels, pigments (Fig. 26-3), nerves, muscle fibers, and lymph spaces. The dermis is made of two layers. The outer layer, called the stratum spongiosum, consists of loose connective tissue and contains (1) pigment bodies that give the frog its spotted pattern (pigments may also be present in the epidermis), (2) small, spherical mucous glands that pour out a slimy secretion upon the surface of the skin, (3) larger spherical poison glands that secrete a whitish, acrid fluid for protection, and (4) numerous sensory and tactile papillae (just below the epidermis) for sensory purposes. The inner layer, called the stratum compactum, consists of dense connective tissues in which the the fibers run somewhat parallel to the surface of the skin and among which are blood vessels. The smooth, scaleless, hairless skin functions as an organ of respiration, gives protection, and is used for sensory purposes. The pigment bodies are responsible for some protective coloration.

The bony endoskeleton (Fig. 26-4) consists of an axial skeleton (skull and vertebral column) and an appendicular skeleton (the pectoral girdle with its forelimbs and the pelvic girdle with its hind limbs). The frog has no ribs. Most of the bones of the skull, except those of the upper and lower jaws and the hyoid bone to which the tongue is attached, form the brain case (cranium). The brain and spinal cord connect through a large opening (foramen magnum) at the base of the cranium. The cranium articulates with the first vertebra (atlas) by means of a pair of rounded prominences, the occipital condyles. The pair of prootid bones, one bone on either side of the posterior part of the cranium, forms the rounded auditory capsule that encloses the inner ear. The dorsal roof of the cranial cavity consists of two bones, the frontoparietals, each formed by the fusion of a frontal and a parietal bone. At the anterior end of the brain case is the tubular sphenethmoid bone. A pair of vomer bones helps to form the ventral wall of the olfactory sacs and also helps to form the roof of the mouth. The vomer bones bear vomerine teeth on the ventral surface. The upper jaw (maxilla) consists of a pair of premaxillae, a pair of maxillae, and a pair of quadratojugal bones. The first two pairs of bones bear teeth. The lower jaw (mandible) is the only part of the two jaws that moves. The hyoid apparatus consists of a large flat plate of cartilage in the floor of the mouth cavity. Rods of cartilage and bone extend anteriorly and posteriorly from its central part.

The vertebral column consists of nine vertebrae and a bladelike, posterior urostyle. A typical vertebra consists of (1) an oval, basal centrum (for articulation), (2) a neural arch, through which the spinal cord passes, (3) a single dorsal spine (neural spine), attached to the neural arch, and (4) a pair of transverse processes (except on the atlas), which extend laterally for the attachment of muscles. The articulating processes at each end of the neural arch are called zygapophyses. Ligaments hold the vertebrae together but allow a certain amount of movement.

The pectoral girdle (L. *pectus,* breast), to which the forelimbs are attached, is attached to the vertebral column by muscles. The sternum (breastbone) is located on the ventral median line and is composed of a number of bones and cartilages. The ventral part of the pectoral girdle consists of an anterior clavicle and a posterior coracoid. Other smaller bones help make up this part of the girdle. The dorsal part of the girdle is composed of the bony scapula, dorsal to which is the cartilaginous suprascapula. The glenoid fossa is the cavity with which the humerus of the forelimb articulates. The radioulna of the foreleg is a fusion of the radius and ulna bones. The wrist consists of six carpal bones, and the hand is supported by five metacarpals. Distal to the hand are the bones (phalanges) of the digits.

The pelvic girdle (L. *pelvis,* cavity), with which the hindlimbs articulate, is attached to the transverse processes of the ninth, or sacral, vertebra. The girdle is composed of a pair of long ilium bones (pl. ilia), a pair of ischium bones, and a pair of pubis bones. These three pairs of bones articulate so that a cavity is formed (acetabulum) with which the femur of the hind limb joins. The anterior part of the acetabulum is formed by the ilium and the posterior part by the ischium, whereas the ventral part is formed by the cartilaginous pubis. The tibiofibula is a fusion of the tibia and fibula in man. The tarsals (ankle bones) are arranged in two rows, the proximal one consisting of long bones, the astragalus and calcaneus. The distal row contains a series of smaller bones. Distal to this are the five elongated metatarsals (foot). The five toes (digits) are composed of a total of fourteen phalanges. On the tibial side of the first toe there is a rudimentary

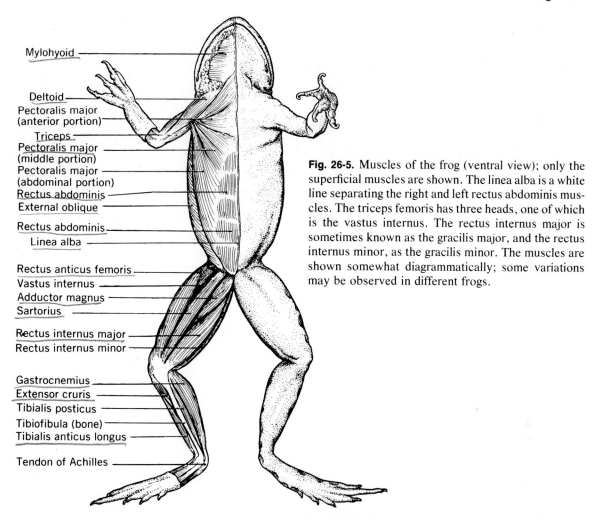

Mylohyoid

Deltoid
Pectoralis major
(anterior portion)
Triceps
Pectoralis major
(middle portion)
Pectoralis major
(abdominal portion)
Rectus abdominis
External oblique

Rectus abdominis
Linea alba

Rectus anticus femoris
Vastus internus
Adductor magnus
Sartorius

Rectus internus major
Rectus internus minor

Gastrocnemius
Extensor cruris
Tibialis posticus
Tibiofibula (bone)
Tibialis anticus longus

Tendon of Achilles

Fig. 26-5. Muscles of the frog (ventral view); only the superficial muscles are shown. The linea alba is a white line separating the right and left rectus abdominis muscles. The triceps femoris has three heads, one of which is the vastus internus. The rectus internus major is sometimes known as the gracilis major, and the rectus internus minor, as the gracilis minor. The muscles are shown somewhat diagrammatically; some variations may be observed in different frogs.

digit called the calcar or prehallux. There are no claws.

MOTION AND LOCOMOTION

Well-developed muscles are present in the body, appendages, and head (Fig. 26-5). Other muscles move the lower jaw, aid in breathing, obtain foods, and produce sounds by means of the vocal apparatus. The muscles attached to the skeleton are called skeletal muscles, each of which has an origin that is the more fixed end and an insertion that is the more movable end. Pul-

sating lymph "hearts" (two near the third vertebra and two near the end of the vertebral column) force lymph into a branch of the renal portal and internal jugular veins.

INGESTION AND DIGESTION

Living insects, worms, and similar organisms are captured by a rather sticky, extensile tongue, attached at its front end. The tongue is thrown forcibly forward by the rapid filling of a lymph space beneath it. The large mouth cavity bears cone-shaped maxillary teeth on the upper jaw

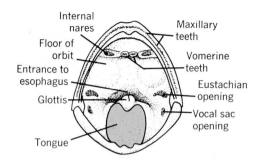

Fig. 26-6. Mouth of a frog, opened to show details.

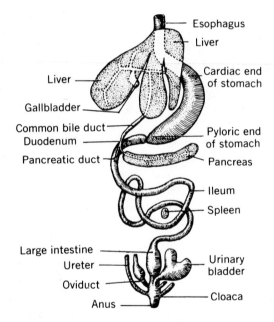

Fig. 26-7. Digestive system of the frog with associated organs.

(Fig. 26-6). The two vomer bones in the roof of the mouth bear vomerine teeth. A constricted, horizontal slit separates the mouth cavity from the esophagus (Gr. *oisophagos*, gullet). The stomach is crescent-shaped and is composed of a larger, anterior cardiac part and a smaller, posterior pyloric part (Gr. *pylorus*, gatekeeper), which connects with the coiled small intestine.

The small intestine consists of an anterior duodenum and a much-coiled ileum (L. *ileum*, groin), which widens into the large intestine. The large intestine connects with the saclike cloaca, which also receives tubes from the kidneys and reproductive system. The cloaca empties to the exterior through the anus.

The pancreas (Gr. *pan*, all; *kreas*, flesh) is a much-branched, tubular organ that lies between the stomach and the duodenum. It passes its alkaline digestive juices into the common bile duct (Fig. 26-7). The large, reddish, trilobed liver secretes an alkaline bile, which is carried to the gallbladder. From the gallbladder the bile enters the duodenum through the common bile duct.

Physiology of digestion

Digestion breaks down complex, insoluble foods, such as proteins, fats, and carbohydrates, into simple, soluble compounds capable of being absorbed by the cells and assimilated into living protoplasm.

The various foods are acted upon by specific enzymes that hasten the conversion processes without being used up themselves.

In the mouth there is no mastication, digestion, or enzymatic action. In the esophagus certain glands produce an alkaline mucous secretion that is mixed with the acid gastric juice secreted by the glands in the walls of the stomach. The cardiac end of the stomach has long, tubular, branched, deeply set glands for the secretion of mucus. The pyloric end of the stomach has short, tubular, shallow glands for secreting gastric juice, which contains the enzyme pepsin and about 0.4% hydrochloric acid (HCl). In other words the reaction is as follows:

$$\text{Proteins} + \text{H}_2\text{O} \xrightarrow[\text{Pepsin}]{\text{HCl}} \text{Soluble peptones}$$

After the partially digested foods pass through the pyloric valve from the stomach into the duodenum, they are mixed with the alkaline pancreatic juice. The alkalinity of the pancreatic juice results from sodium carbonate (Na_2CO_3).

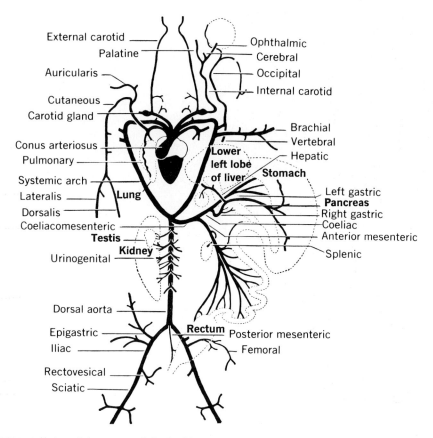

Fig. 26-8. Arterial system of the bullfrog (ventral view).
(From Potter, G. E.: Textbook of zoology, ed. 2, St. Louis, The C. V. Mosby Co.)

The three specific enzymes of the pancreatic juice are amylase, trypsin, and lipase.

The hepatic cells (Gr. *hepar*, liver) of the tubular glands of the liver secrete a greenish bile that is mixed with the pancreatic juice in the common bile duct before they enter the duodenum. Certain bile enzymes convert fats, when in an alkaline environment, into a soapy emulsion capable of passing through the intestinal walls into the blood and lymph systems. The liver also stores glycogen, which is changed by certain liver enzymes into usable sugar when needed.

The production and roles of intestinal juices and their enzymes in the frog are probably similar to those in higher animals. Possibly starches may be converted into sugars in the intestine. The various types of food acted upon by specific enzymes in their proper environments are eventually absorbed by the cells of the intestine, passed into the lymph and blood vessels, and transported to body tissues to be utilized.

CIRCULATION

The heart, located within the thin, saclike pericardium, is three-chambered, consisting of two

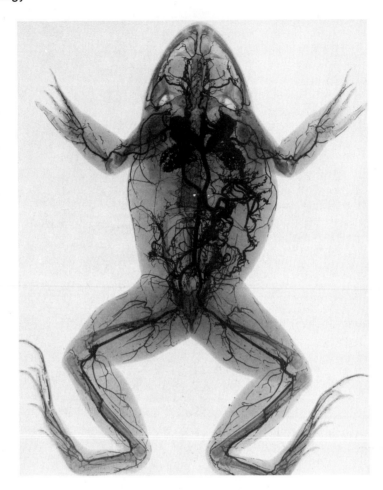

Fig. 26-9. Arterial system of a bullfrog that has been injected with colored latex (x-ray view). Observe the flexures of arteries at the elbow and knee joints to permit free movements. *(Courtesy Carolina Biological Supply Company, Burlington, North Carolina.)*

thin-walled auricles (right and left) and one muscular, cone-shaped ventricle. A thick-walled, tubular conus arteriosus (truncus arteriosus) arises from the base of the ventricle. A thin-walled, triangular sinus venosus, located on the dorsal side of the heart, is connected with the right auricle. In the adult frog the blood is pumped from the ventricle into the conus arteriosus, which has many branches (Figs. 26-8 and 26-9).

After passing from the arteries into thin-walled capillaries, the blood is returned from the various tissues and organs of the body by a system of veins (Fig. 26-10). The right and left pulmonary veins return the oxygenated (aerated) blood from the right and left lungs to the left auricle. The blood from all others parts of the body is returned to the sinus venosus through three large veins known as (1) the posterior vena cava (postcaval) and its branches, (2) the right anterior vena cava (right precaval) and its branches, and (3) the left anterior vena cava (left precaval) and its

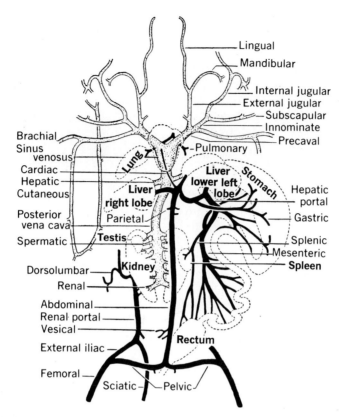

Fig. 26-10. Venous system of the bullfrog (ventral view).
(From Potter, G. E.: Textbook of zoology, ed. 2, St. Louis, The C. V. Mosby Co.)

branches. The blood from the sinus venosus enters the right auricle. The right and left auricles send their blood into the one ventricle, which forces its mixture of oxygenated blood (from the left auricle) and nonoxygenated blood (from the right auricle) into the conus arteriosus through pocket-shaped semilunar valves.

Frog blood consists of the following: (1) Oval, biconvex, and nucleated red blood corpuscles (erythrocytes) contain hemoglobin. Hemoglobin unites temporarily with oxygen in the lungs and skin to form oxyhemoglobin, which in turn gives up its oxygen to cells and tissues when or where it is needed. (2) Ameboid white blood corpuscles

(leukocytes), which are able to move independently and are of different sizes, pass through the walls of blood vessels and tissues. They destroy bacteria and other organisms by ingesting them, thus serving to prevent infections. (3) The spindle cells assist in the clotting of blood upon their disintegration. Blood corpuscles originate principally in the marrow of the bones and in the spleen. (4) The plasma or liquid part of the blood carries foods, carbon dioxide, wastes, proteins, and mineral salts. Blood coagulates, especially after injuries, to form a clot, which includes fibrin, red and white corpuscles, and tissue cells.

RESPIRATION

In the earlier tadpole stages external gills are present, but these are later covered to form internal gills, which communicate with the exterior through a small opening. The internal gills are eventually absorbed, and typical lungs develop in the air-breathing adult frog. In the adult frog respiration takes place through the skin and lungs, but probably more through the skin. During hibernation the lungs are inactive, yet skin respiration continues, even though the rate may be reduced. In lung respiration the air is admitted into the mouth cavity from the outside through the external nares (nostrils) and then through the slit-like glottis (Gr. *glotta,* tongue) into the short, tubular larynx (Gr. *larynx,* voice box). From the larynx the air passes into the trachea (windpipe) and finally into the thin-walled, saclike, paired lungs. The lungs are ovoid, distensible, and internally divided by folds (septae) into a number of compartments known as alveoli (L. *alveolus,* small cavity) to increase the surface exposed to the air. Thin-walled capillaries line the inner surfaces of the alveoli and permit the exchange of oxygen and carbon dioxide between the air in the lungs and the blood in the circulatory system. The amount of exchange of these gases depends upon the concentration of each on either side of the lung and blood vessel membranes. Air is forced into the lungs through the slitlike glottis by closing the nares and contracting the floor of the mouth. Air is expelled from the lungs through the glottis into the mouth cavity by the contraction of the muscles of the body walls. It may be expelled or drawn into the mouth through the nares by closing the glottis and alternately raising or lowering the floor of the mouth. Sounds may be produced by forcing air back and forth through the glottis.

EXCRETION

Some wastes are excreted by the frog skin and intestine, but many are removed from the blood by a pair of elongated kidneys in the dorsal ab-

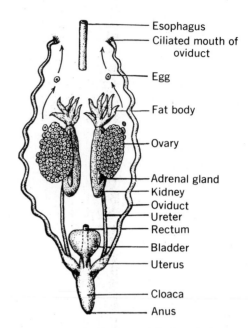

Fig. 26-11. Urogenital system of the female frog (ventral view).

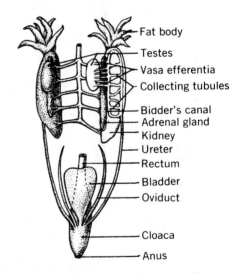

Fig. 26-12. Urogenital system of the male frog (ventral view). On the right, the testis has been moved and the kidney dissected to show the internal tubes. Sperms pass from the testes through the vasa efferentia into Bidder's canal, from which they pass out through the ureter. Part of the circulatory system and adrenal (ductless) glands are also shown. Note the poorly developed rudimentary oviduct in the male.

dominal cavity (Figs. 26-11 and 26-12). Internally, a kidney contains a number of malpighian bodies, each consisting of an enclosing membrane known as Bowman's capsule, which surrounds a coiled mass of thin-walled capillaries known as a glomerulus (L. *glomus*, ball). Wastes are collected from the blood in the glomeruli and carried by uriniferous tubules to collecting tubules, to the tubular ureter, and finally to the saclike cloaca. From the cloaca the urine may be stored in the thin-walled, distensible urinary bladder, which voids only at certain intervals. Ciliated, funnel-shaped nephrostomes in the ventral part of the kidney open into the coelom, from which wastes may be obtained and later eliminated.

COORDINATION AND SENSORY EQUIPMENT

The nervous system may be divided into (1) a central nervous system, consisting of the brain and spinal cord, (2) a peripheral nervous system, consisting of ten pairs of cranial nerves (Table 26-1) and ten pairs of spinal nerves, and (3) a sympathetic nervous system, consisting of nerves and ganglia that supply the internal (visceral) organs (Figs. 26-13 and 26-14).

The brain has the following structures: (1) two small, fused olfactory lobes for the sense of smell, (2) two large, elongated cerebral hemispheres (cerebrum), (3) two large optic lobes for the sense of sight, (4) the midbrain, (5) the small, narrow cerebellum, and (6) the wide medulla oblongata, which connects with the enlarged portion of the spinal cord. When the brain (except the medulla oblongata) is removed, the frog is still able to breathe, jump, swim, swallow food, and use its sense of equilibrium.

On the ventral side of the brain the following structures are distinguishable: (1) the optic chiasma, of the crossing of the optic nerves, (2) the pituitary body (hypophysis), and (3) the infundibulum, a bilobed extension of the diencephalon.

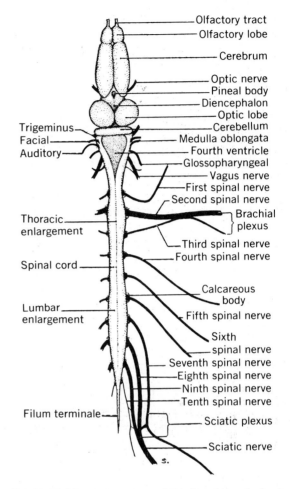

Fig. 26-13. Nervous system of the frog (dorsal view). The optic lobes rest on the midbrain just posterior to the diencephalon.
(From Potter, G. E.: Textbook of zoology, ed. 2, St. Louis, The C. V. Mosby Co.)

The spinal cord is composed of a central mass of gray matter (principally nerve cells) in the shape of the letter H and an outer mass of white matter made up of nerve fibers. The hollow central canal extends throughout the entire cord and may be seen in the middle of the crossbar of the H in a cross section of the spinal cord. The central canal connects anteriorly with the cavities (ventricles) of the brain. The spinal cord has two sur-

Table 26-1 Cranial nerves of vertebrates

Number	Name	Origin	Distribution	Function
I	Olfactory	Olfactory lobe	Mucous membrane lining nose	Sensory (smell)
II	Optic	Second vesicle of forebrain (diencephalon)	Cells of retina of eye	Sensory (sight)
III	Oculomotor	Ventral part of midbrain	Superior, inferior, internal recti; and inferior oblique muscles of eye	Motor (eye movement)
IV	Trochlear	Dorsal part of midbrain	Superior oblique muscle of eye	Motor (eye movement)
V	Trigeminal	Laterally from medulla (hindbrain)	Face, tongue, and mouth, and to muscles of jaws or mandibles	Sensory and motor
VI	Abducens	Ventral part of medulla	External rectus muscle of eye	Motor
VII	Facial	Laterally from medulla	Muscles of face, roof of mouth, hyoid	Motor (principally)
VIII	Auditory	Laterally from medulla	Cells of semicircular canal and other parts of inner ear	Sensory (hearing and equilibrium)
IX	Glossopharyngeal	Laterally from medulla	Membranes and muscles of tongue and pharynx	Sensory and motor
X	Vagus	Laterally from medulla	Heart, lungs, pharynx, stomach, intestine visceral arches	Sensory and motor
XI*	Spinal accessory	Laterally from medulla	Muscles of shoulder	Sensory and motor
XII*	Hypoglossal	Ventral part of medulla	Tongue and neck muscles	Motor

*The XI and XII pairs are absent in fishes and amphibia.

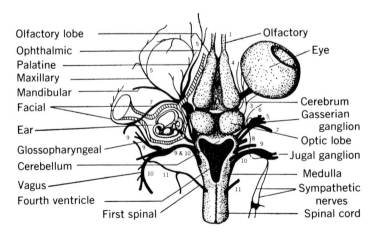

Fig. 26-14. Brain and cranial nerves of the bullfrog *(Rana catesbeiana)* (somewhat diagrammatic dorsal view). **1** to **10**, cranial nerves; **11**, first pair of spinal nerves. Certain cranial nerves show some of their branches.

(From Atwood, W. H.: Comparative anatomy, ed. 2, St. Louis, The C. V. Mosby Co.)

rounding membranous meninges, the outer dura mater and the inner pia mater.

There are ten pairs of spinal nerves, each arising from the gray matter of the spinal cord by a dorsal and ventral root. The union of these two roots at the side of the cord forms a spinal nerve. Each spinal nerve passes out between the bony arches of adjacent vertebrae.

The sympathetic nervous system consists of two main trunks that parallel the spinal cord, one on either side. Each trunk has ganglia or enlargements where the ten pairs of spinal nerves unite with it.

The skin, by way of its sensory nerve endings, receives tactile, chemical, heat, and light stimuli. The eyes resemble the eyes of other vertebrates. There are three eyelids: the rather motionless upper lid and the lower lid, which is fused with the third eyelid, or nictitating membrane (L. *nictare,* to beckon). The lens permits objects, especially moving objects, to be seen at definite distances. The pupil contracts and regulates the amount of light that enters the eye. The sensitive retina (L. *rete,* net) within the eye is stimulated by light and transfers the impulses to the optic nerve, which carries them to the brain to give the sensation of sight. The eyes lie in orbits (sockets) at the side of the skull and are moved by six eye muscles known as the external and internal recti, the superior and inferior recti, and the superior and inferior oblique.

The tympanic membrane of the outer ear communicates with the inner ear by a bony columella that vibrates with the sound stimuli received. The auditory nerve carries the impulses to the brain, where the sensation of hearing is really produced. There are no external ears. The middle ear communicates with the mouth cavity by means of the eustachian tube, which aids in equalizing air pressures on the eardrums. The inner ear also contains organs of equilibrium.

The olfactory sense is located in a pair of nasal cavities lined with folds of sensitive, epithelial, nasal membranes. The external nares (anterior nares) connect with the nasal cavity. The internal nares (posterior nares) connect the nasal cavity with the mouth cavity. The nares in amphibia and other vertebrates (above the fishes) are used for both respiratory and olfactory purposes. The olfactory nerves connect the epithelial nasal membranes of the nasal cavities with the olfactory lobes of the brain. The elevated papillae of the mouth and tongue contain organs of taste, especially if foods and chemicals are in solution.

REPRODUCTION

The sexes are separate (diecious) (Figs. 26-11 and 26-12). The sperms of the male originate in paired, small, oval testes. The sperms pass through the vasa efferentia into the kidneys, into the ureter by means of Bidder's canal, into the cloaca, and out through the anus. The eggs originate in the large paired ovaries and later break out through the ovary walls into the coelom. From here the eggs find their way into the much-coiled, paired oviducts, the funnel-shaped openings of which are located near the anterior edge of the abdominal cavity. Each oviduct leads into a thin-walled uterus that empties into the cloaca. The cloaca leads to the anus. The eggs are supplied with a layer of gelatinous food, which is produced by glandular cells of the oviduct. There are no copulatory organs. A yellowish hand-shaped fat body is located anterior to each reproductive organ and serves for food storage. The embryology of the frog was considered in Chapter 6.

<table>
<tr><td>

Some contributors to the knowledge of
THE FROG

Carolus Linnaeus (1707-1778)

He first used the word "amphibia" for a general group of more or less aquatic vertebrates.

A. M. Marshall

He authored one of the first texts, *The Frog* (1882).

A. Ecker

He wrote the *Anatomy of the Frog* (1889).

H. Gadow

He described the biology and natural history of amphibia in a volume in the *Cambridge Natural History* (1901).

G. B. Howes

He discussed frogs in some detail in his *Atlas of Elementary Zootomy* (1902).

Joseph Leidy (1823-1891)

He studied the worm parasites of frogs (1904).

M. Dickerson

He authored a widely used text called *The Frog Book* (1906).

Samuel J. Holmes

He wrote a fine book called *The Biology of the Frog* (1930).

G. Kingsley Noble

Noble, of the American Museum of Natural History, authored an excellent volume called *Biology of the Amphibia* (1931).

Franz Werner

He gave an extensive account of amphibia in Kükenthal's *Handbuch der Zoologie* (1931).

</td></tr>
</table>

Review questions and topics

1 List the characteristics that place the frog in the phylum Chordata, subphylum Vertebrata, and class Amphibia.
2 Explain the significance of (1) protective coloration, (2) internal bony skeleton, (3) closed system of arteries, capillaries, and veins, (4) three-chambered heart, (5) lymph hearts and lymph, (6) paired appendages with digits, (7) well-developed skeletal muscles that function in opposition, and (8) red blood corpuscles that carry oxygen.
3 Describe the structure and functions of the following for the frog: (1) integument, (2) locomotion, (3) ingestion and digestion, (4) circulation, (5) respiration, (6) excretion, (7) coordination and sensory organs, and (8) reproduction. In what specific ways do these show improvements over comparable structures and phenomena in lower animals?
4 Explain: (1) origin and insertion of a muscle, (2) pectoral and pelvic girdles, (3) auricle and ventricles, (4) vasa efferentia and ureter, (5) oviduct and ureter, and (6) aorta and vena cava.
5 Describe the structures and functions of the various parts of the frog digestive system, including the physiology of digestion in detail.
6 Explain how frog blood is carried from the heart to the lungs and skin.
7 Explain the structure and functions of the blood cells of the frog.
8 When excretory functions are localized in a pair of kidneys, must there be a comparable development of the circulatory system to take wastes to them?
9 Explain the role of the cirulatory system in the respiratory process in the lungs and skin.
10 Discuss the improvements of the structure and functions of the nervous system and sensory organs over those of lower animals.
11 List the number, names, origin, distribution, and functions of the ten pairs of cranial nerves in amphibia. In what ways do reptiles, birds, and mammals differ from amphibia and fishes in regard to cranial nerves?

Selected references

Cochran, D. M.: The new field book of reptiles and amphibians, New York, 1970, G. P. Putnam's Sons.
Frazer, J.: Amphibians, New York, 1973, Springer-Verlag New York, Inc.
Moore, J. A., editor: Physiology of the Amphibia, New York, 1964, Academic Press, Inc.
Muntz, W. R. A.: Vision in frogs, Sci. Amer. **210:**110-119, 1964.
Noble, G. K.: Biology of the Amphibia, New York, 1931, McGraw-Hill Book Company.
Oliver, J. A.: The natural history of North American amphibians and reptiles, Princeton, New Jersey, 1955, D. Van Nostrand Co., Inc.
Stuart, R. R.: The anatomy of the bullfrog, Chicago, 1940, Denoyer-Geppert Co.

PART FIVE

The biology of man

The human form
Human metabolism
Human control mechanisms
Human reproduction

27 The human form

Integument
Skeleton
Motion and locomotion
   Skeletal muscle

The form and the function of the body are results of the repeated division and specialization of a fertilized egg. The resulting embryo contains trillions of cells, each part of a particular tissue and organ. These function as part of the following systems.

Integumentary (skin) system—protection, support, heat regulation, absorption, excretion, stimuli reception

Skeletal system—support, protection, posture, motion, locomotion, manufacture of blood corpuscles (by bone marrow), transmission of sound waves (by ear bones)

Muscular system—locomotion, movement of body parts, such as the stomach, intestines, heart

Digestive system—ingestion, digestion, absorption of foods, elimination of wastes

Circulatory system—transportation of foods, wastes, oxygen, heat, carbon dioxide, endocrine hormones

Respiratory system—supply of oxygen and elimination of carbon dioxide and other wastes

Excretory system—excretion and elimination of metabolic waste materials

Nervous and sensory system—reception of stimuli, transmission and interpretation of impulses for purposes of correlation, movement, locomotion, behavior, secretion; centers of sight, hearing, taste, smell, equilibrium

Endocrine (ductless) gland system—production of hormones for correlating and regulating body processes

Reproductive system—production of sex cells for the continuation of the species

INTEGUMENT

The functions of the human skin and its accessory structures may be summarized briefly as follows: (1) regulation of heat, (2) excretion of wastes, (3) protection against injury, harmful light rays, loss of water, and disease-producing organisms, (4) prevention of the absorption of various deleterious materials, (5) assistance in normal respiration, (6) supplying of information about the surroundings through the various types of sensory organs, and (7) production of hairs, nails, glands, and teeth.

The two layers of the human skin are (1) the external epidermis (cuticle) and (2) the deeper dermis (corium) (Fig. 27-1). The epidermis is composed of stratified squamous and columnar epithelium and contains no blood vessels but has fine nerve fibrils. The hairs, nails, and numerous glands are modified epidermis. When people ''peel'' after a sunburn, the epidermis is shed.

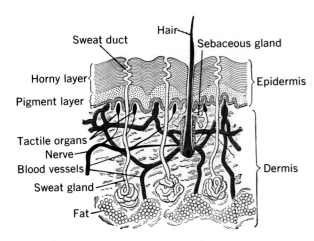

Fig. 27-1. Human skin (cross section).
(From Guyer, M. F.: Animal biology, New York, Harper & Row, Publishers.)

Hair is formed when the pigmented malpighian layer of the epidermis extends downward into the dermis to form tubelike hair follicles. The cells at the base of a follicle produce a hair, which is fused epidermal cells supplied with a horny, protein material called keratin (Gr. *keras,* horny). When a hair is being formed, it first appears as a tiny elevation below the skin's surface. As more is formed, the hair is pushed from the skin's surface. The part of the hair within the follicle is called the root, and the remainder is called the shaft. All skin is provided with follicles, except the palms of the hands, the soles of the feet, and the last portions of the fingers and toes. The consistency of hairs depends upon the structure of the follicle; a round follicle (in cross section) gives rise to straight hair, an oval follicle to curly hair, and a rather flat, ribbon-shaped follicle to wavy (kinky) hair. Hair color is determined by the quantity and quality of the pigments present and their relation to the transparent air within the hair. The base of each hair is supplied with a nerve and with blood vessels for nourishment. Muscle fibers in the dermis are attached to the hair follicle for hair movement.

Sebaceous (oil) glands are formed by the invagination of the malpighian layer of the epidermis. They are usually associated with hair follicles, being especially numerous on the face and scalp. The oils pass from the gland into the hair follicle and then to the skin surface, thereby keeping the hair and skin from becoming dry and preventing undue evaporation or absorption of water and other liquids by the skin.

Nails are formed from closely packed epithelial tissues along a furrow at the base of the nail. The nail is formed by a fusion of clear, dead, horny, keratinized cells.

The dermis of the skin is thicker than the epidermis and has (1) an upper papillary layer, containing numerous papillae (elevations) that increase the surface for nerves and tactile organs, blood vessels, hair follicles, and sebaceous glands, and (2) a lower reticular layer, containing yellow elastic and white fibrous connective tissues, which contain adipose tissue ("fat") and sweat glands.

Sweat glands are present in all human skin but are most numerous under the arms and on the forehead, soles of the feet, and palms of the hands. These coiled, tubular glands are located in the dermis and empty their excretions through pores on the skin surface. Over 2 million sweat glands in the entire skin eliminate more than a quart of liquid per day under normal conditions. Perspiration eliminates body wastes and regulates body heat through the evaporation of water. Heat that is produced in various tissues, especially muscles, is distributed by the blood throughout all parts of the body, thus producing an average, normal body temperature of 98° to 99° F. Since the skin is well supplied with blood, it can act efficiently as a heat-regulating mechanism. Enlargement (dilation) of blood vessels and relaxation of the muscle fibers allow more blood to lose more heat, whereas a contraction of blood vessels and muscles has the opposite effect.

The dermis is attached to the deeper tissues by connective tissue, called subcutaneous tissue. In some parts of the body the skin is tightly attached to the deeper tissues, but in other parts it is loosely attached to permit more freedom of movement. The dermis contains many sensory

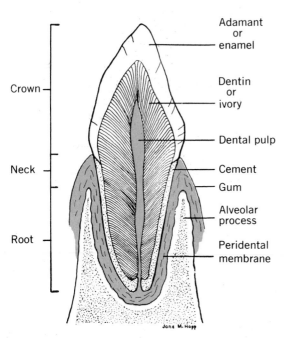

Adamant
or
enamel

Dentin
or
ivory

Dental pulp

Cement

Gum

Alveolar
process

Peridental
membrane

Crown

Neck

Root

Jane M. Hopp

Fig. 27-2. Human incisor tooth (vertical section).
(From Francis, C. C.: Introduction to human anat-
omy, ed. 6, St. Louis, 1973, The C. V. Mosby Co.)

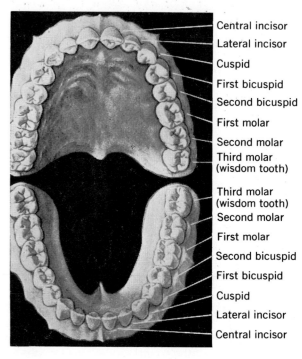

Central incisor
Lateral incisor
Cuspid
First bicuspid
Second bicuspid
First molar
Second molar
Third molar
(wisdom tooth)
Third molar
(wisdom tooth)
Second molar
First molar
Second bicuspid
First bicuspid
Cuspid
Lateral incisor
Central incisor

Fig. 27-3. Cast showing the thirty-two permanent human teeth.
(Courtesy American Dental Association.)

nerve endings for the reception of such stimuli as
heat, cold, pain, pressure, and touch. Receptors
are present in most areas of the skin but are much
more concentrated in some regions than in
others; for example, tactile (touch) receptors are
more numerous on fingertips than on the back of
the hand. On the inner surface of the hands and
fingers and on the soles of the feet there are many
minute ridges that increase friction and form distinctive fingerprint and footprint patterns.

Teeth are derived embryonically from epithelial tissues and are embedded in the upper
and lower jaws. The part of the tooth (Fig. 27-2)
above the gum is called the crown and is covered
with hard enamel. The remainder of the tooth is
composed of softer dentin with its central canal
(pulp cavity), which in turn contains blood vessels and nerves. The teeth are anchored in their
sockets by a cement that also protects the dentin.

Man, like other mammals, has a temporary
(baby) set of teeth, twenty in number, which
appears between the ages of 6 months and 2½
years. The permanent set of thirty-two (Fig. 27-3)
is composed (on each side of each jaw) of two
flat, sharp incisors (front teeth) for cutting foods
as they overlap, one pointed canine (eye tooth or
cuspid) for tearing foods, two broad-surfaced
premolars (bicuspids) for grinding, and three
large molars for grinding. The premolars have
two surface elevations, and the larger molars
have four or more elevations for grinding. The
last pair of molars (wisdom teeth) may not erupt
until later in life, or not at all. The normal dental
formula for man is as follows:

$$I\frac{2}{2}; \ C\frac{1}{1}; \ P\frac{2}{2}; \ M\frac{3}{3} \text{ (for each jaw)}$$

Know these

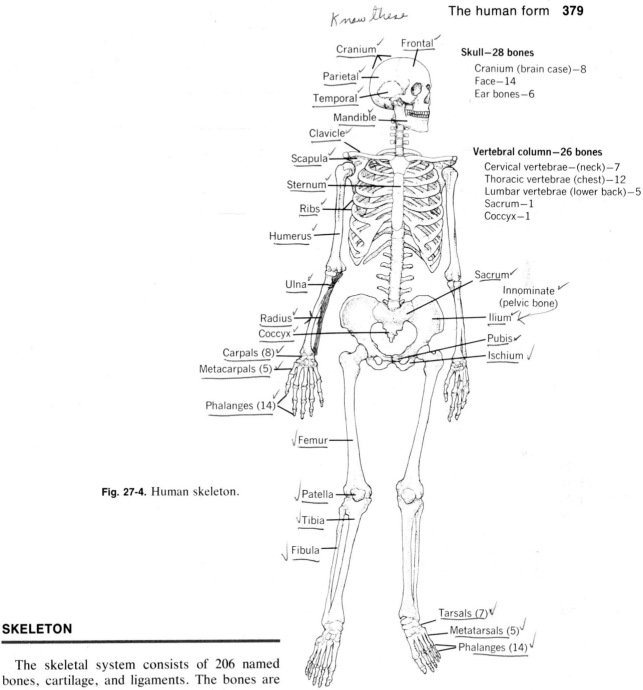

Skull—28 bones
Cranium (brain case)—8
Face—14
Ear bones—6

Vertebral column—26 bones
Cervical vertebrae—(neck)—7
Thoracic vertebrae (chest)—12
Lumbar vertebrae (lower back)—5
Sacrum—1
Coccyx—1

Cranium
Frontal
Parietal
Temporal
Mandible
Clavicle
Scapula
Sternum
Ribs
Humerus
Ulna
Radius
Coccyx
Carpals (8)
Metacarpals (5)
Phalanges (14)
Femur
Patella
Tibia
Fibula

Sacrum
Innominate (pelvic bone)
Ilium
Pubis
Ischium
Tarsals (7)
Metatarsals (5)
Phalanges (14)

Fig. 27-4. Human skeleton.

SKELETON

The skeletal system consists of 206 named bones, cartilage, and ligaments. The bones are illustrated in Figs. 27-4 and 27-5, and the names, and numbers are given in table form in Fig. 27-6. Bones are classified concerning shape as follows: (1) long (arms and legs), (2) short (wrist), (3) flat

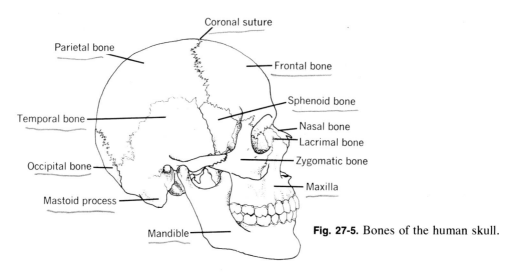

Coronal suture

Parietal bone

Frontal bone

Sphenoid bone

Temporal bone

Nasal bone

Lacrimal bone

Occipital bone

Zygomatic bone

Maxilla

Mastoid process

Mandible

Fig. 27-5. Bones of the human skull.

(shoulder blade, patella), and (4) irregular (vertebrae). It will be noted that the teeth are not listed as part of the skeleton but are included with the integument because of their epithelial origin.

Study of a human skeleton will show that there are several types of joints, each with its specific functions. Joints are classed as (1) immovable (irregular, dovetail connections [sutures] of the bones of the cranium) and (2) movable, with movements of various types for specific purposes. Movable joints may be further classed as (1) ball-and-socket (femur and pelvic girdle, humerus and pectoral girdle), (2) hinge (femur and tibia, humerus and ulna), (3) sliding (most vertebrae), and (4) rotating (radius and ulna).

Most bones originate in the embryo as cartilage; hence they are known as cartilage bones, in contrast to the membrane bones formed by the gradual ossification of soft, fibrous, membranous tissues (skull bones). Certain parts of the skeleton remain as cartilage, such as the external ear, tip of the nose, tip of the breastbone, areas between the vertebrae, and articulatory surfaces of movable joints. The great strength, elasticity, and reduced friction of cartilage make it an efficient part of the skeleton.

Ligaments are bands of tissue that connect the bones and support the interior organs. The liga-

ments bind the body parts together into an efficient structure.

Briefly stated, the functions of the human skeleton are (1) to form a framework to support other organs and give posture to the body, (2) to give protection to vital organs, such as the brain, spinal cord, heart, and lungs, (3) to form solid attachments for muscles so that they may act as a system of levers in motion and locomotion, (4) to store certain mineral reserves, (5) to form blood corpuscles by the bone marrow, and (6) to transmit sound waves as accomplished by the hammer, anvil, and stirrup bones of the ears.

MOTION AND LOCOMOTION

The bones of the skeleton are passive structures to which the active skeletal muscles are attached. By means of these muscles, body parts are moved, or the body as a whole is moved from place to place.

A muscle functions by contracting. This is initiated by the transmission of a nerve impulse, which may be voluntary or involuntary. The muscles of the skeleton are under control of the will, and those of the viscera (stomach, intestines) and heart are not.

The fine structure of the muscles differs and

Human skeleton*

Axial (80)
- **Skull**
  - **Cranium** (brain case) (8)
    - **Occipital** (base of skull) (1)
    - **Parietal** (top of head) (2)
    - **Frontal** (forehead) (1)
    - **Temporal** (above ears) (2)
    - **Ethmoid** (back of nose) (1)
    - **Sphenoid** (back of eye) (1)
  - **Face** (14)
    - **Mandible** (lower jaw) (2)
    - **Maxilla** (upper jaw) (2)
    - **Palate** (2)
    - **Malar or zygomatic** (cheek)
    - **Lacrimal** (inner orbit) (2)
    - **Inferior turbinate** (nose) (2)
    - **Vomer** (nasal septum) (1)
    - **Nasal** (bridge of nose) (2)
- **Ear bones** (6)
  - **Malleus** (hammer) (2)
  - **Incus** (anvil) (2)
  - **Stapes** (stirrup) (2)
- **Vertebral column** (26)
  - **Cervical** (neck) (7)
  - **Thoracic** (chest) (12) with ribs
  - **Lumbar** (lower trunk) (5)
  - **Sacral** (sacrum) (1)†
  - **Coccygeal** (caudal or tail) (1)‡
- **Hyoid** (base of tongue) (1)
- **Sternum** (breastbone) (1)
- **Ribs** (24)

Appendicular (126)
- **Pectoral girdle** (shoulder)
  - **Clavicle** (collarbone) (2)
  - **Scapula** (shoulder blade) (2)
- **Arms**
  - **Humerus** (2)
  - **Ralius** (2)
  - **Ulna** (2)
  - **Carpals** (wrist) (16)
  - **Metacarpals** (hand) (10)
  - **Phalanges** (fingers) (28)
- **Pelvic girdle** (hip) (2)§
- **Legs**
  - **Femur** (thigh) (2)
  - **Tibia** (shin) (2)
  - **Fibula** (2)
  - **Tarsals** (ankle and heel) (14)
  - **Metatarsals** (foot) (10)
  - **Phalanges** (toes) (28)
- **Kneecap** (patella) (2)

*This does not include the variable number of **sesamoid bones** (ses' a moid) (Gr. *sesamon*, sesame seed; *eidos*, like), embedded in the tendons of the hand, knee, and foot, or the **wormian bones** (wur' mi an) (after Worm, a Danish anatomist), which are isolated bones in the sutures or joints, especially those of the skull.

The figures in parentheses give the number of bones of each type.

†Five bones fused.

‡Four bones fused.

§Three bones (ilium, ischium, and pubis) fused.

Fig. 27-6. Outline of bones in human skeleton.

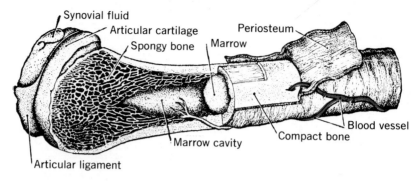

Fig. 27-7. Structure of a long bone and joint (diagrammatic).
(From Neal and Rand: Chordate anatomy, New York, McGraw-Hill Book Company.)

thus, skeletal muscle is also known as striated, visceral muscle as smooth, and heart muscle as cardiac. Under the microscope striated muscle shows a series of fine lines or striae, and it lacks a distinct cellular structure, having nuclei in a scattered pattern. By contrast, smooth muscle is composed of a series of distinct, spindle-like cells, each with its own nucleus. Cardiac muscle appears to have the fine striation and also individual nuclei and cellular structure. A series of branchings and joinings is also characteristic of cardiac muscle.

Skeletal muscle

Muscles functioning in movement are attached to two bones across a joint; contraction then pulls one bone toward (or away from) the other. The muscle end attached to the immovable bone is called the origin, while the muscle end attached to the movable bone is called the insertion. Muscles that move a part away from the median line are called abductor muscles; those that move a part toward the median line are called adductor muscles. The terms flexor and extensor indicate the types of movements muscles effect. The biceps of the forearm is a flexor (flexes or bends the forearm), and the triceps of the forearm is an extensor (extends or straightens the forearm). These muscles act as opposing pairs (Fig. 27-8).

Man has more than 600 skeletal muscles (Fig. 27-9), to which scientific names are applied in

various ways: (1) in relation to the structure, or bone, with which they are associated (*triceps brachii* muscle on the back of the upper arm, or brachium), (2) in relation to the number of "heads" with which they originate (*triceps brachii*, meaning three "heads," or *biceps brachii*, with two "heads" and located on the front of the upper arm), (3) in reference to the shape of the muscle (*deltoid*, the delta-shaped muscle of the top of the shoulder), (4) in reference to the direction in which they run (*external oblique*, a strong muscle of the abdominal wall that lies obliquely), (5) in reference to their loca-

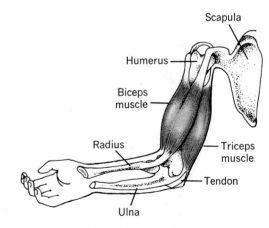

Fig. 27-8. Human biceps and triceps muscles. These are opposing muscles. Contraction of the biceps raises the forearm. Contraction of the triceps lowers the forearm.

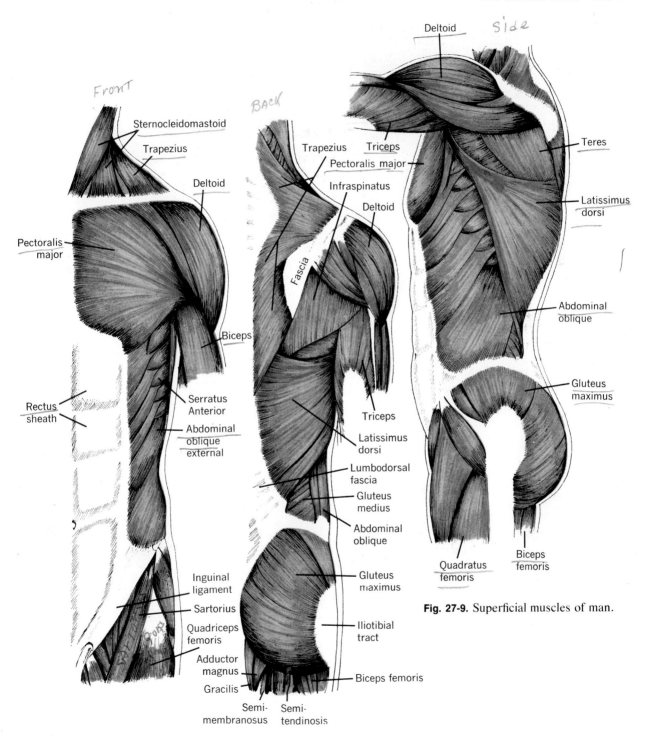

Front

Side

Deltoid

BACK

Sternocleidomastoid

Trapezius

Deltoid

Pectoralis
major

Biceps

Serratus
Anterior

Rectus
sheath

Abdominal
oblique
external

Inguinal
ligament

Sartorius

Quadriceps
femoris

Adductor
magnus

Gracilis

Semi-
membranosus

Semi-
tendinosis

Trapezius

Triceps

Pectoralis major

Infraspinatus

Deltoid

Fascia

Triceps

Latissimus
dorsi

Lumbodorsal
fascia

Gluteus
medius

Abdominal
oblique

Gluteus
maximus

Iliotibial
tract

Biceps femoris

Deltoid

Teres

Latissimus
dorsi

Abdominal
oblique

Gluteus
maximus

Quadratus
femoris

Biceps
femoris

Fig. 27-9. Superficial muscles of man.

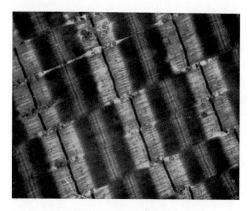

Fig. 27-10. Striated muscle from the rabbit (enlarged 24,000 diameters by the electron micrograph). Each diagonal ribbon is a thin section of myofibril (muscle fibril) with its dense A-bands bisected by less dense H-zones, and lighter I-bands bisected by more dense, narrow Z-lines, or membranes.
(From Huxley: Sci. Amer., November, 1958.)

tion (*external intercostals,* superficial muscles between the ribs, and *internal intercostals,* deeper muscles between the ribs), (6) in reference to their function (*adductor longus,* adducts the thigh toward the median line), and (7) in reference to the length and size of the muscle (*peroneus longus,* the large muscle attached to the fibula bone and *peroneus brevis,* the smaller muscle attached to the fibula bone).

Muscle structure

A medium-sized muscle contains approximately 10 million cells (fibers), so that over 6 billion muscle cells are contained in the total musculature. Each striated, skeletal muscle is composed of numerous bundles, and each bundle consists of hundreds of delicate fibrils, each having of diameter of 10 to 100 μm.

Fibrils are composed of a series of alternate dark and light areas, thus giving a banded (striated) appearance to skeletal muscles. Each skeletal muscle cell contains several nuclei. Skeletal muscles are voluntary (under the control of the will), distinctly striated, and have a rather rapid rate of action and a relatively high rate of

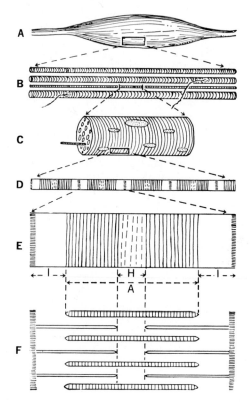

Fig. 27-11. Striated muscle dissected and represented by schematic drawings. Muscle, **A,** is made up of muscle fibers, **B,** which appear striated (banded) in the light microscope. Small branching structures at the surface are "end plates" of motor nerves, which signal the fibers to contract. Single muscle fiber, **C,** is made up of myofibrils, beside which lie cell nuclei and mitochondria. In the single myofibril, **D,** striations are resolved into a repeating pattern of dark and light bands. Single unit of this pattern, **E,** consists of Z-line, then I-band, then A-band which is interrupted by H-zone, then next I-band, and finally next Z-line. Electron micrographs show that the repeating band pattern is a result of the overlapping of thick and thin filaments, **F.**
(From Huxley: Sci. Amer., November, 1958.)

fatigue compared with other types of muscles.

The contractile structure of a muscle fiber is made up of long, thin elements called myofibrils, each about 1 μm in diameter. A myofibril has cross striations like the fiber of which it is a part.

These striations arise from a repeating variation in the density, or concentration, of protein along the myofibrils (Figs. 27-10 and 27-11). The striations can be seen in isolated myofibrils. Under high magnification of a light microscope there is a regular alternation of dense bands (A-bands) and lighter bands (I-bands). The central region of an A-band is frequently less dense than the rest of the band and is known as the H-zone. The I-band is bisected by a narrow, dense line known as the Z-line (or Z-membrane). From one Z-line to the next, the repeating units of the myofibril may be summarized as Z-line, I-band, A-band (interrupted by the H-zone), I-band, and Z-line.

When examined with an electron microscope, the myofibril is found to consist of still smaller filaments, each of which is 50 to 100 angstrom units in diameter (1 angstrom unit is 1 ten-thousandth of a micrometer). When properly prepared thin sections of muscle are observed with an electron microscope, the myofibrils are seen to consist of two kinds of filaments, one of which is twice as thick as the other. Each filament is arranged in register with other filaments of the same kind, and the two arrays overlap for part of their length. This overlapping gives rise to the crossbands of the myofibril, the dense A-band consisting of overlapping thick and thin filaments, the lighter I-band consisting of thin filaments alone, and the H-zone consisting of thick filaments alone. Halfway along their length the thin filaments pass through a narrow zone of dense material that comprises the Z-line.

The two kinds of filaments are linked together by an intricate and well-ordered system of cross bridges, which may play an important role in muscle contraction. The bridges project outward from a thick filament at rather regular intervals of 60 to 70 angstroms, and each bridge is 60 degrees around the axis of the filament with respect to the adjacent bridge. The bridges thus form a helical pattern that repeats every six bridges along the filament. This pattern joins the thick filament to each one of its six adjacent, thin filaments once every 400 angstroms.

Muscle contraction

Knowledge of the band structure of striated muscle made it apparent that changes in the pattern during contraction should give insight into the molecular nature of the process. It has been found that over a wide range of muscle lengths, during both contraction and expansion, the length of the A-bands remains constant. On the other hand, the length of the I-bands changes in accord with the length of the muscle. The length of the A-band is equal to the length of the thick filament. But the length of the H-zone increases and decreases with the length of the I-band, so that the distance from the end of one H-zone through the Z-line to the beginning of the next H-zone remains approximately the same. This distance is equal to the length of the thin filaments, so they, too, do not alter their length appreciably. From all of this it is concluded that when a muscle changes length, the two sets of filaments slide past each other (sliding-filament phenomenon (Fig. 27-11).

When a muscle is treated with an appropriate salt solution and examined with a light microscope, it is observed that the A-bands have disappeared. Such a salt solution will remove myosin (a muscle protein), which demonstrates that the thick filaments of the A-band are composed of myosin. If the myofibril is treated to extract its actin (another protein), a large part of the material in these segments is removed, which indicates that the thin filaments of the I-bands are composed of actin. It is likely that the physical expression of the combination of actin and myosin is found in the bridges between the two kinds of filaments; thus, it seems that the sliding movement is mediated by the bridges.

It is suggested that the cross bridges seem to form a permanent part of the myosin filaments; probably they are those parts of the myosin molecules that are involved directly in the combination with actin. Since the number of myosin molecules in a given volume of muscle is surprisingly close to the number of bridges in the same volume, it is suggested that each bridge may be part of a single myosin molecule.

How can the bridges cause contraction? One theory is that the bridges may oscillate back and forth and hook up with specific sites on the actin filament. The bridges then could pull the filament a short distance and return to their original location, ready for another pull. One would expect that each time a bridge went through such a cycle a phosphate group would be split from a single molecule of adenosine triphosphate (ATP), which would supply the energy for the cycle. When the muscle has relaxed, it may be supposed that the removal of phosphate groups from the ATP has ceased, and that myosin bridges can no longer combine with actin filaments, thus permitting the muscle to return to an uncontracted condition.

ATP is one of the most important substances in muscle. The terminal phosphate group of this compound is linked to the structure by an energy-rich bond. Rupture of this bond produces inorganic phosphate and adenosine diphosphate (ADP) with the release of a large amount of energy that can be used in performing work.

$$ATP \rightarrow ADP + H_3PO_4 + Energy$$

Myosin is the actual enzyme or is at least associated with the catalytic breakdown of ATP to ADP. The reaction is activated by calcium ions and inhibited by magnesium ions. In resting muscle the enzymatic activity is inhibited, and ATP maintains the extensibility of the fiber. When stimulated, the enzymatic activity breaks down ATP to ADP, with the sudden release of energy for contraction. The chemical interaction of ATP, myosin, and actin results in the closer interdigitation of the thick and thin filaments, with consequent muscle shortening. The ATP is subsequently restored, to be used later.

MUSCLE STIMULATION

Around each muscle fiber is an electrical, polarized membrane, the inside of which is about one tenth of a volt negative with respect to the outside. If the membrane is depolarized temporarily, the muscle fiber contracts; it is by this means that muscle activity is controlled by the nervous system. An impulse traveling over a motor nerve is transmitted to the muscle membrane at the motor end plate or neuromuscular junction. It then passes deep into the myofibril by way of a system of fine T-tubules, which cross the area where actin and myosin overlap. At this point the tubules come in contact with the sarcoplasmic reticulum, which runs parallel to the myofibril. An impulse travels by way of the T-tubules to the reticulum, which then releases calcium ions that in turn initiate the muscle contraction. Then a wave of depolarization, the "action potential," passes down the muscle fiber causing it to twitch. When nerve impulses arrive in rapid succession through the motor nerve, the twitches run together, and the muscle maintains its contraction as long as the stimulation continues (or until the muscle becomes exhausted). The muscle relaxes automatically when nerve stimulation ceases.

As long as a man is conscious and his muscles are not actively contracting, the muscles are not completely relaxed but are constantly partially contracted. This phenomenon is known as tonus ("tone") (Gr. tonos, tension). Tonus, by a series of nerve impulses, is responsible for body posture and keeps muscles in readiness for actual contraction. Severing of the nerve to a skeletal muscle eliminates tonus. During tonus only a small fraction of the muscle fibers are involved, and it is thought that individual fibers work in relays, thus giving them a chance to recover between contractions.

An entire muscle cannot contract maximally, but a single fiber can respond only maximally or not at all. This is explained by the all-or-none law and can be demonstrated by dissecting out a single muscle fiber and subjecting it to repeated stimuli of increasing intensity, starting with those too weak to cause contraction. Response occurs only when a certain level of stimulus strength is reached, and at that time the fiber will contract completely. Stimuli of even greater intensity do not cause any greater contraction. The nature and amount of contraction of an entire muscle depend upon the number of fibers that are contracting and upon their contracting simultaneously or alternately.

When a muscle is given a single stimulus, such as a single electric shock, it responds with a single, quick twitch that lasts about 0.05 second in a human muscle and about 0.1 second in a frog muscle. A single twitch of a frog muscle consists of three periods: (1) a latent period (0.01 second), or the interval between stimulus application and the start of visible contraction, (2) the contraction period (0.04 second), during which the muscle contracts and does work, and (3) the relaxation period (0.05 second), during which the muscle relaxes (returns to its original length). After a twitch the muscle uses oxygen and gives off carbon dioxide and heat at a rate greater than that during rest, suggesting a recovery period (several seconds) in which the muscle is restored to its original condition. If stimuli are repeated so that successive contractions occur before the muscle has recovered from the previous stimuli, the muscle becomes fatigued, and the twitches grow fewer and finally stop. Sufficient rest will allow a fatigued muscle to regain its ability to contract. Normal contractions of muscles, however, do not occur as single twitches but as sustained contractions caused by a series of nerve impulses (stimuli) reaching them in rapid succession. Such sustained contraction is called tetanus (Gr. *tetanos,* stretched), and the stimuli occur so rapidly that relaxations cannot occur between successive contractions. In most normal contractions the various muscle fibers are stimulated in rotation, so that although individual fibers contract and relax, the muscle as a whole remains contracted partly. In the weak contraction of a muscle only a small percentage of the contained fibers receives nervous stimuli, whereas in a stronger contraction, more fibers contract at the same time. If a muscle has contracted many times, it will exhaust its supplies of glycogen and organic phosphates and produce lactic acid, so that it can contract no longer.

VISCERAL MUSCLES

Many unnamed, mononucleated, smooth muscles are part of the internal organs, such as the esophagus, stomach, and intestines. These muscles are involuntary, have a rather slow, rhythmic rate of action, and do not fatigue easily. The churning and mixing movements of the digestive system, called peristalsis, and the rhythmic contractions of the uterus during childbirth are examples of their activity.

CARDIAC MUSCLE

Numerous indistinctly striated, branched, cardiac muscles compose the walls of the heart and blood vessels. Because of the branches or connections between some adjacent cells, there may be more than one nucleus per cell. Cardiac muscles are involuntary, have a variable rate of action, and under normal conditions do not fatigue easily.

Review questions and topics

1 What are the systems of the human body and their functions?
2 Describe the integument of man and its various functions.
3 Describe the human skeleton and its functions.
4 Define the following bones: sesamoid, wormian, patella, phalanges, hyoid, pelvic girdle.
5 Discuss the band structure of striated muscles, including changes during contraction.
6 Explain how energy is released when muscles perform.

7 Explain the all-or-none law and its significance.
8 Discuss the periods when a muscle responds to a single stimulus.
9 Describe the following types of muscles: skeletal, visceral, and cardiac.
10 Define: myofibril, fatigue, myosin, actin, adenosine triphosphate (ATP), motor end-plate, "action potential," tonus, tetanus, and sliding-filament phenomenon.

See Chapters 29 and 30 for related references on human biology.

28 Human metabolism

The material presented in this chapter deals with the "logistics" of living. While it must be emphasized that no living process takes place in a vacuum—the interaction of all systems is what constitutes life—for convenience we will consider only certain systems here.

These are, in general, the various systems that deal with the obtaining of food, its processing, distribution and use, and related activities that contribute to or result from this energy cycle. The following areas will be discussed: (1) foods and nutrition, (2) respiration, including breathing and cell respiration, (3) excretion, and (4) circulation and related areas.

 DIGESTION

In the strict sense a food is a substance that can be broken down by the organism to release energy. This definition would apply primarily to fats, carbohydrates, and proteins, and their related substances. A broader definition would include other materials not synthesized by the cells but still necessary for metabolism. Water, vitamins, and inorganic chemicals would be part of an expanded list of basic nutrients.

Because of their size and chemical make-up, most large organic molecules cannot be absorbed until they have been digested. They are then absorbed in the form of smaller molecules. Often these same small food molecules are introduced through the veins of a person too weak to eat

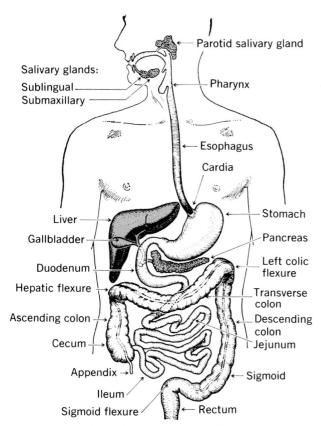

Salivary glands:
Sublingual
Submaxillary

Parotid salivary gland
Pharynx
Esophagus
Cardia
Liver
Gallbladder
Duodenum
Hepatic flexure
Ascending colon
Cecum
Appendix
Ileum
Sigmoid flexure
Stomach
Pancreas
Left colic flexure
Transverse colon
Descending colon
Jejunum
Sigmoid
Rectum

Fig. 28-1. Human digestive system with accessory organs. (The small intestine is shown shortened for clarity.)
(From Schottelius, B. A., and Schottelius, D. D.: Textbook of physiology, ed. 17, St. Louis, 1973, The C. V. Mosby Co.)

solid food. It is important to note that the actual molecules available to the cells are the same after digestion as those introduced in intravenous feeding.

Digestion, as such, is primarily a process of enzymatic hydrolysis, resulting in a change of the size of large organic food molecules. This results, then, in the splitting of food molecules by water under the influence of the appropriate enzyme.

Digestive system

The digestive system in man is basically a long tube, of varying diameter, that provides the proper environment for the chemical change of food

molecules. It consists of the mouth, esophagus, stomach, duodenum, small intestine, colon or large intestine, rectum, and anus. The liver, gallbladder, and pancreas may be considered as associated digestive structures (Fig. 28-1).

The ducts of the three pairs of salivary glands open into the mouth and provide saliva, which contains enzymes important in carbohydrate digestion. After food has been swallowed, it passes through the esophagus into the stomach. The rate of movement depends on the fluidity of the food and the action of peristalsis, the squeezing activity of smooth muscle. Anyone who has taken a cold drink on a hot summer's day is aware of the speed with which liquids pass through the esophagus and into the stomach.

The stomach is an expanded, sacklike structure with its larger area at the anterior end. It tapers down to form a constriction at the pylorus, the site of the pyloric sphincter. The glands of the stomach secrete gastric juices containing hydrochloric acid (HCl) in ion form, and enzymes, in an inactive state. The primary action in the stomach is that of protein digestion. It is interesting to note that the enzymes are active only at the low pH provided by the H^+ ions. Contraction of the smooth muscles in the stomach wall provides a vigorous churning or mixing of the food and gastric juices. Waves of contraction push the food masses forward toward the pylorus, which periodically opens, and the semiliquid food, called chyme, squirts through the valvelike sphincter and into the duodenum. The chyme is mixed with fluids from the intestinal mucosa, liver, gallbladder, and pancreas. Fluids from the pancreas and gallbladder enter through the common bile duct.

In the duodenum and small intestine the process of digestion is finally completed. Carbohydrate digestion, started in the mouth, proceeds to the point of producing simple sugars. Protein digestion, started in the stomach, is completed and results in amino acids. Fat is emulsified by bile, producing smaller particles with a larger surface area; these are then digested to produced glycerin and fatty acids. All of these are absorbed by cells lining the small intestine. The rate of absorption varies with the size and shape of molecule and heat of food.

Unabsorbed materials in a fluid state pass into the large intestine where water is absorbed and the residue concentrated as the feces. Bacterial activity may produce some vitamin precursors at this time. The entire residue then moves into the rectum and is later eliminated by defecation.

Control of digestion

Before discussing details of the digestive process, we shall consider another important factor—the control of the entire process.

The common sensation we have on entering a bakery—the pleasurable smell of cakes and pastry—results in a flow of saliva into the mouth. We are, in a sense, ready to begin eating. This same flow of saliva may be produced by such things as the sight of food, the odor of food, and even the memory of food. We might add to this, perhaps, the sound of the "sizzle" of a steak frying, and of course, the contact of food in the mouth.

These examples emphasize the variety of factors that may initiate the flow of saliva. This secretion is controlled by nerve impulses from the brain. As food is being chewed, other nerve impulses stimulate the flow of gastric and pancreatic juices.

As food is swallowed and actually enters the stomach, the stomach lining is stimulated to release into the blood a hormone, gastrin, which then influences further gastric secretion. Contact by the food entering the duodenum helps stimulate the flow of intestinal juices. The low pH of food entering the duodenum activates a hormone, secretin, which passes through the blood to the pancreas, which then releases pancreatic juices. The same hormone may stimulate the liver to release bile. Previously secreted bile has been stored in the gallbladder. The presence of fats in the food acts as the stimulus for the release of another hormone, cholecystokinin, which is carried by the blood to the gallbladder, which

contracts to release stored bile. Finally, the passage of foods further along the intestine initiates the release of still another hormone, enterogastrone, whose action is the inhibition or turning off of the gastric phase of activity.

The process of digestion is controlled by a variety of psychologic, nervous, and chemical factors, all of which provide the proper digestive fluids in the order of their activity, finally stopping the activity when no more food is present.

As we are aware, the malfunction of any of these processes may lead to complications, such as gastric ulcers.

Digestive process

Digestion is primarily the result of enzymatic hydrolysis. This may be shown by the following scheme:

$$\text{Foods} + H_2O \xrightarrow[\text{pH, ions}]{\text{Enzyme}} \text{Absorbable products}$$

Table 28-1 summarizes the digestive processes. The following scheme shows the general outlines of digestion of the primary foods.

$$\text{Complex carbohydrates} + H_2O \xrightarrow{\text{Amylase}} \text{Disaccharides}$$

$$\text{Disaccharides} + H_2O \xrightarrow{\text{Disaccharases}} \text{Monosaccharides}$$

$$\text{Proteins} + H_2O \xrightarrow{\text{Pepsin}} \text{Polypeptides}$$

$$\text{Polypeptides} + H_2O \xrightarrow{\text{Peptidases}} \text{Amino acids}$$

$$\text{Fats} + H_2O \xrightarrow{\text{Lipase}} \text{Fatty acids} + \text{Glycerol}$$

Fate of absorbed foods

With the exception of fats, which pass into the lacteals and lymphatic circulation, most absorbed foods move into the capillaries of the intestinal villi and then pass by way of the hepatic portal system to the liver. In this, the largest organ of the body, hundreds of metabolic activities take place.

Among these processes are: (1) glucose-glycogen activities, (2) amino acid breakdown with the formation of urea, (3) storage of proteins, minerals, and vitamins, (4) formation of plasma proteins, such as albumin and fibrinogen, and (5) some fatty acid metabolism.

CARBOHYDRATE METABOLISM

The liver is the primary storage site of carbohydrate in the body. The formation of glycogen serves both to store glucose and to regulate its content in the blood. The amount of glucose normally present in the circulating blood is approximately 0.1%. A higher concentration is present in the blood leaving the intestine immediately after a meal. As this quantity of glucose reaches the liver, much is taken out of circulation and converted to glycogen. The process is controlled by insulin, a hormone secreted by the pancreas. When the supply of circulating glucose is low, glycogen breakdown occurs, and the level is raised. In the stress reaction the hormone adrenalin promotes the breakdown of glycogen and leads to an increased supply of glucose in the circulating blood.

Diabetic patients have abnormal quantities of insulin in the blood. If too little is present, glycogen is not formed, and glucose accumulates in the blood and is excreted through the kidneys. When too much insulin is present, the supply of circulating glucose drops as it is converted to glycogen. This produces a type of cell "starvation" and may result in the person going into a coma.

In general, carbohydrates (1) form the principal energy source available to cells, (2) form a storage product, and (3) supply a basis for synthesis of such substances as nucleotides. When carbohydrates are not available, fats and proteins may break down to form glucose.

AMINO ACID METABOLISM

Amino acids are not stored as such in the body, but they are required in the daily diet where their chief function is the formation of proteins. These include enzymes, antibodies, tissue proteins, blood proteins, and cell membranes.

As amino acids are used in metabolism, the amino group, —NH_2, is removed and forms

Table 28-1 Summary of digestive processes in man

Region	Secretion and source	Enzyme	pH	Substances acted upon	Products formed
Mouth	Saliva (from salivary glands)	Amylase	Neutral	Starches	Maltose
Stomach	Gastric juice (from gastric glands)	Pepsin*	Acid	Proteins	Polypeptides
Small intestine	Pancreatic juice (from pancreas)	Amylase	Alkaline	Starches	Double sugars
		Lipase	Alkaline	Fats	Fatty acids and glycerol
		Trypsin	Alkaline	Proteins and polypeptides	Peptides and amino acids
	Intestinal juice (from glands of small intestine)	Peptidase	Alkaline	Peptides	Amino acids
		Maltase	Alkaline	Maltose (malt sugar)	Glucose
		Lactase	Alkaline	Lactose (milk sugar)	Glucose and galactose
		Sucrase	Alkaline	Sucrose (can sugar)	Glucose and fructose
		Enterokinase (not a true digestive enzyme)†	Alkaline	†	†

*When acting with hydrochloric acid (HCl).
†Activates the inactive trypsinogen of the pancreatic juice, changing it to active trypsin. (See information under Substances acted upon and Products formed.)

ammonia (NH_3), a potential poison. Detoxification occurs as the $-NH_2$ is converted to urea, which is then excreted by the kidneys.

Amino acids may also be utilized as food and converted to fat or carbohydrates.

FAT METABOLISM

Fats form an important component of all cell membranes, as part of a lipoprotein complex. They may be readily converted to glucose and undergo respiration, hence, they are a major storage food. Other kinds of foods present may be converted to fats and stored. Thus, any type of excess food may be stored as fat.

Mineral salts

Mineral salts are inorganic substances that, in combination with other food constituents, promote the formation and maintenance of various body structures and functions. Minerals not only aid in metabolic processes but also actually help to form certain body tissues and fluids. The daily excretion in feces, urine, and perspiration of about 30 gm. of mineral salts must be replaced. The absence of the proper amount of minerals in a diet is more serious than lack of the proper amounts of carbohydrates, fats, or proteins and when prolonged, causes the use of the body reserve of salts, with possibly serious consequences. Fortunately, many common foods are rich in minerals, but actual deficiencies of calcium, iron, and iodine may occur at times. About fifteen elements are known to be essential as mineral salts in the human diet; some are required only in traces.

In addition to the formation of muscles and bones, minerals are important as the ions that are involved in various activities, such as respiration, heartbeat, and nerve impulse transmission.

Water

About two thirds of the human body is water. It is the medium in which chemicals are dissolved and in which chemical reactions occur. Water is essential for the separation of the molecules of carbohydrates, fats, and proteins during diges-tion. It helps to from the blood plasma and lymph. Water equalizes body heat and removes metabolic wastes. Over 2 quarts of water are lost daily, the specific amount changing with such variables as climate and activities. This loss must be replaced by liquid foods and water-containing foods, such as fruits and green vegetables.

Vitamins

Vitamins are organic compounds that are essential for a number of living processes. They differ chemically and cannot be formed by the body cells, although bacteria in the intestine may synthesize certain types. It was found in 1912 that animals could not survive on a diet of purified carbohydrates, fats, and proteins alone, but that "accessory growth factors" were essential. Vitamins were originally named because they were considered to be amines ($-NH_2$) that were essential to life (L. *vita*, life). Some vitamins, originally considered as a single unit, are now known to be complexes of vitamins, for example, original vitamin A (now divided into A, D, and E) and vitamin B (divided into over a dozen kinds).

Lunin (1888), in Switzerland, fed mice a diet of carbohydrates, fats, proteins, and inorganic salts that he had isolated or synthetically prepared in the laboratory. The quantities of these ingredients were thought to be in the same proportions as those in milk. When were fed this diet, they died, but when milk was added to the prepared diet, the mice lived. He concluded that milk contained an unknown substance essential to life.

In 1913, McCollum and Davis, and Osborne and Mendel experimentally proved the presence of a fat-soluble dietary substance in butterfat that promoted growth and well-being of rats. The substance was first called "fat-soluble A" factor.

Probably all plants and animals require vitamins of some types, but their specific requirements vary. Vitamins serve as integral parts of coenzymes for the fundamental enzyme reactions in all protoplasm. Plants normally synthesize vitamins, but animals do not ordinarily.

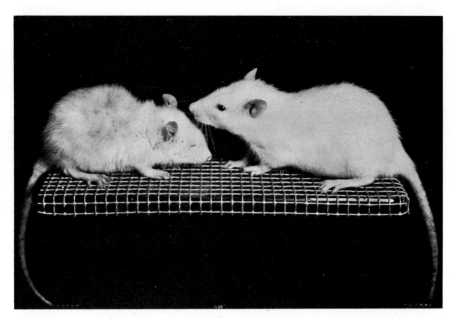

Fig. 28-2. Vitamin A deficiency in white rats. The animals were litter mates, 21 days old at the start of the experiment. For 33 days the animal at left received a diet containing all nutritive substances except vitamin A, whereas the animal at right received all nutritive substances plus a sufficient amount of vitamin A: Note the eye symptoms (xerophthalmia) in the vitamin A-deficient rat.
(Courtesy Upjohn Co., Kalamazoo, Michigan; from Schottelius, B. A., and Schottelius, D. D.: Textbook of physiology, ed. 17, St. Louis, 1973, The C. V. Mosby Co.)

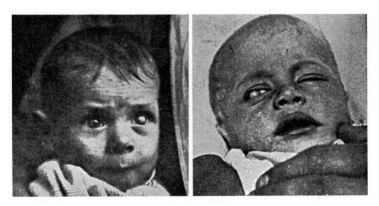

Fig. 28-3. Xerophthalmia, an eye disease caused by a dietary deficiency of vitamin A; the eye becomes dry and a layer of horny tissue forms upon the cornea.
(From Harris: Vitamins in theory and practice; courtesy Cambridge University Press and The Macmillan Company.)

Following is a brief account of important facts concerning vitamins.

Vitamin A

Vitamin A is found primarily in fatty foods of animal origin, hence is fat-soluble. Examples are fish-liver oils, cream, butter, milk, and egg yolk. We derive much of our vitamin A from a pigment, carotene, found in green-, red-, or orange-colored plants. The beta type of carotene ($C_{40}H_{56}$) is a water-soluble precursor, or provitamin, which is transformed into vitamin A chiefly by the liver. Vitamin A is not affected by light or heat but is destroyed by ultraviolet light or oxidation.

Vitamin A aids normal functioning of epithelial tissues (skin, eyes, eyelids, and digestive and respiratory systems). The insoluble proteins of the most superficial layer of the skin become transformed into a special hornlike, protective form known as keratin. Abnormal keratinization impairs the function of the involved tissues. When sweat and oil glands are affected, the skin becomes dry, scaly, and pimpled (acne).

Night vision (dim light vision) depends upon the presence in the rods of the eye retina of a pigment called visual purple. When stimulated by light, the rods and cones generate nerve impulses, which create a sensation of light in the brain. Bright light entering the eye quickly bleaches the visual purple, which accounts for the temporary blindness when going from bright light into dim light; but, in reduced light, the visual purple normally regains its color quickly. This is known as dark adaptation and permits seeing dimly lighted objects that were temporarily invisible. Dark adaptation depends on vitamin A. This vitamin is one of many factors associated with growth, and a deficiency may lead to increased susceptibility to infections, especially of the respiratory tract.

Vitamin B complex

Of the numerous water-soluble vitamins grouped together as the vitamin B complex, only a few are known to be of importance to man. The B complex vitamins are of fundamental impor-

tance to life; they are found in all living organisms. They are associated with metabolic processes throughout the body, as is shown by the fact that the amount of these vitamins required is largely determined by the amount of energy expended in work and the maintenance of body temperature.

Generally speaking, the absence of these vitamins may lead to lowered general well-being, increased fatigue, and emotional disturbances, such as depression and abnormal irritability. The B complex vitamins play important roles as coenzymes in the many successive chemical changes in the glucose and glycogen molecules. Since these chemical changes occur in all types of cells and are necessary for life, their absence is of great importance.

Thiamine (B₁)

Good food sources of thiamine include cereal grains, beans, peas, nuts, wheat germ, and pork. Vitamin B_1 is soluble in water and is not destroyed by cooking.

Vitamin B_1 acts within living cells as an essential part of carbohydrate metabolism. Vitamin B_1 is converted to cocarboxylase and acts with essential enzymes. In vitamin B_1 deficiency not all of the carbohydrate is used, and pyruvic acid builds up in the body.

Vitamin B_1 is not stored in quantity in the body, hence a daily intake is necessary. The body takes in only the amount needed at the time, and any excess is destroyed or excreted. Serious, prolonged deficiency of vitamin B_1 leads inevitably to beriberi (multiple peripheral neuritis). Symptoms include inflamed peripheral nerves, which become painful and degenerate, heart changes, muscle weakness, and swellings caused by excess fluids. The disease can be fatal.

Partial deficiency may cause loss of appetite, fatigue, poor sleep, irritability, and digestive system disorders.

Riboflavin (B₂)

Vitamin B_2, or riboflavin, is slightly water soluble. Good food sources include milk and its

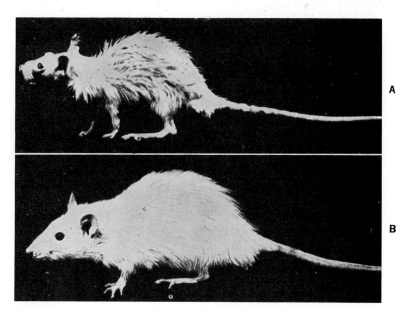

Fig. 28-4. Results of riboflavin (vitamin B₂) deficiency. **A,** Rat, 28 weeks old, received no riboflavin and weighed 63 gm.; **B,** after receiving food rich in riboflavin for 6 weeks the same rat weighed 169 gm.
(Courtesy Bureau of Human Nutrition and Home Economics, U.S. Department of Agriculture.)

products, eggs, meats, legumes, green leaves, and whole-grain cereals. Riboflavin is essential to life. We have no special storage tissues in our bodies for riboflavin, yet a certain level is maintained in various places, especially in the liver and kidneys.

Riboflavin is a vital link in carbohydrate metabolism. It is a constituent of many enzyme systems. In certain animals the requirement of riboflavin may be met by synthesizing it through the activities of bacteria within the intestine.

Within cells riboflavin is attached to a phosphate group, where it is known as riboflavin-5-phosphate, or flavin mononucleotide. This in turn may be attached to still another essential substance, adenylic acid, forming flavin-adenine dinucleotide (FAD). Either nucleotide then is attached to a protein, forming an enzyme, and takes part in important oxidation-reduction reactions.

Deficiencies of riboflavin occur in man in several ways. Symptoms include inflammation of eyes and skin, and nerves and blood are also affected. In riboflavin-deficient rats there is stunted growth, hair loss, inflamed and cataractous eyes, and even death (Fig. 28-4).

Nicotinic acid (niacin)

Nicotinic acid is another B complex vitamin. It is said to be formed by bacterial activity within the intestine. Good food sources include yeast, roasted chicken, beef liver, tuna fish, halibut, peanuts, lean meats, poultry, and enriched flour.

Nicotinic acid deficiency gives rise to pellagra ("skin seizure"), a disease characterized by lack of vitality and strength, loss of appetite, indigestion, diarrhea, pain, skin eruptions and sometimes severe mental disturbances. Formerly, the cause of these symptoms was thought to be nicotinic acid deficiency alone, but it is known that a lack of thiamine, riboflavin, and possibly other B vitamins are also involved, thus making pellagra a multiple-deficiency disease.

A diet lacking the essential amino acid tryptophan produces effects similar to those caused by nicotinic acid deficiency. Tryptophan can also be substituted for nicotinic acid in abolishing most of the pellagra symptoms. Although milk and eggs are poor sources of nicotinic acid, they are excellent pellagra preventives because of their supply of tryptophan.

Combined with a highly complex phosphorus compound, nicotinic acid forms nicotinamide-adenine-dinucleotide (NAD) and nicotinamide-adenine-dinucleotide-phosphate (NADP), often called coenzymes I and II, which play important roles in carbohydrate catabolism.

Pyridoxine (B₆)

Pyrodoxine is another B complex substance that seems to be essential for animals. Good food sources include liver, eggs, fish, lettuce, celery, lemon, whole wheat, and milk.

Many ill effects have been attributed to the lack of vitamin B_6, such as growh impairment in young experimental animals and body weight loss in adults. Skin lesions are common, with anemia in some instances. This vitamin plays an important part in the chemical reactions involving amino acids, apparently acting as a coenzyme.

Pantothenic acid

Pantothenic acid is present in all living cells. Common sources include eggs, meat, liver, fresh vegetables, sweet potatoes, cane molasses, and milk. If diets contain sufficient eggs, meat, and milk, the requirements of this substance apparently are met.

Pantothenic acid functions as one of the important coenzymes of metabolism (coenzyme A of the Krebs cycles; see Fig. 28-12). It seems to be necessary for normal skin and nerves, and its presence is probably needed for many fundamental chemical reactions in cells.

Biotin

Biotin is a substance that plays important roles in living processes. Good sources include liver, kidney, eggs, milk, and most fresh vegetables. Balanced human diets supply sufficient amounts of biotin.

Only a most unbalanced diet would evoke such symptoms of biotin deficiency as skin and tongue lesions. Biotin, acting as a coenzyme, takes part in many enzyme reactions, including carbon dioxide utilization, deamination (splitting of the amine group [NH_2] from the amino acid), and oxidation of intermediate products in protein

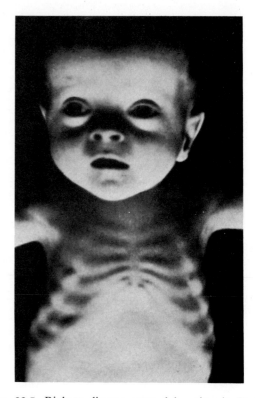

Fig. 28-5. Rickets disease caused by vitamin D deficiency, showing the square head and enlarged junctions between the ribs and rib cartilages. *(From Jeans, P. C., and Marriott, W. McK.: Infant nutrition, ed. 4, St. Louis, The C. V. Mosby Co.)*

catabolism. It is involved also in carbohydrate and fat metabolism.

Folic acid

Folic acid occurs in plant and animal tissues. Good sources include liver, asparagus, broccoli, spinach, yeast, soybeans, and egg yolk.

Folic acid is usually formed in sufficient amount in the human intestine to meet daily needs. It is necessary for the growth and formation of blood cells and for certain metabolic processes. Lack of folic acid causes anemia in man and, in chickens, such conditions as kidney hemorrhages and bone deformities.

Choline

Choline is a complex nitrogenous substance that performs a number of vital functions. Choline aids in fat storage and usage. When absent, the liver becomes filled with fat. Choline also aids in forming phospholipids (phorphorus-containing fat). Lecithin, a common phospholipid in the body, is formed when one of three fatty acids is replaced by choline and phosphoric acid in a molecule of a neutral fat (such as palmitin). Lecithin is essential in cells, where it decreases surface tension, causing it to adhere to the protoplasm's surface. It forms a constituent of the plasma membrane and contributes to its selective permeability—a highly important phenomenon in cell physiology.

Choline is used in the construction of acetylcholine, a compound formed by the union of choline and acetic acid. Acetylcholine plays roles in nerve impulse transmission, skeletal muscle contraction, and cardiac inhibition.

Cobalamin (B_{12} group)

Cobalamin is sometimes referred to as the "erythrocyte maturation factor." It is unique in that it contains the element cobalt. It is manufactured by certain fungi within the intestine. Common sources include beef liver, milk, egg yolk, oysters, and yeast. Its lack causes pernicious anemia, poor growth, and wasting of tissues.

Ascorbic acid (vitamin C)

Ascorbic acid is water-soluble substance that is very important as an antioxidant. Common sources include citrus fruits, tomatoes, potatoes, cabbage, and spinach.

Vitamin C is essential for maintaining healthy bones and for forming collagen, an important skin protein, tendons, cartilage, and connective tissues, where it synthesizes "cell cement." It is required for the normal functioning of blood vessels, promotes healing of wounds and fractures, and is essential for tissue respiration, in which foods are burned to provide energy and heat.

Lack of vitamin C causes scurvy in man and certain other animals. Scurvy symptoms include great fragility of capillary walls, fracture of bones, swollen, spongy, bleeding gums, loose teeth, painful joints, and slow healing wounds.

Vitamin D (calciferol)

Vitamin D is a fat-soluble vitamin that is very limited in its distribution. It is never found in natural foods of plant origin. Normal diets probably contain very little, if any, of it. It is quite stable, can withstand boiling, and is not oxidized readily. The most abundant sources are liver oil and body oils of such fishes as salmon, halibut, mackerel, cod, and swordfish. Egg yolk may contain some vitamin D, but the amount depends on the irradiation of the hen and the amount of vitamin in her food.

It is doubtful that man needs much vitamin D in foods because the irradiation of his skin (by ultraviolet light) changes the precursor (ergosterol) in the skin into this vitamin.

Vitamin D regulates calcium and phosphorus metabolism. Its lack causes rickets (Fig. 28-5), characterized by soft bones, bowed legs, beaded ribs, defective teeth, and enlarged wrists, knees, and ankles.

Vitamin E

Vitamin E is a fat-soluble substance that is abundant in vegetable oils. Common sources include green leaves, wheat germ, yeast, vegetable oils (corn, cottonseed, soybean, and peanut), meats, and eggs. Vitamin E is considered necessary for cell maturation and differentiation in vertebrates and for nuclear growth and activity.

Vitamin K

Vitamin K is a fat-soluble compound, and its absorption into the blood is dependent upon bile. Common sources include green leaves, spinach, soybean oil, kale, liver, and egg yolk.

Vitamin K assists in prothrombin formation in the liver, necessary for blood coagulation. It is manufactured by bacterial action within the in-

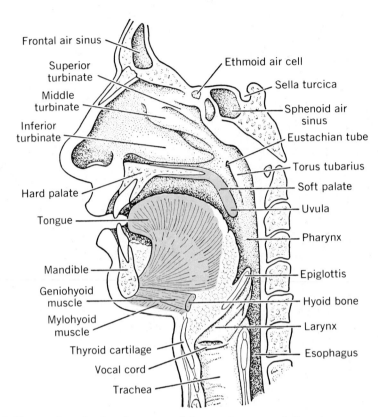

Fig. 28-6. Face and neck, showing parts of the respiratory and digestive systems (sagittal section). The common cavity where the two systems cross is known as the pharnyx. The leaf-shaped epiglottis prevents the food from entering the larnyx from the pharnyx. *(From Schottelius, B. A., and Schottelius, D. D.: Textbook of physiology, ed. 17, St. Louis, 1973, The C. V. Mosby Co.)*

testine. A lack of vitamin K results in an increased tendency for hemorrhages. Vitamin K is sometimes given before surgical operations and to pregnant women shortly before delivery.

RESPIRATION

The term "respiration" has been used in several ways to indicate some feature of the overall process of oxygen activity in the organism. We will modify its use and describe the entire process under the following headings: (1) breathing, the movement of air from the atmo-

sphere into the lungs, (2) gas exchange, the movements of oxygen and carbon dioxide between the lungs and the blood, and between the blood and the tissues, (3) gas transport, the various processes involved in carrying these gases, and (4) cellular respiration, the immediate activity of individual cells in what is called aerobic respiration. This last process will be considered later in some detail.

Breathing system

The *breathing system* is composed of the nose, pharynx, larynx (voice box or Adam's apple),

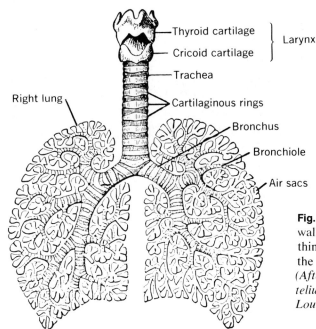

Thyroid cartilage ⎤
Cricoid cartilage ⎦ Larynx

Trachea

Right lung

Cartilaginous rings

Bronchus

Bronchiole

Air sacs

Fig. 28-7. Human respiratory system (section). The walls of the air sacs are dilated to form alveoli, whose thin walls are covered with a network of capillaries for the exchange of gases.
(After Dalton, from Schottelius, B. A., and Schottelius, D. D.: Textbook of physiology, ed. 17, St. Louis, 1973, The C. V. Mosby Co.)

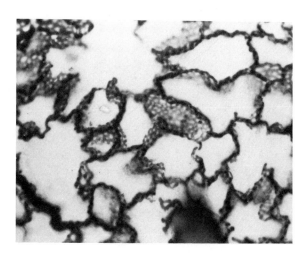

Fig. 28-8. Capillary network from the alveoli of the human lung.

trachea (windpipe), bronchi, and lungs (Figs. 28-6 and 28-7).

The nose is divided by a partition or septum, to form two wedge-shaped cavities, lined by a highly vascular, mucous membrane, the upper layer of which is ciliated. Sinuses in the bones are associated with the nasal cavities, so that inflammations may spread to the sinuses easily. The lateral surface of each nasal cavity has three light, spongy, bony projections called conchae, which make the upper part of the nasal passages very narrow. The nose has the following functions: (1) to remove dust and other foreign materials by hair, cilia, and mucus (secreted by goblet cells), (2) to give warmth and moisture to the inhaled air, (3) to detect odors by means of

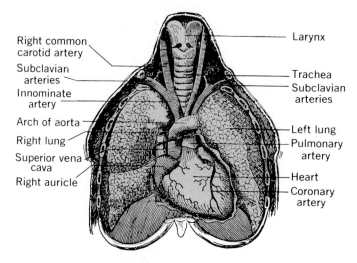

Right common carotid artery

Subclavian arteries

Innominate artery

Arch of aorta

Right lung

Superior vena cava

Right auricle

Larynx

Trachea

Subclavian arteries

Left lung

Pulmonary artery

Heart

Coronary artery

Fig. 28-9. Organs of the human thoracic cavity.
(After Ingals, from Turner, C. E.: Personal and community health, ed. 10, St. Louis, 1956, The C. V. Mosby Co.)

the olfactory nerve endings in the upper passages, and (4) to act as a sounding board for the voice.

The pharynx is a common cavity that connects the nasal cavities with the larynx and the mouth with the esophagus. This is the site of the sore throat associated with a cold that is often caused by breathing unfiltered air through the mouth.

The larynx is a cartilaginous box that forms the prominence in the midline of the front part of the neck. Within the laryngeal cavity are two folds of mucous membrane, extending from front to back but not quite meeting in the middle. Embedded in the edges of these folds are fibrous and elastic ligaments that constitute the vocal cords, because they function in voice production as air passes between them. Above the vocal folds are two smaller folds that do not aid in voice production but protect the larynx during swallowing, help keep the true vocal folds moist, and assist in holding the breath. The opening between the true vocal folds is the glottis. The size of the glottis and the tension of the vocal folds regulate the tone produced. The glottis is protected above by

a leaf-shaped fibrocartilage, the epiglottis, which prevents food from entering the trachea. The trachea is a membranous tube about 4 inches long, located in front of the esophagus. The walls are strengthened by sixteen to twenty cartilaginous, C shaped structures. It extends from the lower end of the larynx to the two branches, or bronchi. Each bronchus divides and subdivides; the smallest branches are the bronchioles. Each bronchiole terminates in a series of saclike alveoli. The thin-walled alveoli are surrounded by thin-walled capillaries through which the exchange of gases occurs (Fig. 28-8).

The lungs are cone shaped and lie in the thorax (Fig. 28-9), separated by the thick mediastinum, which contains the heart, larger blood vessels, and trachea. The left lung is smaller and longer than the right because the heart occupies part of this space. Each lung is enclosed in a serous sac called the pleura, consisting of an outer layer, or parietal pleura, which adheres closely to the diaphragm and the walls of the thorax, and a visceral pleura, which covers the lungs. The two pleurae are separated by a thin layer of serum to reduce friction. Inflammation of the pleura is called pleurisy.

Breathing process

Breathing may be defined as the rhythmic inhalation of air into the lungs and the exhalation of carbon dioxide and other gases from them. During inhalation the size of the thorax is increased by the contraction of the breathing muscles, the diaphragm and rib muscles, thereby decreasing the pressure within the lungs and allowing the greater pressure of the external air to force air into the lungs until the pressures are equalized. The decrease in pressure actively enlarges the size of individual alveoli, much as a balloon is enlarged by air rushing into it. During exhalation the size of the thorax is decreased by the relaxation of these muscles, thus forcing out a quantity of the air from the lungs. About 500 ml. of air is moved with each breath. With a breathing rate of approximately sixteen times a minute, this means that an average of 8 liters of new air enters the lungs each minute.

An increase in the force of breathing results in moving more air per breath. The maximum amount of air involved is about 4,000 ml. and is called the vital capacity of the person. As with other physiologic activities, many factors influence the vital capacity.

The air eventually comes in contact with the alveolar lining, where the following take place: (1) loss of an estimated 5% of the oxygen from the inhaled air, (2) gain of about 4% of carbon dioxide, (3) gain of approximately 1% of nitrogen, (4) saturation of the expired air with moisture (about 1 pint daily), (5) warming the expired air to nearly that of the blood (98.6° F.), thereby losing body heat, and (6) transfer of oxygen and carbon dioxide through the thin-walled air sacs of the lungs (Table 28-2).

Breathing control

The coordination of both chemical and nervous stimuli regulates the breathing rate. In general, as the tissues are more active, they produce more

Table 28-2 Composition of inspired and expired air

	Inhaled air (percent)	Exhaled air (percent)
Oxygen	20.96	15.8
Carbon dioxide	0.04	4
Nitrogen	79	80.2

carbon dioxide, which is carried by blood to the brain where certain cells of the medulla oblongata are stimulated. This causes an impulse to be transmitted via the phrenic nerve to the diaphragm and rib muscles, which then contract, enlarging the thoracic cavity. The resulting enlargement of the alveoli stimulates stretch receptors, and the impulse is transmitted by the vagus nerve back to the medulla, stopping the impulses of the phrenic nerve. The muscles relax, reducing the cavity, and the body exhales. Seconds later the process is repeated.

Gas exchange

The result of the preceding activity is a change in gas concentration in the alveoli. Very simply, it is here that oxygen moves into the blood and carbon dioxide into the lungs. A similar diffusion takes place wherever there are differing gas concentrations. Thus every cell is a site of gas exchange. The following transfers are the result of diffusion involving high and low concentration of gases.

Blood entering the lungs through the pulmonary artery has previously passed over the tissues. As the blood passes through the capillaries, a system of differing concentrations of gases, separated by a membrane, is set up. The blood loses carbon dioxide to the alveoli and picks up oxygen from them. Thus there is a higher oxygen and lower carbon dioxide content in the blood leaving through the pulmonary veins and passing to the various tissues. As a result of respiration, there is a buildup of carbon dioxide and a utilization of oxygen in the various cells. Therefore the blood and cells provide another system of diffusion, with carbon dioxide moving out of the cells and oxygen moving into the cells. The blood then moves back to the lungs.

Gas transport

Oxygen is carried in the blood by hemoglobin in the red blood cells. The combination is known as oxyhemoglobin (HbO_2). Briefly, this combination is stable in places where there is a high concentration of oxygen, as in the capillaries of the lungs. It breaks down, releasing the oxygen in places where there is a low concentration of oxygen, as in the tissues. The rate of release of oxygen to the cells is therefore controlled by the rate of usage by the cells.

Red blood cells are also important in the transport of carbon dioxide from tissues to the lungs. Approximately one third of the carbon dioxide combines with hemoglobin, forming carboxyhemoglobin ($HbCO_2$) in much the same way that oxyhemoglobin is formed. The other two thirds of the carbon dioxide combines with water in the red blood cells, forming carbonic acid and then a bicarbonate ion. In the lung capillaries these compounds break down releasing carbon dioxide, which diffuses into the alveoli.

Cellular respiration

Respiration is the release of energy by the chemical breakdown of foods. This process takes place in the mitochondria, and the energy is trapped in molecules of adenosine triphosphate (ATP). The energy in ATP is then used by the cell in such activities as synthesis, secretion, contraction, and active transport. Molecules of the energy-rich ATP are generally formed and used within the same cell.

The source of the energy in the original foods is photosynthesis. The energy is passed through a series of processes involving plants, herbivores, and carnivores, finally becoming concentrated, in the high-energy phosphate bonds of ATP, by a series of biologic oxidations involving electron transfers.

From various investigations it is evident that the fundamental reactions in respiration in all cells are quite similar. The steps by which glucose is converted to pyruvic acid are the same in man, plants, and animals, and the remainder of breakdown reactions are similar if oxygen is available.

For convenience, the entire process of respiration will be considered under the three headings of glycolysis, pyruvic acid breakdown, and citric acid cycle.

Glycolysis

The breakdown of glucose, with the release of energy and formation of ATP, fits the preceding description of cell respiration.

The process begins with the phosphorylation or addition of a phosphate group —(P) (actually, —PO_3H_2), of glucose. It ends with the formation of pyruvic acid. During the glycolytic cycle (Fig. 28-10), the 6-carbon glucose has split, producing two 3-carbon triose phosphates (PGAL). Through a series of enzymatic reactions these 3-carbon molecules are oxidized by splitting off hydrogen, producing an energy-rich phosphate bond. The high-energy phosphate, ~(P), is is now transferred to a molecule of adenosine diphosphate (ADP), forming the energy-rich ATP. The process of oxidation, high-energy ~(P) formation, and transfer to ADP is repeated and ends in the formation of pyruvic acid.

It should be noted that in this cycle the following three things are outstanding: (1) phosphorylation, providing the —(P) that will later become ATP, (2) oxidation and the removal of hydrogen from the triose, resulting in a high-energy bond, and (3) the formation of ATP, by transfer of this energy-rich ~(P) to ADP.

PHOSPHORYLATION

By consulting Fig. 28-10 it can be seen that four separate phosphorylations take place. In the two at the 6-carbon level the source of the —(P) is ATP. At the 3-carbon level the source is the inorganic —(P) of the cell. Although all four of these phosphates will eventually be raised to the high-energy phosphate, ~(P), and form ATP,

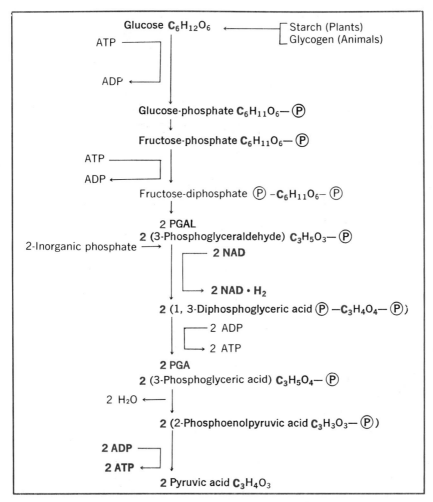

Fig. 28-10. Glycolytic cycle, in which glucose is converted into pyruvic acid in living organisms. The net yield is 2 ATP plus 2 NADH$_2$.

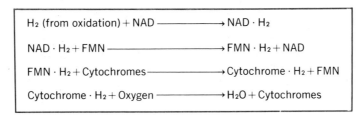

Fig. 28-11. Hydrogen transport sequence.

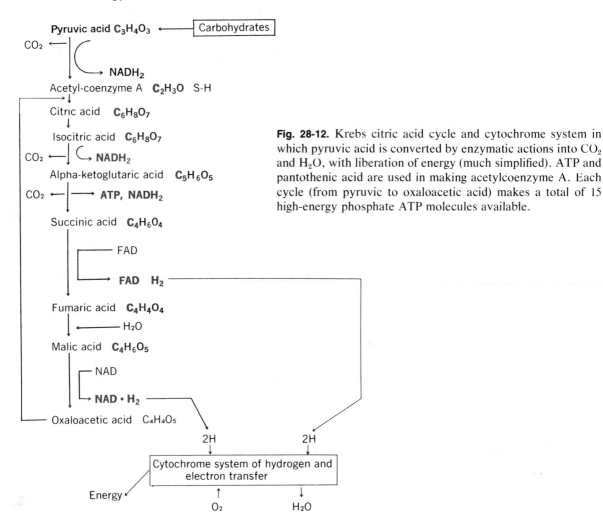

Fig. 28-12. Krebs citric acid cycle and cytochrome system in which pyruvic acid is converted by enzymatic actions into CO_2 and H_2O, with liberation of energy (much simplified). ATP and pantothenic acid are used in making acetylcoenzyme A. Each cycle (from pyruvic to oxaloacetic acid) makes a total of 15 high-energy phosphate ATP molecules available.

the net gain will be 2 ATP. The other two ATP molecules merely replace the starting molecules.

OXIDATION

Of the several types of energy-rearranging reactions during respiration, two of these—dehydrogenation and dehydration—take place during glycolysis. This removal of hydrogen and water can be seen in Fig. 28-10. These and a third, decarboxylation (removal of carbon dioxide), also take place in the later stages of respiration.

The immediate result of oxidation is the removal of a food fraction (hydrogen, water, carbon dioxide) and the resulting formation of a high-energy bond.

ATP FORMATION

As mentioned earlier, ATP is formed by the transfer of the high-energy $\sim\text{P}$ from the oxidized triose phosphate to ADP. A second, and more plentiful, source of ATP is available when the type of oxidation involves the removal of hydrogen. The hydrogen does not simply dissipate into the cell contents but combines temporarily with a series of hydrogen acceptors, which are then said to be "reduced."

The first of these reducing agents is a nucleotide derivative of the B vitamin nicotinic acid, which has been discussed previously in connection with photosynthesis. This carrier is nicotinamide-adenine-dinucleotide or NAD. As an acceptor in its reduced form it is $NAD \cdot H_2$.

In a cell that uses oxygen this hydrogen may be transferred further through a "bucket brigade" of acceptors. The final acceptor is oxygen, and the result of the final transfer is water.

These acceptors include a second nucleotide, flavin mononucleotide (FMN) or its dinucleotide (FAD), and the cytochrome series mentioned with photosynthesis. The entire sequence may be summarized as in Fig. 28-11.

It should be noted that after each transfer the previous acceptor is freed and may again function as an acceptor or reducing agent for further oxidation. Also, and of more importance at this time, for every hydrogen molecule transferred three molecules of ATP are formed. Each transfer has served as an additional oxidation and produced more energy that is used to form more high-energy $\sim \textcircled{P}$.

The net energy available as a result of glycolysis is thus seen to be 8 ATP, two from the oxidation of the triose and six from the hydrogen-transfer sequences.

Pyruvic acid breakdown

The result of the sequence of pyruvic acid breakdown is the formation of carbon dioxide and a 2-carbon fraction, combined with a coenzyme, called acetyl-coenzyme A. The process involves many steps and, as in glycolysis, depends on oxidation, hydrogen acceptors, nucleotides, and vitamin derivatives, as well as other cell components. Coenzyme A is a derivative of the B vitamin, pantothenic acid.

Carbon dioxide is formed by the breakdown of the 3-carbon pyruvic acid ($C_3H_4O_3$) to form a 2-carbon molecule. This then combines with coenzyme A, splitting off hydrogen and forming acetyl-coenzyme A. The hydrogen then enters the transfer cycle, forming an additional 6 ATP (three from each of the pyruvic acid molecules resulting from the original glucose breakdown).

Citric acid cycle

The series of reactions in the citric acid cycle were postulated by the English biochemist H. A. Krebs and is called the Krebs citric acid cycle (tricarboxylic acid cycle), because citric acid ($C_6H_8O_7$) is the first substance formed in the series (Fig. 28-12).

The 2-carbon actyl-coenzyme A unites with 4-carbon oxaloacetic acid to form 6-carbon citric acid. Then successive enzymes break down the citric acid, step by step, through eight different intermediate compounds to oxaloacetic acid, which then combines with another molecule of acetyl-coenzyme A, to continue the cycle.

In the cycle, carbon dioxide is given off, hydrogen atoms are removed, and the electrons of the hydrogen atoms are transferred by cytochromes to the oxygen, which then unites with the hydrogen ions to form water.

As the two molecules of acetyl-coenzyme A, derived from the two molecules of pyruvic acid, are metabolized in the Krebs cycle and cytochromes, about 24 energy-rich phosphate compounds are formed.

The carbon and oxygen atoms are removed together from a substrate molecule by decarboxylation, with thiamine pyrophosphate as a necessary cofactor. One such decarboxylation process occurs as 3-carbon pyruvic acid is converted to 2-carbon acetyl-coenzyme A. Two decarboxylation processes occur in the Krebs cycle as 6-carbon citric acid is converted into 4-carbon oxaloacetic acid.

The Krebs cycle is the final common pathway for the oxidation of amino acids and fatty acids, as well as for carbohydrates. The fatty acids usually found in tissues contain 16 and 18 carbons

Aerobic   **Net yield**

1. Glucose ————————————————→ Pyruvic acid

   C_6 $2\,C_3$ 8 ATP

2. Pyruvic acid ————————————————→ Acetyl-coenzyme A

   $2\,C_3$ $2\,C_2 + 2\,CO_2$ 6 ATP

3. Acetyl-coenzyme A————————————→ Carbon dioxide

   $2\,C_2$ Krebs cycle ———————————→ $4\,CO_2$ 24 ATP

   _____

                                            Total   38 ATP

Anaerobic

1a. Glucose ————————————————→ Triose phosphate

1b. Triose—(P) + NAD $\dfrac{\text{Oxidation}}{\text{Dehydrogenation}}$ → Pyruvic acid + NAD · H_2 2 ATP

                                   ↗ Lactic acid (in animals)

1c. Pyruvic acid + NAD · H_2 ——————   0 ATP

                                   ↘ Alcohol + CO_2 (in plants)

   _____

                                            Total   2 ATP

Fig. 28-13. Summary of cellular respiration.

Table 28-3 Excretion of human wastes

	Primary	Secondary
Kidneys	Water and soluble salts	Carbon dioxide and heat
Lungs	Carbon dioxide (12 cubic feet daily)	Water (250 ml. daily) and heat
Skin	Water, salts, carbon dioxide, and heat	Dead skin, nails, etc.
Alimentary canal	Solids and secretions	Water, carbon dioxide and other gases, salts, and heat

in a long chain. The long chains are split into 2-carbon acetyl-coenzyme A (Fig. 28-12), and the latter enter the Krebs cycle by uniting with oxalo-acetic acid.

Certain amino acids, through enzyme actions, can be converted to pyruvic acid, and others are converted to other members of the Krebs cycle. By various means the amino groups are removed, and the carbon chains of the amino acids eventually enter the Krebs cycle to be oxidized to yield energy, carbon dioxide, and water.

Anaerobic respiration

If the "bucket brigade" of the hydrogen transfer system is blocked by a lack of any of the factors, pyruvic acid may serve as an acceptor. While this is a temporary activity in oxygen-using organisms, it should be noted that a group of anaerobic, or non–air users, survives on the small amounts of energy available by glucose breakdown (Fig. 28-13). The ATP formed by glucose breakdown alone is only two molecules, in contrast to the 38 formed when the six carbons of glucose are finally converted to 6-carbon dioxide.

EXCRETION

Excretion may be considered as the elimination from the body of the waste products of metabolism. Organs involved in this process include the kidneys, lungs, and skin. Although it is sometimes included in the list, the intestine, strictly speaking, is not an excretory organ, since the products of defecation have not been previously absorbed (Table 28-3).

The main products excreted by the lungs are carbon dioxide and water. The condensing of the breath on a cold surface is a good example of this water loss (up to 1 pint per day).

The many sweat glands in the skin account for further loss of water, carbon dioxide, and mineral salts.

The kidneys

Two bean-shaped kidneys, located at the back of the abdominal cavity, one on each side of the vertebral column, select wastes from the blood and pass them through the ureters to the urinary bladder where wastes are stored (Fig. 28-14). Each kidney consists of an outer cortex and an inner medulla. When examined microscopically, the cortex contains approximately one million globelike, renal corpuscles, or nephrons, each

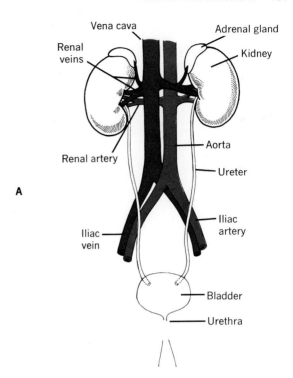

A

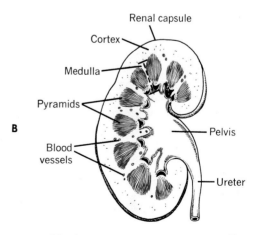

B

Fig. 28-14. The human excretory system. **A,** General anatomy. **B,** Section through a kidney.

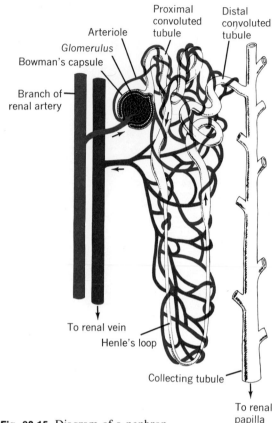

Proximal convoluted tubule

Distal convoluted tubule

Arteriole

Glomerulus

Bowman's capsule

Branch of renal artery

To renal vein

Henle's loop

Collecting tubule

To renal papilla

Fig. 28-15. Diagram of a nephron.

of which is composed of (1) a coiled mass, or glomerulus, of thin-walled capillaries arising from the renal arteries, and (2) a thin, double-walled, enclosing glomerular capsule (Bowman's capsule), which is the beginning of a renal tubule (Fig. 28-15). The convoluted renal tubules travel irregularly and empty into the straighter collecting tubes, which in turn pass the urine into the basin-like pelvis of the kidney then out through the ureter. The muscle contractions in the walls of the ureters cause the urine to pass toward the muscular bladder, located in the pelvic cavity. The bladder normally holds about 1 pint, and the contraction of its three layers of muscles forces the urine to the exterior through the tubular urethra.

The structure of the individual renal corpuscle is such that the walls of the tubules and the surface of the blood vessels are in contact at two sites. The first site is in the capsule itself, where the blood vessels form the highly branched glomerulus. The second is in the capillary network that surrounds the renal tubule. The dual kidney function of filtration, followed by reabsorption, takes place at these areas of contact.

Blood pressure in the glomerulus is such that most of the fluid and molecules smaller than proteins are forced out of the vessels and into the capsule itself. This capsular fluid then passes through the renal tubules and comes in contact with the area surrounded by capillaries. Here, most of the water is reabsorbed, and the ions and molecules are selectively removed.

While approximately 170 quarts of fluid pass through the kidneys each day, only about 1 quart leaves the body as urine. This urine is primarily water, with varying amounts of urea, carbon dioxide, and ions, depending on the type of diet. Thus the kidneys, formerly thought to function only in excretion, also control the water content, ion and mineral balance, and pH of the blood.

CIRCULATION

The function of the circulatory system is to carry blood through all tissues of the body. The system itself is composed of (1) the heart, a biologic pump, (2) blood vessels, consisting of arteries, capillaries, and veins, and (3) blood, the circulating fluid.

While the interaction of heart, vessels, and blood results directly in circulation, the system may also be considered to function in all of those activities influenced by the materials it carries.

The heart

The heart is functionally a pump, forcing blood through arteries in a double path—the pulmonary circulation to the lungs and the systemic circulation to the rest of the body. Its main structure, the muscular myocardium, is organized into four

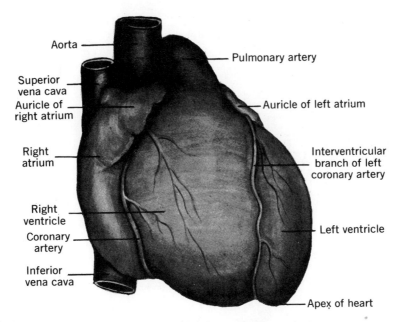

Fig. 28-16. Human heart, showing some blood vessels (front view). *(From Francis, C. C.: Introduction to human anatomy, ed. 6, St. Louis, 1973, The C. V. Mosby Co.)*

chambers, the right and left atria and the right and left ventricles. The chambers on each side are connected through atrioventricular valves. Each atrium and each ventricle has an epithelial lining, the endocardium. The entire heart is enclosed in a saclike pericardium.

Blood from the systemic circulation flows through the venae cavae into the right atrium, which contracts, forcing it through the right atrioventricular (tricuspid) valve into the right ventricle. Contraction of the ventricle forces blood through the pulmonary semilunar valve into the pulmonary artery and then to the lungs. Blood from the lungs reenters by way of the pulmonary veins into the left atrium, then through the left atrioventricular (mitral) valve, to the left ventricle. Contraction of the ventricle forces blood through the aortic semilunar valve and into the systemic circulation.

It should be understood that blood enters both sides of the heart at the same time and leaves through the pulmonary artery and aorta at the same time.

Heartbeat

The simultaneous contraction of the atria, followed by the simultaneous contraction of the ventricles, is controlled by a series of unique structures. The heartbeat begins in the sinoatrial node or "pacemaker," located in the right atrium between the venae caval openings. A wave of contraction, similar to the ripples in a pond, spreads out through both atria, stimulating an atrioventricular (A-V) node located in the wall separating these chambers. The contracting impulse then spreads rapidly through the right and left atrioventricular bundles (bundles of His), located in the wall between the ventricles. Each bundle terminates in a series of branching fibers (Purkinje fibers) in the outer walls of the ventricles.

The speed of transmission through these fibers

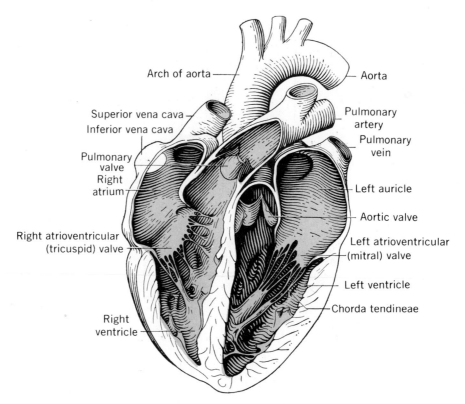

Fig. 28-17. Human heart (longitudinal section).
(From Haggard: Man and his body, New York, Harper & Row, Publishers.)

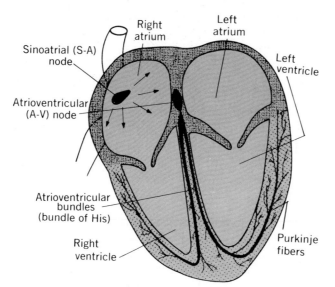

Fig. 28-18. Neuromuscular conducting system in the human heart (arteries omitted).

Table 28-4 Human blood

	Red blood corpuscles (erythrocytes)	White blood corpuscles (leukocytes)	Blood platelets
Nucleus	No nucleus when mature	Always nucleated	None
Shape	Flat, biconcave disks	Variable	Spherical
Motility	None (may bend)	Ameboid movement	None
Diameter	$7.7\mu m$	8 to $20\mu m$	$3\mu m$ (average)
Number	5,000,000 per cubic millimeter	5,000 to 9,000 per cubic millimeter	250,000 per cubic millimeter
Where formed	Red bone marrow	Red bone marrow, lymphoid tissue, and spleen	Bone marrow
Where lost	Liver and spleen	Liver and lumen of intestine	Disintegrate rapidly, which may explain variations in number
Length of life	125 days	Variable, few hours to months	4 days
Functions	Carry oxygen and aid in transportation of carbon dioxide	Protect against bacteria, produce antibodies, repair tissues	Clotting of blood

initiates a uniform contraction of the ventricles, forcing blood out of the heart at the rate of approximately 72 times per minute.

The rate of contraction is influenced by such factors as activity, emotion, and diet. Impulses from a "heart-rate center" in the brain inhibit the pacemaker through the vagus nerve or accelerate it through the spinal nerves. The interaction of these nerves controls the beating of the heart and therefore the supply of blood throughout the body.

The blood

Blood is a liquid tissue that consists of clear, faint-yellow plasma in which are suspended the red blood corpuscles (erythrocytes), the various types of white blood corpuscles (leukocytes), and the blood platelets (Fig. 4-10). Blood forms about one thirteenth of the total body weight and in an average man totals about 6 liters (over 6 quarts). It is somewhat viscous, slightly heavier than water, and slightly alkaline (pH of 7.35) (Table 28-4).

Erythrocytes constitute about 50% of the blood volume. When mature, they are without a nucleus and consist of an elastic framework known as the stroma and hemoglobin. Hemoglobin consists of a protein and an iron-containing compound, the latter being responsible for the chemical affinity for oxygen. When hemoglobin carries oxygen, it is known as oxyhemoglobin. Anemia (Gr. *an-*, not; *haima*, blood) is a condition in which there is a decrease in the number or erythrocytes, in the amount of hemoglobin, or in both. These conditions may result from impaired blood formation, increased destruction of erythrocytes, or both.

Leukocytes, because of their ameboid movements, are able to escape from the blood vessels and penetrate into the body tissues. Leukocytes may be classified as (1) granulocytes, with distinguishing granules in the cytoplasm, and (2) agranulocytes, without granules in the cytoplasm. Because of the variations in the lobes of the nuclei of the granulocytes, they are sometimes referred to as polymorphonuclear leukocytes (Gr. *polys*, many; *morphe*, form). They act as phagocytes (Gr. *phagein*, to eat; *kytos*, cell) by engulfing bacteria, cell fragments, and foreign materials. The granulocytes are classified in three groups according to the type of granules in their cytoplasm: (1) eosinophils, which stain readily by

Table 28-5 Summary of leukocytes

Type	Percent of total leukocytes	Diameter in microns	Characteristics (with Wright's stain)
Granulocytes			
Neutrophils	60 to 70	10 to 12	Fine, light blue cytoplasmic granules; 3- to 5-lobed nucleus
Eosinophils	1 to 4	12	Bright red cytoplasmic granules; 2-lobed nucleus
Basophils	0.5	10	Large, dark purplish blue cytoplasmic granules; irregular nucleus; often S-shaped
Agranulocytes			
Lymphocytes	25	8	Thin layer of nongranular, robin's-egg blue cytoplasm; large, bright purple nucleus
Monocytes	5 to 10	12 to 20	Thick layer of nongranular cytoplasm; large horseshoe- or kidney-shaped, purple nucleus

eosin (acid) stains; (2) neutrophils, which stain by neutral dyes; and (3) basophils, which stain well with basic stains (Table 28-5).

The agranulocytes lack cytoplasmic granules and are classified in two groups, lymphocytes and monocytes; both are formed in the lymphoid tissue. Lymphocytes contribute to the repair of wounds by connective tissue formation. They are important in the process of immunity, since they are antibody carriers. Monocytes are active in phagocytosis, engulfing bacteria and cell debris.

Platelets (Gr. *platys*, flat) of mammals are small and colorless, and their various sizes average about 3μm. They are roundish or oval biconvex disks that look spindle shaped when seen in profile. In man they usually average approximately 250,000 per cubic millimeter. When they leave the blood vessels, they tend to adhere to one another and to surfaces that they contact. They originate in the bone marrow and are associated with the process of blood clotting.

Functions of blood

The functions of blood are: (1) respiratory, transporting oxygen to the tissues and carbon dioxide from them, (2) excretory, carrying waste materials from the tissues to the organs of excretion, (3) nutritive, transporting sugars, amino acids, fats, and ions from the intestine to the tissues, (4) regulatory, providing a fluid environment for cells and osmotic balance, maintaining

pH by buffer action, distributing hormones to all cells, and controlling body temperature by water movement and sweating, and (5) protective, by the phagocyte action of white blood cells, by antibody action, and by the clotting mechanism.

Blood vessels

Arteries are vessels whose walls are rather thick, contractile, and elastic. They consist of (1) an inner layer of endothelial cells and elastic tissue, (2) a middle or intermediate layer of muscle and elastic tissue, and (3) an external layer of elastic tissues. Arteries carry blood away from the heart, and veins carry blood back toward the heart.

Veins are vessels whose structure is similar to that of arteries, being composed of three layers, but whose walls are thinner and less elastic because of a poorly developed middle layer and the lack of much muscle and elastic tissue. Certain veins, especially those of the legs, have a series of semilunar valves to prevent the backflow of blood. In general, the systemic veins parallel the systemic arteries, frequently having the same names as the arteries.

Capillaries are thin-walled vessels that form a network connecting the arteries and veins. The walls are a single layer of flat endothelial cells. The capillaries are so numerous that there is hardly any body part that does not contain them. The exchange of materials takes place through

Arteries

Veins

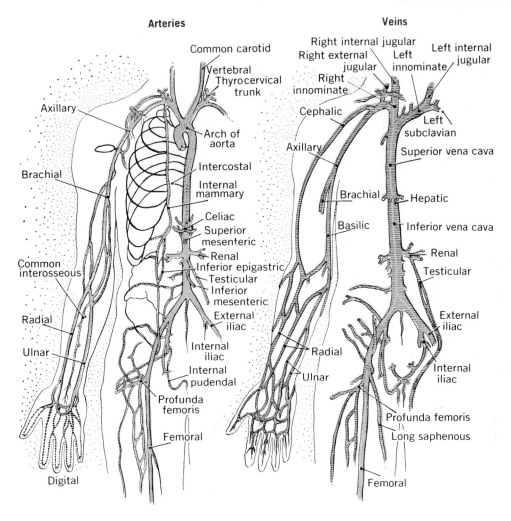

Fig. 28-19. Main blood vessels of man (diagrammatic).
(From Braungart, D. C., and Buddeke, R.: An introduction to animal biology, ed. 7, St. Louis, 1968, The C. V. Mosby Co.)

them because of their thin walls and the slow movement of the blood within. An average capillary is about 8μm in diameter. Compare this with the diameter of a red blood corpuscle.

Blood plasma

Blood plasma is the fluid part of the blood and is approximately 90% water. It serves as the basic carrier for the cells and other materials. Plasma proteins form about 8% of the remaining fraction, and inorganic ions, foods, wastes, and hormones account for the rest.

Most of the protein is albumin, which is important in maintaining the osmotic pressure of the blood, globulins, which are essential to antibody formation and immunity, and fibrinogen, which functions in clotting of the blood.

The ion concentration is important in maintaining pH and all metabolism as indicated earlier.

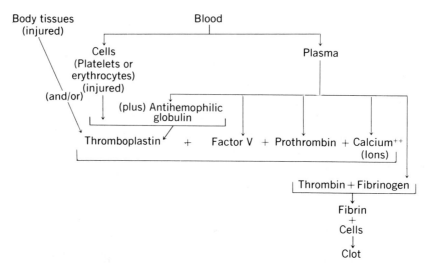

Fig. 28-20. Summary of the salient factors in blood coagulation. Many details are not given, but some are considered in the text. Thromboplastin activity can come from two sources: one from injured body tissues, the other from blood alone. There is lipoprotein present in platelets and in erythrocytes. With the release of this substance it reacts with antihemophilic globulin to form powerful thromboplastin activity. Blood cell lipoprotein may be called thromboplastic cell component (TCC), and plasma antihemophilic globulin may be called thromboplastic plasma component (TPC). In physiologic blood clotting the platelets play a more important role than do the erythrocytes.
(Courtesy Dr. George Y. Shinowara, New York University-Bellevue Medical Center, New York, New York.)

Clotting of blood

Loss of blood is known as hemorrhage; this initiates the following protective activities in the body: (1) clotting of blood at the site of the injury, (2) decrease in the general blood pressure, (3) contraction of the small vessels of the skin, muscles, and intestines in order to supply the vital parts of the body, (4) increase in the blood volume by the contraction of the spleen, which normally contains a large quantity of blood, and (5) passage of water and salts from the tissues into the capillaries because of increased osmotic pressure.

Shortly after blood leaves blood vessels it loses its normal fluidity and becomes jellylike through a process known as coagulation, or clotting. Maintaining the fluidity of the blood while it flows through normal blood vessels and yet preserving its inherent coagulability is a marvelous phenomenon of vital importance.

When a quantity of freshly drawn blood stands for a few minutes, a meshwork of microscopic, thread-like fibrils is formed throughout the blood mass. The fibrils entangle blood corpuscles and transform the original liquid into a gel. Then the fibrils shrink and press out of the gel a slightly yellowish fluid, the blood serum. This shrinkage continues with the formation of more fluid, and eventually the gel grows firmer to form a clot, composed of fibrils and entangled blood corpuscles.

When the stringy material is washed properly, it is found to be composed of a white, fibrous, elastic, protein substance. Since there is no insoluble protein in the blood plasma, what is its origin and what causes its formation? The red and white corpuscles do not contribute to this, because after their removal, the remaining plasma still has the ability to coagulate. Thus, the insoluble fibrin of the clot must originate from some soluble protein, such as serum albumin, serum globulin, or from a substance in the blood called fibrinogen (Fig. 28-20). By proper techniques these three can be separated. The serum albumin and globulin have been found not to be associated with blood coagulation. Fibrinogen, however, after being obtained free from the other proteins and when dissolved in a salt solution, will coagulate in a way similar to that of plasma. Thus in blood coagulation soluble fibrinogen is changed into insoluble fibrin. As long as blood remains in normal blood vessels, no fibrin is formed, and the fluidity of the blood is maintained.

When it is realized that approximately 35 compounds take part in clot formation, it is understandable that this phenomenon is one of the most complicated chemical processes in the body. Many theories have been proposed concerning the roles of the various chemicals involved, but we shall limit our consideration to a summary of a few salient facts that have been generally accepted.

The conversion of the soluble protein, fibrinogen, into the insoluble protein, fibrin, is caused by an enzyme called thrombin (Gr. *thrombos*, clot). In blood, thrombin is present in an inactive form, called prothrombin. For the formation of prothrombin an adequate supply of vitamin K is essential. Since bile is required for absorption of vitamin K from the intestine into the blood, a bile deficiency will reduce the amount of prothrombin and hence cause deficient clotting, even if sufficient amounts of the vitamin are present in the intestine. Certain types of hemorrhages can be prevented by the administration of vitamin K before operations.

Injury to tissues or blood vessels, or the disintegration of blood platelets, gives rise to the formation of a substance called thromboplastin. In the presence of calcium ions this transforms the inactive prothrombin into active thrombin.

It is highly important that shed blood should coagulate, but it is equally desirable that the fluidity of blood be maintained so long as it remains in blood vessels normally. Two important factors in this connection are intact blood platelets and intact blood vessels, especially the endothelial lining of the latter. Injury to blood vessels not only liberates thromboplastin but also increases the destruction of platelets. Since platelets live only about 4 days, they constantly supply thromboplastin.

The formation of a clot within a blood vessel that is not severed is called a thrombus. This may be caused by an injury of the vessel wall from a blow or from toxins of bacteria that injure the blood platelets. If a part of a thrombus circulates in the vessels, it is called an embolus, which, if it should block circulation to a vital part, has serious consequences.

Under normal conditions clotting is inhibited by the action of two enzyme-inactivating substances, antithrombin and heparin.

Functions of human lymph

The composition of the lymph is similar to that of the blood plasma. It ranges from colorless to a yellowish color, has an alkaline reaction, contains no blood platelets, clots slowly and not firmly, may contain a few red blood corpuscles, and contains lymphocytes.

The lymph is derived from (1) the blood plasma by filtration through the thin walls of the capillaries and (2) secretions of the endothelial cells that line the numerous capillaries.

The functions of human lymph are summarized as follows. (1) It bathes all parts of the body not reached directly by the blood, thus supplying foods and oxygen and receiving carbon dioxide and wastes. There is a continuous interchange between the blood plasma and the lymph through the processes of osmosis and diffusion. (2) It aids in the fight against foreign materials, such as bacteria and protozoans. (3) It helps to equalize body temperature. (4) It helps to regulate the acid-alkaline balance of the various body parts. (5) It helps

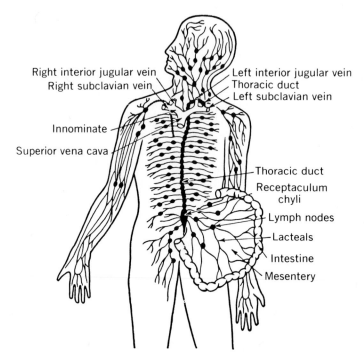

Fig. 28-21. Lymph system and parts of certain veins of the upper part of the body. Lymph from the abdominal organs and lower limbs flows into the thoracic duct, which empties into the left subclavian vein. Lymphatics from the left arm and left sides of the thorax, neck, and head also empty into the thoracic duct. Lymph from the right arm and right sides of the thorax, neck, and head flows into the right subclavian vein.
(From Zoethout, W. D., and Tuttle, W. W.: Textbook of physiology, ed. 11, St. Louis, The C. V. Mosby Co.)

to collect and transport fatigue products that are the result of cellular activity. (6) It aids in transporting enzymes and other secretions to various body parts (Fig. 28-21).

Human lymph is found in various places and consequently has a variety of functions. The principal locations are (1) in the lymph ducts and their enlargements, the lymph nodes, (2) in tissue spaces (tissue sinuses) or cavities in various tissues, (3) in the pleural cavity (around the lungs), (4) in the pericardial cavity around the heart, (5) in the peritoneal cavity (abdominal cavity), (6) in the perineural cavities (spaces between the various linings of the brain and spinal cord),

and (7) in the lacteals or lymphatics, which originate in the small, finger-like villi of the intestine. Fats are absorbed from the intestine by the lacteals and eventually placed in the bloodstream.

Human blood groups
Blood types

The typing of human blood is based on the presence or absence of certain substances in the red blood cells or plasma. The substance in the cells is a type of antigen called an agglutinogen. The plasma substance is a type of antibody called an agglutinin. When the proper kinds of each are

Table 28-6 Human blood groups

Blood group	Agglutinogen (antigen) in red blood corpuscles	Agglutinin (antibody) in blood plasma	May receive blood from group	May give blood to group	Percent of the white population of the United States represented by each blood group*
O	None	a and b	O	O, A, B, AB	41
A	A	b	O, A	A, AB	45
B	B	a	O, B	B, AB	10
AB	A and B	None	O, A, B, AB	AB	4

*Different populations will show other percentages.

mixed, clumping, or agglutination, of the red blood cells results.

The name of the blood type is derived from the kind of agglutinogen (antigen) present in the red blood cells. In the A-B-O system, type A has agglutinogen A, type B has agglutinogen B, type AB has both A and B, and type O has neither A nor B.

Since a mixture of plasma from a person of type A and red blood cells from a type B will result in a clumping of cells, it is apparent that type A plasma contains the agglutinins against type B cells. Table 28-6 shows the scheme in detail. Note that type O may be given by transfusion to all blood types, and type AB may receive from all types. This, of course, assumes that all precautions have been taken. Under normal circumstances the same type would be transfused.

Since type O blood contains agglutinins a and b and type AB contains agglutinogens A and B, an apparent contradiction exists in this "universal donor–universal recipient" scheme. This is solved when it is demonstrated that the agglutinins lose their effect when diluted, as they are during transfusion.

Other blood groups, such as the Rh and MN systems, depened on similar sets of agglutinogens and agglutinins.

BODY DEFENSES

The mere presence of microorganisms in air, food, milk, and water does not produce a disease in all instances. Among the most important deterrents to disease production in man are the various body defenses.

Defenses (first line)

The main defenses of the body are: skin, which acts as a mechanical barrier, nose, in which mucus collects the organisms that cilia move toward the exterior, after which the enzyme lysozyme destroys the bacteria, and sneezing and coughing occur, eyes, from which organisms are washed away mechanically by tears, which also contain the enzyme lysozyme, mouth, a mucous membrane acting as a mechanical barrier, stomach, which contains the acid of the gastric juices, intestine, which contains a mucous membrane acting as a mechanical barrier and in which the antagonistic action of other organisms occurs, and urethra, in which the action of urine takes place.

Defenses (second line)

The body's second line of defenses consists of (1) the production of inflammations, in which many organisms that have penetrated the deeper tissues are trapped and destroyed (inflammations are usually characterized by redness and swelling because of increased supplies of blood; increased temperature, because of increased metabolic activity in the area, by pain, because of abnormal activities, and by pus formation in later stages because of the destruction of microorganisms and tissues); and (2) phagocytic action of certain white blood corpuscles, in which the phagocytic cells engulf and destroy microorganisms in the various body tissues, in the bloodstream, and lymph nodes.

Defenses (third line)

Other defensive reactions of the body that depend upon present or previous contact with the infectious organisms, or their products, are known as immunologic reactions. Pathogenic organisms may injure the hosts, which they attack in two ways: (1) they consume or destroy vital tissues directly, or (2) they produce substances that poison host tissues in the immediate vicinity of the organisms or are carried to other parts of the body. These poisonous substances, called toxins, are either produced by living organisms as they grow and reproduce, or are liberated when organisms die and disintegrate. Some toxins are general in their action, affecting a variety of host tissues, while others are more selective in their action, affecting heart muscle, nervous tissue, or other specific body tissues. Some toxins are proteins, while others are of unknown chemical composition.

Any substance that stimulates the body of an animal to produce a specific antibody is called an antigen. When animals are attacked by certain pathogenic bacteria or their toxins, they produce substances that neutralize the poisonous effects of the toxins or that inactivate or destroy the bacteria. These protective substances, called antibodies, are of several types, among which are antitoxins, agglutinins, bacteriolysins, and opsonins. Antitoxins are substances that neutralize specific toxins and are formed in animal bodies in direct response to the toxins. Anti-

toxins are specific toxins and are formed in animal bodies in direct response to the toxins. Antitoxins are specific in their action; for example, diphtheria antitoxin reacts only with diphtheria toxin and not with the toxin of the tetanus organism. Agglutinins are antibodies that agglutinate (clump) the specific organisms that acted as antigens in their formation in a body. The agglutination may occur in the body or outside, where it can be used for identifying specific organisms for diagnosis of certain diseases. For example, if the blood serum of an animal immunized against typhoid organisms is mixed with a suspension of typhoid organisms, the latter are agglutinated. Bacteriolysins are antibodies that dissolve the specific organisms that have stimulated the body cells to produce them. Opsonins, meaning "to prepare food for," are antibodies that weaken bacteria so that they are more readily destroyed by the phagocytes.

If the production of antibodies by the host is sufficiently extensive and rapid, pathogenic bacteria are destroyed and their toxins neutralized, and the animal may recover from the disease. An animal that has formed sufficient antibodies is therefore immunized against the specific disease caused by the type of bacterium that led to the production of the antibodies. This immunity often persists for the life of an individual animal that has experienced and has recovered from a particular disease. Thus, one attack of measles or typhoid fever usually renders the victim immune to these diseases indefinitely.

It is not always necessary for an animal to have a disease in order to be immunized against it. One method of immunization involves the administration of vaccines, which are preparations of dead or weakened bacteria or viruses. When such vaccines enter the body, they stimulate the formation of antibodies, which in turn may immunize the animal for certain periods of time. Such antibodies are specific for the organisms from which the vaccines were prepared. Vaccines are used in immunizing against diseases such as smallpox, poliomyelitis, typhoid,

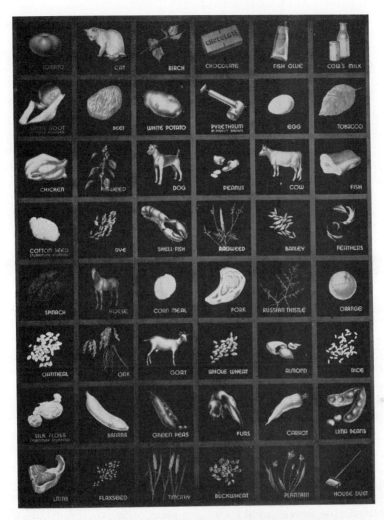

Fig. 28-22. Forty-eight causes of allergic reactions.
(Courtesy Lederle Laboratories, Pearl River, New York.)

cholera, and bubonic plague. Since vaccines stimulate the animal body to produce its own antibodies, this type of immunity is similar to that acquired as a consequence of having that disease.

The condition of active immunity after either vaccination or having the disease is the result of antibody production by the individual himself. A varying period of time must elapse before immunization is effective.

On occasion, and especially prior to the development of newer antibiotics, the severity of some particular diseases necessitated the development of passive immunity. This is the result of receiving a serum containing antibodies from a person or animal who had active immunity previously. Passive immunity has the advantage of being effective immediately; however, it does not last very long.

ALLERGIES

Tissue cells that have responded to stimulation by an antigen are immunologically different from tissue cells that have not so responded. The antigenically stimulated cells often retain antibodies for some period of time. Later, when a specific antigen again contacts such cells, they react in a distinctive manner and are said to be hypersensitive, or in an allergic state (Gr. *allos,* changed; *ergon,* activity). When cells have produced antibodies, the allergic reaction may be vigorous, especially if some time has elapsed between the first contact with the sensitizing antigen and that which causes the later allergic reaction. Sometimes normal cells that have not been stimulated actively by an antigen may passively acquire the antibodies necessary for certain types of allergic reactions by merely absorbing them from the blood.

In some types of allergies the antibodies remain associated closely with the cells that produce them and are called allergic reagins. The antigens that stimulate their production often differ from substances commonly considered as antigens and are called allergens. Common allergens include certain cosmetics, some plastics, sulfonamides and other drugs, certain dyes on fabrics, and certain antibiotics (Fig. 28-22). Hypersensitivity to any of them may occur. There are hundreds of allergic reactions, and their specific characteristics depend on the kind and location of tissues involved, the nature and dosage of the allergen, the nature of the reagin, and the degree of hypersensitiveness.

While the term allergy may include all reactions resulting from tissue hypersensitiveness, some classifications exclude those reactions that do not occur naturally and have unique properties. For example, if a guinea pig is injected with 1 ml. of a foreign protein, such as horse serum, in about 3 weeks the guinea pig is hypersensitive to a second injection of horse serum and will exhibit a striking response, called an anaphylactic reaction (Gr. *ana,* against; *phylaxis,* protection). Shortly after the injection of even a small second dose of the horse serum the guinea pig may display such symptoms as uneasiness, coughing, scratching of the nose, gasping for breath, urination, defecation, finally cessation of breathing, and death. If the injection is not fatal, the animal may recover within 2 hours. Most symptoms in anaphylaxis are caused by contractions of smooth muscles, such as are found in lungs, bladder, and intestines. Common allergic reactions include the so-called hay fevers (caused by pollens), the early stage of the common cold, asthma, food allergies (caused by such things as eggs and strawberries), cosmetic rashes, and certain industrial dust reactions.

To test for an allergy, a small quantity of sterile solution of a particular protein is injected into (not under) the patient's skin. If the patient is allergic to this protein, a large, localized inflamed area results because of the allergen-reagin (antigen-antibody) reaction. An allergic tendency is inheritable.

Review questions and topics

1 Describe the various parts of the human digestive system, including the specific functions performed by each.
2 Discuss the digestive process and the control of digestion.
3 Discuss the metabolism of carbohydrates, amino acids, and fats.
4 Discuss the roles of the following in nutrition: mineral salts, water, and vitamins.
5 List the characteristics and functions of each of the important vitamins.
6 Explain the terms breathing and respiration, including gas transport and cellular respiration.
7 Discuss the various parts of the breathing system and the function of each part.
8 Explain in detail the process of glycolysis, including phosphorylation, oxidation, and ATP formation.
9 Explain pyruvic acid breakdown and Krebs citric acid cycle (tricarboxylic acid cycle).
10 Explain anaerobic respiration.
11 Discuss the 2-carbon acetyl-coenzyme A and its function in respiration.
12 Explain how the various wastes are eliminated by the human body.
13 Describe the structure of the human circulatory system and the functions of the various parts.
14 Discuss the functions of the human blood, including the characteristics and specific functions of its various parts.
15 Explain in detail the functions of each correlated phenomenon in the clotting of blood.
16 Describe the human lymph system and the functions of lymph.
17 Discuss each of the human blood groups and their significance.
18 Define agglutinogen (antigen) and agglutinin (antibody) and describe where each is found.
19 Discuss in detail the body defenses against microorganisms.
20 Discuss allergies and their significance.

See Chapters 29 and 30 for related references on human biology.

29 Human control mechanisms

All living protoplasm is necessarily irritable or subject to stimulation. A stimulus is any external or internal substance, material, or condition that affects a cell or group of cells, thereby setting up a change known as a response. General types of stimuli can be chemical, electrical, thermal, or mechanical; general types of responses can be moving, secretory, thermal, or chemical. The responsive mechanisms of man are complex and varied. The three steps involved are as follows: (1) a special structure called a receptor must be stimulated; (2) there must be some method of conduction of the effects of stimulation to (3) a specialized structure called an effector, which must respond in some way.

While the actual coordinating substances in man are quite varied, the more obvious mechanisms are found in the nervous and endocrine systems.

NERVOUS SYSTEM

Organization

The human nervous system will be considered under the headings of (1) the central nervous system, which consists of the brain and spinal cord and their related nerves, (2) the autonomic nervous system, which is associated with involuntary control; and (3) the sense receptors, which are in contact with the environment.

Neurons

The basic unit of the nervous system is the individual neuron, or nerve cell. This is a living, metabolizing cell that is specialized to receive and transmit nervous impulses. Although the structure may vary, an individual cell has an irregular shape with several protoplasmic extensions. According to the direction of impulse transmission, these extensions are referred to as dendrites, leading toward the cell body, and axons, leading from it. The length of these extended fibers varies tremendously.

An individual fiber such as an axon has an outer fatty covering called the myelin sheath, surrounded by a cellular neurilemma (Fig. 29-1). These apparently serve as nutritive, protective,

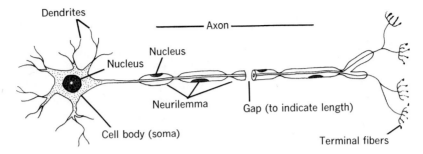

Fig. 29-1. Diagram of a nerve cell, or neuron.

and insulating sheaths. Nerve fibers usually terminate in a series of branching fibers.

Based on the direction of impulse transmission, neurons are called (1) sensory, leading from sense organs to the brain or spinal cord, (2) motor, leading from the brain or spinal cord to a muscle or gland acting as an effector, and (3) interneurons, connecting the other two in the brain or spinal cord.

Nerve impulses

How do nerves carry a message by what is commonly called a "nerve impulse?" Luigi Galvani, professor of anatomy at the University of Bologna, accidentally discovered that if a frog's leg touched a piece of iron, an electric current was propagated (1786). Sixty years later it was proved that nerve and muscle cells actually possess electrical charges and are capable of generating an electric current.

To measure a nerve impulse, two pairs of electrodes are placed on a dissected nerve. One pair (the "transmitter") stimulates the nerve, and the other pair (the "receiver") is applied at various points to ascertain the strength and velocity of the signal. The signal, which travels at a constant velocity, arrives at the receiver after a delay that represents the time required to travel along the nerve between the two electrodes. In the fastest nerve fibers of warm-blooded animals this velocity is about 300 feet per second.

A nerve fiber, or axon, is a long, thin tube that is attached to a microscopic nerve cell, or neuron. These fibers vary in diameter from less than

thirty thousandths to one thirtieth of an inch. In limbs of large animals the fibers may be several feet long.

The explanation of the electrical activity of the fiber lies in its chemistry and that of the tissue fluid around it.

There are two types of nerve fibers: one is enclosed by a myelin sheath and the other is almost bare (nonmyelinated) and freely exposed to the surrounding tissue fluid. In both however, there seems to be a type of membrane that acts as a resistant barrier to the movement of ions.

The nerve membrane is extremely delicate and is intimately associated with signal transmission along the fiber. This membrane is probably a very thin surface layer of fatty material, only one or two molecules thick, thus making it invisible with high-powered magnifications. The surface membrane displays selective permeability, allowing certain chemical substances to pass through much more readily than others.

Under normal conditions the ion content of the fluids outside and inside the membrane would be the same. The selective permeability of the membrane appears to prevent this equilibrium, however, especially with regard to positive sodium (Na^+) ions, which are "pumped" out of the fiber as fast as they enter. This results in relatively more positive ($+$) ions outside the fiber than inside it—a condition known as polarization (Fig. 29-2).

Stimuli such as electricity, pressure, or chemicals upset the selective permeability of the fiber membrane. This in turn permits the inrush of

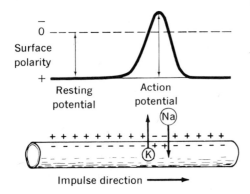

Fig. 29-2. Action potential and "sodium theory" of nerve impulse transmission. The action potential wave (top) spreads along the surface of the nerve fiber (bottom). During the rise of action potential, sodium ions (Na) enter the fiber and make it positive; during the resting state of the nerve, outward pressure of potassium ions (K) keeps the interior of the fiber negative. It is believed that the so-called resting cell may operate a "sodium pump" that constantly drives out sodium ions as fast as they enter. This may explain how a nerve cell manages to hold down its concentration of sodium ions against the combined forces of diffusion and electrical pressure.
(From Katz, B.: The nerve impulse, Sci. Amer. 187: 55-64, 1952.)

positive sodium ions (Na^+) and causes depolarization, with more positive ions inside than outside. This spreads to the neighboring areas, resulting in more depolarization, and generates the nerve impulse. In other words the nerve impulse is a spreading wave of depolarization caused by the temporary upset of the selective permeability of the fiber membrane.

Synapses and impulse transmissions

Individual neurons are placed end to end, forming long pathways. The axon tip of one neuron "connects" functionally with a dendrite of the next. This is not a structural connection because the fiber terminals are separated by a microscopic gap, called a synapse (Gr. *synapsis*, union).

The tip of an axon usually terminates in a series of brushlike fibers. Recent studies with electron

microscopy have shown that the ends of the fibers are swollen into synaptic knobs. The knobs contain many small sacs called synaptic vesicles that hold the transmitter substances. Apparently the action of the nerve impulse is to force the emptying of the synaptic vesicles into the gap between the axon and dendrite. After discharging the transmitter substances, the vesicles move back into the synaptic knob where they are refilled (Fig. 29-3).

Two hormones, epinephrine (also called adrenalin) and acetylcholine, play important roles in synaptic transmission. Epinephrine is secreted by the axon terminals of some fibers of the sympathetic nervous system. Such epinephrine-producing fibers are often said to be adrenergic. Other fibers of the sympathetic system, and of the so-called parasympathetic nervous system and possibly the nerve fibers of the central nervous system, secrete a second hormone, acetylcholine, a rather simple chemical that is normally present in many parts of the body and has important physiologic functions. Acetylcholine brings about impulse transmission across many neural synapses. Acetylcholine-secreting fibers are called cholinergic.

In these synapses a potent enzyme, cholinesterase, is present, which splits acetylcholine into acetyl and choline fractions, thus making the hormone ineffective. If cholinesterase were absent, acetylcholine would remain in the synapse and would stimulate dendrites constantly. If two impulses entered a synapse in quick succession, acetylcholine produced by the first would still be effective when the second produced its own hormone. Hence, the impulses would merge, and the result would be one long, extended response and not two distinct responses. Thus cholinesterase keeps the impulses separated. There is just sufficient time for acetylcholine to stimulate dendrites once, and then the hormone is immediately destroyed.

The consequences of impulse transmission in synapses caused by chemicals include the following. (1) A diffusion through a synapse takes a

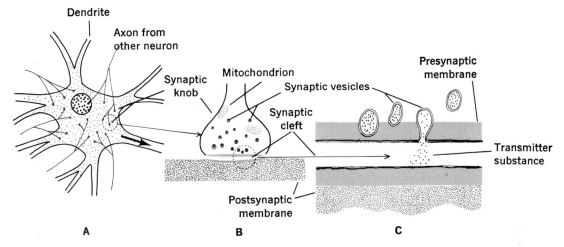

Fig. 29-3. A synapse. **A,** Motor neuron with many synaptic knobs of other neurons. **B,** Synapse between knob on axon tip and cell membrane of motor neuron. **C,** Detail of synaptic vesicles discharging transmitter substance.

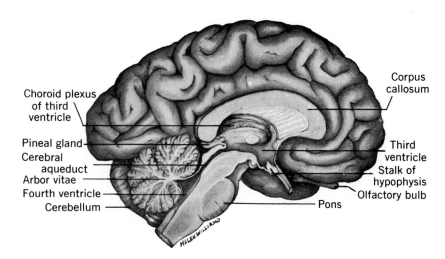

Fig. 29-4. Human brain, showing the medial aspect of the left half (sagittal section). The convoluted cerebrum (cerebral hemisphere) is shown above the corpus callosum, and the spinal cord is shown below the pons. The hypophysis is also called the pituitary gland. *(From Francis, C. C.: Introduction to human anatomy, ed. 6, St. Louis, 1973, The C. V. Mosby Co.)*

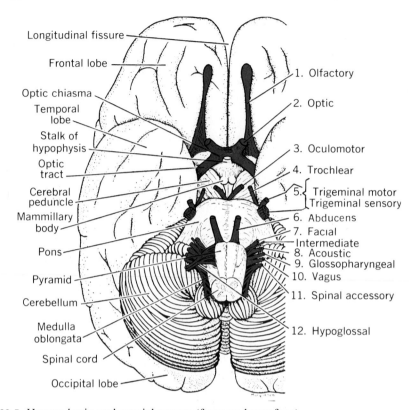

Longitudinal fissure

Frontal lobe

Optic chiasma

Temporal lobe

Stalk of hypophysis

Optic tract

Cerebral peduncle

Mammillary body

Pons

Pyramid

Cerebellum

Medulla oblongata

Spinal cord

Occipital lobe

1. Olfactory
2. Optic
3. Oculomotor
4. Trochlear
5. Trigeminal motor / Trigeminal sensory
6. Abducens
7. Facial
   Intermediate
8. Acoustic
9. Glossopharyngeal
10. Vagus
11. Spinal accessory
12. Hypoglossal

Fig. 29-5. Human brain and cranial nerves (from undersurface). *(After Morat, from Schottelius, B. A., and Schottelius, D. D.: Textbook of physiology, ed. 16, St. Louis, 1969, The C. V. Mosby Co.)*

relatively longer time than does impulse transmission within a nerve fiber. Thus a reflex through many synapses requires more time than is expected on the basis of impulse speeds through fibers alone. (2) Different types of impulses within fibers probably release different concentrations of hormones from axon terminals. If a certain terminal secretes an insufficient amount of hormone, it may be unable to activate an adjacent dendrite. However, if each of several axons at a synapse secretes subthreshold amounts of hormone at the same time, the total may suffice to initiate an impulse in the adjacent dendrite. This type of synaptic summation probably plays an important role in normal nervous activity. (3) Synapses direct the impulses in only

one direction, from axon toward dendrites. Only axon terminals are sensitive to these hormones. (4) Synapses fatigue fairly easily, whereas nerve fibers rarely become tired. When axon terminals function intensely, they may exhaust their hormone-secreting ability temporarily, and synaptic transmission may slow down even more, or stop completely for some time.

Central nervous system
The brain

The human brain (Figs. 29-4 and 29-5) consists of (1) cerebrum, (2) cerebellum, (3) midbrain, (4) medulla oblongata, and (5) pons varolii. The cerebrum, which is the largest and most prominent part of the brain, consists of the right and left

Central nervous system {

Brain {
Cerebrum, which is large, ovoid, convoluted, and made of 2 hemispheres with 5 lobes
Cerebellum, which is smaller, oval, nonconvoluted but with smaller furrows (sulci)
Midbrain, which is short and connects the cerebellum with the pons varolii
Medulla oblongata, which is pyramid shaped and continues with the spinal cord
Pons varolii, which is in front of the cerebellum between the midbrain and the medulla oblongata and which connects the parts of the brain
}

Cranial nerves (12 pairs) and their end organs
Spinal cord for reflexes and pathways to and from the higher nervous centers
Spinal nerves (31 pairs) and their end organs
}

Autonomic nervous system {
Sympathetic, which has centers, ganglia, and plexuses in the cervical, thoracic, and lumbar regions of the spinal cord
Parasympathetic, which consists of the centers and ganglia of the cranial and sacral parts of the autonomic system
}

Fig. 29-6. Summary of the human nervous system.

cerebral hemispheres. Each hemisphere is divided by sulci into five distinct areas known as lobes (frontal, parietal, temporal, and occipital lobes, and the insula, which is invisible from the surface). The outer layer of the cerebrum, known as the cortex, has numerous, foldlike convolutions that greatly increase the surface area. Specific regions of the cerebral cortex include the following: motor area, sensory areas (heat, cold, pain, touch, light pressure, and muscle sense), auditory area, visual area, olfactory area (taste and smell), and speech area. Beneath the gray cortex of the cerebrum is a mass of nervous tissue known as white matter.

The brain contains cavities (ventricles) as follows: (1) two lateral ventricles, one in each cerebral hemisphere, (2) the third ventricle behind the lateral ventricles and connected with each by an opening called the foramen of Monro, and (3) the fourth ventricle in front of the cerebellum and behind the pons and medulla, connected with the third ventricle by a small canal called the aqueduct of Sylvius. The coverings of the brain are called meninges and are the same as for the spinal cord (dura mater, outer layer; arachnoid, middle layer; pia mater, inner layer). Thin layers of fluid separate the various coverings.

Functions of the cerebrum, in addition to those

already mentioned, are as follows: it governs all man's mental activities (reason, will, memory, intelligence, higher feelings, and emotions); it is the seat of consciousness, the interpreter of sensations, and the originator of voluntary acts; it acts as a control on many reflex acts that originate involuntarily (weeping, laughing, defecation, urination).

The cerebellum lies at the base or posterior part of the brain. The outer cerebellar cortex is made of gray matter, which is not convoluted but is traversed by numerous furrows (sulci). All functions of the cerebellum are below the level of consciousness, the main function being the reflex control of skeletal muscle activities.

The midbrain connects the cerebral hemispheres with the cerebellum and pons. Two pairs of round elevations, the corpora quadrigemina, act as centers for auditory and visual reflexes. Important pathways to and from other parts of the brain pass through the midbrain.

The pons lies in front of the cerebellum and above the medulla. Its fibers connect the two halves of the cerebellum and join the medulla with the midbrain.

The medulla oblongata lies between the pons and the spinal cord; structurally it is much like the spinal cord. The fourth ventricle of the brain is located within the medulla and connects with the central canal of the cord. The medulla contains such vital centers as cardiac, respiratory, and vasoconstrictor centers, the latter for the control of arterial pressure.

Cranial nerves

Nerve impulses pass to and from the brain by way of the twelve pairs of cranial nerves. In some of these, such as the optic nerve, impulses pass only in the direction of the brain and are called sensory nerves. In others, called motor nerves (such as the facial nerve), impulses pass from the brain. Still others carry impulses both to and from the brain; these are mixed nerves, for example, the vagus nerve. The following is a listing of the cranial nerves*:

   I. Olfactory—sense of smell
  II. Optic—sense of sight
 III. Oculomotor—control of the following eye muscles, ciliary; inferior oblique; superior, inferior, and internal (medial) recti; sphincter of the iris of the eye
 IV. Trochlear (pathetic)—superior oblique muscle of the eye
  V. Trigeminal—sensory to the head; motor for the muscles of mastication
 VI. Abducens (abducent)—external (lateral) rectus muscle of the eye
VII. Facial—motor to the face and scalp; sensory to the tongue; secretory to the submaxillary and sublingual (salivary) glands of the mouth
VIII. Auditory (acoustic)—to the cochlear part of the ear for hearing; to the semicircular canals of the ear for equilibrium
 IX. Glossopharyngeal—motor to the pharynx; sensory to the tongue, mucous membranes of the pharynx, tonsils, eustachian tube, and tympanic cavity of the ear; secretory to the parotid gland (salivary) of mouth
  X. Vagus (pneumogastric)—sensory to the larynx, trachea, lungs, esophagus, stomach, small intestine, and part of the large intestine; motor for respiration, heart action, and digestion (inhibits

heart action); secretory for gastric and pancreatic glands
 XI. Accessory—to muscles of the shoulder
XII. Hypoglossal—to the tongue.

Spinal cord

The spinal cord consists of a central canal surrounded by a core of gray matter, which is surrounded by white matter. The gray matter in cross section resembles the letter H, the two forward projections are called anterior columns and the two backward projections, the posterior columns. The spinal cord serves as a center for spinal reflexes and as a pathway to and from the brain. The white matter of the cord has (1) long ascending tracts to transmit afferent impulses from the spinal nerves to the brain and (2) long descending tracts to transmit efferent impulses from the motor centers of the brain to the anterior columns of the cord to control muscular movements. A summary of the spinal nerves is given in Table 29-1.

Reference to Fig. 4-13 will show the general relationship of neurons, spinal nerves, and spinal cord. Dendritic fibers of a sensory neuron lead back to the spinal cord by way of a spinal nerve, which branches just before entering the spinal cord. The dorsal root has an enlargement, called a ganglion, which contains the cell bodies of the sensory neurons. The sensory axon then passes into the cord where it forms a synapse with an interneuron. The entire interneuron is located in the gray matter of the spinal cord. A second synapse contacts a motor neuron whose cell body is also in the cord. The motor axon leaves by way

Table 29-1 Spinal nerves

Spinal nerve	Number of pairs
Cervical (neck)	8
Thoracic (thorax)	12
Lumbar (back)	5
Sacral (pelvis)	5
Coccygeal (tail)	1
TOTAL	31 pairs

*See also Table 26-1.

of the ventral root, which rejoins the dorsal root forming the intact spinal nerve. The spinal nerve branches at its terminus, supplying the various organs of that area.

Reflex pathways

A nervous reflex involves the transmission of an impulse, initiated in a receptor, causing a response in an effector. This system involves receptors, sensory neurons, the spinal cord or brain, motor neurons, and an effector. The familiar knee jerk reflex is one of the simplest examples.

Usually, a small hammer is used to strike the patellar tendon below the kneecap. This stretches the quadriceps muscle (in front of the thigh) and results in an impulse being carried by way of a sensory neuron to the spinal cord. Here, a synapse is made with a motor neuron, sending an impulse back to the quadriceps, which contracts, ''jerking'' the lower leg.

Other reflex patterns can be more complex than this, involving one or more interneurons. In some cases a particular stimulus may result in a variety of responses, the impulse being carried over several interneurons in the spinal cord. At the other extreme several sensory neurons have to be involved before a single response will take place. If the knee jerk may be considered an example of a simple reflex pathway, consider the type of pathway involved in an action such as balancing on one foot.

Autonomic nervous system

The autonomic nervous system consists of the sympathetic system (thoracolumbar), which has centers, ganglia, and plexuses in the cervical, thoracic, and lumbar regions of the spinal cord, and the parasympathetic (craniosacral) system, which consists of centers and ganglia of the cranial and sacral parts of the autonomic system. The involuntary control of the viscera by this system is essential to normal metabolism.

The two divisions are usually considered to be antagonistic to each other, since they control opposite effects in various organs. See Table 29-2 for some examples.

The neuron units of the autonomic or visceral systems differ from the other parts of the central nervous system in that there are usually two motor units involved. Thus a second cell body is located outside the brain or spinal cord in a ganglion. Fibers from this innervate the visceral organs.

A reflex arc in the autonomic system would involve: (1) a sensory neuron, with a cell body in the dorsal root ganglion of the spinal nerve, (2) a preganglionic neuron, with a cell body in the spinal cord and axon, leaving to terminate in a ganglion outside the cord, and (3) postganglionic fibers, from a cell body in the ganglion, leading to the particular parts of the heart, lungs, and intestine.

Perhaps the major difference between the sympathetic and parasympathetic neurons is that the postganglionic fibers usually secrete different neurohormones. Like the fibers of the central system, preganglionic neurons secrete acetylcholine and are called cholinergic fibers. Postganglionic fibers of the parasympathetic system are primarily cholinergic also. Sympathetic postganglionic fibers, however, secrete adrenalin and are called adrenergic.

Sense receptors

As indicated previously, a neuron may be depolarized by a variety of stimuli. In many cases a

Table 29-2 Some functions of autonomic nerves

Organ	Sympathetic	Parasympathetic
Iris muscles (pupil of eye)	Dilates	Constricts
Heart rate	Increases	Slows
Bronchi	Dilates	Constricts
Intestine	Decreases peristalsis	Increases peristalsis
Bladder	Fills	Empties
Sphincters	Contracted	Relaxed

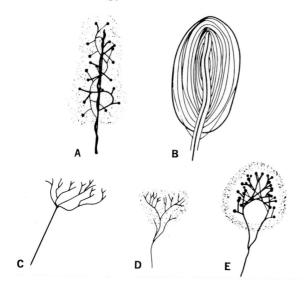

Fig. 29-7. Simple sense receptors. **A,** Meissner's corpuscle (touch); **B,** Pacini's corpuscle (pressure); **C,** free nerve ending (pain); **D,** Ruffini's corpuscle (warmth); **E,** end bulb of Krause (cold).

particular type of stimulus results in an impulse sooner than another type. Accordingly, then, this efficient stimulus is the basis of the designation of certain sensory neurons as receptors for a particular type of stimulus.

Simple receptors

The sensory nerves of the skin transmit sensations of pressure, pain, heat, and cold from the specific sense organs to the proper parts of the central nervous system to be interpreted (Fig. 29-7). Pressure receptors of the skin, sex organs, and internal organs give us a sense of touch and automatic regulation of muscular coordination. The skin contains about 500,000 pressure receptors, which are especially abundant on the fingertips and the tip of the tongue.

The endings of the afferent nerve fibers in muscles are branched terminal arborizations (internal-pressure receptors), which frequently extend over the ends of the muscle fibers. Such receptors are stimulated by pressure caused by muscular contractions. However, afferent nerve

fibers have other types of endings in muscles, such as spirals wound around muscles, tiny circular end plates, or club-shaped terminal organs. These receptors also have an effect on sensation. This conscious reaction is known as muscle sense, or kinesthetic sense (Gr. *kinein,* to move; *aisthesis,* perception), and through it we know whether muscles are contracted or moving, or whether a joint is bent.

Thermoreceptors in man are heat receptors, stimulated by a temperature higher than that of the skin, and cold receptors, stimulated by temperatures lower than that of the skin. The thermoreceptors of the tongue are particularly sensitive, those of the face less so, and those of the trunk and limbs least sensitive.

Complex receptors

Complex receptors are similar to the preceding simple receptors in that they are based on single sensory neurons. They differ in the number of cells associated in the particular sense organ and in their somewhat complex supporting tissues.

Chemoreceptors

Chemoreceptors are stimulated by suitable concentrations of definite chemical substances.

TASTE

Taste (gustatory) receptors are groups of specialized epithelial cells, located chiefly on the tongue, which give the primary taste qualities of sweet, sour, salty, and bitter.

The human taste buds are the end organs of nerve filaments arising from the trigeminal, facial, and glossopharyngeal nerves (cranial nerves).

The importance of smell to taste is well known, as demonstrated by foods losing their taste when one has a cold with a stuffy nose.

SMELL

Smell (olfactory) receptors are specialized epithelial cells in the mucous membranes of the nose for the detection of several types of odors.

Although there is still much doubt about the correct explanation for the sense of smell, recent investigations have suggested that the actual stimulus for an impulse in an olfactory fiber is caused by the shape of the particular molecule. The terminal fibers in the nasal epithelium have a series of pits, or depressions, into which a particular molecule may fit. Thus, a variety of substances having the same general molecular shape would produce the sensation of similar odors. A list of suggested primary odors appears in Table 29-3. Complex odors would result from a combination of these, stimulating a variety of receptors.

Table 29-3 Primary odors

Primary odor	Example
Camphoraceous	Moth balls
Musky	Skunk
Floral	Roses
Pepperminty	Mints
Ethereal	Ether
Pungent	Vinegar
Putrid	Rotten eggs

VISION

The visual apparatus consists of the eyes, their muscles that control movement, the tear glands, and the eyelids (Fig. 29-8). The eye itself is a spherical body formed of three coats. The outermost covering, the sclera, is a fibrous, protective layer, which is transparent in front, forming the slightly bulging cornea. In the back it continues as the covering of the optic nerve. The second, or middle layer, is pigmented and contains many blood vessels, giving it a dark color. This choroid coat continues in the front of the eye as the iris. The adjustable opening in the iris is called the pupil. A lymphlike fluid, the aqueous humor, fills the space on either side of the iris. Just behind this is the lens, held in place by ligaments. The innermost layer is the retina, the actual light receptor area. This does not continue in the front of the eye. The interior of the eyeball is filled with a clear, jellylike vitreous humor.

The amount of light entering the eye is determined by the iris muscles, which in turn control the diameter of the pupil. These muscles are under autonomic control and dilate or constrict

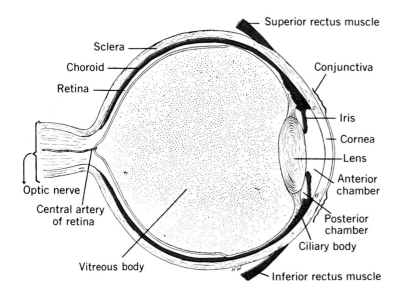

Fig. 29-8. Human eyeball (diagrammatic vertical section).
(From Francis, C. C.: Introduction to human anatomy, ed. 6, St. Louis, 1973, The C. V. Mosby Co.)

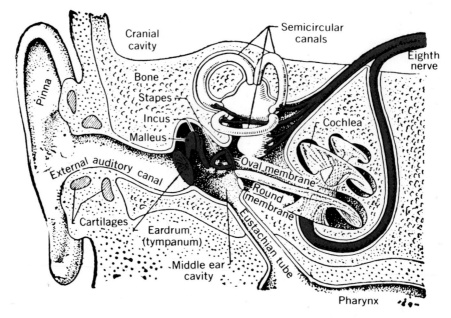

Fig. 29-9. Human ear (diagrammatic section). Ear bones are the malleus (hammer), incus (anvil), and stapes (stirrup).
(From Storer, T. I.: General zoology, New York, McGraw-Hill Book Company.)

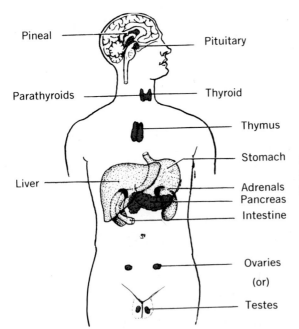

Fig. 29-10. Approximate locations of certain endocrine (ductless) glands. The liver shows the relative positions of other organs.

with varying light intensity. The actual fine focusing of the light waves on the retina is determined by the shape of the lens, which is controlled by muscles and is alternately bulging or slender, depending upon whether the object being viewed is near or far.

The retina is a multilayered structure that is specialized for light reception and impulse conduction. Most light receptor cells are sensitive to black and white light and are called rod cells. A smaller number of cells, called cones, function in color vision. There is evidence that separate cones exist that are sensitive to red, blue, and green light. For both rods and cones the action of a light wave appears to split a particular pigment molecule, resulting in a stimulus to a sensory neuron that leads to the brain by way of the optic nerve. In the rods this substance is called rhodopsin, or visual purple. Vitamin A is necessary for its formation.

SOUND

The human auditory apparatus (Fig. 29-9) consists of (1) an external ear, with its auditory canal, which has a membranous tympanum (eardrum) at its inner end, (2) the middle ear, with its eustachian tube connecting it with the pharynx to equalize air pressure, the middle ear bones— hammer or malleus (L. *malleus,* hammer), anvil or incus (L. *incus,* anvil), and the stirrup or stapes (L. *stapes,* stirrup), and the two openings of the middle ear into the inner ear, which are known as the fenestra vestibuli (ovalis) and the fenestra cochleae (rotunda), (3) the internal ear, with its vestibule, its cochlea (shaped like a snail shell), and the three semicircular canals, which serve the purpose of equilibrium, and (4) the auditory, or acoustic, nerve leading from the internal ear to the central nervous system.

Sound receptors are located in the coiled cochlea of the inner ear and are composed of vibrating "hair cells" that are stimulated by vibratory pressure waves brought to them from the ossicles (ear bones) of the middle ear. Sound vibration frequencies which range from about

16,000 to 20,000 per second, are detectable. Gravity receptors consist of three pairs of membranous semicircular canals (labyrinth) within the inner ear. The canals are so arranged that one lies in a plane at right angles to both planes of the other two canals in its labyrinth and is in a plane parallel to that of one canal in the labyrinth of the opposite ear. An enlarged ampulla near the end of each canal contains most of the receptor organs. Their receptor cells bear flexible hairs projecting into the internal lymph. Movements create currents in the lymph of the canals, thus stimulating the hairs, which send impulses to the brain.

ENDOCRINE SYSTEM

The structure and functions of various organs in the human body are also affected by substances produced in other organs and transmitted primarily by the blood. This chemical coordination is brought about by specific chemical substances known as hormones (Gr. *hormaein,* to excite). These hormones are manufactured in certain organs from ingredients brought to them by the blood and are carried away by the blood instead of through ducts or tubes (Figs. 29-10 and 29-11).

Endocrine glands and their secretions that contain the specific hormones are influenced by such factors as the quantity and quality of foods brought to them by the blood, the action of hormones from other endocrine glands, and the action of certain parts of the nervous system affecting the medulla of the adrenals. Endocrine glands were formerly studied separately, but recent work has shown the great interdependence of many of them, and this new approach to their study has helped in obtaining a more correct picture. Some organs, such as testes, ovaries, and the pancreas, may function as both ductless and duct glands. These organs produce some substances that move through special tubes, such as the pancreatic duct, and other substances that are carried away in the blood. Some glands, such as the pituitary, thyroid, and adrenals, function only

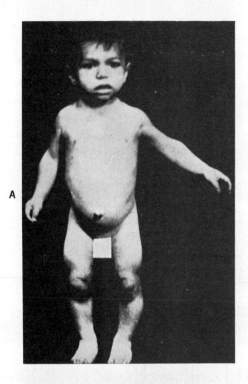

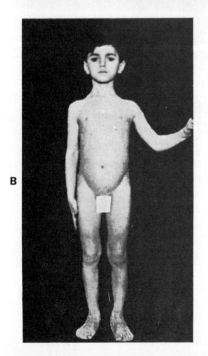

as ductless glands. Several produce a number of hormones with more of less specific functions, which complicates the problem of their investigation.

Pituitary gland

The bilobed pituitary gland (Fig. 29-12) produces many hormones that influence many parts of the body, and consequently it is often called the master control gland of the body. In man it weighs about 0.5 gm. and is about the size of a pea. It lies between the roof of the mouth and the floor of the brain (hypothalamus) in a depression of the sphenoid bone. it is a double gland, consisting of (1) the anterior lobe, which together with the intermediate part (pars intermedia), arises embryologically from a pouch on the roof of the mouth and (2) the posterior lobe, which arises from the brain. This gland thus has a dual origin and quite diverse functions. It is also called the hypophysis. The various hormones and some of their functions are summarized in Table 29-4.

When the pituitary is experimentally removed from young animals, growth stops and sexual maturity fails to develop. If it is removed from adults, male and female reproductive organs atrophy, as do the adrenal gland's cortex and the thyroid. If pituitary extracts are administered to normal young animals, they become giants who are sexually mature at an early age, and the sex glands, adrenal cortex, and thyroid enlarge because of increased secretions. An underactive pituitary during growth causes a small, properly proportioned midget. Oversecretion of the growth-stimulating hormone, after normal growth ceases, causes acromegaly (Gr. *akron*, tip; *megalou*, large), with grossly enlarged hands, feet, facial bones, and broad long jaws.

Fig. 29-11. A, Hypothyroidism in a 6-year-old child; **B,** same individual 2 years after the beginning of replacement treatment. *(Courtesy Dr. Jacob Rosenblum. From Wolf: Endocrinology in modern practice, Philadelphia, W. B. Saunders Company.)*

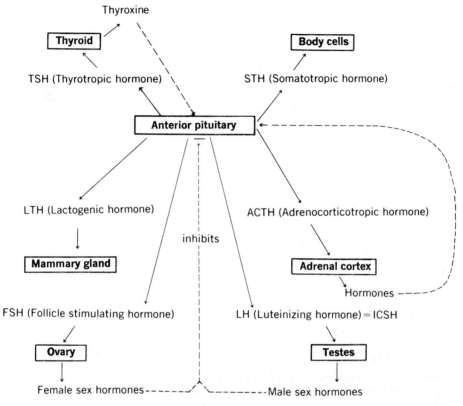

Fig. 29-12. Hormonal activity of the anterior pituitary. Broken lines indicate that the pituitary hormone is inhibited.

Adrenal glands

One soft, cup-shaped adrenal gland is attached to the top of each kidney and is composed of (1) an outer cortex and (2) a darker, inner medulla. The two parts differ concerning embryologic origin, structure, and functions. Each adrenal gland weighs about 5 gm.

Chemically all of the hormones secreted by the cortex are called steroids. The surgical removal of adrenal glands from animals results in decreased sodium and chlorine in the blood, a loss of water with decreased blood plasma (hence lowered blood pressure), and increased potassium in the blood. Insufficient cortical secretion results in Addison's disease. Symptoms include muscular weakness, low blood pressure, digestive upsets, and bronzed skin caused by the depositing of a pigment called melanin (Gr. *melas,* black).

Thyroid gland

The paired, shield-shaped lobes of the thyroid gland, weighing about 25 gm., are connected by an isthmus and located in front of the trachea just below the voice box, or larynx. They are present in all vertebrates.

The thyroid contains a protein (thyroglobulin) that possesses a large amount of iodine. From the protein a hormone, thyroxine, can be obtained that contains about 65% iodine. A deficiency of

Table 29-4 Summary of endocrine glands, hormones, and functions

Gland	Hormone	Function
1. Pituitary		
a. Anterior lobe	1. Growth-hormone (STH) (somatrotropin)	1. Regulates growth of body cells
	2. Follicle-stimulating hormone (FSH)	2. Regulates growth of ovarian follicles; regulates spermatogenesis in seminiferous tubules
	3. Interstitial cell-stimulating hormone (ICSH), or luteinizing hormone (LH)	3. Forms corpus luteum (ovaries); stimulates interstitial cells to produce male sex hormone
	4. Lactogenic hormone (prolactin)	4. Stimulates lactation (milk production)
	5. Thyroid-stimulating hormone (TSH)	5. Stimulates thyroid gland
	6. Adrenocorticotropic hormone (ACTH)	6. Controls cortex of adrenal glands
b. Posterior lobe	1. Vasopressin; antidiuretic hormone (ADH)	1. Stimulates smooth muscles, urinary bladder, and gallbladder; constricts small arteries; decreases volume of urine secreted
	2. Oxytocin (Pitocin)	2. Contracts muscles of uterus; ejects milk from mammary glands
2. Adrenals (suprarenals)		
a. Cortex	1. Cortisone (hydrocortisone, glucocorticoids)	1. Controls food metabolism (glucose, proteins, and fats); suppresses inflammation
	2. Aldosterone (desoxycorticosterone [DOCA], mineralocorticoids)	2. Controls secretion of salt and water; regulates sodium-potassium metabolism
	3. Androgens (androsterone and others)	3. Similar to testosterone; stimulates masculine traits as beard and deep voice
b. Medulla	1. Epinephrine	1. Increases blood pressure, blood sugar and heart rate; inhibits gastrointestinal tract; hastens blood coagulation; decreases glycogen in liver
	2. Norepinephrine	2. Effects similar to adrenalin but with wider vasoconstrictor influences
3. Thyroid	1. Thyroxine	1. Controls rate of metabolism, growth, and development
	2. Triiodothyronine	2. Similar to thyroxine
	3. Calcitonin	3. Controls blood calcium
4. Parathyroids	1. Parathormone	1. Regulates metabolism of calcium and phosphorus
5. Islands of Langerhans of pancreas	1. Insulin	1. Controls metabolism of glucose (sugar) and fats; lack of insulin causes diabetes mellitus
	2. Glucagon	2. Controls breakdown of glycogen in liver; effect on carbohydrate metabolism opposite that of insulin
6. Thymus	"Thymic hormone"	Aids in forming lymphocytes and antibodies

Table 29-4 Summary of endocrine glands, hormones, and functions—cont'd

Gland	Hormone	Function
7. Pineal	Suspected of secreting a hormone influencing Na⁺	
8. Testes	1. Testosterone 2. Androsterone	Controls secondary sex traits, including larger muscles, body hair, beard, and deepened voice
9. Ovaries		
a. Graafian follicles	1. Estrogen	1. Controls secondary sex traits menstrual cycle, mammary glands, and sex changes at puberty
b. Corpus luteum	1. Progesterone	1. Regulates menstruation, mammary glands, and pregnancy
10. Placenta	1. Chorionic gonadotropin	1. Regulates growth of corpus luteum during pregnancy
	2. Relaxin	2. Relaxes pelvic ligaments during childbirth
11. Gastrointestinal mucosa		
a. Stomach	1. Gastrin	1. Stimulates gastric juice secretion
b. Small intestine	1. Secretin	1. Stimulates pancreatic juice secretion
	2. Enterogastrone	2. Inhibits gastric secretion
	3. Cholecystokinin	3. Stimulates bile secretion

thyroxine causes a reduction in the basal metabolism rate (BMR) of 30% to 50% of normal, whereas excess thyroxine causes increased oxygen usage, increased heat production, and increased metabolic wastes.

HYPOTHYROIDISM

Underactivity or degeneration of the thyroid, resulting in a deficiency of thyroxine, produces a disease in adults called myxedema (Gr. *myxa,* mucus; *eidema,* swelling). This condition is characterized by a low metabolic rate (decreased heat production); slow pulse; mental and physical lethargy; waxy, puffy skin caused by mucus in the subcutaneous tissues; usually a normal appetite which, because of the lowered metabolic rate, results in food storage and a tendency toward obesity; and usually a falling of the hair. When hypothyrodism (Fig. 29-11) is present from birth, the disease is called cretinism, and the child

has low intelligence and does not mature sexually. Successful treatments can occur if started early. Another type of hypothyroidism is produced by insufficient iodine in the diet, resulting in thyroxine deficiency. In an attempt to compensate the thyroid enlarges, producing simple goiter, in which the gland may be slightly or greatly enlarged with symptoms resembling those of myxedema but of a milder nature.

HYPERTHYROIDISM

Hyperthyroidism is a condition that results either from the overactivity of a normal-sized gland or an increase in its size, with a highly increased metabolic rate (increased heat production) and excessive perspiration. Other symptoms include loss of weight (foods are quickly used), high blood pressure, irritability and nervous tension, weakness and involuntary trembling of muscles, protrusion of the eyeballs, known as exophthalmos (Gr. *ex,* out; *oph-thalmos,* eye) which gives the name exophthalmic goiter to this condition.

Parathyroid glands

There are usually two pairs of parathyroid glands, about the size of a pea, which are embedded in the thyroid. When the parathyroids are removed surgically from an animal, death follows in a few days unless the hormone (parathormone) is administered. Symptoms in animals after removal include cramps, involuntary muscle twitching, and convulsions. This condition, called tetany, results from an increased irritability of nerves and muscles caused by a decrease of calcium in body fluids and blood. If calcium solution is injected into a tetanized animal, the convulsion stops.

In case of increased parathyroid functioning caused by gland enlargement or a disease such as a tumor, the blood calcium increases. Part of this calcium may come from bones, rendering them soft and easily fractured. Muscles become painful, decrease in size, and are less irritable than normal. The excess blood calcium may be deposited in various organs.

Islands of Langerhans of the pancreas

The flat, elongated pancreas lies in the curvature between the stomach and small intestine. Groups of certain cells within the pancreas secrete the hormone, insulin, whereas other cells secrete a different hormone called glucagon.

Insulin was extracted from the pancreas by two Canadians, Banting and Best, in 1922, and has been used successfully in treating sugar diabetes (diabetes mellitus), which is a disease that has been known since the first century A.D. In sugar diabetes there is excess sugar in the blood and urine, causing excessive excretion of urine, and producing dehydration and thirst. The patient becomes progressively weaker and may die eventually unless treated. If too much insulin is administered, the blood-sugar level may be so lowered as to produce insulin shock, and unconsciousness may result unless the blood-sugar level is restored quickly.

Thymus gland

The thymus gland lies in the upper part of the chest near the lower part of the trachea. It is rather large during childhood but regresses at puberty (12 to 17 years of age).

Recent studies have indicated a possible endocrine function of the thymus in mice. Apparently, a substance is formed in the thymus that passes to the lymph glands and initiates the formation of a type of lymphocytes called plasma cells. These cells are antibody-producing cells and are the basis of immunity.

Pineal gland

The small, cone-shaped pineal gland between the two cerebral hemispheres is located at the base of the brain (thalamus) and posterior to the pituitary. Exact functions of this gland are not well established. It appears to be involved in light-controlled reactions, pigment functions, and aldosterone secretion.

Testes

Two ovoid bodies, the testes (L. *testis,* testicle), are suspended in the scrotum (see Fig. 30-1), and their interstitial cells produce endocrine hormones; the seminiferous tubules of the testes produce sperms. Castration (removal of the testes) results in improperly developed secondary sex traits. A eunuch (GR. *eune,* couch; *ehcein,* to hold) is a man who was castrated before maturity and as a result is beardless and has a high-pitched voice. Injection of testosterone into females causes some secondary male sex traits to develop. The male hormone also partially influences male sexual behavior.

Ovaries

Two bean-shaped ovaries, about 1½ inches long, near the female uterus (see Fig. 30-2) contain an outer layer of germinal epithelium for forming eggs within the graafian follicles. After a fluid-filled follicle breaks and releases its egg, it fills with yellowish cells, which constitute the corpus luteum (L. *corpus;* body; *luteum,* yellow). Between the onset of menstruation and the menopause one or more follicles mature and release an egg (ovulation), once each month. The ovarian hormones and their functions are summarized in Table 29-4.

Placenta

The placenta (L. *placenta,* flat cake) is a flattish organ that attaches the developing embryo to the uterine wall of the mother, supplying it with nourishment.

During the first eight weeks of pregnancy the placenta secretes a hormone, chorionic gonadotropin, which stimulates the corpus luteum to continue secretion of its hormones. In the third and fourth months, as the amount of chorionic gonadotropin decreases, the placenta begins to produce progesterone and estrogens.

Chorionic gonadotropin secreted in the urine has been used as an almost unfailing test for pregnancy. A small quantity of urine of a pregnant woman is injected into an immature female mouse, rat, or rabbit, where it stimulates the reproductive organs and results in estrus ("heat") and the characteristic changes in the genital tract.

Gastrointestinal mucosa

The mucous membranes lining the stomach and small intestine produce several hormones that are summarized in Table 29-4. One hormone, cholecystokinin (Gr. *chole,* bile; *kystis,* bladder; *kinesis,* to move), causes the gallbladder to empty its stored bile into the small intestine.

Interrelationships of hormones

Studies of several hormones and their functions reveal an involved interrelationship among the various endocrine glands, such as shown by the pituitary, adrenals, and reproductive organs.

The hormones are exceedingly powerful agents, and in some instances their activities cover practically the entire body. What regulates the amount of each hormone produced? How do they interact so well and normally in most cases? In some instances the body requirements for a particular hormone will regulate its production automatically. Under normal conditions an increase in blood sugar stimulates the production of insulin, and the production is decreased when the need has been met.

The pituitary and its "target glands" serve as a good example of endocrine relationship (Fig. 29-12). The anterior pituitary secretes the following hormones: (1) thyroid-stimulating hormone (TSH), which stimulates the thyroid gland to secrete thyroxine, which controls cell respiration and oxygen use as well as inhibiting further TSH release, (2) somatotropic hormone (STH), which promotes protein formation and body growth, (3) adrenocorticotropic hormone (ACTH), which stimulates the production of several hormones, including one that inhibits further production of ACTH, and (4) gonadotropin hormones, which are associated with sex and reproduction in females. The follicle-stimulating hormone (FSH), the luteinizing hormone (LH), and the lactogenic hormone (LTH) individually and collectively control the development of female sex characteristics, ovary maturation, egg production and release, milk production, and the release of hormones including estrogens and progesterone, which in turn influence further female development and inhibit the anterior pituitary.

These same FSH and LH, called the interstitial cell-stimulating hormone (ICSH), influence the development of male characteristics and especially the supporting interstitial cells of the testes and the seminiferous tubules that produce sperms. The testes in turn produce the androgens, such as testosterone, which, in addition to supporting masculine characteristics, also inhibit the pituitary.

Some contributors to the knowledge of the
BIOLOGY OF MAN

Hippocrates (460-370 B.C.)

A Greek who presented methods of medical treatments and the responsibilities of a physician to the patient and to the profession. His code, called the "Hippocratic Oath," is still followed by doctors. He made a science of medicine and is called the "Father of Medicine."

Herophilus (300 B.C.?)

A Greek who worked in Alexandria, Egypt, and was called the "Father of Anatomy." He is said to have dissected animals and human bodies. He wrote a book on anatomy, recognized the brain as the center of nervous activity, and distinguished between the kinds of nerves. At that time Alexandria was a great center of Greek learning in Egypt.

Galen (Claudius Galenus) (130-200 A.D.)

A Greek who was born in Asia Minor but received his medical training in Alexandria, Egypt, and then did much of his studying in Rome. In order to solve problems of anatomy and physiology he resorted to experiments and observations by dissections. For a period of 500 years after the time of Herophilus the dissection of human bodies was prohibited, so that Galen dissected animals primarily. His dissections showed in detail the brain, spinal cord, and nerves. His book on dissection, entitled *Anatomical Procedure,* originally consisted of sixteen volumes. His work was the authority in anatomy and physiology for 1,000 years, until the Renaissance.

Leonardo da Vinci (1452-1519)

The great Italian painter who believed that experience and direct observations, not classical authority, were the source of knowledge. He studied anatomy because of its relationship to painting, and he made comparisons between human anatomy and that of animals.

Vesalius

Andreas Vesalius (1514-1564)

A Belgian anatomist who published a large treatise, *On the Structure of the Human Body* (1543), based on dissections and direct observations rather than on the previous assumptions of so-called authorities. He went to Paris at the age of 16 years to study medicine, where in dissection he excelled over the barber dissectors who demonstrated the structures. He taught anatomy at the University of Padua, Italy, and was called the "Father of Modern Dissective Anatomy." He stressed the importance of direct observations and discovered that the right and left ventricles of the heart are completely separated from each other. *(Historical Pictures Service, Chicago.)*

Michael Servetus (1511-1553)

A Spanish anatomist, religious philosopher, and mystic who postulated the circulation of the blood, later proved by William Harvey. He was burned at the stake for his antireligious views.

Hieronymus Fabricius (153?-1619)

An Italian anatomist at Padua who discovered the valves in veins—knowledge that was later used by his student, William Harvey.

William Harvey (1578-1657)

An English physician who proved that blood circulates through the body in arteries and veins, although he did not see capillaries because he had no microscope. He proved that the heart was muscular and pumps actively and that veins controlled the direction of flow.

Giovanni Borelli (1608-1679)

An Italian physiologist who established physiology in what was later to become an effort to describe organic functions in mechanical physiochemical terms. He taught Marcello Malpighi.

Richard Lower (1631-1691)

An English physician who first published an account of a successful blood transfusion, but it is probable that unsuccessful transfusions had been tried as early as 1492.

Stephen Hales (1677-1761)

An English physiologist and botantist who conceived experiments in animal and plant physiology, which served as a connection between physiologic studies of the seventeenth century and the nineteenth century.

Luigi Galvani (1737-1798)

An Italian naturalist who studied the roles of electricity in organisms, particularly those in muscular movements.

Edward Jenner (1749-1823)

An English physician who discovered that vaccination with cowpox would immunize man against the more serious smallpox (1798), thus ushering in a new era in preventive medicine.

William Beaumont (1785-1853)

An American surgeon and physiologist who performed famous experiments on digestion using the incompletely healed stomach of a wounded patient. His work accelerated the establishment of the modern physiology of digestion.

Thomas Addison (1793-1860)

An English physician who contributed greatly to establishing modern physiology of glandular secretions, especially those of the adrenal glands.

Johannes Müller (1801-1858)

A German who laid the broad outlines for present-day physiology, using chemistry and physics as tools. He stressed comparative physiology by studies of the functions in lower and higher animals.

Eijkman

Ivan P. Pavlov (1849-1936)

A Russian physiologist who performed many experiments with the digestive system and whose book, *Digestive Glands* (1897), is a classic in digestive studies. He received a Nobel Prize in 1904. He proposed the concept that acquired reflexes play a role in the nervous reaction patterns in animals (1910).

Christian Eijkman (1858-1930)

A Dutch physiologist who produced experimental polyneuritis in fowls by feeding them polished rice. He called attention to rice hulls as being the source of an agent for preventing human beriberi (1897). This later led to the discovery of the antineuritic vitamin B_1. *(Historical Pictures Service, Chicago.)*

Frederick G. Hopkins (1861-1947)

An English biochemist and physiologist who made historic studies of muscular chemistry, cellular respiration, and biochemical dietary deficiencies.

Edward C. Kendall (1886-1972)

An American biochemist who first isolated the hormone thyroxine from the thyroid gland in crystalline form (1914) and isolated cortisone from the adrenal glands. ACTH (adrenocorticotropic hormone) was isolated from the pituitary gland (1943).

Frederick G. Banting (1891-1941) and Charles H. Best (1899-1978)

Banting and Best, Canadians, extracted the hormone insulin from the pancreas (1921), thus paving the way for the treatment of diabetes mellitus. This work was based on results of the removal of the pancreas from dogs by Mering and Minkowski (1889). *(Historical Pictures Service, Chicago.)*

Banting

Review questions and topics

1 Explain coordination in man by (1) nervous system and (2) chemical substances.
2 Describe the structure and function of (1) the various kinds of receptors, (2) the various kinds of effectors, and (3) the different types of conductors.
3 Explain in detail the theory regarding the initiation and conduction of nerve impulses.
4 List the more important structural and functional characteristics of the parts of the human nervous system.
5 Describe the structure and important functions of each endocrine gland.
6 Describe the anatomy and physiology of the male and female reproductive systems.

Selected references

Axelrod, J.: Neurotransmitter, Sci. Amer. **230**:58-71, June, 1974.
Barrington, E. J. W.: The chemical basis of physiological regulation, Glenview, Ill., 1968, Scott, Foresman and Company.
Bentley, P. J.: Comparative vertebrate endocrinology, Cambridge, 1976, Cambridge University Press.
Case, J.: Sensory mechanisms, New York, 1966, The Macmillan Company.
Comroe, J. H.: The lung, Sci. Amer. **214**:56, 1966.
De Coursey, R. M.: The human organism, ed. 4, New York, 1974, McGraw-Hill Book Company.
Dienhart, C. M.: Basic human anatomy and physiology, Philadelphia, 1973, W. B. Saunders Company.
Eccles, J.: The synapse, Sci. Amer. **212**:56-66, 1965.
Francis, C. C.: Introduction to human anatomy, ed. 7, St. Louis, 1975, The C. V. Mosby Co.
Glassman, E., editor: Molecular approaches to pshychobiology, Belmont, Calif., 1967, Dickenson Pub. Co., Inc.
Greisheimer, E. M.: Physiology and anatomy, ed. 8, Philadelphia, 1963, J. B. Lippincott Co.
Guyton, A. C.: Function of the human body, ed. 3, Philadelphia, 1969, W. B. Saunders Company.
Huxley, H. E.: The mechanism of muscular contraction, Sci. Amer. **213**:18-27, 1965.
Jacob, S. W., and Francone, C. A.: Structure and function in man, Philadelphia, 1974, W. B. Saunders Company.
Le Baron, R.: Hormones, New York, 1972, Pegasus.
Montagna, W.: The skin, Sci. Amer. **212**:56-66, 1965.
Turner, C. D., and Bagnara, J. T.: General endocrinology, ed. 6, Philadelphia, 1976, W. B. Saunders Company.
Vander, A. J., Sherman, J. H., and Luciano, D. S.: Human physiology: The mechanisms of body function, New York, 1970, McGraw-Hill Book Company.
Wood, J. E.: The venous system, Sci. Amer. **218**:86-94, 1968.
Wurtman, R. J., and Axelrod, J.: The pineal gland, Sci. Amer. **213**:50-60, 1965.

30 Human reproduction

Sexual maturity in humans is signalled by the onset of puberty. Secondary sexual characteristics are developing—the boy becomes a man and the girl becomes a woman. The boy begins to develop pubic and axillary hair, whiskers, and a deeper voice. The girl begins to accumulate the stores of body fat that give her womanly contours. Her breasts begin to develop, and pubic and axillary hairs also begin to appear. Sometimes during puberty there is also a growth spurt.

Strictly speaking, sexual maturity and puberty refer to the possibility of fertilization and conception. In common usage puberty may refer to all of the developmental changes, including attainment of secondary sex characteristics, which culminate in sexual maturity. When all of the physical, sexual, mental, and perhaps attitudinal changes are considered, this transition period is aptly called adolescence.

The onset of puberty and the subsequent sexual maturity are initiated and controlled by hormones secreted by the pituitary gland. Although these hormones have other functions, the follicle stimulating hormone (FSH) serves to stimulate the start of growth in primary follicles in the female and seminiferous tubules in the male. The second hormone, luteinizing hormone (LH), acts with FSH to complete follicle development and ovulation and also to secrete ovarian hormones called estrogens, which control female characteristics.

In males this same hormone is called interstitial cell stimulating hormone (ICSH). Here it acts on the cells between the sperm forming tubules. These cells are part of the interstitial tissue, which secretes hormones controlling male characteristics. All of these hormones have been discussed earlier.

MALE REPRODUCTIVE SYSTEM

Male reproductive organs consist of the paired testes, epididymis, seminal ducts (vas deferens), and seminal vesicles, which join the vas deferens to form the ejaculatory ducts. These open into the urethra which leads out through the penis.

The paired testes are located in the saclike scrotum (Fig. 30-1). The functioning parts of the testes are the seminiferous tubules, which produce the spermatozoa and the interstitial cells between them, which produce male hormones.

443

Each of the tubules leads into the highly coiled epididymis. We have already discussed the details of spermatogenesis in humans (see Fig. 5-8). The process takes place inside the seminiferous tubules. The tubular walls are lined with cells that are capable of continuing division throughout the sexual life of the male.

When one of these cells, called a spermatogonium, divides by mitosis, the two resulting cells are of different types. One of them forms another spermatogonium and continues to divide in the same way. The second cell forms a potential sperm cell called a primary spermatocyte. This cell and others like it divide by meiosis to form two secondary spermatocytes, which divide forming four spermatids. Each of these develops into a mature spermatozoan or sperm cell. The sperm cells are developed toward the cavity of the tubule.

Sperm formed in seminiferous tubules of each testis pass into the epididymis and from there to the sperm duct or vas deferens. (This is the tube cut in a vasectomy.) The sperm ducts pass through the body wall into the abdominal cavity. Here they loop over the pubic bone, pass over and behind each side of the bladder where they are joined by the ducts of the seminal vesicles. The common tube on each side is now called the ejaculatory duct. The ducts come together to join the urethra, which then leads to the outside through the penis.

The prostate gland lies just below the bladder, surrounding the junction of the ejaculatory ducts and the urethra. Both it and the seminal vesicles form much of the fluid part of the semen or seminal fluid. The average amount of semen ejaculated is about 3 ml., containing about 3 million sperm. Both the quantity and quality of the fluid and sperm cells are important in influencing fertility.

The human penis consists of three masses of spongy erectile tissue, the corresponding nerves and blood vessels, and the enclosing skin. The lower tissue mass surrounds the urethra as it passes through the penis. The possibility of sperm deposition in the female depends on the filling of the spongy tissue with blood. When this happens, erection of the penis takes place. This is caused by filling of the spongy tissue with blood and the constriction of the veins leading away from the tissues. Since the blood is under pressure, the spongy tissue remains firm and the penis remains erect. When the pressure on the veins is released, blood flows out of the spongy tissue and the penis is again limp.

FEMALE REPRODUCTIVE SYSTEM

The female reproductive system consists of several parts specialized for a variety of functions. The organs are: a pair of ovaries, a pair of oviducts (or fallopian tubes), the uterus (or womb), the vagina, and the external genitals (or vulva) (Fig. 30-2).

The basic part of the system is the ovary. Each ovary is a small oblong body slightly over an inch long. Eggs are produced here and later released. (Details of the functioning ovary will be covered later.) The eggs are then picked up in the funnel-like openings of the oviducts or fallopian tubes. These tubes are lined with cilia that sweep the egg toward the uterus. The tubes open on either side of the uterus.

The muscular uterus or womb is somewhat pear shaped. It lies internally between the bladder and the rectum. The oviducts lead into the thicker upper end. Its narrower lower end, called the cervix, opens into the vagina. The inner lining called the *endometrium* is highly glandular and goes through a cyclic buildup, which terminates in menstruation if fertilization and embryo implantation do not take place. Normally, the hollow center of the uterus is almost closed, but it swells greatly during pregnancy.

The vagina is a narrow muscular tube about 4 inches long. Like the uterus it is capable of extreme dilation or stretching. During childbirth its width increases to permit the passage of the baby. The vulva, or external genitals, consists of several folds of tissue called the labia. A mound

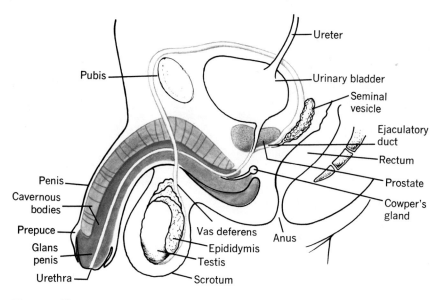

Fig. 30-1. Human male reproductive system.

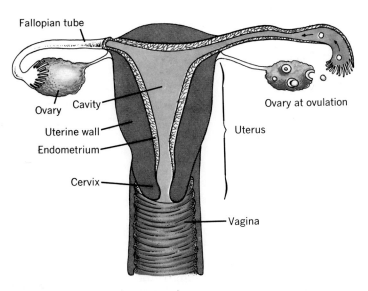

Fig. 30-2. Human female reproductive organs.

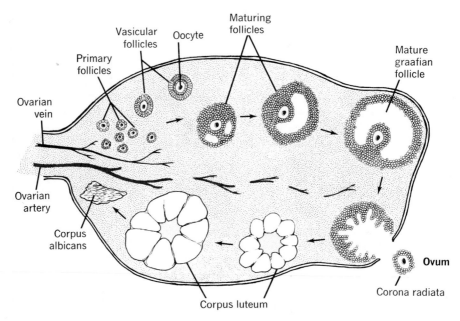

Fig. 30-3. Human ovary. Cycle starts with primary follicles and continues clockwise.

of fatty tissue, the *mons veneris,* extends over the pubic bone just above the vulva.

The sequence of events leading to pregnancy involves: oogenesis, the formation of the egg (see Fig. 5-8), ovulation, the release of the mature egg, fertilization by sperm, and actual implantation of the embryo into the expanded uterine wall. If fertilization does not take place, the engorged endometrium is shed during menstruation.

The functioning ovary

At birth the ovary of a baby girl contains over one quarter of a million primary follicles. These were formed during embryonic life, and no more will be formed. Each primary follicle was formed by mitosis of embryonic cells that enlarged and became oocytes. Other cells in the ovary formed a follicular layer around them. Only one in every thousand (about 250 to 300) will reach maturity and release an ovum or egg cell. The rest will degenerate during the adult years.

After the onset of puberty and until menopause is reached, the ovary goes through cyclic changes related to maturation and ovulation. These changes are initiated by the release from the pituitary gland of a hormone FSH or *follicle stimulating hormone.* This hormone is carried by the blood. It acts by stimulating an increase in the size of the ovary itself and of the primary follicles, which have been in an interphase since birth.

A number of the primary follicles begins to grow as the oocytes enlarge and the surrounding follicle cells increase in number, forming a layered structure. The surrounding ovary cells form a cellular capsule, the *theca,* around the follicle (Fig. 30-3). As this growth continues, the cells of the theca and the follicle cells secrete a fluid into small cavities between the cells. About this time the scretion of another pituitary hormone, LH or *luteinizing hormone,* increases. LH acts to enhance the effect of FSH and results in more rapid growth of follicle cells and greater secretion of fluid. As the cells continue to grow and secrete fluid, the cavity enlarges and the oocyte and some surrounding cells are pushed to one side. This maturing follicle is sometimes referred to as a graafian follicle.

A mature graafian follicle consists of: a capsule, the theca; a lining of follicular cells, the granular membrane; and a fluid filled cavity. The oocyte lies at one side surrounded by a gelatinous sphere and embedded in a mass of follicle cells. As the follicle grows, its diameter increases to 12 mm. (½ inch). It now lies just under the surface of the ovary and appears similar to a blister ready to burst. During each monthly cycle 20 to 30 primary oocytes begin the maturation process but usually only one of them completes development and ovulates. Shortly before ovulation there is a large increase in the amount of LH released by the pituitary. This, together with an increase in follicular fluid pressure, causes the follicle and nearby ovarian wall to rupture. As the fluid escapes, it carries the ovum and some follicle cells with it.

The events leading up to ovulation appear to be controlled by an interaction of hormones of the pituitary gland and the ovary. The release of FSH influences both the ovary itself and the primary follicle to grow. As they grow they secrete hormones called estrogens into the blood. Small amounts of estrogens increase these secretions of FSH which, in turn, stimulate more follicle growth. This increases estrogens in circulation and the increase leads to pituitary secretion of LH, which works with FSH to produce still more release of estrogens. At the peak of estrogen secretion the effect is to inhibit FSH and cause a sudden surge of LH production. The LH surge leads to ovulation sometime during the next day.

The ovum is picked up by the end of the uterine tube and is moved toward the uterus. If sperms are present in the tube, fertilization and pregnancy may follow.

Shortly after ovulation the follicle walls collapse. The cells of the theca and granular layer gradually change cytoplasmic composition and take on a yellowish appearance. The follicle is now called a *corpus luteum.* The corpus luteum lasts about 2 weeks. It begins a secretory function after the amounts of FSH and LH have diminished as a result of large amounts of estrogens secreted by the follicle cells.

The diminishing amounts of LH initiate corpus luteum secretion of large amounts of estrogens and progesterones for about a week. After this time LH is almost gone, and the secretion of estrogen and progesterone begins to decrease. The diminishing secretion of estrogen and progesterone leads to menstruation about a week later. The decrease also removes the inhibiting effect on the pituitary, and FSH begins to increase. This stimulates the primary follicles and the cycle starts again. Although it is common practice to indicate the onset of menstruation as day one of the cycle, the beginning of the follicle cycle occurs about 5 days later.

Menstruation

The monthly shedding of the uterine lining is called menstruation (L. *menses,* months). The cyclic changes in the lining are controlled by changes in the concentration of estrogen and progesterone. Estrogen serves to stimulate the buildup of the uterine lining, the endometrium. Progesterone then acts on this lining to convert it to a highly vascular tissue.

The release of the ovum is followed by an increased secretion of progesterone from the developing corpus luteum. The increase in progesterone stimulates the increased growth referred to previously. Roughly 1 week after ovulation the corpus luteum decreases its secretion of both estrogen and progesterone. In another week the corpus luteum is completely regressed, hormonal secretion stops, and the endometrium begins to disintegrate and menstruation starts. About 70 ml. of fluid, about one half of which is blood, are passed in the menstrual flow.

The entire process is dependent on events taking place in the ovary and is evidenced by changes in the uterus. As the estrogen content decreases during the menstrual phase, its inhibiting influence on the pituitary also decreases.

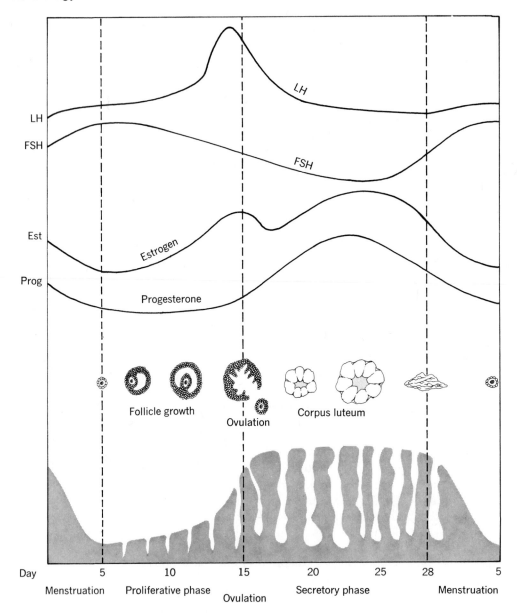

Fig. 30-4. Events of the pituitary (LH, FSH), ovary (estrogen, progesterone, and follicles), and uterus (growth and shedding of lining) during the menstrual cycle.

This results in a release of follicle stimulating hormone (FSH). FSH then acts on the primary follicles, which in turn begin to secrete estrogen. The resulting estrogens stimulate the endome- trium, which begins its *proliferative* buildup. The proliferative phase lasts until ovulation, when se- cretion of progesterone by developing corpus lu- teum begins to rise.

The period of the corpus luteum influence is called the *secretory* phase. In this the progesterone acts on the lining, prepared under the influence of estrogen, to convert it into highly glandular secretory tissue.

When it is recalled that the secretory phase begins with ovulation, the highly vascular condition is explained as a preparation to receive a fertilized ovum. If the ovum is not fertilized, the corpus luteum regresses, estrogen and progesterone levels drop, and menstruation begins.

It is evident that the processes of pituitary secretion of FSH and LH, the follicle changes and estrogen secretion, the corpus formation and secretion of estrogen and progesterone, and the uterine changes are all highly interrelated. See Fig. 30-4 for a diagram of these relationships.

FERTILIZATION AND LIFE BEFORE BIRTH

During copulation semen containing millions of sperms is deposited in the vagina near the cervical opening of the uterus. The sperms swim through the uterus and along the fallopian tubes. They remain capable of fertilizing an ovum for about 2 days. If an egg is expelled from the ovary during this time, it may be fertilized while in the fallopian tube. After union of the egg and sperm, their nuclei also combine. This restores the paired condition of the chromosomes. The fertilized egg is now called a zygote.

About 30 hours after fertilization the zygote begins a series of divisions called cleavage. The two cells from this first division divide again in about 10 hours, and this cleavage results in 16 to 32 cells by the third day. The mass of cells resulting from this slowly moves along the fallopian tube toward the uterus. The 32-cell stage forms a compact mass called a morula. As further development takes place, the outer layer of cells begins to divide faster than the inner cell mass.

After about 6 days the cell mass has developed a hollow ball shape with a small mass of cells inside (Fig. 30-5). This is called the blastocyst stage. In the uterus the blastocyst adheres to a sticky substance secreted by the endometrium. Combined action from the blastocyst and from the endometrium results in an erosion of the uterus and a sinking of the blastocyst. The point of erosion soon heals and a slight pimple-like swelling is the only evidence of blastocyst implantation. By the twelfth day after fertilization implantation is completed.

Meanwhile, the inner cell mass of the blastocyst continues to divide, forming two small cavities in a figure **8** shape separated by a double layer of cells. This double layer, called the embryonic disk, forms the embryo. The upper sphere becomes the amnion, a fluid-filled sac that protects and supports the developing embryo. The upper region of the lower yolk sac becomes the roof of the alimentary canal.

By the end of the second week, the outer layer of cells of the entire blastodisk and an inner lining layer begin to form many tiny fingerlike outgrowths called villi (Fig. 30-6). Under a microscope the external appearance is that of a hairy sphere. These outgrowths form the chorion, which grows into the uterine lining and begins the formation of the placenta. Both embryo and mother contribute to placental formation.

Between the third and fourth week of gestation, the blastocyst has enlarged to form a bulge on the uterine's surface. The embryo soon pushes into the cavity of the uterus. The original stalk of the inner cell mass forms the umbilical cord and is continuous with the chorion and placenta. About two thirds of the embryonic disk forms the future head, and the remainder forms the neck, trunk, and tail. The two cellular layers of the embryonic disk consist of (1) the lower endoderm layer (nearest the yolk sac cavity), and (2) the upper ectoderm layer, which gives rise to the brain, spinal cord, and outer skin. Between the ectoderm and endoderm, on either side of the median line and originating from both, two groups of cells are formed: (1) the membranous mesoderm (mesothelium), and (2) a loosely arranged meshwork of cells called the mesenchyme. The ectoderm, mesoderm, and endoderm

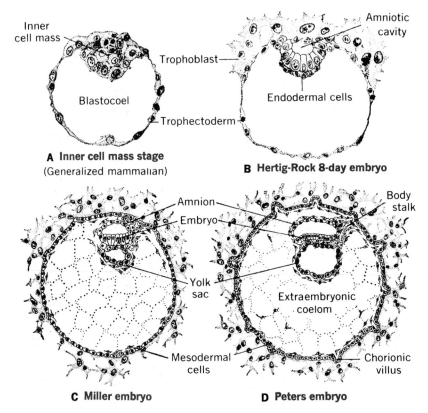

Fig. 30-5. Human embryos during the first 2 weeks after fertilization. **A,** Generalized embryo just prior to implantation in the uterus. **B,** Hertig-Rock embryo (8 days) with differentiation within the inner cell mass; that part in contact with the uterine mucosa forms specialized, flame-shaped, trophoblastic extensions. **C,** Miller embryo (12 days), with the amnion and yolk sac recognizable. **D,** Peters embryo (13 to 14 days) showing the extraembryonic coelom lined with mesodermal cells. These embryos are named after those who described them originally.

(From Patten, B. M.: Human embryology, New York, McGraw-Hill Book Company.)

are known as the three primary germ layers because from them arise all the tissues and organs of the future organism.

The detailed description of the embryologic origin of each tissue and organ cannot be given here, but a few typical examples will be sufficient. As the embryo develops, the upper part of the yolk sac forms the tubular primitive mid- and hindgut, with its outgrowth, the liver, when the embryo is about 3 weeks old. The saclike "heart" begins to beat soon after this time. By the third week the yolk sac has numerous "blood islands" for developing the embryonic vitelline circulation. The embryonic disk infolds (invaginates) to form a troughlike groove (neural groove), the open upper side of which later

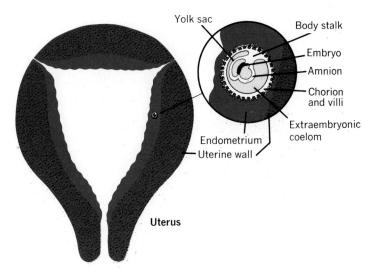

Fig. 30-6. Human embryo shortly after implantation.

Table 30-1 Early embryology and germ layer derivatives

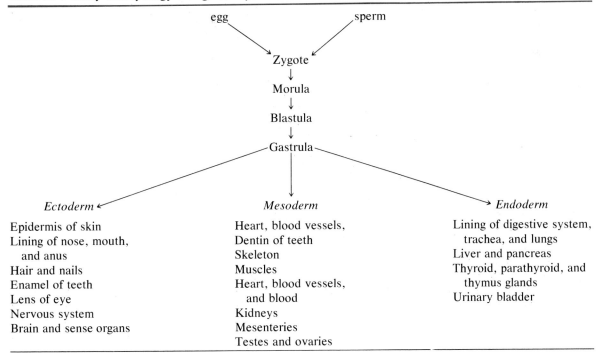

Ectoderm	Mesoderm	Endoderm
Epidermis of skin	Heart, blood vessels,	Lining of digestive system,
Lining of nose, mouth,	Dentin of teeth	trachea, and lungs
and anus	Skeleton	Liver and pancreas
Hair and nails	Muscles	Thyroid, parathyroid, and
Enamel of teeth	Heart, blood vessels,	thymus glands
Lens of eye	and blood	Urinary bladder
Nervous system	Kidneys	
Brain and sense organs	Mesenteries	
	Testes and ovaries	

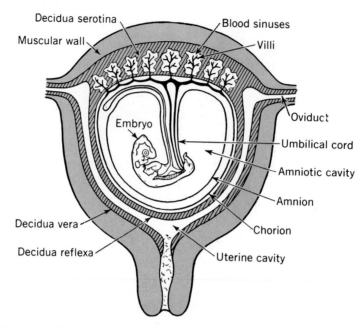

Fig. 30-7. Uterus at about the seventh week of pregnancy (diagrammatic). *(From Schottelius, B. A., and Schottelius, D. D.: Textbook of physiology, ed. 16, St. Louis, 1969, The C. V. Mosby Co.)*

closes to form a hollow tube (neural tube). From the anterior end of the tube will develop the various parts of the brain and cranial nerves, whereas the remainder forms the spinal cord with its spinal nerves.

By the end of the first month after fertilization, all major systems of the embryo have begun to form. None of them is completely functional. The whole embryo is about ¼ inch in diameter, and the woman may not be aware that she is pregnant.

Before 5 weeks the embryo shows a head with rudimentary eyes, an external tail, and a neck with four pairs of gill arches and four pairs of incompletely formed slits, which somewhat resemble the gill-bearing arches of a fish. There are numerous blood vessels here but no true gills. From the gill arches arise muscles used in chewing food, the middle ear bones, the hyoid bone (at the base of the tongue), certain facial nerves and

muscles, and part of the cartilage of the larynx and its muscles. From the slits between the arches arise such structures as the eustachian tubes, the external ear passage, part of the tonsils, the thymus, and the parathyroids.

By the end of the second month, the embryo has developed its basic form and is producing bone cells to replace the softer embryonic cartilage (Fig. 30-7). It is now referred to as a fetus.

Fetal development continues throughout the next 7 months until the actual birth of the child (Fig. 30-8). All of the tissues and organs of the fetus have resulted from specialization of the original primary germ layers.

From the ectoderm come (1) the epidermis and its derivatives, such as hair, nails, glands, and lenses of the eyes, (2) nervous tissues including the neuroglia, (3) the epithelium of the organs of special sense, of the mouth and its oral glands, of the hypophysis, of the anus, and the amnion, (4)

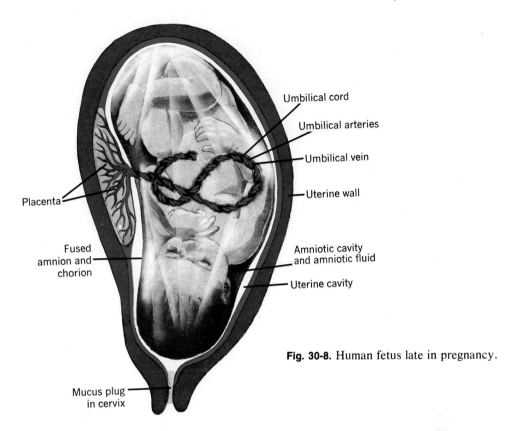

Umbilical cord

Umbilical arteries

Umbilical vein

Uterine wall

Amniotic cavity
and amniotic fluid

Uterine cavity

Placenta

Fused
amnion and
chorion

Mucus plug
in cervix

Fig. 30-8. Human fetus late in pregnancy.

the chorion, and (5) the smooth muscles of the iris (eye), and the sweat glands.

From the mesoderm come (1) the epithelial lining of the pericardium, pleura, peritoneum, and urogenital system, (2) striated muscles, (3) smooth muscles, (4) the notochord, (5) connective tissues, including cartilage and bone, (6) bone marrow, (7) blood, (8) the lining (endothelium) of the blood vessels and lymph system, (9) lymphoid organs, and (10) the cortex of the suprarenal gland.

From the endoderm come (1) the epithelium of the pharynx and its derivatives, the thyroid, parathyroids, thymus, tonsils, and auditory tube, (2) the digestive tract, including the liver and pancreas, (3) the respiratory tract, including the lungs, trachea, and larynx, (4) the bladder, (5) the prostate, (6) the urethra, (7) the yolk sac, and (8) the allantois.

Review questions and topics

1 What is meant by the term "puberty"?
2 What initiates the onset of puberty?
3 Discuss the hormones controlling puberty and the structures they influence.
4 List the parts of the male reproductive system.
5 List the parts of the female reproductive system.
6 Review the process of egg formation. See Chapter 7.
7 Review the process of sperm formation. See Chapter 7.
8 Trace the path of a sperm cell from its place of formation to its deposition in the female. Indicate all structure through which it passes.
9 Trace the path of an egg cell from its initial formation to fertilization and implantation in the uterus.
10 Discuss the hormone control of ovulation.
11 Discuss the hormone control of menstruation.
12 Distinguish between a graafian follicle and a corpus luteum.
13 What is the "LH surge"?
14 Discuss the events in the uterus during a complete menstrual cycle.
15 Discuss fertilization and early development of the fertilized egg.
16 Review the early stages of cleavage.
17 Define: ovum, zygote, blastocyst, embryo.
18 How is the umbilical cord formed?
19 Discuss the origin of the primary germ layers of the embryo. What tissues are derived from these layers?
20 When does an embryo become a fetus?

Selected references

Balinsky, B. I.: An introduction to embryology, ed. 4, Philadelphia, 1975, W. B. Saunders Co.

Beach, F. A., editor: Sex and behaviour, New York, 1965, John Wiley & Sons, Inc.

Berrill, N. J., and Karp, G.: Development, New York, 1976, McGraw-Hill Book Co.

Demarest, R. J., and Sciarra, J. J.: Conception, birth and contraception: a visual presentation, New York, 1969, Blakiston.

Jacobs, S. W., and Francone, C. A.: Structure and function in man, Philadelphia, 1974, W. B. Saunders Co.

LeBaron, R.: Hormones, New York, 1972, Pegasus.

Masters, W. H., and Johnson, V. E.: Human sexual response, Boston, 1966, Little, Brown and Co.

Rugh, R., and Shettles, L.: From conception to birth: the drama of life's beginnings, New York, 1971, Harper and Row, Publishers.

Turner, C. D., and Bagnara, J. T.: General endocrinology, ed. 5, Philadelphia, 1971, W. B. Saunders Co.

Volpe, E. P.: Human heredity and birth defects, New York, 1971, Pegasus.

PART SIX

Heredity and evolution

31 Genetics

For thousands of years, mankind has been aware of the transmission of physical characteristics from parent to offspring. It has been only in the last one hundred years, however, that we have had any explanation of how this transmission works, and only in the last decade or so have we really begun to understand it.

Today the names of Mendel, Watson, and Crick are at least familiar to most people, and terms such as DNA and genetic code are not completely unknown. Conferences on the genetic effects of radiation or the possible control of heredity in man are not at all unusual.

The present chapter will describe the work of Mendel and the visible results of heredity, while the following chapter will deal with DNA, genes, and the genetic code.

GENERAL CONSIDERATIONS

Genetics (Gr. *genesis,* origin) is the study of how hereditary materials are transmitted through successive generations, and further, how the resultant traits have been influenced not only by their genetic inheritance, but also by internal and external environmental conditions.

Studies related to genetics have been conducted in many ways. Among the methods used are (1) the experimental crossing method, in which organisms of known genetic composition are mated and the offspring observed, (2) the pedigree method, whereby individuals in the same family line are studied and similarities and differences are analyzed, (3) the cytogenetic method, involving the study of cells, nuclei, and chromosomes, and (4) biochemical genetics, in which the relation of DNA and RNA to cell metabolism is considered. In some studies, the cell is subjected to various physical forces such as atomic radiation, x-rays, ultraviolet light, and various chemicals, and the results are then interpreted.

MENDEL'S LAWS

Gregor Mendel, an Austrian monk, gave the first scientific interpretation of the heredity mechanism in 1865 after 8 years of experimental work with garden peas. These interpretations led to the formulation of his famous laws and laid the foundation for future studies in genetics. Although the laws do not explain all types of inheritance, wherever they apply they are as valid today as when they were formulated.

There has been some speculation about why Mendel was successful in his experiments. He produced hybrids among different types of peas and studied their heredity in successive generations. His method of study differed from that of previous workers, because he singled out and studied a particular trait of a plant instead of attempting to study the inheritance of the whole individual; for example, he studied seed color, plant height, flower location, and pod color. Fortunately, he also chose for study organisms that had rather simple methods of inheritance. He counted all of the thousands of individuals that were produced and kept complete pedigree records, making sure that he knew the ancestry of each individual plant and the traits displayed by each of its ancestors and offspring. He carefully made the particular artificial pollinations desired and prevented all undesired pollinations by variables such as winds or insects. In each generation where contrasting (alternative) traits appeared, for example, where both tall and short plants appeared among the offspring from a single cross, he counted the number of each type obtained. He then interpreted the data and formulated conclusions now known as Mendel's laws (mendelism).

Mendel experimentally crossed two pea plants with alternative traits. The resulting hybrids, each resembling one parent or the other, were then crossed with each other. In the hybrids he recognized the expressed trait and referred to it as the *dominant* trait; the other that was latent and did not express itself he called the *recessive* trait. When he crossed the hybrids, both traits reappeared in the next generation in a ratio of approximately three dominants to one recessive. The result based on outward appearance is called the *phenotype* ratio. This may be resolved into a *genotype* ratio, which is based on the actual genic compositions of the various individuals. When an individual contains two genes that are alike (as TT or tt), that individual is homozygous (Gr. *homo-*, same), whereas an individual with two different members of a pair (Tt) is heterozygous (Gr. *heteros*, different).

The laws that Mendel formulated may be summarized as the law of unit characters, the law of dominance, the law of segregation, and the law of independent assortment.

Law of unit characters

Genes* occur in pairs and control the inheritance of traits as a unit.

Law of dominance

One gene of a pair may mask or inhibit the expression of the opposite member of that pair. In an individual pea plant that has one gene for tallness (T) and one for shortness (t), the tall one will dominate the other and express itself. The one that expresses itself is called the dominant gene and the other the recessive gene. The trait so developed is called the dominant trait. When two recessive genes (tt) are together, they will express themselves and exhibit the recessive trait.

Law of segregation

The genes that make up the different pairs are segregated (separated) from each other when gametes are formed in animals, or spores are formed in plants. Only one of each pair of genes goes into a sex cell or into a spore of plants.

---

*Mendel used the word "factor." The term "gene" was not used until 1909.

Parents (P)   Tall (TT) × Dwarf (tt)

Gametes produced   T T   t t

Offspring (F₁)   Tall (Tt)

Gametes produced by both male and female   T t
    individuals of the F₁

Offspring (F₂) when various individuals are crossed and shown
        by so-called checkerboard (Punnett square)

Male gametes

		T	t
Female gametes	T	TT	Tt
	t	Tt	tt

Fig. 31-1. Monohybrid cross in peas.

Law of independent assortment

The genes representing two or more contrasting pairs of traits are distributed independently of one another at the time of gamete formation in animals and at the time of spore formation in plants. The gametes unite at random.

HYBRID STUDIES

When two organisms, homozygous for opposite members of a trait, are crossed, the offspring are heterozygous for that character. These offspring are known as hybrids.

Monohybrid crosses are those in which only one pair of traits is studied, while dihybrid and trihybrid are those in which two or three pairs are considered simultaneously.

By using the pea plant and the following traits with their symbols, the different crosses (monohybrid, dihybrid, and trihybrid) may be illustrated.

T, tall plant
t, short (dwarf) plant
R, round seed
r, wrinkled seed
Y, yellow seed color
y, green seed color

When the dominant (shown by the capital letter) is present, it expresses itself even though the opposite or recessive is present. Two recessive genes must be present together in order to express themselves.

In a *monohybrid cross* we assume the parents to be homozygous dominant and homozygous recessive, for example, tall and dwarf, TT × tt. Each homozygous parent produces only one gamete type, T or t. The offspring F_1 (first filial generation) will all be heterozygous, Tt, having received one gene from each parent. Phenotypically these are all tall.

Parents (P):   Tall (TT) × Dwarf (tt)

Gametes produced: T   T   t   t

Offspring: Tall (Tt)

A cross between offspring results in an F_2 (second filial generation) composed of both tall and dwarf plants. This is because each F_1 parent produces two gamete types, that is, both produce T and t. Since the gametes unite at random, possible F_2 genotypes are TT, Tt, Tt, and tt, hence the phenotypes ratio of three dominant to one recessive (Fig. 31-1).

The events of a **dihybrid** cross are essentially the same as those in a monohybrid cross. We

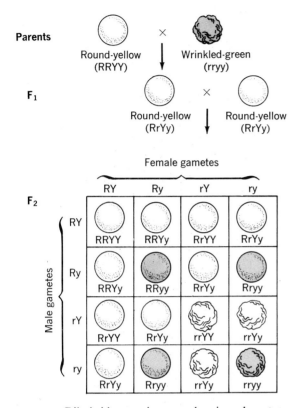

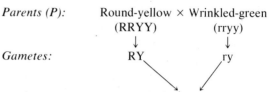

assume that there are homozygous dominant and homozygous recessive parents, for example, round-yellow seeds × wrinkled-green seeds, that is, RRYY × rryy. Each parent produces one gamete type RY or ry. The F_1 are therefore all heterozygous RrYy.

Parents (P):   Round-yellow × Wrinkled-green
              (RRYY)   (rryy)
                 ↓   ↓
Gametes: RY   ry

Offspring (F_1):   Round-yellow (RrYy)

If these are crossed, each parent produces four gamete types: RY, Ry, rY, and ry. Gametes again unite at random with sixteen possible F_2 phenotypes. These are: 9 round-yellow, 3 round-green, 3 wrinkled-yellow, and 1 wrinkled-green (Fig. 31-2). Thus the typical F_2 ratio of a dihybrid cross is 9-3-3-1.

When crossed, F_1 parents each produce 8 gamete types, which uniting at random, produce 64 possible F_2 phenotypes in a ratio of 27-9-9-9-3-3-3-1. It is important to understand that the phenotypes represent different genotypes. See Fig. 31-3 for details.

The same principles may be shown in any other organism in which dominant traits appear. The guinea pig is used as an example in Figs. 31-4 to 31-6. In this animal black is dominant over white,

Fig. 31-2. Dihybrid cross in peas showing phenotype color and shape and genotype-gene makeup of each individual. F_2 in checkerboard shows possible outcomes when a male gamete at left is crossed with a female gamete at top.

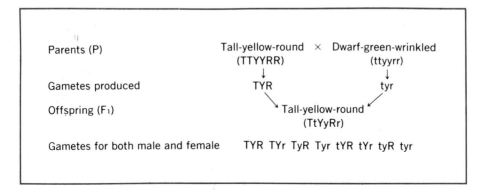

Fig. 31-3. Trihybrid cross in peas.

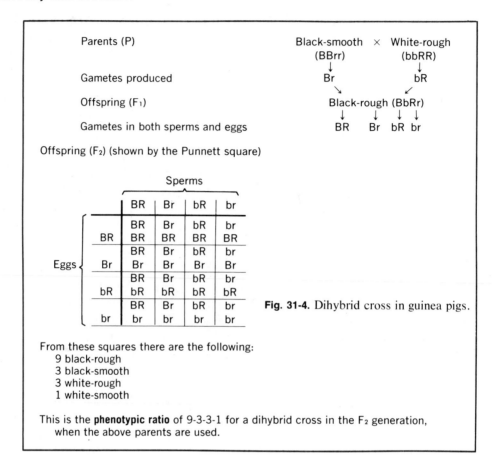

Parents (P)   Black-smooth × White-rough
                                                    (BBrr)   (bbRR)
                                                        ↓   ↓
Gametes produced   Br   bR
                                                          ↘   ↙
Offspring (F₁)   Black-rough (BbRr)
                                                      ↓   ↓   ↓   ↓
Gametes in both sperms and eggs   BR   Br   bR   br

Offspring (F₂) (shown by the Punnett square)

Sperms

	BR	Br	bR	br
BR	BR BR	Br BR	bR BR	br BR
Br	BR Br	Br Br	bR Br	br Br
bR	BR bR	Br bR	bR bR	br bR
br	BR br	Br br	bR br	br br

Eggs {

Fig. 31-4. Dihybrid cross in guinea pigs.

From these squares there are the following:
9 black-rough
3 black-smooth
3 white-rough
1 white-smooth

This is the **phenotypic ratio** of 9-3-3-1 for a dihybrid cross in the F₂ generation, when the above parents are used.

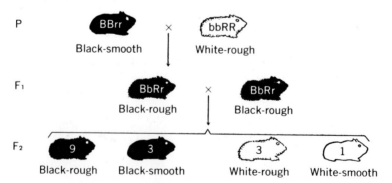

P   BBrr   ×   bbRR

Black-smooth   White-rough

F₁   BbRr   ×   BbRr

Black-rough   Black-rough

F₂   9   3   3   1

Black-rough   Black-smooth   White-rough   White-smooth

FIG. 31-5. Dihybrid cross in guinea pigs when the parents, P, are crossed. All of the F₁ generation is black-rough; when the F₁ generation is intercrossed, the F₂ generation is produced as shown. Genes for the P and F₁ generations are shown.

rough over smooth, and short over long. This illustrates the occurrence of several pairs of traits influencing the same structure.

Incomplete dominance

From studies made so far it has been noted that when opposite members of a pair of genes are present, one or the other is completely dominant. In incomplete dominance (partial dominance, or absence of dominance) the F_1 does not resemble either parent exactly for the trait in question, since neither gene of the pair is completely dominant. An example of this is the four-o'clock flower (Fig. 31-7), in which a homozygous white (rr) is crossed with a homozygous red (RR). The F_1 is neither red nor white, but is pink (Rr). When two F_1 individuals are crossed, there are produced in the F_2: one white (rr), two pinks (Rr); and one red (RR), with a ratio of 1-2-1. In this case the heterozygous (Rr) individuals can be detected visually, which is not possible in cases of complete dominance.

This phenomenon is also illustrated by the blue Andalusian fowl (Fig. 31-8). When a black and a white-splashed individual are crossed, the F_1 shows an intermediate shade of blue Andalusian that is heterozygous. When two blue Andalusian fowls are crossed, the ratio of offspring is: one fourth white-splashed, one half blue Andalusian, and one fourth black, that is, 1-2-1.

Lethal genes

Lethal genes cause developmental or metabolic reactions that result in the death of an organism possessing them. They may affect the organism at any time during life, from the time of fertilization to maturity. The term lethal is usually applied to the killing in early life, whereas sublethal is reserved for describing those conditions that cause death before the reproductive age. Some lethals are dominant; others are recessive. Two kinds of recessive lethals are known. Those that are often called "dominant lethals" are actually recessive genes that have

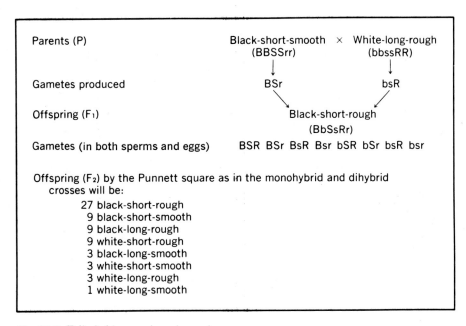

Fig. 31-6. Trihybrid cross in guinea pigs.

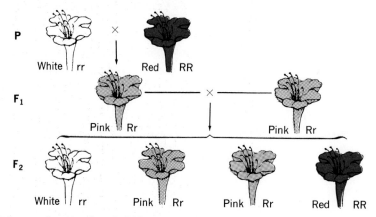

Fig. 31-7. Incomplete dominance in four-o'clock flower *(Mirabilis)*. Homozygous flowers are white or red. Heterozygous flowers are pink.

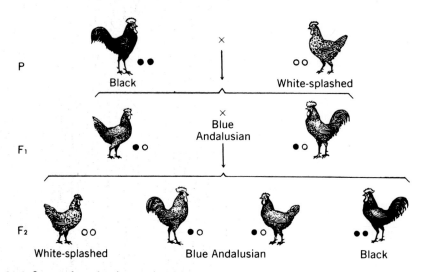

Fig. 31-8. Incomplete dominance in blue Andalusian fowls. When a black fowl is crossed with a white-splashed-with-blue, all the F_1 generation will be blue Andalusian. When the latter are interbred, there are produced one fourth white-splashed-with-blue, one half blue Andalusians, and one fourth black in the F_2 generation. When white-splashed-with-blue of F_2 are interbred, only white-splashed-with-blue are produced. When blue Andalusians of F_2 are interbred, they produce offspring like those resulting from F_1. When black of F_2 are crossed with each other, only blacks are produced. Black dots and circles show the genes involved in each individual.

Table 31-1 Genetic phenotypes of plants

Plant	Dominant	Recessive
Peas		
Plant height	Tall	Short (dwarf)
Pod color	Green	Yellow
Seed color	Yellow	Green
Seed shape	Round (smooth)	Wrinkled
Flower color	Colored	White
Four-o'clock plant		
Flower color	Colored (incomplete)	White
Snapdragon plant		
Flower color	Colored (incomplete)	White
Tomato		
Vine	Tall	Short (dwarf)
Fruit color	Red	Yellow
Summer squash		
Fruit color	White	Yellow
Fruit shape	Disk shaped	Sphere shaped
Corn		
Seed color	Colored	Colorless
Seed shape	Full	Shrunken (wrinkled)
Endosperm (stored food)	Starchy	Sugary (sweet)

Table 31-2 Genetic phenotypes of animals

Animal	Dominant	Recessive
Fruit fly		
Eye color	Red	White
Body color	Gray	Ebony (black)
Wing length	Long	Vestigial (short)
Poultry		
Comb	Rose (as in Wyandottes)	Single (as in Leghorns)
Shank (thigh)	Feathered	Bare
Number of toes	Extra toes	Normal number
Guinea pigs		
Hair length	Short	Long
Coat	Rough	Smooth
Hair color	Colored	White
Cattle		
Leg length	Short	Long
Horns	Hornless (polled)	Horns present
Horses		
Hair color	Gray	Other colors
Running form	Trotting	Pacing

dominant, visible effects in heterozygous individuals of the phenotype. Other recessive lethals have no visible effect in the heterozygous condition. In the truly dominant lethals, death may occur even in the heterozygous condition.

In plants lethal genes may prevent the development of chlorophyll, so that photosynthesis cannot occur. Certain lethal genes in corn produce albinism (lack of cholorphyll), so that the young plant dies after the stored food of the seed is exhausted. Many mutations (changes in a gene that are inheritable) that occur naturally, or are artificially induced by radiations, such as x-rays and radium, are lethal. Lethals may express themselves merely through the nonappearance of certain types of offspring.

A certain type of snapdragon has golden leaves rather than the normal green. When such plants are self-pollinated (self-fertilized), they may give a ratio of offspring of two golden to one green. This ratio is understood when the early seedlings are examined, for there may be three phenotypes with a ratio of one yellow, two golden, and one green. The genes causing the abnormality apparently do not prevent sufficient nutrition when heterozygous (as in golden), but act as lethals (yellow) when homozygous, shown as follows:

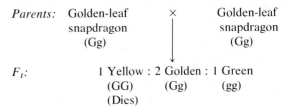

Parents: Golden-leaf   ×   Golden-leaf
          snapdragon   snapdragon
            (Gg)   (Gg)

F_1:   1 Yellow : 2 Golden : 1 Green
              (GG)   (Gg)   (gg)
              (Dies)

The inheritance of a lethal, sex-linked gene in the fruit fly *(Drosophila)* and its effects on the sex ratio are shown in Fig. 31-9. This lethal gene is a recessive and is present on the X chromosome

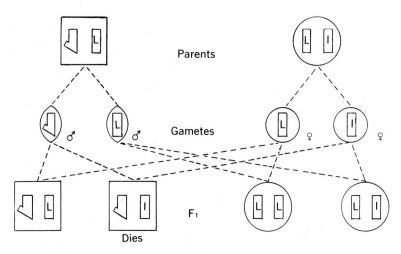

Fig. 31-9. Inheritance of the lethal, sex-linked gene in the fruit fly *(Drosophila)* and its effect on the sex ratio. **I**, Lethal gene; **L**, normal, nonlethal allele. Males are represented by squares, females by circles. In F_1 (first filial generation) one half of the males fail to develop, giving a sex ratio of two females to one male: Note that both parents are normal but that the female carries the lethal gene, which she later contributes to one half of her sons, who consequently never develop.

(sex chromosome). Note that both parents appear normal, but the female carries the lethal gene (one), which she contributes to one half of her male offspring, which consequently never develop. In the F_1 one half of the males fail to develop, giving the adult sex ratio of two females to one male.

In mice yellow (Y) dominates gray (y) color. If crosses are made as follows, the lethal results are shown:

Parents: Yellow mouse × Yellow mouse
  (Yy) (Yy)

F₁: One-fourth : One-half : One-fourth
  yellow yellow gray
  (YY) (Yy) (yy)
  (Die before birth)

To explain the early deaths of one fourth of the mice it is surmised that when the lethal gene (Y), is present in duplicate (homozygous), it will cause death, which is not true in the heterozygous (Yy) condition. When matings as described are made, litters are about three fourths normal size, but the uteri of the pregnant mothers show normal numbers of embryos. About one fourth of these embryos die before birth.

Included in the human lethals and sublethals are the following. (1) Amaurotic idiocy (progressive mental deterioration and blindness with death usually within 5 years) is caused by two recessive genes acting together but is not fatal in the heterozygous condition. Parents of an affected child may have good health and possess superior intelligence. There is no known treatment for heterozygous persons. (2) A rare condition of shortened fingers, called brachydactyly (brak i -dak′ ti ly) (Gr. *braxys,* short; *daktylos,* digit), is caused by a dominant lethal gene. Homozygous offspring have grossly abnormal skeletal defects and die before birth. (3) Cancer of the eye retina in early childhood, called retinoblastoma (Gr. *blastos,* young; *oma,* tumor), is caused by a dominant lethal. Early surgical removal of the affected eye before cancer spreads may be successful. (4) The human sublethal hemophilia (Gr. *haima*, blood; *philos*, loving) is a sex-linked condition in which the blood of males clots so slowly that death may result from minor hemorrhages. (This condition is discussed under sex-linked traits.) There seem to be other types of hemophilic diseases caused by other genes, and the methods of inheritance are different. (5) The early death of a certain percentage of babies may be caused by the action of certain Rh blood types and is caused by multiple genes. This condition is called erythroblastosis fetalis and affects about 1 in 200 or 500 pregnancies and may be fatal. Some human abortions occur because of the expulsion of a dead fetus resulting from lethal homozygous recessive genes.

Multiple genes and interaction of genes

Many traits in plants and animals are determined by the interaction of multiple genes (more than one pair). The specific methods of inheritance vary, but the following examples may give a general idea of such genetic phenomena. When a so-called quantitative trait (one with various degrees of expression) is the result of the interaction of more than one pair of genes, the latter are known as multiple genes.

The Swedish geneticist, Nilsson-Ehle, found three pairs of interacting genes in certain types of red wheat. These possessed incomplete dominance, and various degrees of trait expression were produced because of the cumulative effects of the genes in question. Deepest red wheat is represented by R_1R_1 R_2R_2 R_3R_3, and white wheat by r_1r_1 r_2r_2 r_3r_3. The more genes represented by the capital letters in any individual plant, the darker the red. This is shown as follows:

Parents: Deepest red wheat × White wheat
$(R_1R_1$ R_2R_2 $R_3R_3)$ | $(r_1r_1$ r_2r_2 $r_3r_3)$

F₁: Medium red wheat
$(R_1r_1$ R_2r_2 $R_3r_3)$

F₂: When two of the F_1 are crossed, the F_2 ratio is:

1 deepest red (6 red genes)
6 very deep red (5 red genes)
15 deep red (4 red genes)
20 medium red (3 red genes)
15 pale red (2 red genes)
6 very pale red (1 red gene)
1 white (no red genes, but 6 white genes)

Another, but somewhat different, example of multiple gene inheritance is illustrated by the flower color is sweet peas (Fig. 31-10). When two dominant genes located in different pairs of chromosomes interact, they may complement each other (both present and produce a visible effect). When two strains of white-flowered sweet peas with the genic composition shown here are crossed, the F_1 will be purple flowers. Purple flowers are represented by at least one C gene and at least one P gene. White flowers are represented by at least one C and pp, or by cc and at least one P or by ccpp. This may be shown as follows:

Parents: White-flowered × White-flowered
sweet pea | sweet pea
(CCpp) | (ccPP)

F₁: Purple-flowered sweet pea
(CcPp)

F₂: When two of the F_1 are crossed, the F_2 ratio is:

9 purple (at least one C and at least one P)
3 white (at least one C and pp)
3 white (cc and at least one P)
1 white (cc and pp)

This phenotype ratio is 9 purple-7 white.

Another example of multiple genes (two pairs) is shown in the inheritance of the combs of chickens. In this case pea comb is represented by at least one P and rr, rose comb by pp and at least one R, walnut comb by at least one P and at least one R, and single comb by pprr. The following crosses show a homozygous rose comb crossed with a homozygous pea comb with the genic contents and ratios:

Parents: Rose comb × Pea comb
              (ppRR)   (PPrr)
Gametes: pR   Pr

F₁: Walnut comb
                    (PpRr)
F₂: When two walnut-combed individuals
       of the F₂ are crossed, the F₂ shows:
          9 walnut comb (at least one P and at
            least one R)
          3 pea comb (at least one P and rr)
          3 rose comb (pp and at least one R)
          1 single comb (pprr)

This shows that not only is a new comb trait produced in the F₁, but when intercrossed a still different type, namely single comb (pprr), is produced.

There are many human genetic traits, each of which results from the interaction of more than one pair of genes. Many of our so-called quantitative traits that have variable degrees of expression are in this category. An example is the production of skin color in which Negroes differ from whites in two pairs of genes, which interact cumulatively and show incomplete dominance. This is explained by: Negro, AABB; dark mulatto, AABb or AaBB; medium mulatto, AaBb, AAbb, or aaBB; light mulatto, Aabb or aaBb; white, aabb. Using these symbols, if a pure Negro (AABB) and a pure white (aabb) are crossed, the F₁ offspring are medium mulatto (AaBb). Another crossing may be shown in the accompanying diagram.

Parents: Medium mulatto × Medium mulatto
               (male)   (female)
               (AAbb)   (AaBb)
Gametes: Ab   AB Ab aB ab

F₁: 1 dark mulatto (AABb); 2 medium mulat-
       toes (AAbb) (AaBb), and 1 light mulatto
       (Aabb). This ratio is based on the one kind
       of male sperm and four different types of
       eggs in the female.

The actual process of color determination is much more complex and may involve more gene pairs.

Multiple alleles

Another multiple method of inheriting a human trait may be illustrated by the blood groups A, B, AB, and O, in which three alternative genes (triple alleles) are responsible. It is known that when blood from certain individuals is mixed this results in an agglutination (clumping) of the red blood corpuscles. The four types of human blood are known as A, B, AB, and O. Agglutination of human blood depends upon the presence of (1) the specific substance known as the agglutinogen (a type of antigen) in the red blood corpuscles and (2) the specific substance known as the agglutinin (a type of antibody) in the blood plasma. Both are necessary for blood to agglutinate, so naturally both cannot occur in the same person or his blood would agglutinate in his blood vessels. The two inheritable agglutinogens are known as A and B in man. Hence, the following human blood groups are possible: an individual of group A has agglutinogen A in his red blood corpuscles, group B has agglutinogen B, group Ab has both agglutinogens A and B, and group O has neither agglutinogen. Whichever agglutinogen an individual has in his blood corpuscles, the corresponding agglutinin is absent in his blood plasma. When an agglutinogen is absent in his erythrocytes, the corresponding agglutinin is present in his plasma.

Of the triple alleles, each person has only one pair of these genes, which consists of any one of the six possible combinations as follows:

Blood group	Genes and genotype
A	$I^A I^A$, or $I^A i$
B	$I^B I^B$, or $I^B i$
AB	$I^A I^B$
O	ii

In the above both I^A and I^B are dominant over gene i, and group O, being a recessive, must breed true. Group AB is a hybrid, whereas groups A and B may either breed true or be hybrids.

In the inheritance of the preceding groups gene I^A produces agglutinogen A; gene I^B produces agglutinogen B; and gene i produces no ag-

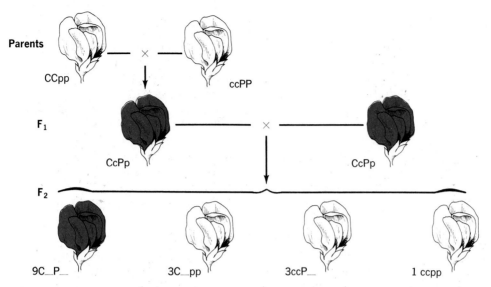

Fig. 31-10. Multiple gene inheritance in sweet pea flowers. Purple flowers show up only when at least one C and one P genes are present. White flowers appear whenever C or P is lacking. Note that in the F_2, only purple flowers have both a C and a P.

glutinogen. When both I^A and I^B are present, an individual of the AB group results. Blood types are inherited specifically and do not change from the embryo to old age, hence blood tests may be used in certain cases of disputed parentage. Such tests cannot prove that a certain man is the father of a particular child but whether he could or could not have been. In other instances a certain child with a specific blood group could not have been conceived by certain individuals with their specific blood group.

Additional agglutinogens, called M and N, are known in human erythrocytes, and these are inherited independently of other blood groups. They may be useful in identifying bloods but usually need not be considered in blood transfusions.

LINKAGE AND CROSSING-OVER

Linked traits are developed from a linear series of genes that are associated (linked) together on the same chromosomes. Each gene occupies a particular position (locus) on a specific chromo-

some, and linkage tends to maintain this sequence. Linked genes have a tendency to be passed on to the next generation. However, linkage does not prevent segments of chromosomes, with their genes, from separating and crossing over (Fig. 31-11) during synapsis (temporary union of pairs of homologous chromatids at the time of meiosis). Crossing-over might be defined as the mutual exchange of blocks of homologous genes located on two members of a pair of chromatids.

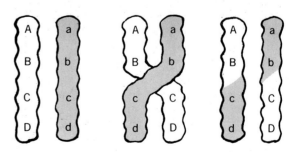

Fig. 31-11. Crossing-over of genes from one homologous chromatid to another during synapsis (diagrammatic).

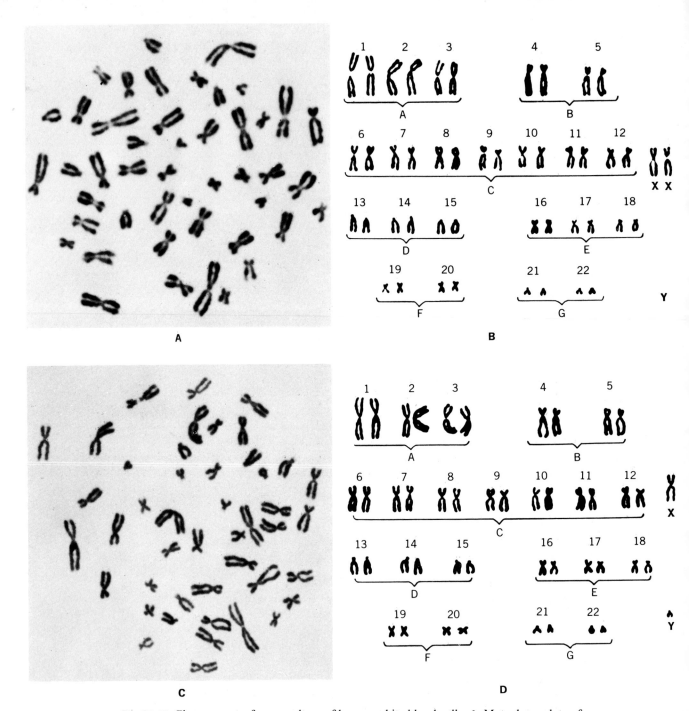

Fig. 31-12. Chromosomes from a culture of human white blood cells. **A,** Metaphase plate of a normal female; **B,** Karyotype (paired chromosomes) made by photographing the plate at **A,** cutting out individual chromosomes, and arranging them in pairs according to the Denver system; **C,** metaphase plate of a normal male; **D,** Karyotype made from **C.**
(Courtesy Willard Yarema and Cytogenetic Laboratory of St. Elizabeth Hospital, Dayton, Ohio.)

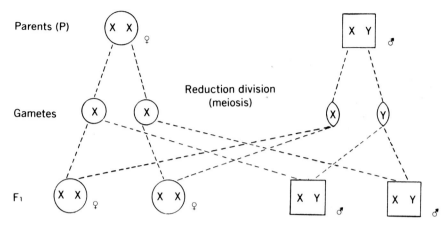

Fig. 31-13. Inheritance of sex. In addition to a specific number of autosomes the cells of the female (♀) have two X chromosomes (sex chromosomes), whereas the cells of the male (♂) have an X chromosome and a Y chromosome.

Under uniform environmental conditions the linked genes in a uniform stock of organisms cross over with a rather definite frequency. The exchanged segments consist of homologous pieces of chromatids belonging to homologous chromosomes, which result in two types of visible cross-overs (recombinations) that occur in equal numbers, and also two types of non-crossovers that occur in equal numbers.

If all genes in all chromosomes were permanently linked, and no crossing-over were possible, variations in organisms would be limited. Crossing-over has the effect of increasing variations. Linkage tends to restrain a complete disorganization of the natural loci of genes, yet crossing-over permits recombinations, thereby permitting a greater variety of total traits with which the organism can attempt to adapt itself to its living conditions.

Linkage and crossing-over were first studied in England by Bateson and Punnett by experimenting wih sweet peas. Thomas H. Morgan, in the United States, developed these two phenomena in 1910 by experimenting with fruit flies (Drososphila).

The frequency of crossing-over between two groups of genes depends for the most part upon the distance between the genes on the chromosome. In general, the farther apart two genes lie, the greater the percentage of crossing-over, whereas two genes that lie nearer each other have less opportunity for crossing-over. A crossover in a particular region tends to inhibit the occurrence of another crossover nearby through a process called interference.

INHERITANCE OF SEX AND THE SEX RATIO

Sex is an inheritable trait, as are many other characteristics of human beings. Each normal human body cell contains twenty-two pairs of autosomes and a pair of sex chromosomes. In females the sex chromosomes are two similar X chromosomes, whereas in males the sex chromosomes are an X chromosome and a dissimilar Y chromosome (Fig. 31-12). When human sex cells are formed, through meiosis each sex cell (whether male or female) receives the single, haploid number of chromosomes. Thus, by union of two sex cells, the resulting zygote again has the

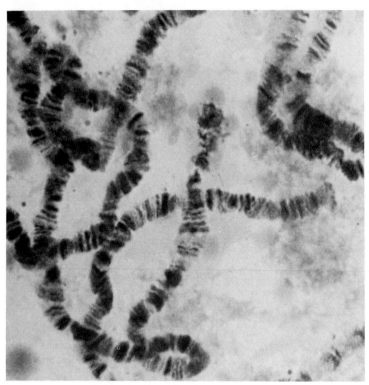

Fig. 31-14. Giant chromosomes from the salivary gland of the fruit fly *(Drosophila melanogaster).*
(Courtesy General Biological Supply House, Inc., Chicago, Illinois.)

diploid number. The sex of the offspring is determined by which of the two types of sperms (X or Y type) unites with the egg (Fig. 31-13).

From cytologic studies of fruit flies we find that each somatic (body) cell contains three pairs of autosomes and one pair of sex chromosomes (Fig. 31-14). The female has a pair of similar sex chromosomes called X chromosomes, whereas the male has one X chromosome and one Y chromosome. When female sex cells are produced, each one contains three autosomes and one X chromosome. When male gametes are produced, one type contains three autosomes and one X chromosome, whereas the other type contains three autosomes and one Y chromosome. Since the two types of male gametes are produced in equal numbers and if each type has an equal chance of fertilizing an egg, the ratio of male to female (sex ratio) should be approximately 50-50. *Drosophila melanogas-*

ter, the common fruit fly, has the XX-XY type of sex chromosome–sex determining mechanism in which the male is the heterogametic sex with XY. It is known that in this species sex determination is a result of a balance of female determiners on the X chromosome and of male determiners on the autosomes; the Y chromosome is inert, although necessary for male fertility.

Sex is determined at the time of fertilization. In general, the methods of gamete production, fertilization, and sex determination are somewhat alike in fruit fly and man.

Mammals, with the exception of a few species, have an XX-XY type of sex mechanism in which the male is the heterogametic sex with XY. In 1959 it was shown that mammalian sex determination is not quite like the *Drosophila* type in certain respects, but the Y chromosome, instead of being inert, is strongly male determining. The male-determining ability of the Y chromosome in the mouse was demonstrated by combined genetic and cytologic evidence. At about the same time there were parallel findings in the human species.

The fact that the Y chromosome in mammals is strongly male determining suggests that it is not genetically empty, yet no other genes are definitely known to be located on the Y chromosome, although some are suspected.

Fig. 31-15 shows a nucleus of a cell taken from the lining of the mouth of a normal human female. The dark dot at the right side of the nucleus is called a Barr body, after its discoverer. It represents the second X chromosome in the female nucleus. The sex of developing embryos may be determined as early as twenty-one days after conception by culturing embryonic fluid and examining the cells for these Barr bodies.

Extra sex chromosomes

An error of meiosis, called nondisjunction, results in unequal numbers of chromosomes entering the gametes. With regard to sex chromosomes then, the following may result: a gamete with no sex chromosomes, a gamete with both an X and a Y chromosome, and a gamete with two X chromosomes.

If any of these abnormal gametes would unite with a normal egg or sperm, the resulting embryo could be: XO, YO, XXY, XYY, or XXX. All of these except YO have been found. (O means that the chromosome is missing.)

Two of these abnormal situations have been studied extensively. The first of these, the XO condition, is called Turner's syndrome and produces females who are sterile and may show other abnormalities. It is interesting to note that the same syndrome exists in mice but that the female mouse remains fertile. The second case occurs in males with an extra X or Y chromosome. This is called Klinefelter's syndrome and the men are sterile with some degree of femaleness. Studies of this abnormality suggest that there is a connection between the condition and aggressive antisocial behavior. In 1968 in

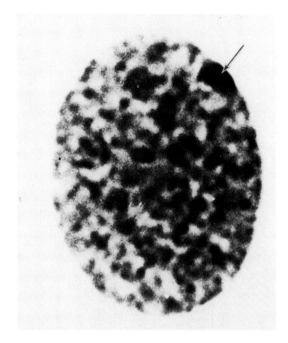

Fig. 31-15. Nucleus of epithelial cell from the mouth of a normal human female. The dark spot, called a Barr body (as indicated by the arrow), is a chromatin mass of the second X chromosome.
(From Redding, A., and Hirshhorn, K.: Guide to human chromosome defects. In Bergsma, D., editor: Birth defects, Original Article Series, vol. 4, 1968, The National Foundation.)

Australia, a man was acquitted of murder on the basis of "supermaleness" due to an XYY genetic makeup. The implications of this finding in the world's courts are interesting to ponder.

Sex-linked traits

Genes for some traits are located on the sex chromosomes. Naturally, the particular distribution of these sex-linked genes determines the appearance of such traits according to the sex of the individuals. Over fifty human sex-linked traits have been found.

One of these is a type of color blindness in which the individual is unable to distinguish red from green. The recessive gene, c, for this trait is

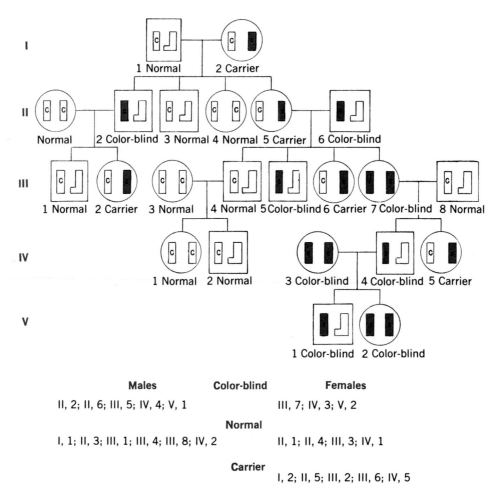

Fig. 31-16. Pedigree for human sex-linked color blindness caused by the recessive gene, c, on X chromosome; normal is C. Squares represent males; circles represent females. X chromosome is straight; Y chromosome is hooked.

carried on the X chromosome, while the Y chromosome lacks a corresponding dominant gene. This explains the different ratio of color blindness in the two sexes (Fig. 31-16).

A color-blind man (XcY) producing children with a normal woman who is not color-blind (XX) passes the gene (Xc) to his daughters. They carry the gene but have normal color vision because of their second, normal X chromosome. These fe-males produce eggs with either (X) or (Xc) chromosomes. Assuming the father of their children has normal color vision, it is possible to produce normal girls (XX), carrier girls (XXc), normal boys (XY), and color-blind boys (XcY).

If a carrier girl married a color-blind male, the possibility of a color-blind girl is present (XcXc), along with a carrier female, normal boy, and color-blind boy.

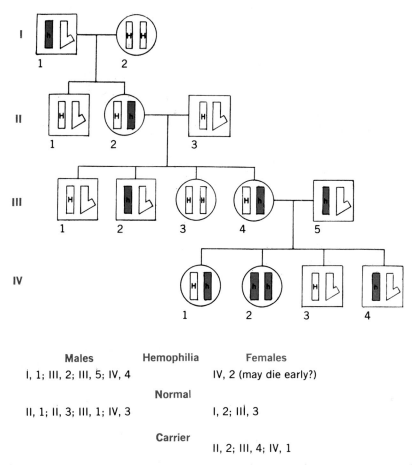

Fig. 31-17. Pedigree for human sex-linked hemophilia caused by the recessive gene, h; normal is H. Squares represent males; circles represent females. X chromosome is straight; Y chromosome is hooked.

Another human sex-linked trait is hemophilia (Fig. 31-17) caused by a recessive gene, h, linked on the X chromosome. Even minor injuries may result in excessive bleeding caused by very slow clotting of blood. The disease may be present in all social strata, although much publicity has been given to its presence in the royal Romanoff and Bourbon families in Europe. Many hemophiliacs die before maturity and many others choose not to produce children so that only a few family pedigrees are recorded. Few

cases of hemophilic girls are known, and this is usually interpreted as resulting from the homozygous condition (hh), which usually causes death of the embryo before birth. Incomplete evidence for this comes from interrupted pregnancies in such marriages, and from the fact that hemophilic men have fewer than the expected percentage of daughters.

If a hemophilic man (h) lives to produce children and marries a normal woman (HH), all of his daughters will be carriers (Hh) and all of his sons

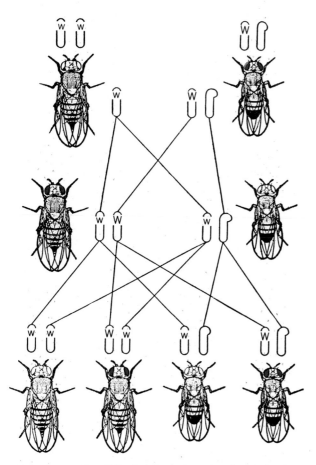

will be normal (H), because the sons inherit their only X chromosome from their mother. If the same hemophilic man (h) marries a carrier woman (Hh), one half of his sons will be normal (H) and one half will be hemophilic (h); and theoretically, one half of his daughters will be carriers (Hh) and one half will be hemophilic (hh). It will be noted that the method of inheritance of this disease is the same as for the inheritance of color blindness, since both are sex linked on the X chromosome.

Very few human traits show absolute Y linkage (do not appear in females and are not transmitted to them). A possible Y-linked trait is a special type of webbed toe (webbed skin between the second and third toes). A different type of webbed toe exists in both males and females, with a preponderance in males.

In the fruit fly the gametes are heterozygous in males and homozygous in females. White eyes are sex-linked traits that are recessive to normal red eyes. When a white-eyed female (Fig. 31-18) is crossed with a red-eyed male, the F_1 males are white eyed and the females red eyed. In the F_2 one half of the individuals of each sex are white eyed and one half are red eyed.

In the reciprocal cross in which a red-eyed female (Fig. 31-19) is crossed with a white-eyed male, all the offspring (both male and female) of the F_1 are red eyed; all the females of the F_2 are also red eyed. One half of the males of the F_2 are red eyed, and the other half are white eyed. Compare and contrast these two types of crosses to understand how sex-linked traits develop.

Fig. 31-18. Sex-linked inheritance in the fruit fly *(Drosophila* sp.). The red-eyed male (upper right) and white-eyed female (upper left) are crossed to produce a male and female of the F_1 generation. Gene W for red eyes is carried on the X chromosome of the male, whereas the curved male Y chromosome does not carry an eye-color gene. Gene w for white eyes is carried on the X chromosome of the female. When members of the F_1 generation are crossed, four kinds of the F_2 generation are produced (lower group). Note that a male has a larger tip of black on his abdomen than the female. Contrast these results with those in Fig. 31-19. *(From Morgan, T. H.: Evolution and genetics, Princeton, New Jersey, Princeton University Press.)*

Sex-influenced traits

Sex-influenced traits are inherited from genes on the autosomes (not on the sex chromosomes), but they are influenced or modified by the sex hormones of the male and female gonads and are responsible for differences in trait expressions in the two sexes, even though the genes may be the same in both.

An example is the inheritance of horns in sheep. Neither sex of Suffolk sheep bears horns,

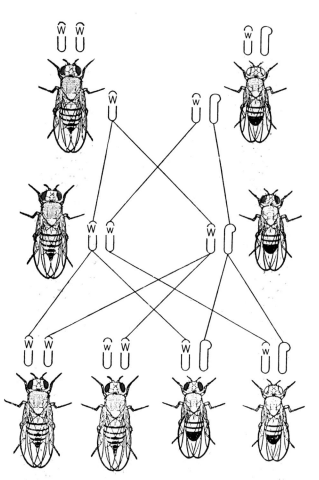

Fig. 31-19. Sex-linked inheritance in the fruit fly (*Drosophila* sp.). A white-eyed male (upper right) and a red-eyed female (upper left) are crossed to produce a male and female of the F_1 generation. Gene W for red eyes is carried on the X chromosome of the female. Gene w for white eyes is carried on the X chromosome of the male. Curved male Y chromosome does not carry an eye-color gene. When members of the F_1 generation are crossed, members of the F_2 generation are produced, as shown in the lower line: Note that the male has a larger tip of black on his abdomen than the female. Contrast these results with those in Fig. 31-18. *(From Morgan, T. H.: Evolution and genetics, Princeton, New Jersey, Princeton University Press.)*

but both sexes of Dorset sheep always bear horns. When Suffolks and Dorsets are crossed, the male offspring are horned and the females are hornless. When offspring of the F_1 are interbred, the F_2 ratio is obtained in which the males show three horned to one hornless and the females show three hornless to one horned.

Parents: Suffolk (hornless) × Dorset (horned)
           (hh)   ↓   (HH)
                (Hh)
F_1:   Hornless females and horned males
F_2: If 2 of the F_1 are crossed, the following ratios result:
    Males: 1 horned : 2 horned : 1 hornless
         (HH)   (Hh)   (hh)
  Females: 1 horned : 2 hornless : 1 hornless
         (HH)   (Hh)   (hh)

The gene (H) for horns expresses itself in heterozygous males, whereas its allele expresses itself in females.

Another example is the type of inheritable baldness in humans called pattern baldness, in which the hair recedes around the temples, then becomes thinner on the top of the head, finally leaving a fringe of hair low on the head. Analyses of pedigrees show that the genotype BB results in baldness in both sexes, whereas the genotype Bb results in baldness in men but normal hair in women. The genotype bb results in normal hair in both sexes. Many pedigrees show bald women, and this may be caused by a quantity of male sex hormone in their blood that is sufficient to produce baldness in women of genotype BB but insufficient to cause it in women of genotype Bb. Evidence for such phenomena has been obtained from adrenal cortex abnormalities, because this endocrine gland produces a small amount of male sex hormone. Secondary sex traits in general are examples of sex-influenced traits.

PLANT BREEDING

Many types of cultivated plants have been developed by pedigree culture. In pedigree culture individual plants are selected and seeds are

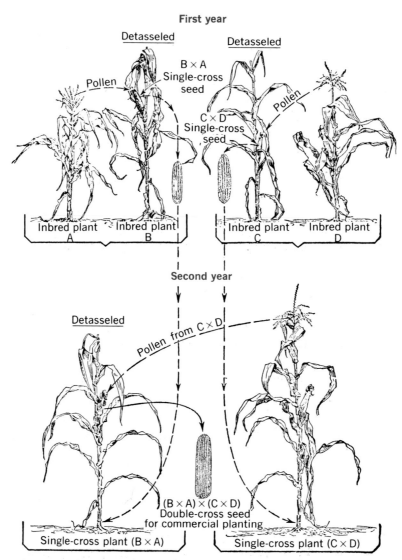

Fig. 31-20. Method of crossing inbred corn plants and the resulting single-cross to produce a double-cross hybrid seed (diagrammatic). The four plants, **A, B, C,** and **D,** are inbred for several generations; then strain **A** is crossed with strain **B** (**A** furnishes pollen and **B** is detasseled). Strains **C** and **D** are crossed similarly. Then the product of these two single-cross lines is crossed to produce a double-cross seed used in commercial plantings.
(From Richey, F. D.: The what and how of hybrid corn, Farmers Bull. 1744, U.S. Department of Agriculture.)

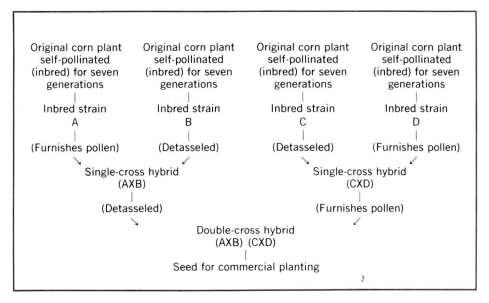

Fig. 31-21. Summary of steps used in the production of hybrid corn. (Also see Fig. 31-20.)

planted. An accurate record (pedigree) is kept of the offspring of each selected individual, and after several generations, the best strain is preserved. Hence, selection is based on hereditary behavior rather than outward appearance. If the original plant were homozygous for its selected trait, the new type of plant will breed true. Pedigree culture is well adapted for plants that normally self-pollinate, and it has been employed successfully with wheat, oats, beans, peas, tobacco, and potatoes. Pedigree culture is useful in preserving mutants, individuals that appear spontaneously and show some striking variation. If isolated, nearly all mutants breed true. Many new horticultural varieties have arisen as mutants. Examples include plants bearing seedless fruits, double flowers, variegated leaves, drooping branches.

Hybridization is a method that is used extensively in bringing together the traits of two different parents to form a new combination. Because such a desirable combination depends on chance, usually a great many crosses of the same kind have to be made, with the hope that in some of the offspring the desired combination of traits will occur. Hybridization is valuable for plants where vegetative propagation is possible, because with no sexual reproduction there is no segregation of genes, so the traits of the hybrid are preserved for a long time. Where vegetative propagation cannot be used, the hybrid offspring may be diverse and exhibit many new trait combinations. If some of these are desirable, they may be isolated and raised under the pedigree culture method until a pure-breeding race is secured.

Hybridization often produces increased vigor. Many hybrids reach a larger size than their parents, grow more rapidly, have larger parts, or have greater resistance to adverse conditions. Hybrid corn (Figs. 31-20 and 31-21) is now grown extensively in the United States. Seeds for planting are produced by crossing two or more inbred strains that have been developed by self-

Fig. 31-22. Some domesticated plants derived from the cliff cabbage found in Europe. The wild ancestor is sketched diagrammatically at lower left. (Not shown to true scale.) *(Courtesy W. Atlee Burpee Co., Philadelphia, Pennsylvania.)*

pollinating selected plants for about seven generations. These inbred strains decrease in vigor, but through constant selection they become uniform for their desired traits. When finally crossed, hybrids are produced that excel the parents in yield, size, and vigor. The yield of grain is often one third greater than that of the original corn plants from which the inbred strains were developed. If seeds from the hybrids are planted, the quality is markedly deteriorated.

Because corn hybridization is one of the most recent and successful examples, it might be considered briefly. Corn is normally cross-pollinated. When self-pollinated for at least seven successive generations, the corn plants tend to become progressively smaller and less productive. Eventually, when two such self-pollinated corn plants are crossed, the resulting hybrid is more productive and larger than the ancestors. The procedure, in brief, is as follows: (1) inbreeding (self-pollination) for about seven successive generations tends to produce homozygous strains; (2) two such homozygous strains that possess the desired traits are then cross-pollinated to produce the F_1 hybrids known as single-cross hybrids. However, such seeds are not sold, primarily because the yield is usually low and grains are of variable size; (3) two single-cross hybrids, produced from different homozygous strains, are now crossed, producing double-cross hybrids, which produce higher yields of uniformly larger seeds (Fig. 31-20).

Many genetic improvements in plants have been made, including such common examples as: fiber length in cotton, sugar content of melons, yellow color of peaches, seedless grapes, improved tobacco plants, and resistance of plants to diseases (wheat stem rust, corn blight, oats smut, and tomato wilt), and to insect pests.

Most of our important species of present-day cultivated plants were domesticated by early man. He recognized certain valuable properties, or parts, in the wild ancestors. For some time he collected these plants in the wild state; later he cultivated particular species and noted variations in their heredity that had also occurred in the wild state. Under conditions of cultivation he selected seeds or vegetative parts for propagation, and these were more widely used. Continued variations accompanied by selections for hundreds of years resulted in cultivated varieties that differed from their wild ancestors in many ways. For example, our present cultivated varieties of head cabbage, kohlrabi, cauliflower, broccoli, and Brussels sprouts were all derived from a mustardlike wild ancestor (cliff cabbage) in Europe (Fig. 31-22).

Some contributors to the knowledge of
GENETICS

J. Gottlieb Koelreuter (1733-1806)

A German who demonstrated the existence of sexes in plants and produced a plant hybrid by crossing two species of tobacco.

Gregor (Johan) Mendel (1822-1884)

An Austrian scientist and monk who described the inheritance of garden peas in a paper entitled ''Experiments in Plant Hybridization'' (1866). He formulated Mendel's laws by scientifically interpreting the results of experimental crossings. *(Historical Pictures Service, Chicago.)*

Mendel

Hugo de Vries (1848-1935)

A Dutch botanist who proposed the ''Mutation Theory'' based upon experiments with the evening primrose. However, most of the changes he observed were not gene mutations but a result of such other phenomena as chromosome changes and unusual combinations.

William Bateson (1861-1926) and R. C. Punnett

Bateson and Punnett, in England, performed experiments on the color of sweet peas and also on the comb shape of chickens. They discovered unusual ratios in these traits and attributed them to the action of two different genes. They also made observations that led to the discovery of the linkage and crossing-over of genes (1905). Bateson postulated the production and transmission of discontinuous variations, and the results on the origin of species (1894).

C. E. McClung (1870-1946)

An American zoologist who established the role of chromosomes in sex determination in grasshoppers (1902), which later led to the discovery of similar phenomena in other animals.

Friedrich Miescher (1844-1895)

A Swiss biochemist who discovered nucleic acids (1871), but whose work attracted little interest until the 1930s when several cell biologists, especially T. Caspersson, appreciated the importance of nucleic acids in cell physiology and applied new methods for their study.

Walter S. Sutton (1876-1916)

An American who explained fully (1903) the operation of Mendel's laws on the basis of chromosome behavior, derived largely from chromosome behaviors as explained by the German biologist Boveri.

Alfred H. Sturtevant (1891-   )

An American zoologist who made a chromosome map (1913) on which genes were located approximately on chromosomes by a scientific interpretation of the crossing-over percentages between chromosomes.

Herman J. Muller (1890-1967)

An American Nobel Prize winner (1946) who was the first to show (1927) that mutations could be artificially induced by radiations.

Thomas Hunt Morgan (1866-1945)

An American biologist and Nobel Prize winner (1933) who contributed to the knowledge of the mechanism of heredity. He reported the first gene mutation (white eye) in the fruit fly *(Drosophila)*. *(Historical Pictures Service, Chicago.)*

Morgan

Albert F. Blakeslee (1874-1954)

An American botanist who applied the drug colchicine to dividing cells of plants, causing a doubling of the number of chromosomes with altered traits in the treated organisms (1927).

B. O. Dodge (1872-1960)

An American mycologist who suggested the importance of the pink bread mold *(Neurospora),* in genetic studies (1930).

G. W. Beadle (1903-   ) and E. L. Tatum (1909-   )

Beadle and Tatum, in California in 1941, studied the action of genes in the pink bread mold *(Neurospora).* It was concluded that the primary work of a gene is the production of a specific enzyme, which controls a living process.

Tracy M. Sonneborn (1905-   )

An American biologist who discovered plasmagenes (cytoplasmic genes) in *Paramecia* (1943) that could reproduce and initiate mutation.

Adolph Butenandt (1903-   )

A German biochemist and Nobel Prize winner (1939) who with the biologist Alfred Kühn experimented with the flour moth *(Ephestia)* and worked out a chainlike process of gene actions (1952) caused by the action of specific enzymes.

James D. Watson (1928-   ) and Francis H. Crick (1916-   )

Watson of Harvard University and Crick of Cambridge, England, found (1953) that the molecules that control heredity and growth in organisms are made of deoxyribonucleic acid (DNA) and consist of two long chains of atoms linked together and twisted spirally. They, with Maurice Wilkins, received Nobel Prizes in 1962.

Arthur Kornberg (1918-   )

An American biochemist who with his co-worker, Severo Ochoa, received a Nobel Prize (1959) for work on the "Biologic Synthesis of Deoxyribonucleic Acid (DNA)." They found that an isolated enzyme catalyzes the synthesis of this nucleic acid in response to directions given from preexisting DNA.

Robert W. Holley (1922-   ), H. Gobind Khorana (1922-   ), and Marshall W. Nirenberg (1927-   )

These three Americans, working independently, explained how genes work. Their work led to formulation of the genetic code and its action in synthesis. They received the 1968 Nobel Prize in Physics.

Review questions and topics

1 Define genetics in your own words.
2 Describe the structure, composition, and functions of genes.
3 Define and give examples of: alleles, loci of genes, dominant, recessive, homozygous, heterozygous, synapsis, phenotype ratio, and genotype ratio.
4 State and give examples of Mendel's laws.
5 Explain and give examples of monohybrid crosses, dihybrid crosses, and trihybrid crosses.
6 What is the value of the checkerboard (Punnett square) in studying genetics?
7 Explain with examples complete dominance and incomplete dominance.
8 Discuss multiple genes and interaction of genes, describing each type with examples of each.
9 Discuss lethal genes and the results of their presence, including examples.
10 Explain the phenomena of linkage and crossing-over, with the results that follow in each case.
11 Explain how sex is determined genetically in plants and animals. What has this to do with heredity? Explain with examples sex-linked traits and sex-influenced traits.
12 Discuss the significance of cytoplasmic inheritance, including recent work that has been done in this field.
13 Problems in genetics:
   Solve the following problems in peas, using the following symbols: T, tall plant; t, dwarf plant; Y, yellow seed; y, green seed.
   a. Work out the entire monohybrid cross, using the proper symbols; carry through to the F_2 generation in each case: (1) homozygous tall × homozygous dwarf; (2) heterozygous tall × heterozygous tall.
   b. Work the following dihybrid crosses as in a: (1) homozygous tall-yellow × dwarf-green; (2) heterozygous tall-yellow × heterozygous tall-yellow; (3) homozygous tall-yellow × heterozygous tall-yellow.
   c. Work the following problems in guinea pig inheritance, using the following traits and symbols: R, rough coat of hair; r, smooth coat of hair; B, black hair; b, white hair.
      (1) Work the entire monohybrid cross in the following by using the correct symbols; carry through to the F_2 generation in each: (a) homozygous rough × homozygous rough; (b) homozygous rough × smooth; (c) heterozygous rough × heterozygous rough; (d) heterozygous rough × smooth.
      (2) Work the entire dihybrid cross, using the proper symbols; carry through to the F_2 generation in each: (a) Homozygous black-rough × homozygous black-rough; (b) homozygous black-rough × white-smooth; (c) heterozygous black-rough × heterozygous black-rough; (d) white-smooth × white-smooth.
14 What are the phenotype ratios and genotype ratios in each of the preceding problems?

Selected references

Beadle, G. W.: Biochemical genetics, Chem. Rev. **37:**15-96, 1945.

Burns, G. W.: The science of genetics: an introduction to heredity, ed. 2, New York, 1972, The Macmillan Company.

Herskowitz, I. H.: Principles of genetics, New York, 1973, The Macmillan Company.

Hexter, W., and Yost, H. T., Jr.: The science of genetics, Englewood Cliffs, N. J., 1976, Prentice-Hall, Inc.

Pai, A. C.: Foundations of genetics: a science for society, New York, 1974, McGraw-Hill Book Company.

Sinnott, E. W., Dunn, L. C., and Dobzhansky, T.: Principles of genetics, ed. 5, New York, 1958, McGraw-Hill Book Company.

Stern, C.: Principles of human genetics, ed. 2, San Francisco, 1960, W. H. Freeman & Co., Publishers.

Sutton, H. E.: An introduction to human genetics, New York, 1965, Holt, Rinehart & Winston, Inc.

Watson, J. D.: Involvement of RNA in the synthesis of proteins, Science **140:**17-26, 1963.

Winchester, A. M.: Genetics, Boston, 1965, Houghton Mifflin Company.

32 Genes and gene action

The term "gene" was originated by W. Johannsen of the University of Copenhagen in 1909. He also made the distinction between heritable and nonheritable traits, largely as a result of his work on the genetics of pure lines of beans. For most purposes we can consider a gene as (1) a unit of biochemical action, or a minimum section of a DNA molecule that controls metabolic activity in a cell, (2) a unit of mutation, or a minimum segment of a DNA molecule that has the inherent ability to mutate, or change, thus altering a heritable trait, and (3) a unit of crossing-over, or a minimum segment of a chromosome that during certain stages in the life cycle of an organism, can cross over (transfer) to a neighboring chromosome.

DEVELOPMENT OF THE DNA CONCEPT

The best approach to understanding the function of the gene is through the history and study of deoxyribonucleic acid (DNA). Although DNA was first isolated from the cell nucleus in 1869 by a Swiss chemist, Friedrich Miescher, its recognition as the active principle of the gene is a fairly recent development. In 1914, another chemist, Robert Feulgen, developed a selective staining procedure that he used to identify DNA in a mass of ground-up cell components. Ten years later he used the same technique to show that the DNA of intact cells is concentrated in the chromosomes. Several investigators later demonstrated that the amount of DNA was fairly constant in all of the body cells of a particular organism. The eggs and sperms of these organisms, however, contained only one half as much DNA as the somatic cells. This suggested that DNA might be a component of genes.

Still later, in 1928, an English bacteriologist, Frederick Griffith, made an intriguing discovery while working with the bacteria that cause pneumonia. He studied two different strains of pneumococcus, one virulent, or disease causing, and

483

A Type *A*. Mycelium with conidia (spores)   **B** Type *a*. Mycelium with ascogonium

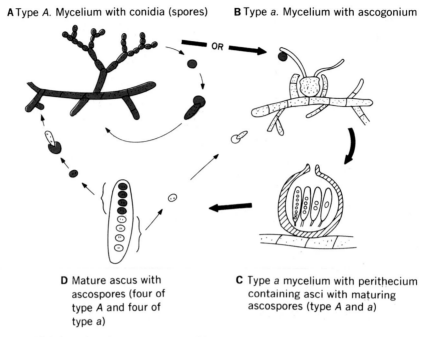

D Mature ascus with
ascospores (four of
type *A* and four of
type *a*)

C Type *a* mycelium with perithecium
containing asci with maturing
ascospores (type *A* and *a*)

Fig. 32-1. Pink bread mold *(Neurospora)* life cycle. Note that alternate methods of ascus formation are possible.

the other nonvirulent. The virulent strain was covered with a thick capsule, and the other was noncapsulated. When injected into mice, the virulent, capsulated strain caused death, while the other strain was harmless. Mice injected with heat-killed capsulated forms were not harmed.

However, when Griffith injected mice with a mixture of heat-killed virulent forms and live harmless forms, the mice died. When the mice were examined for the presence of bacteria, only live capsulated forms were present. Somehow the live, noncapsulated forms had been transformed into live, capsulated forms which were virulent. Griffith and others repeating his work believed that some heat resistant substance from the killed virulent strain had entered the non-virulent noncapsulated forms and transformed them into the capsulated form.

It was not until 1944 that Avery, MacLeod, and McCarty showed that this transforming principle

was actually DNA. They demonstrated that bacteria could be changed from harmless to virulent merely by supplying them with DNA extracted from a virulent strain.

This series of experiments, along with many others, showed that DNA is found in chromosomes, that it can be transmitted to other cells, that it acts in the same manner as the chromosomes during meiosis in that the quantity is halved, and that it is the hereditary substance.

DEVELOPMENT OF THE GENE THEORY

Once it was demonstrated that DNA was, indeed, the genetic principle the obvious question was, "How does it work?" For the answer to this we are indebted to George W. Beadle and E. L. Tatum and their work with the pink bread mold, *Neurospora crassa,* in the early 1940s. This fungus, belonging to the class Ascomycetes, has a short life cycle, is propagated rather easily, and produces large populations of a given genotype.

Reproduction occurs asexually by spores called conidia, or simply through unspecialized fragments of the mycelium. The vegetative hyphae are segmented, with each segment containing many haploid (N) nuclei (Fig. 32-1). Sexual reproduction occurs only when there is a union of cells of opposite mating types, known as A and a. The resulting fusion nucleus is the only diploid (2N) stage. Nuclear fusion and meiosis occur within the saclike ascus. The ascus is sufficiently narrow so that the divisions of meiosis occur in sequence, and the resulting nuclei, do not slip past each other but hold their original positions. The meiotic divisions give rise to four nuclei, with each in its particular position dividing once by mitosis, thus producing eight nuclei in a linear series in each ascus.

While still within the ascus, the eight linearly arranged ascospores provide an accurate record of what has happened at meiosis because (1) the position of a particular ascospore can be traced to the actual position of a nucleus in meiosis, as determined by the orientation of separating chromosomes on a spindle, and (2) all products of a single meiosis are retained within one structure and hence cannot be confused with products of other meiotic divisions.

With careful dissection each of the eight ascospores can be removed singly from the ascus yet retain its original position in the sequence. Each isolated ascospore can be cultured separately (in agar tubes), and the genetic characteristics of each such derived culture can be traced to segregation of genes at meiosis. If eight of these culture tubes are lined up in the same sequence as their corresponding ascospores in the ascus, it can be observed directly which alleles went to which nuclei at meiotic segregation.

Beadle and Tatum took advantage of the growth patterns of *Neurospora,* which forms a mycelium with haploid (N) nuclei. Any expressed trait was caused by one gene and not confused by

possible dominants or recessives. *Neurospora* is very easy to grow, requiring only a sugar, some inorganic salts, and one vitamin (biotin) as nutrients. From this minimal medium of nutrients, it can then synthesize any other required substances.

Beadle and Tatum treated asexual spores with radiation and germinated them on an enriched medium containing additional vitamins, amino acids, and sugars. When all spores had produced growth, bits of each colony were transferred to minimal media. Some of the spores did not grow, indicating that they had lost the capacity to synthesize some essential substance presumably because of earlier radiation treatment.

Next Beadle and Tatum studied the original culture that had been used for the minimal media trials. Samples from the enriched medium were transferred to several tubes, each containing minimal medium plus either vitamins, amino acids, or sugars. By observing the growth characteristics on the new media, they were able to devise further tests to determine exactly which substance was required for growth by that particular strain of *Neurospora*. They reasoned that if a colony of *Neurospora* now required a particular amino acid that it had been able to synthesize prior to radiation treatment, some change had taken place in the genetic structure. In order to demonstrate that this was truly a genetic mutation, they crossed the new strain with the wild type (nonradiated) *Neurospora*. In the resulting ascus, four of the spores germinated to form wild type colonies while the other four formed colonies requiring supplemental amino acids.

Beadle and Tatum had therefore demonstrated not only that mutations could be induced and transmitted as genetic traits but also that each mutation resulted in loss of the ability to synthesize one type of molecule. Since the synthesis was controlled by an enzyme, they proposed that a gene works by controlling the synthesis of one enzyme.

This one gene–one enzyme hypothesis has been slightly modified by further studies but the

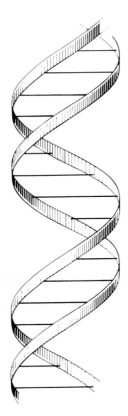

Fig. 32-2. The double helix first proposed by Watson and Crick as a model of double stranded DNA. The helix is composed of alternating molecules of sugar and phosphate. The cross bars represent the base pairs.

Fig. 32-3. Replication of DNA. As the original double stranded DNA uncoils, new complementary strands are synthesized.

work of Beadle and Tatum, for which they were awarded Nobel Prizes, is recognized as the basis of the gene theory.

SYNTHESIS OF PROTEINS

Roles of DNA and RNA

Once it was established that DNA was the genetic principle and that a gene controlled protein synthesis, the next task was to determine how this worked. Before going into this, however, the chemistry of DNA should be reviewed (see Chapter 7).

This molecule, or rather macromolecule, is composed of large numbers of nucleotides, which in turn are composed of a five-carbon sugar, phosphoric acid, and a nitrogenous ring molecule. The D in DNA refers to the sugar deoxyribose, just as the R in RNA refers to a similar sugar, ribose.

The nitrogenous compounds are double ring purines and single ring pyrimidines. Five kinds of these nitrogen bases are found in nucleotides. Two purines, adenine (A) and guanine (G) are found in both DNA and RNA. Three pyrimidines occur, cytosine (C) in both DNA and RNA, thymine (T) only in DNA, and uracil (U) only in RNA. Both DNA and RNA then are composed of four types of nucleotides that differ in the sugar and in the nitrogen base present. (See Table 7-4.)

THE DOUBLE HELIX

In 1950 it was shown by E. Chargaff that a consistent feature of the DNA molecule was that the amount of adenine always equaled the amount of thymine, and the guanine equaled the cytosine. By 1953 Maurice Wilkins was able to show by use of x-ray diffraction that the DNA molecule had a very regular and characteristic shape.

Later in 1953 James Watson and Francis Crick were able to put this information together to show that DNA was, in fact, two strands of DNA bonded together in the now famous "spiral staircase" or more accurately, double helix (Fig. 32-2). (Details and personal recollections of the search for DNA are given in Watson's book, *The Double Helix.**)

In their model, Watson and Crick showed that the "steps of the staircase" were always composed of a purine bonded to a pyrimidine. This explained the earlier finding that the adenine content was always equal to the thymine and the guanine to the cytosine. The helical railings of the staircase were formed by the sugar and phosphate joined together. The significance of these findings earned the Nobel Prize for Watson, Crick, and Wilkins.

Once they had constructed their model, Watson and Crick suggested that the double helix structure had all of the criteria to account for mitosis and gene duplication. A double helix could unwind by breaking the hydrogen bonds between the base pairs of A-T or G-C, and thus provide two single strands (Fig. 32-3). Each nucleotide on each strand would then attract its complementary nucleotide. Thus, an adenine nucleotide would attract a thymine nucleotide and a guanine would attract a cytosine nucleotide. In this manner each of the original strands would determine the structure of a new member of a double strand.

In 1957, this suggestion was proved true by the work of another recipient of the Nobel Prize, Arthur Kornberg. He showed that DNA, if provided with enzymes, adenosine triphosphate (ATP), and a quantity of nucleotides in a test tube, would synthesize these into new DNA molecules. This work was confirmed in living cells by M. Meselson and F. Stahl. They grew cells in media containing nitrogen bases with different isotopes of nitrogen and analyzed the distribution in later generations (Fig. 32-4).

---

*Watson, J. D.: The double helix; a personal account of the discovery of the structure of DNA, New York, 1968, Atheneum Publishers.

We are now ready to return to the problem of the control of protein synthesis by DNA.

FUNCTION OF DNA

Deoxyribonucleic acid (DNA) is formed in the nucleus of cells. (In bacteria and other Monera, it is formed in an equivalent area.) Its synthesis requires the presence of previously existing DNA, the enzyme DNA-polymerase, ATP as an energy source, and appropriate nucleotides and nucleotide precursors, and the proper cell environment, including pH, temperature, and ions.

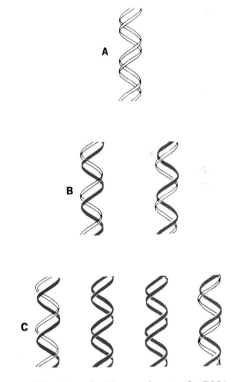

Fig. 32-4. Meselson-Stahl experiment. **A,** DNA from cells grown on isotope nitrogen for many generations. **B,** After one cell division in a nutrient containing normal nitrogen, one half of each new double helix contained normal nitrogen and the other contained heavy isotope nitrogen. **C,** When the cells divided again, three fourths of the strands were normal nitrogen.

The nucleotide precursors are: inorganic phosphate —℗, deoxyribose sugar and the nitrogen bases adenine, thymine, guanine, and cytosine. These units form the four kinds of desoxyribose nucleotides.

DNA in a living cell exists as a double helix—a twisted double strand of polymerized deoxyribonucleotides. The two strands are held together by weak bonds between the nitrogen bases of the complementary (paired) nucleotides (Fig. 32-5). In each pair a specific purine is always linked to a specific pyrimidine. The pairings then are always adenine-thymine, guanine-cytosine, thymine-adenine, and cytosine-guanine.

When new DNA is formed in a cell, a previously existing double helix unwinds. This results in two single strands of DNA at the point of unwinding. The two strands do not have to completely unwind before a new complementary strand is started by each of them.

Each nucleotide, on each of the unwinding strands, attracts its complementary nucleotide. Bonds are formed between A-T, G-C, T-A, and C-G, in whatever order they occur on the original strand. The new nucleotides then polymerize to form a new complementary DNA strand. This is under the control of the enxyme DNA-polymerase and depends on energy from ATP. The result is the formation of two molecules of double stranded DNA in the form of the typical double helix. The entire process is called DNA *replication* (Fig. 32-6).

Ribonucleic acid (RNA) is also formed in the nucleus of cells (or at least in a chromosomal area, as in bacteria). As in the synthesis of DNA, its formation requires previously existing DNA, ATP, and enzyme, and appropriate nucleotides.

The enzyme required is RNA-polymerase. The nucleotides are similar, but not identical, to those in DNA. These nucleotides contain the sugar *ribose* instead of deoxyribose. The nitrogen bases are adenine, guanine, cytosine, and uracil in place of thymine.

As in DNA replication the double strand starts to unwind. The complementary ribose nucleo-

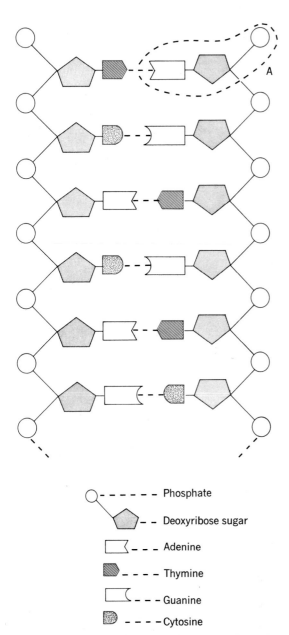

Phosphate

Deoxyribose sugar

Adenine

Thymine

Guanine

Cytosine

Fig. 32-5. Double stranded DNA, unwound to show base pairs. A nucleotide is composed of phosphate, sugar, and base. The nucleotide adenosine monophosphate (AMP) is shown at **A.**

tides are attracted to the DNA strand. The bonds between nucleotides are now thymine-adenine, guanine-cytosine, cytosine-guanine, and adenine-uracil (note that in RNA thymine is replaced by uracil). The action of RNA polymerase now combines the new strand of RNA in the sequence demanded by the DNA sequence. This is called RNA *transcription* (Fig. 32-7).

If a series of nucleotides in the DNA strand is T-A-G-C-A-T- and so on, then the complementary series in RNA is A-U-C-G-U-A- and so on. Very long strands of both DNA and RNA are known. Note that although DNA is usually double stranded, RNA is formed in single strands.

RNA action. Three different functional types of RNA are formed by DNA action. All of these may be found in the cytoplasm. One of these, rRNA, forms the ribosomes, the actual site of protein assembly in the cytoplasm. A second type, tRNA, is involved in transferring amino acids to the assembly site. The third type, mRNA, acts as a *messenger,* carrying the sequence required for synthesis of a specific protein.

Polypeptide synthesis. Cytoplasm contains the

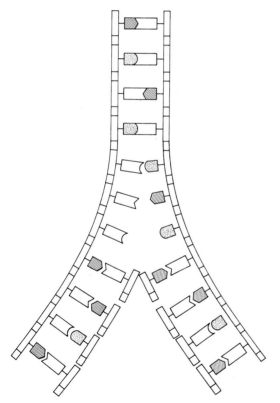

Fig. 32-6. DNA replication. Double strand unwinds and base pairs separate. Each strand attracts new complementary nucleotides from the cytoplasm. New base pairs line up and the "backbone" unites. The result is two new double strands of DNA.

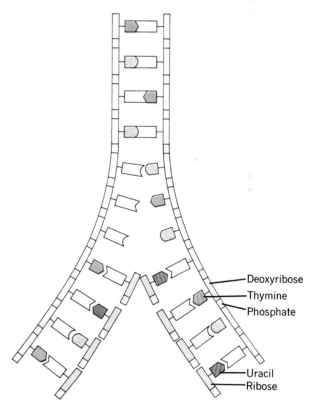

Fig. 32-7. RNA transcription. Original double strand of DNA unwinds and base pairs separate. Each strand attracts complementary *ribose* nucleotides. In RNA formation, thymine is replaced by *uracil* and deoxyribose by *ribose*.

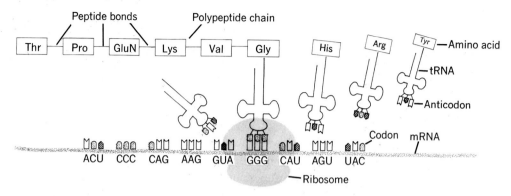

Fig. 32-8. RNA translation. Each triplet on the mRNA chain is a code for a complementary triplet on a tRNA carrying a specific amino acid. When the triplets are joined, the amino acids form a bond. The previous tRNA is then released and returns to the cytoplasm. The next tRNA, with its amino acid, then moves in. This continues until a "stop" codon is reached and the polypeptide is released.

various chemical compounds necessary for metabolism. These include the amino acids and enzymes required for protein formation. Proteins and smaller polypeptides are formed along the surface of an mRNA–ribosome complex. Amino acids are carried from the cytoplasm to this complex by tRNA molecules. Each type of amino acid has a specific tRNA molecule specialized for its transfer. A specific enzyme controls the specific amino acid–tRNA combination. Just as each kind of tRNA combines with a specific amino acid, it also binds only with a specific area on the mRNA strand.

The union of various amino acids takes place as one tRNA binds to its specific site on the mRNA, then is joined by a second tRNA attracted to the next site on the mRNA. As the respective amino acid molecules come into proximity, a peptide bond forms between them. The first tRNA molecule is released and goes back to the cytoplasmic pool. As more tRNA brings more amino acids to the mRNA complex, more peptide bonds are formed and the polypeptide grows. When the last amino acid has been delivered, the entire polypeptide is released into the cytoplasm. The total process of polypeptide formation is referred to as *translation* (Fig. 32-8).

The entire process might be summerized as follows: DNA, by transcription, forms mRNA, which carries the message spelled by the nucleotide sequence to ribosomes; tRNA carries specific amino acids and binds to the message site; the message is translated as a specific protein or polypeptide (Fig. 32-9).

More specifically, the sequence of nucleotides of DNA determines the sequence of complementary nucleotides in mRNA. This in turn determines the sequence of tRNA and the resulting amino acid combination. All of this in turn is determined by the *genetic code*.

THE GENETIC CODE

The question of how a cell "knows" that the next amino acid in a polypeptide should be one instead of another is puzzling. The answer lead to the establishment of what is called the genetic code.

By a series of ingenious experments, Robert W. Holley, Har Gobind Khorana, and Marshall W. Nirenberg, working separately, determined that a sequence of nucleotides on mRNA "codes" for a particular amino acid. The first of these discoveries, made by Nirenberg, showed

in nucleus:

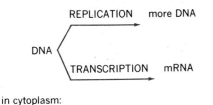

Fig. 32-9. Polypeptide synthesis. In the nucleus DNA may produce more DNA or messenger RNA. In the cytoplasm mRNA moves to the ribosomes and codes for various tRNA-amino acids, which form the polypeptides.

in cytoplasm:

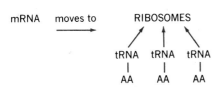

that the sequence of three adenine nucleotide, -AAA-, on a DNA molecule, resulted in three uracil -UUU- nucleotides on mRNA. This, in turn, was the "codon" for the amino acid *phenylalanine*. Similarly, a sequence of CAA on the DNA molecule produced GUU on mRNA. This proved to be the codon for the amino acid *valine*. CTT on DNA produced GAA, which coded for the amino acid *glutamic acid*.

The sequence involved in the formation of a single polypeptide composed of these three amino acids would start with the DNA molecule. This DNA would have a sequence of nine nucleotides as follows: A-A-A-C-A-A-C-T-T. Transcription would result in an mRNA molecule that contained the complementary sequence U-U-U-G-U-U-G-A-A. After moving to the ribosome site, this sequence would be translated as the codon for the amino acid sequence phenylalanine–valine–glutamic acid.

Further studies involving all of the possible three nucleotide combinations revealed the specific codes for all twenty of the amino acids. Two amino acids, methionine and tryptophan, have only one triplet codon. All of the others are coded for by two or more triplets (see Table 32-1).

In 1968 Holley, Khorana, and Nirenberg received a Nobel Prize for their pioneering studies in this field.

Anticodons and tRNA

As indicated previously, specific tRNA molecules act as the carriers that bring amino acids to the mRNA ribosome complex. Apparently the tRNA has a complementary set of nucleotides that "recognizes" the message of the codon. A sequence of G-U-U on mRNA would bind with an *anticodon* sequence C-A-A on a tRNA molecule carrying the amino acid valine. All of the codons for amino acids would provide the binding site for specific anticodons on tRNA molecules.

Several tRNA molecules have been synthesized in the laboratory. The structure appears similar to a three-leaf clover with one side of the stem slightly longer than the other. This long side is the attachment point for the amino acid. The anticodon nucleotides are in the middle loop.

The actual polypeptide sequence then might proceed in the following way. The anticodon of tRNA for the first amino acid binds to the first codon triplet on mRNA. The second anticodon triplet for the next amino acid binds to the next

Table 32-1 The genetic code*

Amino acid	Nucleotide triplets code
Alanine (Ala)	GCU, GCC, GCA, GCG
Arginine (Arg)	AGG, CGU, CGC, CGA, CGG
Asparginine (AspN)	AAU, AAC
Aspartic Acid (Asp)	GAU, GAC
Cysteine (Cys)	UGU, UGC
Glutamic Acid (Glu)	GAA, GAG
Glutamine (GluN)	CAA, CAG
Glycine (Gly)	GGU, GGC, GGA, GGG
Histidine (His)	CAU, CAC
Isoleucine (Ileu)	AUU, AUC, AUA
Leucine (Leu)	UUG, UUA, CUU, CUC, CUA, CUG
Lysine (Lys)	AAA, AAG
Methionine (Met)	AUG
Phenylalanine (Phe)	UUU, UUC
Proline (Pro)	CCC, CCU, CCA, CCG
Serine (Ser)	UCU, UCC, UCA, UCG, AGU, AGC
Threonine (Thr)	ACU, ACC, ACA, ACG
Tryptophan (TRYP)	UGG
Tyrosine (Tyr)	UAU, UAC
Valine (Val)	GUU, GUC, GUA, GUG
Stop (punctuation)	UAA, UAG, UGA

*A = adenine, C = cytosine, G = guanine, U = uracil.

codon. The amino acids form a peptide bond. This frees the *first* tRNA, which returns to the cytoplasm. A third tRNA brings a third amino acid to its appropriate position, and a second peptide bond is formed. This frees the *second* tRNA, which returns to the cytoplasm. The process continues in this manner producing a lengthening polypeptide chain. It continues until a "stop" codon is reached. The polypeptide is then released.

CONTROL OF GENE ACTION

Although detailed information of gene action and control is not available for all kinds of organisms, a few systems have been studied extensively. Most of our knowledge is based on a bacterium, *Escherichia coli*. This is easy to obtain in large numbers, requires little space, and can be grown in a tube or flask containing precisely de-

fined nutrients. It is not surprising that it is probably the most widely studied organism known to biology.

Because of the work of two Frenchmen, Francois Jacob and Jacques Monod, and others after them, we have some idea of the working of a *gene* control system in *E. coli*. From earlier workers it is known that a function of DNA is to provide the template, that is, the genes, which determine the production of messenger RNA. This is mediated by a specific enzyme called *mRNA-polymerase*. Once mRNA is produced, it moves to ribosomes where polypeptides are formed. The site on the DNA molecule that determines the mRNA, and therefore the structure of the polypeptide, is called a *structural gene*.

When *E. coli* is grown in a medium containing the sugar lactose, it produces several enzymes in large quanitities. One of these, which helps to get the lactose inside the cell, is *beta (β)-galactoside permease.* Another, *β-galactosidase,* acts to split lactose into two smaller sugars, glucose and galactose. These sugars are then involved in further metabolic activities. When lactose is present, the average cell of *E. coli* contains up to 3000 molecules of *β*-galactosidase. What is somewhat surprising, however, is that when the bacterium is grown on a different sugar, it contains only 2 or 3 molecules of *β*-galactosidase per cell. Something must control the action of the structural genes that produce the mRNA for *β*-galactosidase. Jacob and Monod postulated that this was an *operator gene* located next to the structural genes for the two enzymes. The combination of operator gene and adjacent structural genes was called an *operon.*

Jacob and Monod further postulated the existence of a *repressor* substance, which inactivated the operator gene. This substance, since shown to be a protein, was thought to be synthesized as a result of the action of a *regulator* gene.

Jacob and Monod then formulated a model for control of gene action. This states that a regulator gene (by mRNA-ribosome + RNA action) forms a repressor structure that prevents an operator gene from activating structural genes. In this manner enzymes are not formed when they are not needed. If molecules, which block the action of the repressor, are present in the cytoplasm, then the operator gene turns on the activity of the structural genes. The structural genes then form mRNA, which lends to enzyme formation.

In *E. coli* then, a *lac-operon* was involved. When lactose was absent, the repressor substance blocked the lac-operon. When lactose was present, it acted as an *inducer* for its own enzymatic breakdown by blocking the repressor, thus freeing the operator gene and the adjacent structural genes. In 1965 Jacob and Monod received the Nobel Prize for their work in developing the operon model of gene control.

Later work revealed that other types of operons exist. In one system, *E. coli* is capable of synthesizing all of the 20 or more types of amino acids it requires. Chemical analysis of these cells shows that they contain all of the enzymes necessary for this synthesis. If these cells are grown in a medium that contains the amino acids, the enzyme content of the cells is sharply reduced. In other words, the presence of the external substance stops the enzyme synthesis. This is in contrast to the lactose system, which increased the enzyme synthesis.

GENE CONTROL IN HIGHER ORGANISMS

While very little is known about gene control in eucaryotes, it is thought that the control is in the cytoplasm rather than in the DNA.

DEFINITION OF GENES

We will conclude with a further definition of the gene as the minimum section of a DNA molecule that can be transcribed in formation of a molecule of mRNA. The gene functions in the formation of polypeptides when the mRNA moves to ribosomes, attracts the coded tRNA–amino acids, and is translated as a particular amino acid sequence.

Review questions and topics

1 Distinguish between virulent and nonvirulent. Does there seem to be a connection between capsulated bacteria and virulence?
2 Trace the story of DNA from discovery to identification as the heredity substance.
3 What properties of *Neurospora* made it ideal for the work of Beadle and Tatum?
4 How did Beadle and Tatum state the way a gene works?
5 Review the chemistry of DNA and of RNA.
6 What are the component parts of a nucleic acid?
7 What are the genetic functions of DNA?
8 Starting in the nucleus, trace the components involved in the synthesis of a polypeptide.
9 What is the significance of the genetic code?
10 Distinguish between the various types of RNA.

Selected references

Clark, B. F. C., and Marcker, K. A.: How proteins start, Sci. Amer. **218:**36-42, 1968.
Crick, F. H. C.: The genetic code: part III, Sci. Amer. **215:**55-60, 1966.
Haynes, R. H., and Hanawalt, P. C., editors: The molecular basis of life. Readings from Scientific American, San Francisco, 1968, W. H. Freeman and Co. Publishers.
Herskowitz, I. H.: Principles of genetics: an introduction to heredity, New York, 1973, The Macmillan Company.
Ingram, V. M.: The biosynthesis of macromolecules, New York, 1965, W. A. Benjamin, Inc.
Kornberg, A.: The synthesis of DNA, Sci. Amer. Oct. 1968.
Lerner, I. M.: Heredity, evolution and society, San Francisco, 1968, W. H. Freeman and Co. Publishers.
Levine, L.: Biology of the gene, ed. 2, St. Louis, 1973, The C. V. Mosby Co.
Pai, A. C.: Foundations of genetics: a science for society, New York, 1974, McGraw-Hill Book Company.
Watson, J. D.: The double helix; a personal account of the discovery of the structure of DNA, New York, 1968, Atheneum Publishers.
Watson, J. D.: Molecular biology of the gene, ed. 3, New York, 1976, W. A. Benjamin, Inc.

33 Evolution

THEORIES OF THE ORIGIN OF LIFE

Abiogenesis

Until the seventeenth century people believed in abiogenesis (Gr. *a-*, not; *bios,* life; *genesis,* origin), which is the spontaneous generation of living organisms, especially lower types, from nonliving substances. For example, field mice were thought to arise spontaneously from mud of the Nile River, and flies were thought to originate from dirt, manure, and decaying meats.

Empedocles (495-435? B.C.), a Greek philosopher, suggested that parts of animals had arisen spontaneously and separately from the earth and had assembled themselves at random into entire animals, thus explaining why certain organisms were so odd. Anaximander (611-547 B.C.) thought that air imparted life to all living things. Aristotle (384-322 B.C.) believed that eels (fishes) arose spontaneously from nonliving materials. He believed that complex forms of life evolved from simpler ones, although he had no clear idea of an activating mechanism.

For hundreds of years many persons believed that each species of plant and animal was created separately and remained unchanged thereafter. There seemed to be little conflict between this belief and the acceptance of spontaneous generation of flies, fleas, and mice at that time. Kircher thought that animals arose through the action of water on the stems of plants. Von Helmont (1577-1644), a Flemish physician, thought that house mice were generated from pieces of cheese placed in bundles of rags.

One reason why abiogenesis was accepted was because the complex life cycles of many animals and plants were not known, and the microscopic stages of living organisms had never been seen. This lack of accurate information was a natural setting for such a theory. Experimental methods of attacking such problems were unknown.

Joseph Needham, in 1749, believed that he had demonstrated the spontaneous origin of minute living organisms in infusions that he had boiled a few minutes in corked flasks. By boiling he thought that he had killed all living matter from which living organisms could arise later. However, the living organisms that later appeared in the flasks arose from living materials such as cysts and spores that had withstood the brief boiling temperature.

Biogenesis

Abbé Spallanzani (1729-1799), an Italian, stated that Needham's conclusions were incorrect because they were derived from incomplete sterilization of the infusions in the flasks. Spallanzani improved on Needham's experiments by sealing the necks of the flasks with a blast lamp to prevent the entrance of anything, including air. He boiled the infusions for hours instead of a few minutes and found that under such conditions no living organisms could be found even after long periods of time (1776). Others stated that free oxygen, which is essential for life processes, was excluded in the experiments, thus preventing the possibility of spontaneous generation.

To answer the latter point investigators during the first half of the last century showed that thoroughly sterilized infusions never developed living organisms even when sterile air was admitted. The air was sterilized by heating or passing through acids to remove the suspended "dust particles" on which living substances might be attached.

Francisco Redi (1621-1697), an Italian scientist, demonstrated by a simple experiment of protecting decaying meat with a fine gauze from contamination by flies, that the maggots of these flies did not arise spontaneously from the dead meat, but that they developed from living eggs deposited by flies. Redi's results showed that "life must arise from living organisms."

Louis Pasteur (1822-1895), a French microbiologist, by means of a special type of flask showed that the source of life, which appeared in infusions exposed to the air, was the air itself, or rather the life existing in the air. Thus it was contended that much of the dust of the air contained microorganisms in a somewhat dormant condition that were ready to become active when suitable environmental conditions, such as proper food, temperature, and moisture, were encountered.

Biochemical theory

The rather intriguing notion that life rose from nonliving matter dates from the early Greeks. Suggestions concerning how this may have happened, and some data to support the theories, have been available only in recent years as a result of the work of many biochemists, among whom A. I. Oparin is outstanding. His work aided that of Dr. H. Urey, who in turn stimulated his graduate student Stanley Miller. Miller, like others after him, succeeded in establishing a primitive atmosphere in which some of the conditions present on the earth in its early history were duplicated.

The exact details of the formation of the earth are not known, and several conflicting theories exist, but there is some agreement on the course of development of complex molecules.

There were a number of elements in the atmosphere of early earth; among them we are most concerned with carbon (C), hydrogen (H), nitrogen (N), and oxygen (O) because of their prevalence in present-day organisms. The presence of large amounts of hydrogen resulted in the formation of compounds of water (H_2O), methane (CH^4), ammonia (NH_3), and later hydrogen cyanide (HCN).

These early molecules reacted under the influence of available energy sources, especially ultraviolet light, and also lightning, radiation, and volcanic activity. It is this early reaction that has attracted the attention of recent investigators. They have tried, and in most cases succeeded, to duplicate the various steps in mo-

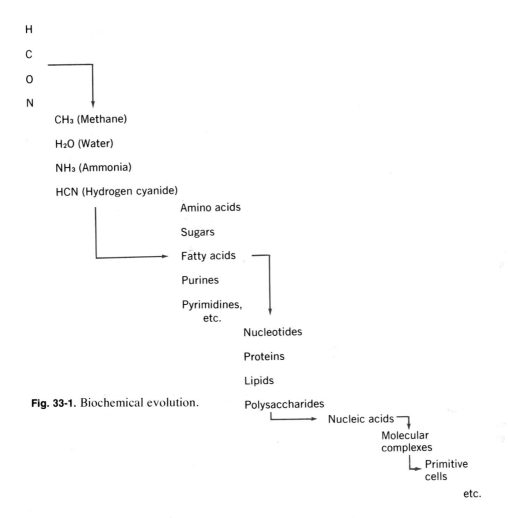

Fig. 33-1. Biochemical evolution.

lecular synthesis in present-day organisms (Fig. 33-1).

In general, it appears that the interaction of methane, ammonia, water, and hydrogen cyanide produced more complex molecules, including simple sugars, amino acids, purines, pyrimidines, and fatty acids.

The further synthesis of these compounds produced more complex molecules. If we consider this process as occurring in the early oceans, or at least in watery accumulations that were being constantly fed by runoff, it is easy to picture the gradual buildup of the inorganic molecules involved in the process.

The laboratory synthesis involving the purine, adenine, and simple sugar resulted in the formation of a common nucleotide, which in a further reaction with phosphoric acid, formed adenosine triphosphate. It is significant that this nucleotide is the principal source of energy in metabolism. It has been suggested that the polymerization of adenosine triphosphate and related nucleotides led to the formation of nucleic acids.

The reaction involving amino acids led to the formation of proteins including primitive enzymes; lipids were formed from fatty acids and glycerol-like molecules.

These and similar reactions resulted resulted in

the formation of many molecules that are important to metabolism. Some of these molecules, or at least the steps in their formation, have been prepared under laboratory conditions similar to the conditions of primitive earth.

The next step in this biochemical evolution appears to have been the gradual accumulation of rather simple groups of nucleic acids, surrounded by complexes of proteins and lipids—what we would recognize as a very primitive "cell."

The properties of the various molecules attracted to this complex would, in turn, determine its activities. There is reason to infer a process of change and development of the primitive complex into the fore-runners of present-day bacteria and protozoa (Fig. 33-1).

The current interest in outer space has greatly stimulated the investigation of primitive environments and the possibility of extraterrestrial life.

EARLY THEORIES OF EVOLUTION

Jean Baptiste Lamarck (1744-1829)

Jean Baptiste Lamarck, a French zoologist, was one of the first to attempt to explain seriously the phenomenon of evolution. Lamarck observed the changes and adaptations of certain parts of animal bodies that were used continuously in certain ways. He noted that certain structures increased in size if used but grew smaller or atrophied if not used. In 1809, in his theory of acquired traits through use and disuse, he concluded that these gains or losses through use or disuse were passed on to offspring in succeeding generations. Apparently, Lamarck never experimented to ascertain whether or not these acquired changes were hereditary. Use and disuse may change certain body traits, but the transmission of acquired traits, as such, to future generations has not been proved. Although the basic idea of his theory is no longer accepted, it did stimulate serious thought about evolution.

Erasmus Darwin (1731-1802)

Erasmus Darwin, the grandfather of Charles Darwin, recognized that the various types of organisms arose from each other, and he stressed the response of the animal to environmental changes as the basis for its modifications.

Sir Charles Lyell (1797-1875)

Sir Charles Lyell, a geologist, helped to foster sound thinking along evolutionary lines through his publication, *Principles of Geology* (1830). He formulated the theory of uniformitarianism, in which he stated that the forces that produced changes in the earth's surface in the past are the same as those that operate upon the earth's surface at present. Such forces over long periods of time could account for all the observed changes, including the formation of fossil-bearing rocks. This concept showed that the age of the earth was millions of years, rather than thousands of years. This great geologic work stimulated and influenced Charles Darwin's own thoughts and the formulation of his theory of organic evolution.

Charles Darwin (1809-1882)

While it is customary to refer to Darwin's presentations as the theory of evolution, he actually made two great contributions in his work, the first of which was so basic that it was easily overlooked. It was simply that evolution has occurred. The second theory was the mechanism by which it occurred, namely, natural selection.

The idea that evolution had occurred was not new with Darwin. His contribution was in stating the theory and then supporting it with massive evidence. He maintained that living species were descendants of similar, slightly different, forms that lived in the past.

Theory of natural selection (1859)

Charles Darwin's theory was based on the observation and study of many plants and animals in many parts of the world. He noted that all living plants and animals constantly displayed variations, great or small, structural or functional. He noted that the offspring of a pair of individuals, either plant or animal, showed they were not all alike. He also observed that, by selecting for breeding purposes those individuals that possessed certain desirable traits, there could be produced new kinds of domesticated animals and plants. He observed fossils, studied comparative anatomy and embryology, was familiar with geographic distribution of species and with geology, and he conducted breeding studies that produced variations in animals and plants.

His interests in nature and his lack of success in any other field led to an appointment as naturalist on a 5-year cruise of HMS Beagle around the world. This gave him the information that led to his belief in the truth of evolution.

Darwin was greatly influenced by Malthus' book *An Essay on the Principles of Population* (1798). Malthus stated that the human breeding potential far outstrips the limited supply of potential resources, leading to competition for the available goods and thus a struggle for existence. Darwin adopted Malthus' ideas of competition and struggle for existence in his theory of organic evolution.

Darwin was also greatly influenced by Alfred Russell Wallace (1823-1913), an English scientist, who had independently reached the same general conclusions concerning the theory of natural selection. However, when Darwin published his book *The Origin of Species* (1859), a year after Wallace had announced his conclusions in a short essay, most of the credit went to Darwin.

Evolution was explained as natural selection —a theory that was based on a few observations and the conclusions drawn from them.

First, it can be shown that the reproductive potential of any species is more than sufficient to replace the existing population. This is true whether in elephants, which require several years to produce offspring, or in flies, which require only days. A geometric progression for the increase is possible.

It can also be shown that this geometric in-

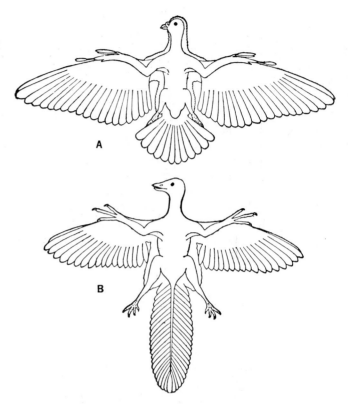

Fig. 33-2. Modern and ancient birds. **A,** A modern bird, the pigeon *(Columba livia).* **B,** *Archaeopteryx,* a reptilelike bird of Upper Jurassic period.
(From Lull, R. S.: Organic evolution, New York. The Macmillan Company.)

crease does not take place. For example, the offspring of a pair of flies, mating in early spring and producing many generations per season, have been calculated to be 1.91×10^{20}, which is more than sufficient to cover the earth with dead flies. Obviously they are not here. It may be concluded, therefore, that not all offspring survive. The next question is, then, which of them do survive?

Since variations among individuals, even close relatives, do exist, it becomes evident that some variations are "favored" over others in the survival race. Darwin believed that those variations that made the organism "better adjusted" were favorable ones and were transmitted to offspring, which then also had a greater survival advantage.

Some of the salient points of Darwin's theory of evolution might be summarized as follows. (1) Organisms of the same species are not all alike, but vary slightly among themselves, and the off-spring of a particular pair of parents tend to vary around the average of these parents. (2) More offspring are produced than can possibly survive. (3) In order to survive these offspring must compete with each other for existence. (4) As a result of this competition, those individuals best adapted to that environment tend to survive (survival of the fittest). (5) New species originate, over a period of many generations, by this natural selection of the best-fitted individuals. According to Darwin, whenever two groups of a plant or animal population are each faced with slightly different environmental conditions, they would tend to diverge from each other and in time would become sufficiently different from each other to form separate species. In a similar manner greater divergencies could arise at later times.

Although Darwin's theory of natural selection gives, in general, a satisfactory explanation of evolution, several parts of it must be altered in the light of more recent biologic knowledge. Darwin did not distinguish clearly between variations induced by environmental factors and those that involved chromosomal materials or al-terations of germ plasm. Many of his types of variations are now known to be nonheritable, thus having little significance in evolution. Only variations that involve genes are inherited and furnish material on which natural selection can act. In fact, the mechanism of genetics was not appreciated until 1903, quite some time after Darwin's death.

NEO-DARWINISM

Modern studies of evolution are primarily the result of the rediscovery in 1903 of the work of Mendel, in which the mechanism of survival along the lines of differential reproduction and the genetic basis of variation was investigated.

Evolution takes place in populations of a particular species, not in its individual members. The breeding population is the potential unit of change. Recognition of this has led to the concept of gene pools, which consist of the sum of the different genes present in the entire population. In time any member of the pool may exhibit any combination of its genes.

A species of any organism would then consist of all the interconnected gene pools. Thus, not only does every member of population A potentially share in its genes, but each member may contribute also to the gene pools of populations B, C, D, and so on, by migration.

Shortly after the rediscovery of Mendel's work, G. H. Hardy, an English mathematician, and W. Weinberg, a German physician, recognized the precision of the preservation of genes in a population. Stated quite simply, the Hardy-Weinberg law is that, under usual conditions, the gene frequencies in a population remain constant from generation to generation.

Evolution occurs when gene frequencies change, thus upsetting the Hardy-Weinberg equilibrium. Although many agencies enter the process, the major factors influencing gene frequencies are (1) mutation, (2) genetic drift, (3) selection or differential reproduction, and (4) migration.

Thus, the Hardy-Weinberg equilibrium is only an indication of gene frequencies and is applicable only when (1) mutations either do not occur or are balanced by back mutations; (2) genetic drift, or changes in gene frequency purely by chance in a small population, does not occur; (3) reproduction is random with no sexual selection; and (4) gene migration into or out of the pool does not take place.

Since it can readily be demonstrated that the preceding changes do occur, evolution can be defined simply as a series of changes in gene frequencies in a population, or gene pool. When a mutation is selected or has a reprodcutive ad-

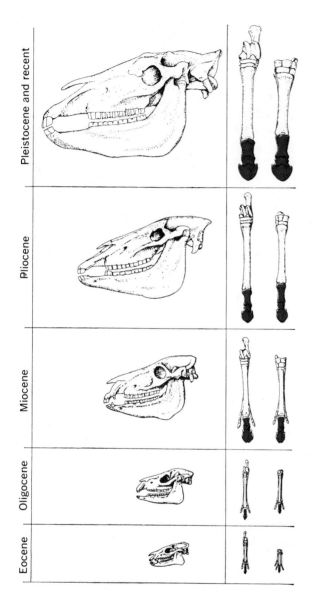

Fig. 33-3. Evolution of the horse. Comparisons of the shape and size of skulls, hindfeet, and forefeet of animals of successive geologic periods, indicating changes that led from a fox-size ancestor in the Eocene period, sixty million years ago, to the modern horse (top): Note the changes from four toes and three toes of the early ancestors through successive stages to the one-toed type with splint bones that do not touch the ground.
(From Crow, J. F.: Ionizing radiation and evolution, Sci. Amer. 201:138-145, 1959.)

vantage, it becomes the norm of that particular population.

EVIDENCE OF EVOLUTION

Evidences of evolution have been observed from such sciences as (1) paleontology, (2) taxonomy (classification), (3) comparative embryol-ogy, (4) comparative anatomy, (5) comparative physiology, (6) biogeography (geographic distribution), and (7) genetics and variations.

Evidence from paleontology

The science that deals with fossil plants and animals is called paleontology (Gr. *palaios,* ancient; *onto-,* being; *logos,* study). Geologists can determine, in most instances with remarkable accuracy, the chronologic succession in time of the various strata composing the earth. The fossils of these strata testify to the order of appearance and disappearance of various types of animals and plants on the earth, the more recent appearing nearer the earth's surface. Three examples, birds, horses, and wheat, may be cited.

EVOLUTION OF BIRDS

Birds seem to have evolved from a reptile-like ancestor, because despite superficial dissimilarities, such as the scaly-skinned, cold-blooded reptile and the feathered, warm-blooded bird, there are many fundamental structural and functional similarities not only between adult reptiles and birds but also between their embryonic stages (Fig. 33-14). Fossil remains of a reptile-like bird *(Archaeopteryx)* show a connecting link between reptiles and birds as they are known today (Fig. 33-2).

EVOLUTION OF HORSES

Much of the evolution of the horse occurred in North America and extends back 60 million years ago to the Eocene epoch in geologic history (Table 33-1). Fossil records indicate that the descent proceeded in a definite direction, but there were undoubtedly many side forms that showed variations from the main line of descent. Many of these forms became extinct. Important progressive changes in the horse included differences in the shape and size of the skull, in the bones of the feet, and in the structure and character of the teeth (Figs. 33-3 and 33-4). There also seemed to be a progressive increase in the size of most types in the direct line of descent.

The first known ancestor of the American horse is *Eohippus* (Eocene epoch), which was about 1 foot high at the shoulder and grazed on forest underbrush. Its forefeet had four toes (and a splint bone), whereas the hind foot had three well-developed toes (and two splints, which represented the first and fifth toes). Another type in the line of descent was *Mesohippus* (Oligocene epoch), which was taller. Each foot had three digits, with the middle one larger and better developed.

Merychippus (Miocene epoch) was probably the direct ancestor of later horses. It was three-toed, with the lateral toes high above the ground. The skull was larger and the lower jaw was heavier than the ancestral forms. *Merychippus* was between 3 and 4 feet high at the shoulder and gave rise to several types, most of which became extinct. However, one that persisted was *Pliohippus* (Pliocene epoch), which was the first one-toed horse. The modern horse, *Equus,* arose from *Pliohippus* during the Pleistocene epoch in North America, and it spread to other continents. *Equus* finally became over 5 feet high at the shoulder, with only one toe on each foot (with two splint bones of former lateral toes). The horse became extinct in North America by the end of the Pleistocene epoch, but certain types persisted in Eurasia to become the ancestors of present-day horses. Wild types still exist in Asia. Early Spanish colonists reintroduced horses to America after the time of Columbus, and some of them became the wild horses of our western plains and of South America.

EVOLUTION OF WHEAT

Wheat was one of the earliest plants cultivated by man. Carbonized kernels of wheat have been found by the University of Chicago archaeologist Robert Braidwood at Jarmo, Iraq, which may have been a birthplace of agriculture nearly 7,000 years ago. There is a remarkable resemblance between the ancient grains and modern grains, the latter having been carbonized to simulate the archeologic speciments. Two types of kernels were found in Jarmo, one almost identical with wild wheat still growing in the Near East, and the other almost exactly like present-day, cultivated wheat of the type called einkorn (Fig. 33-5).

Evidently, in certain types of wheat there has been little change in nearly 7,000 years. However, other types have evolved, so that we have

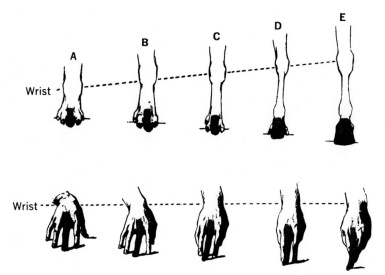

Fig. 33-4. Comparison stages in the elevation of the horse's foot to the tip of the middle toe, as shown by the human hand.
(Courtesy The American Museum of Natural History, New York, New York.)

today a number of distinct species of wheat, all belonging to the genus *Triticum* (L. *tres,* three). Some authorities recognize 14 species, which fall into three distinct groups determined by the number of chromosomes in their reproductive cells. These numbers are, respectively, seven, 14, and 21, or double this number in body (somatic) cells. The chromosome numbers are closely associated with differences in such things as anatomy, resistance to disease, productivity, and baking qualities. The wheats with 14 and 21 chromosomes have arisen from the seven-chromosome wheat and certain related grasses through hybridization, followed by chromosome doubling. Hence, cultivated wheats are an excellent example of evolution, and the evidences collected involved studies of the genetics mechanisms.

Fossil record

A fossil may be defined as any trace, remains, or impression of a plant or animal of past geologic ages. Much of our present knowledge about an-cient life has been gained by a study of these ancient organisms in the various strata of the earth. According to their origin, the rocks of the earth's surface are of two kinds, sedimentary and igneous. Sedimentary rocks, such as limestone, sandstone, and shale, may contain fossils and are formed by the transportation and deposition of small rock particles, by the precipitation of materials from solutions, or by the secretions by certain organisms, as in the case of limestones. Igneous rocks (L. *igneus,* fire), such as volcanic rocks formed by the consolidation of molten lava, are produced as a result of heat and do not con-tain fossils. In sedimentary rock formations, the oldest occur at the bottom of a series of strata and the youngest near the surface. The most ancient fossils thus will be found in the oldest rocks, whereas the most recent fossils will occur in the youngest rocks.

NATURE AND KINDS OF FOSSILS

The following are ways in which animals and plants of the past have left records: (1) by actual

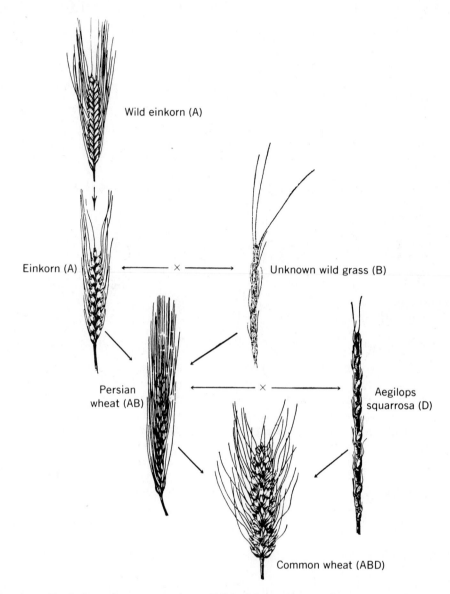

Wild einkorn (A)

Einkorn (A) ← ——— × ——— → Unknown wild grass (B)

Persian
wheat (AB) ← ——— × ——— → Aegilops
squarrosa (D)

Common wheat (ABD)

Fig. 33-5. Evolution of common wheat. Wild einkorn (seven chromosomes, genom **A**) evolved into einkorn, which crossed with a wild grass (genom **B**) and gave rise to Persian wheat (fourteen chromosomes, genom **AB**). Persian wheat crossed with another grass (genom **D**) which resulted in common wheat (twenty-one chromosomes, genom **ABD**): Note the increase in chromosome numbers from seven to twenty-one, and from one genom (**A**) to three genoms (**ABD**). The chromosome numbers are listed as haploid (N), and somatic cells would have diploid (2N) numbers. Genom refers to the complete basic set of different (N) chromosomes.

(From Mangelsdorf, P. C.: Wheat, Sci. Amer. 189:50-57, 1953.)

preservation of the organism, (2) by preservation of the skeletal structures, (3) by natural molds or incrustations, (4) by petrifaction, and (5) by leaving trails and impressions (imprints).

Actual preservation may occur by freezing and preserving in ice or soil. An example is the mammoth discovered frozen and perfectly intact in Siberia, with plants of the same period frozen with it. Many records of mammoths (Fig. 33-6) have been discovered. Remains of organs may be enclosed in rocks, as illustrated by a plant leaf (Fig. 33-7). The remains of animals and plants may be preserved, more or less intact, in tar, amber, or oil-impregnanted soils. Amber is a yel-

Fig. 33-6. Prehistoric mammoths, found frozen and well preserved in Siberia. *(Courtesy Chicago Natural History Museum; from Hickman, Cleveland P., Sr., Hickman, Cleveland P., Jr., and Hickman, Frances M.: Integrated principles of zoology, ed. 5, St. Louis, 1974, The C. V. Mosby Co.)*

Fig. 33-7. Photographs of the remains of fronds of fernlike plants *(Pecopteris)*. **A,** Two halves when opened; **B,** enclosed (upper right) and open (lower right).

Fig. 33-8. Photographs of remains of crustaceans called trilobites (Gr. *tri*, three; *lobos,* lobe), because the body was divided lengthwise into three lobes. Animals were also divided transversely into three body regions. Tri-lobites were marine animals with numerous jointed legs and breathing organs. They were probably able to walk or swim. **A,** Type of genus *Isotelus;* **B,** type of genus *Calymene;* **C,** six specimens of *Phacops,* shown in their rolled condition.

Fig. 33-9. Photographs of brachiopods (Gr. *brachion,* arm; *pous,* foot), so called because of two long "arms" enclosed between the two shells or valves, by means of which food was obtained and respiration was carried on. They were also known as lamp shells because of their resemblance to Roman lamps.

lowish plant resin that was originally soft and may have captured an animal or plant intact. The more volatile materials of the resin disappeared later, leaving the hard amber with its imprisoned organism. Certain organisms may remain intact by being buried in quicksands or swamps.

When the skeletal structure of an organism is preserved (Figs. 33-8 to 33-13), it remains almost in its original condition, except that it has lost most, if not all, of its organic material. Several skeletons of ancient mastodons have been found in different states of preservation. In some in-stances the skeletal remains may have added a chemical, such as carbonate of lime, which makes them heavier and more compact than the original. In other instances the shell-like skele-tons of animals have become more porous and lighter than the originals.

Fig. 33-10. Specimens showing **A,** natural mold of the interior of an animal from which the shell has disap-peared; **B,** original shell of a similar specimen. *(From Cleland, H. F.: Physical and historical geol-ogy, New York, American Book Company.)*

In natural molds or incrustations neither the minute structures nor the materials of the original organism are preserved; merely the general outlines of form and shape are recorded. Animals and plants may be enclosed by incrustations of calcium carbonate or silica, which harden around the buried organism before it decays. The organic materials of animals are removed eventually by decay and percolation of dissolving waters. The cavity that remains eventually retains the general form and shape of the original organism. In some instances the skeleton has disappeared entirely, leaving only a mold of it as a record. The shells of mollusks are covered with sediment, and the soft parts decay. The interior is then filled with sediment. Acidified waters then dissolve the limy shell, leaving only the molds of the exterior and interior. Sometimes the shell is removed, and the space left between the external and internal molds is filled with mineral matter, carried in by percolating waters. In this manner the form of the original skeleton is preserved but not its natural structure.

In petrifaction the original materials of the organism have undergone a certain amount of mineralization. In this case lime, silica, iron oxides, iron pyrites, and other substances have replaced the original materials of the organism, sometimes faithfully retaining its original shape, size, and even minute details. Usually, the older the fossil in time the greater the degree of mineralization. The harder parts of an organism, such as shells, teeth, tusks, bones, and the harder, woody parts of plants, are preserved most frequently. Petrified wood may show the minute structures just as they existed in the living trees, but the cell walls are now formed of the mineral silicon instead of the original cellulose. In this process, as each particle of cellulose disappeared, it was accurately replaced by a particle of silicon, thus retaining the minute details.

Trails and impressions (imprints) might be called "fossils of living organisms," whereas other records are of dead organisms. Animals may leave their trails and imprints, such as footprints. The records must be left in soft materials that later become hardened and preserved. A leaf may leave an impression in soft mud that later hardens. If the material in which the plant or animal impression is made turns to rock, the result is a fossil. This type of fossil usually does not give much information concerning internal details but rather aids in determining the form and shape of the organism.

Fig. 33-11. Prehistoric rhinoceros-like animal *(Brontotherium)* common during late Eocene and Oligocene epochs.
(Courtesy Chicago Natural History Museum; from Hickman, Cleveland P.: Integrated principles of zoology, ed. 2, St. Louis, 1961, The C. V. Mosby Co.)

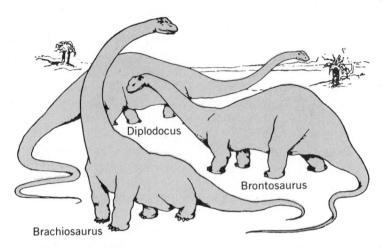

Fig. 33-12. Three enormous dinosaurs (Gr. *deinos*, terrible; *sauros*, lizard), extinct reptiles. *Diplodocus* was more than 80 feet long and weighed 40 tons; *Brontosaurus* was more than 65 feet long; *Brachiosaurus* was about 80 feet long.
(From Atwood, W. H.: Comparative anatomy, ed. 2, St. Louis, The C. V. Mosby Co.)

Fig. 33-13. La Brea tar pool near Los Angeles, California, with entrapped animals. An elephant and two wolves are caught, and a saber-toothed tiger is about to suffer the same fate.
(From Cleland, H. F.: Physical and historical geology, New York, American Book Company.)

CONDITIONS FOR FOSSIL FORMATION

In order for fossils to be formed there must be a rather rapid burial of the organism. This burial usually is accomplished by waterborne sediments. Organisms with hard parts are more likely to fossilize than those with softer ones. This may explain the absence of many fossils of the earliest, simplest organisms. The organism must also remain intact a sufficient length of time to permit fossilization to occur. Air must be excluded in order to prevent oxidation of the organism and bacterial decay before the fossil is formed. During and after formation the fossil must withstand such natural conditions as the elevation or sinking of the earth's strata, pressure and heat, erosion processes, and the slow circulation of waters, especially acid waters, through the fossil.

Terrestrial plants and animals have less chance of becoming fossilized, unless they are placed in water and eventually covered. Most dead land plants and animals will be decomposed quickly on the earth's surface, thus leaving no extensive records. However, if covered with large quantities of volcanic ashes, sand or dust, earth through landslides or earthquakes, or materials from calcareous springs, even terrestrial plants and animals may leave some type of record. If the structures are thin, fragile, easily dissolved, and easily decayed, there is little opportunity to form a fossil.

SIGNIFICANCE OF FOSSILS

Certain types of fossils indicate the boundaries and extent of former lands and waters. There is little land today that has not been below the level of the sea, sometimes repeatedly. This explains why we may find fossils of former marine organisms even on mountains. The Bering Strait between Asia and Alaska is about 35 miles wide and has a maximum depth of 200 feet. Studies of fossils on both sides of the strait show that these two lands were probably connected originally by land that sank beneath the water.

The character of fossils found in certain strata of the earth gives clues concerning their geologic ages and the times those particular sediments that formed these strata were laid down. Hence, certain animal and plant fossils are known as index fossils because through them it is possible to determine particular geologic ages and periods (Table 33-1).

Table 33-1 Major divisions of geologic time (timetable)*

Era	Period	Epoch	Time when each epoch started (millions of (years ago)	Life and general characteristics
Cenozoic (Gr. *kainos*, recent; *zoe*, life)	Quaternary	Recent	$1/40$	Modern man dominant; modern species of plants and animals; climate warmer; end of fourth ice age
		Pleistocene (Gr. *pleistos*, most; *kainos*, recent)	1	Extinction of giant mammals and many plants; development of man; cold and mild climates; four ice ages with glaciers covering much of North America, Europe, and Asia

*This simplified presentation shows the earth's history briefly.

Continued.

Table 33-1 Major divisions of geologic time (timetable)—cont'd

Era	Period	Epoch	Time when each epoch started (millions of years ago)	Life and general characteristics
Cenozoic— cont'd	Tertiary	Pliocene (Gr. *pleion*, more)	12	Modern mammals; rise of herbaceous plants; invertebrate animals similar to modern types; dry, cool weather; volcanoes active; continents elevated
		Miocene (Gr. *meion*, less)	28	Grazing mammals; man apes; temperate types of plants; moderate climates; grasslands and plains developed; forests diminish
		Oligocene (Gr. *oligos*, little)	40	Primitive monkeys and apes; whales; forests widely distributed; temperate types of plants; mild climate; mountains formed
		Eocene (Gr. *eos*, dawn)	60	Modern mammals appear; first horses; subtropical forests; heavy rainfall; North America and Europe connected by land
		Paleocene (Gr. *palaios*, ancient)	75	Former mammals dominant; modern birds; dinosaurs extinct; subtropical plants; temperate to subtropical climates; mountains formed
Mesozoic (Gr. *mesos*, middle; *zoe*, life)	Cretaceous (L. *creta*, chalk)		130	Extinction of giant reptiles and toothed birds; origin of placental mammals; flowering plants arise; gymnosperms decline; mild to cool climates; mountains formed; inland seas and swamps forming
	Jurassic (fine developments in Jura Mountains)		180	Giant dinosaurs and marine reptiles dominant; mammals appear; first toothed birds; angiospermous plants appear; conifers dominant; shallow inland seas
	Triassic (threefold development in Germany)		230	First small dinosaurs; marine reptiles; mammal-like reptiles; conifers dominant; seed ferns disappear; deserts formed
Paleozoic (Gr. *palaios*, ancient)	Permian (extensive in Perm, Russia)		260	Reptiles displace amphibians; many marine invertebrate animals extinct; modern insects; evergreens appear; cold, dry, and moist climate; mountains formed

Table 33-1 Major divisions of geologic time (timetable)—cont'd

Era	Period	Epoch	Time when each epoch started (millions of years ago)	Life and general characteristics
Paleozoic—cont'd	Pennsylvanian (well developed in Pennsylvania) (Upper Carboniferous) (carbon and coal-bearing rocks)		310	Reptiles originate; amphibians; giant insects; spore-bearing trees dominant; extensive coal-forming swamp forests; moist, climate; shallow inland seas
	Mississippian (well developed in Mississippi River valley) (Lower Carboniferous)		350	Amphibians; winged insects; sharks and bony fishes; horsetails and seed ferns abundant; early coal deposits; warm climate; hot, swamp lands; inland seas; mountains formed
	Devonian (common at Devon, England)		400	First amphibians; freshwater fishes; sharks; forest and land plants; wingless insects; corals; primitive horsetails, ferns, and seed ferns; arid land; heavy rainfall; small inland seas; mountains formed
	Silurian (Silures, ancient tribe of Wales)		430	Fishes with lower jaws; first land plants and arthropods; algae dominant; mild climate; European mountains formed
	Ordovician (Ordovici, ancient tribe of Wales)		475	First vertebrates (ostracoderms); brachiopods; trilobites abundant; marine algae dominant; warm, mild climate; oceans increase in size; land submerging
	Cambrian (Latin for Wales)		550	Algae and marine invertebrate animals; first abundant fossils; trilobites and brachiopods dominant; mild climate
Proterozoic (Gr. *proteros*, early)	Upper Precambrian (L. *pre-*, before)		2,000	Bacteria; fossil algae; sponges; soft-bodied animals; start of autotrophic nutrition (making their own food); warm, moist to dry, cold climate; volcanoes active; glaciers; mountains forming
Archeozoic (Gr. *arch-*, beginning; *zoe*, life)	Lower Precambrian		4,500	Life presumed to originate; start of heterotrophic nutrition (obtaining nourishment from organic substances); no fossils found; lava flows; granite formed

A study of animal and plant fossils is important because it often includes the ancestors of modern species. The data obtained often explains relationships among various types of present-day animals and present-day plants. In some instances the ancient types serve to form connections between groups of organisms that at present seem to have no direct connections.

Formerly, the classification of plants and animals was based primarily on living forms, but as the knowledge of fossils increased, the classification was made more accurate by incorporating the data contributed by paleontology. Although they may have disappeared completely, many large groups of organisms of the past, through their fossil records, have thrown light on the relationships between present-day organisms. Other groups of organisms that were originally dominant have diminished in numbers and importance until they are represented by a limited number of types at the present time.

The earliest known organisms were simply constructed. Throughout geologic time there has been a continued evolutionary succession of plants and animals. With each era and period more complex and highly evolved organisms arose, only to be superseded later by newer and more complex groups.

A study of paleontology reveals certain geographic conditions and distributions of organisms of the past. Regions of the world that are united at the present time were once separated by barriers. Mountains have arisen or large land areas have been submerged beneath water. The presence of large numbers of plants at a certain period precludes a certain type of climate so that they might flourish. Studies reveal the former existence of luxuriant vegetation in regions that are at present more or less devoid of those types of plants. Natural phenomena such as floods, glaciers, volcanic eruptions, and earthquakes have affected plant and animal distributions in the past. A change in the atmosphere, water, foods, or soil in the past no doubt influenced the distribution of organisms. The greater and more extensive the changes the greater the effects on organisms.

GEOLOGIC TIME CHART

By accurate studies of the strata of the earth scientists have divided the earth's history into eras (Table 33-1). Each era has been divided into periods, and the periods subdivided into epochs. Each division has specific characteristics, ages, durations, and types of life that were common during those times. The most recent fossils are found in the upper strata, and the ancient are arranged in a series with the most ancient at the bottom.

The relative lengths of the eras and periods may be calculated in two ways. The age can be approximated by the thickness of the sedimentary rocks formed during each period. A definite time is required to form a certain thickness of sedimentary rock of a particular type. From these data it can be estimated how long it would take for a certain thickness to be formed.

The other method is by the radioactive disintegration method. Radioactive elements, such as uranium 238, uranium 235, and thorium, are slowly changed to lead. An analysis of the ratio between the lead and the uranium (or thorium) content of a rock gives its approximate age. Estimated by radioactive methods, the oldest rocks are thought to be over 2 billion years old. The rates at which uranium (and thorium) are converted to lead have been determined approximately as follows:

$$\frac{\text{Amount of lead}}{\text{Amount of uranium}} \times 7,600,000 \text{ years} = \text{Time}$$

Radioactive carbon (carbon 14), which loses one half of its radioactivity in about 5,760 years, has also been used in dating certain specimens. When bones are formed, small amounts of carbon 14 are incorporated, and upon the organism's death the radioactivity is gradually lost. The determination of the amount of radioactivity in bone makes it possible to approximate the time of death. By determining the age of trees killed by

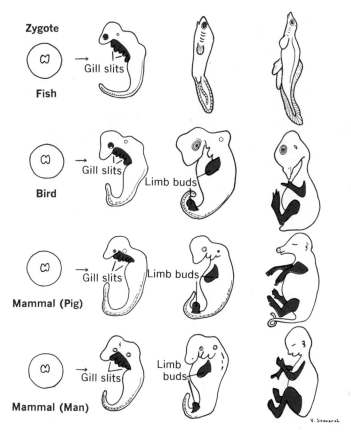

Fig. 33-14. Comparison of four vertebrate embryos showing the parallelism of their development: Note that the early stages of all four are quite similar structurally and that distinctive characteristics appear in later stages. Because of this pattern of similarity, Haeckel formulated his biogenetic law, which stated that "ontogeny (development of the individual) recapitulates (or repeats) phylogeny (evolutionary history of that group)." This view is no longer held in its extreme form, because all stages of ontogeny are modified in evolution. Von Baer's view is that embryonic stages of higher forms are like the embryonic stages of lower animals but not like the adults of those animals. This is considered to be more accurate.
(From Hickman, Cleveland P.: Integrated principles of zoology, ed. 4, St. Louis, 1970, The C. V. Mosby Co.)

the last ice glacier that covered parts of North America, it is estimated that the time of its retreat was about 11,000 years ago. The study of carbon from the charcoal of fires shows that man entered North America shortly after the glacier retreated, and eventually migrated over the continent. By analyzing the wood of trees killed when a volcano exploded it has been estimated that Crater Lake, Oregon, was formed about 4,500 years ago.

Evidence from taxonomy (classification)

Comparative studies of various species of plants and animals reveal great similarities in many instances; in fact, similarities are often so

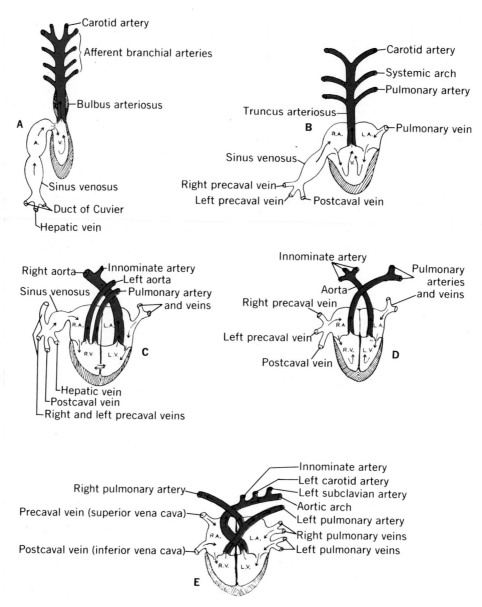

Fig. 33-15. Hearts and blood vessels of vertebrates (somewhat diagrammatic). **A,** Fish *(Pisces);* **B,** frog *(Amphibia);* **C,** turtle *(Reptilia);* **D,** pigeon *(Aves);* **E,** man *(Mammalia).* **A** (within heart) shows the auricle; **V** (within heart) shows the ventricle; **R,** right; **L,** left. The terms auricle and atrium are sometimes used interchangeably.

great that it is difficult to decide where one species with its variations ends and another species with its variations begins. The intergrades (divergent individuals of a certain species) frequently are very similar functionally and structurally to those of closely related species. When we attempt to classify similar types of organisms, we observe the close anatomic and physiologic relationships between many of them.

Evidence from comparative embryology

A comparative study of the embryonic stages through which animals pass reveals a widespread general correspondence of the developmental stages in higher forms with the existing adult stages of lower forms. The history of the embryonic development of an individual frequently corresponds in a general and broad way to the history of the development of the race as a whole (Fig. 33-14).

A study of the embryonic development of a bird's or mammal's heart shows that the various stages through which it develops succeed each other in the same general way, from the two-chambered to the four-chambered condition, as shown when comparisons are made among the lower vertebrates, such as the fishes, up through the amphibia, reptiles, and birds, to mammals (Table 33-2 and Fig. 33-15).

A similar comparative study of the brains (Fig. 33-16), reproductive systems, skeletal systems, and digestive systems of these vertebrates shows a similar condition, in which the organs of higher animals, during their development, pass through stages that correspond in general with the larval or adult conditions of similar organs in the lower

Table 33-2 Chambers of the hearts of vertebrates

	Auricles	Ventricles
Fishes	One; receives blood returning through veins from entire body	One; receives blood from auricle and pumps it through gills on its way to all parts of body
Amphibia (frogs, toads)	Two separate; left receives blood from veins from lungs; right receives blood from veins from all parts of body	One; receives blood from both auricles and pumps mixture through arteries to all parts of body
Reptiles (lower types; lizards, snakes, turtles)	Two separate; left receives blood from veins from lungs; right receives blood from veins from all parts of body	Two partially separated; right receives blood from right auricle; left receives from left auricle; blood from two auricles mixed in partially separated ventricles and pumped to all parts of body
Reptiles (higher types; alligators, crocodiles)	Two separate; left receives blood from veins from lungs; right receives blood from veins from all parts of body	Two completely separated; right receives blood from right auricle; left receives blood from left auricle
Birds (adults)	Two separate; left receives blood from veins from lungs; right receives blood from veins from all parts of body	Two completely separated; right receives blood from right auricle; left from left auricle; left ventricle pumps blood to all parts of body; right ventricle, to lungs
Mammals (adults)	Two separate; left receives blood from veins from lungs; right receives blood from veins from all parts of body	Two completely separated; right and left have same functions as in birds

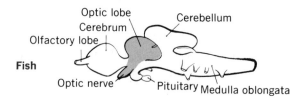

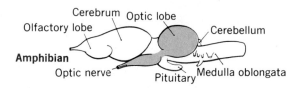

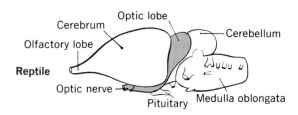

Fig. 33-16. Comparative structure of the principal brain divisions as shown by such vertebrate groups as fishes, amphibia, reptiles, birds, and mammals. Note the progressive enlargement of the cerebrum in higher groups. The pituitary body of the mammal is not shown; the origins of certain cranial nerves are shown. The optic nerve of mammals connects with the visual center of the cerebrum.
(From Hickman, Cleveland P.: Integrated principles of zoology, ed. 3, St. Louis, 1966, The C. V. Mosby Co.)

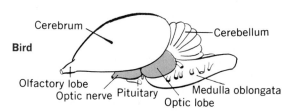

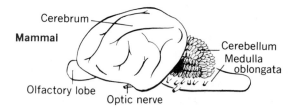

forms. Thus, knowledge of the anatomy of an animal gives a broad and general idea concerning its type of embryonic development.

These similarities of lower and higher types of organisms found by embryonic studies suggest a similar inheritance as a basis and probably an actual relationship between them. The only other alternative is that the same "blueprint," with slight modifications and alterations, was used in the process of specially and individually creating each species.

Evidence from comparative anatomy

Evidences from comparative anatomy include those from gross anatomy and vestigial structures.

From gross comparative anatomy

A detailed comparative study of the anatomy of apparently different types of animals reveals a multitude of similarities that really overbalance the more obvious dissimilarities. For instance, the differences exhibited by the five classes of vertebrates are relatively slight when compared with the many fundamental resemblances that they all possess.

The forelimbs of the frog, bird, cat, horse, and man, for example, are constructed on the same general structural plan and arise in a similar way embryonically. They are thus homologous structures, and such differences or variations as exist are principally the result of the absence of some minor part or the transformation of a certain part, depending on the specific use to which that part has been put. In general, nearly all the bones, muscles, nerves, and blood vessels are constructed and arranged in a homologous manner in the forelimbs of the entire group, from the lower types, frogs, to the higher types, man. The same thing is true for the hind limbs, digestive systems, reproductive systems, and circulatory systems of this series.

From vestigial structures

Man has approximately one hundred vestigial structures that are also represented, and often useful, in lower types.

Illustrations of vestigial structures in man include the following: (1) The vermiform appendix is a remnant of an organ that is useful in certain herbivorous animals and may have had a specific function in man generations ago. (2) The third eyelid in the inner angle of the human eye corresponds to the nictitating membrane, or lid, that moves laterally across the eye in such lower animals as the frog, bird, and dog. (3) Muscles of the external ear are useless for man but are used by lower animals to turn the ear in the proper direction to acquire the sound waves more accurately. (4) The terminal vertebrae (coccyx) are of no value to man, but they are the foundation for the external tail in lower animals. It is interesting to note that the early embryo of man (Fig. 33-14) possesses an external tail that is discarded before the advanced stages are reached. Only occasionally does the external tail persist in the adult man. (5) The lobe of the ear is of no practical benefit to man, although it may have had some function in the past. (6) The point of the ear, known as Darwin's point, on the edge of the upper roll or margin of the human ear, corresponds to the tip of the ear of animals, which holds their ears upright.

Vestigial structures in other animals are illustrated by the following: (1) The splint bones of the legs of the horse are remnants of original toes. (2) The poison glands of certain snakes are modified, specialized salivary glands from which they have evidently developed through descent with change. (3) The gill slits of the embryos of higher vertebrates disappear except one pair, which develops as the eustachian tubes, connecting the pharynx and the middle ear. (4) The milk glands of mammals are modified and specialized sweat glands of the skin. (5) Certain snakes bear small, useless hind limbs that structurally resemble those of other animals in which they are useful.

Evidence from comparative physiology

Since functions and structures are interdependent, one would expect to find fundamental physiologic similarities in organisms with structural similarities. The following examples illustrate this. (1) The blood of closely related organisms is more nearly alike chemically and physiologically than the blood of the more distantly related types. (2) The hormones of closely related organisms are quite similar and in some instances interchangeable. The insulin of the sheep pancreas may be used for an insulin deficiency in man. (3) The crystalline structures of bloods of similar organisms are more nearly alike than those of more dissimilar or unrelated organisms. In general, common properties persist in bloods of closely related types, and variations are greatest in distantly related forms.

Evidence from heredity and variations

Many evidences for evolution have come from the field of genetics. In the discussion of genetics in Chapter 31 the inheritance of variations in traits in living organisms was considered in connection with such phenomena as mutations, recombination of traits when parents with different traits were crossed, and crossing-over. These phenomena might be reviewed advantageously.

A mutant gene that has been altered, thus changing some trait of its bearer, is an important foundation of evolution. Such mutants are known to occur naturally, but they have also been induced experimentally. In each case the resulting observations may have shed some light on the evolution of certain traits in such organisms. H. J. Muller discovered that radiation induces mutations in the fruit fly *(Drosophila),* and more than 30 years ago L. J. Stadler noted similar phenomena in plants. It is hypothesized that heredity information is encoded in various permutations in the arrangement of subunits of the molecule of DNA (deoxyribonucleic acid) in the chromosomes. A mutation occurs when this molecule fails to replicate itself exactly. The mutated gene, and its resulting trait, is reproduced thereafter until its bearer dies without reproducing or until, by chance, the altered gene mutates again.

Evolution of man

Early man has left many interesting and valuable records that enable us to get an idea of his physical and mental traits, achievements, and activities. Skeletons, implements, and tools form the basis for the knowledge of our remote ancestors. Early man did not always bury his dead, so that records do not exist before the Pleistocene epoch.

Pebble-culture man (Australopithecus prometheus)

One of the most interesting sites in the history of early man is in South Africa where the Makapan River has formed a series of limestone caves. In the lower strata are found small stones that have been chipped into crude tools. Geologic studies suggest that this primitive, so-called pebble culture may have flourished approximately 750,000 years ago. Raymond A. Dart found remains of *Australopithecus prometheus,* a smallish, erect-walking creature whose brain was large enough to equip him for making such tools. Since anthropologists define man as a "toolmaking animal," the pebble-chipping *Australopithecus prometheus* may qualify as the most primitive creature that can be called definitely human.

The discoveries of *Zinjanthropus,* by Dr. L. S. B. Leakey, in eastern Africa, may extend the age of early man to 1,750,000 years ago.

Java man (Pithecanthropus erectus)

The skull cap, the left femur, and the lower jaw with three teeth of *Pithecanthropus erectus* were found in Java in 1891 by Eugene Dubois. It is thought that this type of man existed during the first glacial age of the early Pleistocene epoch. His cranial capacity was about 950 ml., which is approximately two-thirds that of an average, modern European, but half again as much as that of a large gorilla. His higher mental functions were limited because of the poorly developed frontal regions of his brain. The centers of taste, touch, and vision were probably well developed,

Fig. 33-17. Skulls of prehistoric man. **A,** Java man *(Pithecanthropus erectus);* **B,** Peking man *(Sinanthropus pekinensis);* **C,** Neanderthal man *(Homo neanderthalensis);* **D,** Cro-Magnon man *(Homo sapiens).* All skulls except **D** have been restored to a certain extent and are somewhat diagrammatic.

Fig. 33-18. Prehistoric men, photographs of restorations. From left: Java man, Neanderthal man, and Cro-Magnon man. *(Courtesy Dr. J. H. McGregor; from Hickman, Cleveland P., Sr., Hickman, Cleveland P., Jr., and Hickman, Frances M.: Integrated principles of zoology, ed. 5, St. Louis, 1974, The C. V. Mosby Co.)*

and he may have used speech of some type. The skull cap was very thick, and his forehead was low, receding, and with massive supraorbital ridges. His skull was narrow, and his jaw projected in an almost snoutlike fashion (Fig. 33-17 and 33-18). His average height was about 5 feet 7 inches. He lived on land and seems to have been more similar to man than any ape. He used sharpened sticks and stones for implements. It is likely that this and the next type, Peking man, were closely related, perhaps subspecies, of *Homo erectus.*

Peking man (Sinanthropus pekinensis)

Skulls, teeth, and brain cases were found near Peking, China, from 1926 to 1928, by D. Black. Peking man is supposed to have lived during the first interglacial age of the early Pleistocene epoch. His cranial capacity was about 1,000 ml.; hence his head was larger than that of *Pithecanthropus.* The brain case shows the brain to be human but small and comparing rather favorably with normal human brains of primitive men of today. The walls of the skull were thick. The forehead was low, receding, and possessed heavy supraorbital ridges. This early man used fire, because charcoal and charred remains of various materials have been found buried with his remains. He used tools and implements of bone and stone. At present, over 2,000 stone implements have been found with the unearthed remains.

Neanderthal man (Homo neanderthalensis)

The skull cap and parts of the skeleton of Neanderthal man were found in the Neanderthal valley near Düsseldorf, Germany. This type of man is thought to have existed during the third interglacial and third glacial ages of the Pleistocene epoch. His cranial capacity was between 1,400 and 1,600 ml. His higher mental faculties were not well developed. The anterior region of his brain was not as highly developed nor as large as that of *Homo sapiens*. Neanderthal man had a low, broad forehead with massive supraorbital ridges. His eyes were large and round, his nose was broad, and he had a receding chin. His knees were bent, and his head was held forward when he stood or walked; his spinal column was slightly curved—characteristics that gave him a peculiar slouching attitude. His skeleton was not over 5 feet 4 inches tall, usually averaging less than 5 feet. His feet and hands were large, and his legs were longer than his arms. Many skeletons of this type of man have been found in England, Belgium, Germany, France, Spain, Italy, Palestine, Syria, Arabia, Iraq, Rhodesia, and China, suggesting a very wide distribution. From their remains it is thought that they lived at the entrance to caves rather than in them; that they used a language; that they used fire for warmth and cooking; that they were great hunters and ate the bone marrow of their captured animals; that they used implements of flint, bone, and unpolished stone; that they believed in a life after death because they buried flint implements and foods with their dead, probably clothing the hairy body in animals skins.

Additional skeletons of Neanderthal man were found in 1957 and 1960 by Columbia University anthropologists in a dry, well-protected cave at Shanidar, 250 miles north of Baghdad, Iraq. Probably these remains represent a "conservative" type of man that became extinct in Iraq about 45,000 years ago, as found by carbon 14 dating. Skeletons were found at different levels, the oldest at a 45-foot depth were estimated to be 70,000 years old. The skeletons were more complete than those usually found, almost entire ones were quite well preserved, and they give physical evidence of the stone-age (Mousterian) culture. Also discovered in a burial place were wild wheat, date palm pollen, and additional evidence of a change from cave to village life.

Cro-Magnon man or modern man (Homo sapiens)

Five skeletons were found in the Cro-Magnon cave in Dordogne, France, in 1868. Cro-Magnon man is thought to have been present during the fourth glacial or ice age of the late Pleistocene epoch, even down to the recent epoch. His cranial capacity was from 1,400 to 1,500 ml., which is equal to if not greater than, that of the average European of today. The anterior part of the brain was large and well developed. The skull was large, long, and narrow. The forehead was high with moderate supraorbital ridges. The face was broad; the jaws, wide; the cheek bones, large; the eyes, large and far apart; the spinal column had four distinct curves. The male averaged 6 feet 2 inches in height, which suggests a strong, athletic race. The chin was well developed. In general, Cro-Magnon men were probably handsome people, comparing quite well with existing races. They lived in caves and rock shelters. They hunted and fished by means of skillfully made harpoons and spears. Many implements and ornaments of bone have been found. They developed an art in which they carved and made drawings in oil. They developed primitive industries in which they used bone more extensively than flint. The Cro-Magnon man is a good ancestor of modern man from a physical and mental standpoint. In the distant future, when the remains of some of us are unearthed, what type of record will we have left, and for what will our civilization be noted?

Review questions and topics

1 Explain why abiogenesis was believed in the past, and when and how it was disproved.

2 Discuss biogenesis, including some early investigators who contributed to an explanation of this phenomenon.

3 Discuss some of the theories of the origin of life on earth. Which one, if any, seems most plausible? Why do you say so?

4 Discuss the various evidences that attempt to explain evolution, including examples of each type of evidence. Which science, if any, contributes the most logical evidence? Why do you say so?

5 Discuss the evolution of horses and common wheat in some detail.

6 What do comparative studies of the hearts and brains of vertebrates suggest?

7 Discuss the role of modern genetics in explaining evolution, including some examples to prove your statements.

8 Of what importance in everyday life is a knowledge of animals and plants of the past and their records?

9 List several reasons why certain softer types of animals and plants have left no fossil records.

10 How are we able to estimate the age of the earth by a scientific study of the fossils in the successive strata of the earth?

11 How have the various estimates of the age of the earth been made? How accurate are these estimates?

12 What is the estimated age of the earth from the Lower Precambrian period down to the present?

13 What percentage of the total age of the earth represents the time that human beings have inhabited the earth?

14 List the various types of fossils and records that ancient man has left and include the specific manner in which each has been preserved.

15 Make a table of the more representative types of ancient man, including the outstanding characteristics of each.

16 Explain how and where records of ancient man are discovered. What is the significance of where these records have been discovered? Are additional records being discovered at the present time? Where? (Read articles on present-day discoveries.)

17 What conclusions might you draw from studies of the sequence of records left by ancient man?

18 Give the general characteristics and boundaries of each of the regions into which the animal world may be divided. What factors might prevent migration from one region to another?

19 How may a certain condition act as a barrier to one type of organism and at the same time act as a method of transportation for another type? Give specific examples.

20 What is the effect of better methods of transportation by man on the distribution of certain types of organisms? Explain in detail giving specific examples.

Selected references

Boughey, A. S., editor: Population and environmental biology, Belmont, Calif., 1967, Dickenson Pub. Co., Inc.

Dobzhansky, T.: Mankind evolving, New Haven, 1962, Yale University Press.

Dobzhansky, T.: The present evolution of man, Sci. Amer. 203:206-217, 1960.

Eiseley, L. C.: Charles Darwin, Sci. Amer. Feb. 1956.

Howells, W. W.: Homo erectus, Sci. Amer. 215:46-53, 1966.

Huxley, P. M.: The confirmation of continental drift, Sci. Amer. April 1968.

Laughlin, W. S., and Osborne, R. H., editors: Human variation and origins. Readings from Scientific American, San Francisco, 1967, W. H. Freeman and Co. Publishers.

Lock, D.: Darwin's finches, Sci. Amer. April 1953.

Mayr, E.: Animal species and evolution, Cambridge, Mass., 1963, Harvard University Press.

Napier, J.: The evolution of the hand, Sci. Amer. 207:56-62, 1962.

Romer, A. S.: The vertebrate story, ed. 4, Chicago, 1959, University of Chicago Press.

Ross, H. H.: A synthesis of evolutionary theory, Englewood Cliffs, N. J., 1962, Prentice-Hall, Inc.

Ruibal, R., editor: The adaptations of organisms, Belmont, Calif., 1967, Dickenson Pub. Co., Inc.

Simpson, G. G.: The major features of evolution, New York, 1953, Columbia University Press.

Simpson, G. G.: This view of life, New York, 1963, Harcourt, Brace & World, Inc.

Solbrig, O. T.: Evolution and systematics, New York, 1966, The Macmillan Company.

Stebbins, G. L.: Process of organic evolution, Englewood Cliffs, N. J., 1971, Prentice-Hall, Inc.

Tax, S., editor: Evolution after Darwin, 3 volumes, Chicago, 1960, University of Chicago Press.

Wallace, B.: Chromosomes, giant molecules and evolution, New York, 1966, W. W. Norton & Company, Inc.

Wallace, B., and Srb, A.: Adaptations, ed. 2, Englewood Cliffs, N. J., 1964, Prentice-Hall, Inc.

Some early contributors to the knowledge of
EVOLUTION AND RELATED TOPICS

Thales (624-548 B.C.)

A Greek who proposed a theory that water (ocean) was the mother of all life.

Anaximander (611-547 B.C.)

A Greek who thought that life arose from a mixture of water and earth and that land forms arose from aquatic types, particularly under the influence of the sun's heat.

Heraclitus (510-450 B.C.)

A Greek natural philosopher who stated that "struggle is life" and "all is flux"—thoughts that are basic to modern ideas in evolution.

Empedocles (495-435 B.C.)

A Greek who theorized that living organisms were generated spontaneously from scattered materials, being attracted by love and hate.

Aristotle (384-322 B.C.)

A Greek who theorized that in the living world there was a gradual change from the simple and imperfect to the more complex and perfect, thus suggesting the idea of evolution.

Saint Augustine (353-439 A.D.)

Saint Augustine interpreted the first chapter of Genesis as stating that in the beginning matter was created with the properties and potentialities to evolve into living and nonliving worlds as we know them today.

Francesco Redi (1621-1697)

An Italian who overthrew the theory of spontaneous generation, which stated that life arose from nonliving materials spontaneously, by discovering that eggs and larvae of insects originated from previous living insects, rather than from nonliving substances.

Buffon

Georges de Buffon (1707-1788)

A French naturalist who excluded the possibility that species have the ability to evolve (change). He probably was influenced by the ancient ideas revived during the Renaissance that discredited such a possibility. *(Historical Pictures Service, Chicago.)*

Charles Bonnet (1720-1793)

A Swiss naturalist and philosopher who first used the term "evolution" but not quite as we do today, and who conceived that organisms could be arranged in a ladder-like, linear series.

Erasmus Darwin (1731-1802)

An English evolutionist who was the grandfather of Charles Darwin. He believed that acquired traits could be transmitted to future generations and he may have influenced Lamarck, who had similar views. *(The Bettmann Archive, Inc.)*

Jean Baptiste Lamarck (1744-1829)

A French biologist who was a student of organic evolution. He believed that environmental influences, and the effects of the use and disuse of body parts, were causes of evolutionary changes—a theory that laid the foundation for the "Theory of the Inheritance of Acquired Traits."

Thomas Robert Malthus (1766-1834)

An English economist whose "Essay on Population," published in 1798, inspired both Darwin and Wallace in their theories of evolution.

Charles Lyell (1797-1875)

A Scottish geologist who laid the foundations for the science of earth structure in his *Principles of Geology* (1830-1832). He is considered an important contributor to the theory of organic evolution because of his influence on Charles Darwin and Alfred Russel Wallace.

Georges Cuvier (1769-1832)

A Frenchman who supported the "Cataclysmic Theory," in which he stated that there had been numerous creations, each of which was followed by a cataclysm that destroyed it, and its place was taken by new forms.

Darwin

Charles Darwin (1809-1882)

Darwin, in a voyage around the world in the sailing ship Beagle, indicated the descent of species by the development of varieties from common stocks. This process entailed a "struggle for existence," which resulted in a "natural selection of species" and a "survival of the fittest." He wrote *The Origin of Species by Means of Natural Selection* (1859).

Thomas Henry Huxley (1825-1895)

An English surgeon who actively supported the views of Charles Darwin and assisted in promoting them extensively.

Wallace

Alfred Russel Wallace (1823-1913)

An Englishman who studied animal geography in the East Indies and concluded that the life of wild animals is a struggle for existence. He worked on the problem of the origin of species and arrived at conclusions concerning evolution that resembled those of Charles Darwin. *(Historical Pictures Service, Chicago.)*

Louis Pasteur (1822-1895)

A French microbiologist and chemist who proved that only living organisms such as bacteria and yeasts could cause fermentations. He proposed a method of preventing this process by heating to a temperature high enough to kill the germs. This method is known as pasteurization. Pasteur's work ended the controversy regarding the possibility of spontaneous generation of living organisms from nonliving materials.

August Weismann (1834-1914)

A German who distinguished between body cells and germ cells and proposed the "Theory of the Continuity of Germ Plasm From Generation to Generation" (1885). He opposed the idea that acquired traits might be transmitted.

PART SEVEN

Organism and environment

Ecology
Behavior and relationships among organisms

34 Ecology

Before getting into the specifics of this chapter it seems necessary to confront the fact that the term "ecology" is misused by a great many people. Words seem to mean what we want them to mean whether we are reading *Alice's Adventures in Wonderland* or the latest report from our favorite politician. This chapter will not deal with the exhilaration of backpacking on a fine day nor with the casual observing of animals in the woods. Neither is it directly involved with such things as "destroying the ecology by oil spills," nor air pollution, strip mining, and endangered species. All of the above and many more appeals such as "Save the Ecology," "Energy not Ecology" are decisions and attitudes of individual people. To say that we will not deal with these topics here is not to state that they are unimportant. It is primarily to state that we will be considering the science of ecology.

By definition, then, *ecology* is the study of the interrelationships between organisms and their environment. The environment includes both the physical components such as temperature, humidity, chemicals, soil, and other physical features of the region as well as other members of the same species and other kinds of organisms in the same area.

Fig. 34-1. A small woodland stream as an example of an ecosystem.
(Photograph by F. S. De-Martino, University of Dayton.)

THE BIOSPHERE

That part of the earth in which life exists is usually called the *biosphere*. Effectively it is the outer land and water envelope on the surface of the earth. It extends at most a few miles above and below sea level. It contains the water, energy, and nutrients necessary to support metabolism of living things. Its major components are: the land, the water in and around it, the oceans, estuaries, and fresh water ponds, lakes and streams, and the atmosphere surrounding and linking these features. When these physical features are considered along with the organism existing in a particular region, they form ecosystems.

ECOSYSTEMS

The term *ecosystem* refers to a natural unit of organisms and their physical environment. With the exception of energy input, an ecosystem is a relatively self-sufficient system. There is a cyclic exchange of minerals and energy between organism and organism and between organism and environment.

The physical environment is important in determining the general appearance of the organisms in an ecosystem, especially with reference to the size and type of vegetation. The species in a particular environment adapt to it by acquiring a somewhat similar body form; for example, fish and mammals in the sea have the same streamlined form. The limiting factors of temperature, light, water, oxygen, minerals, and others, influence the physiologic and structural adaptation of organisms.

Ecosystems vary in size from small ponds to oceans, from a small field to half a continent of a desert or forest (Fig. 34-1).

Physical factors

The physical environmental factors include temperature, light, wind, gravity, soil conditions, pressure, and the presence or absence of natural barriers or of natural methods of dispersal.

Fig. 34-2. Sufficient light filters through the trees, permitting the growth of seedlings. *(Photograph by F. S. De-Martino, University of Dayton.)*

Fig. 34-3. Cliff face plant growth. Even in such a seemingly unfavorable environment there is sufficient water, light, and substrate for a variety of plants. *(Photograph by F. S. De-Martino, University of Dayton.)*

TEMPERATURE

Most organisms have an optimal temperature at which their lives and their metabolic processes are maintained most successfully. They also have a minimal and a maximal temperature below and above which they cannot live. Hence, organisms tend to select those temperatures for which they are best fitted. The freezing of water in which animals live affects them in the following ways: (1) some become inactive during the frozen period; (2) some escape the freezing by burrowing deep in the earth; (3) some die under such conditions, but only after they have produced resistant stages in order to carry on the species. A covering of ice on a body of water not only affects the animals directly but also indirectly by altering the oxygen and food supplies. This may explain why aquatic animals come to holes that are cut in the ice.

LIGHT

Organisms differ in their requirements for light. Some species require light in order to carry on many of their metabolic activities. Indirectly, animals are affected by the presence of plants that require light for their existence. Plants that depend on light supply animals with food, protection, and oxygen and thus influence animal distribution. Different plants vary widely in their light requirements. The amount of light may determine the production of leaves or of flowers or other features. In some cases light is an inhibiting factor. Plants tend to thrive in areas that contain the proper quality and quantity of light to meet their particular needs (Fig. 34-2).

The distribution of certain animals is quite different in daylight than at night. For example, certain insects can be observed only in the daytime (diurnal), and others are found more abundantly at night (nocturnal). In addition to the absence of light, a lower temperature and additional moisture at night influence the distribution of such animals. This is true particularly for those types that have no special abilities to prevent the rapid evaporation of moisture from their body's surfaces. The extremes of life on the desert during the day and during the night are good examples of this.

WIND

The direction and velocity of the wind are factors in the dispersal of organisms. The wind affects animals in various ways: (1) it may aid their dispersal or it may cause injury to them directly; (2) it may stir up sediments in waters or place dust in the atmosphere, thus affecting their respiration; (3) it may influence the oxygen content of the water or atmosphere; (4) it may affect the temperature of the water or atmosphere; and (5) it may bring obnoxious gases that may affect distribution.

Wind affects plants in similar ways: (1) it may assist or prevent pollination; (2) it may help to disperse seeds or spores; and (3) it may distribute carbon dioxide, oxygen, or obnoxious gases that might influence plant activities; and (4) strong winds may uproot or destroy vegetation.

SOIL CONDITIONS

The various physical conditions of different types of soils are important factors in the distribution of living organisms in or on them. Among such factors are included moisture content, degree of aeration, exposure to the sun (heat and light), hardness or looseness, and presence or absence of specific nutrient materials in usable forms (Fig. 34-3). The slope of the soil affects its drainage and erosion, thus influencing the distribution of organisms in or on it. Some animals require certain types of soil for protection or burrowing.

PRESSURE

Pressures in water, soil, and air vary with the depth or altitude. Air pressure is 15 pounds per square inch at sea level and decreases uniformly as one ascends from sea level to higher regions. In higher elevations air pressure may become too low to allow normal respiration in certain animals. Water pressure increases with the depth; in the ocean it is equal to the depth in feet multiplied by 0.434. Thus, at 200 feet the water pressure is approximately 87 pounds per square inch. This pressure influences the vertical distribution in deeper bodies of water because not all organisms are constructed to withstand such pressures.

Natural barriers and methods of dispersal

Different types of living organisms exist in environments that are conducive to their survival. Any natural hindrance to dispersal is known as a barrier. A barrier to dispersal for one species may be a natural method of dispersal (highway) for another species. Water is a natural method of dispersal for fishes, but it may prove to be a natural barrier for terrestrial forms if it is too deep or extensive. Mountains may be natural barriers even for terrestrial organisms because of altitude, snow, ice, and lack of proper vegetation. The absence of plants of certain types from a region may serve as a barrier to animal dispersal because some animals depend upon those plants for food, shelter, and homesites. Floods are barriers to certain forms but serve as methods of dispersal for others. Heavy seeds, which cannot be carried easily by wind or animals, may float on water.

Most animals and plants have particular methods for their dispersal. Dandelion seeds are so constructed that they are carried easily by the wind (see Fig. 17-8). Some types of seeds, such as burs, are distributed by animals by sticking to fur. Seeds may be dispersed by the bird population. If types of dispersal are lacking, a particular plant or animal is limited in its distribution.

Chemical factors

Chemical environmental factors include quality of the soil and water (moisture), oxygen, carbon dioxide, and obnoxious gases, and quality and quantity of usable foods.

QUALITY OF THE SOIL AND WATER

The chemical composition of a soil affects not only animals and plants living in or on it but also the aquatic life living in any water that comes in contact with that soil. Highly acid waters or soils are not ideal for certain organisms, whereas neutral or even alkaline soils may be. The pH of soil or water is an important ecologic factor. The chemical quality of soil or water is an important ecologic factor. The chemical quality of soil or water determines the kind of vegetation growing in or on it; the lack of trace elements is usually the limiting factor. Plants affect animal distribution, because different vegetations supply a variety of foods, protection from the elements, and concealment from enemies. Earthworms may not abound in sandy soils because of the irritations induced by sand grains, a minimum of usable dead plant food, a lack of sufficient moisture, and interference of loose sand in making permanent tunnels.

Plants require an adequate amount of water and soil of the proper quality (chemical composition) to meet their requirements. In some instances the requirements are rather specific, and plants will not be found in areas that do not satisfy those needs. For example, cranberry plants grow primarily in acid soils. Certain plants may be eliminated by altering the pH of the soil or water. A plant such as the dandelion is less specific in its requirements, growing in a variety of soils.

Living organisms require water for various purposes, although the quantity and quality satisfactory for one type may not be suitable for another.

Protective substances and structures prevent excessive evaporation and thus permit such plants and animals to survive in relatively dry areas.

The pollution of waters with wastes and obnoxious materials affects the distribution of animals and plants. Organisms normally living in fresh, clean water cannot survive in a polluted environment. Conversely, organisms thriving in polluted waters do not thrive in fresh clean water. Industrial wastes are responsible for the elimination of fishes, aquatic snails, clams, crustaceans, and certain aquatic plants from many streams.

OXYGEN, CARBON DIOXIDE, AND OBNOXIOUS GASES

All living organisms require a certain quantity of oxygen, and an insufficient supply will affect their distribution. Oxygen is required for the oxidation of foods to supply energy and other materials necessary for their metabolism. Chlorophyll-bearing plants require oxygen, but they may obtain it from the process of photosynthesis. In this process the supply of carbon dioxide may be influential in their distribution. Excess carbon dioxide affects the distribution of terrestrial or aquatic animals; non–chlorophyll-bearing plants, such as bacteria and fungi, may not be affected as much by the carbon dioxide content of their environments.

Obnoxious gases, either naturally or artificially produced, affect the distribution of plants and animals. In fact, certain gases are utilized to combat many undesirable animals such as insects, rats, moles, and gophers. Other gases may interfere with plant respiration, transpiration, and photosynthesis, thus limiting the distribution of vegetation.

HABITATS AND ECOLOGIC NICHES

The *habitat* of a plant or animal is the locality or external environment in which it lives. More than one plant or animal species may live in a particular habitat, which may be large or small. The ecologic *niche* (F. *niche*, place or recess) is the "status" of a plant or animal within an ecosystem or community. An often used analogy is that the habitat is the address of a species and the niche is its profession.

The niche includes all the physical, chemical, physiologic, and biologic factors that an organism needs in order to live. This is influenced by the organism's structure, physiologic responses and abilities, and adaptations. When considering the niche of a particular organism, we must include such things as its food source, the organisms that use it as food, its effects on other organisms, its range of movement, and its effect on the nonliving surroundings.

The place where a species is found is its habitat. Generally we speak of four major habitats: marine—the oceans of the world, estuarine—the bays and tidelands where the saltwater of the ocean meets the tideflats and/or freshwaters, freshwater—the ponds, lakes, and rivers receiving the runoff of rain from the soil, and terrestrial—the land masses, fields, plains, deserts, and mountains of the world (Table 34-1).

Within any major habitat both physical and biotic pressures are operating to cause a species to disperse. Some of these have been mentioned previously. The result of these pressures is the dispersal or spreading away of a species from its place of origin. The dispersal is aided by highways and hindered by barriers.

Biogeographic regions

Because of its adaptation to a particular environment, a species is restricted by its morphology and physiology to those areas where that type of environment exists. However, it may be found that an apparently identical environment in two different and entirely separate places contains different species. In the geologic past the boundaries of sea and land changed, sometimes creating natural barriers, other times providing highways for migration and dispersal. Today we know that many widely separated current types had common ancestors in the past. The fossils of extinct North American camels are the remains of the ancestors of the present day Old World camels as well as the llamas of South America.

The earth has been divided into regions (Fig. 34-4), each with its different characteristics and typical flora and fauna. A brief consideration of these regions is given in Table 34-2. Many mammals of the Palaearctic region are similar to those of the Nearctic region, and many trees and plants are common to both regions. These two regions are so similar that they are sometimes combined into what is called the Holarctic region (Gr. *holo*, whole; Arctic). Animals common to both Nearctic and Palaearctic regions include beavers, deer, hares, foxes, wildcats, and bears.

Table 34-1 Typical habitats and organisms found in them

Habitat	Characteristics of habitat	Typical organisms present
FRESH WATER OR AQUATIC		
Rapid streams	Rapid flow of water; usually hard, firm bottom; usually loose rocks with crevices for protection; usually shallow, hence much light and oxygen	Larvae of mayflies, stone flies, caddis flies, midges (bloodworms); certain snails and fishes (such as darters); usually a minimum of vegetation
Pools (quiet parts of streams)	Slow flow of water; usually soft bottom of mud or sand for burrowing	Larvae of dragonflies, damselflies, midges; clams; various types of fishes, water snakes, and turtles; certain types of vegetation common
Ponds	Slow flow of water; usually soft bottom; many factors depend upon depth	Larvae of dragonflies, damselflies, caddis flies, midges; clams and water snails; crustaceans; fishes, water snakes, and turtles; certain types of vegetation common
Lakes	Wave action depends upon many conditions; bottom may be mud, sand, gravel, loose rocks, or solid, each influencing type of life found; sand and moving gravel undesirable for sessile organisms; sand may interfere with gill-breathing types; lakes larger and deeper than ponds	Types of animals present vary greatly, depending on specific qualities present in different areas of lake; vegetation may be limited because of sand movement, limited nourishment, wave action, and reduced light and oxygen at certain depths
LAND OR TERRESTRIAL		
Open fields	Temperatures usually severe in summer and winter; moisture evaporation high; intense light; great wind action; little protection except in ground	Insects such as beetles, grasshoppers, leaf-hoppers; certain types of spiders, toads, snakes, birds; bees (if flowers present); vegetation usually short
Deserts	Temperatures severe in summer and winter; moisture evaporation very high; intense light; great wind action; limited protection	Insects such as beetles, grasshoppers, leaf-hoppers; certain types of spiders, toads, snakes, birds; vegetation usually sagebrush, cacti, yucca, bunchgrasses
Tundras	Winters long and cold; only upper limits of soil thaw; ground water cold and limited; plant growth season short; often strong winds	Few animals can tolerate the ravages of this polar or subpolar area; vegetation consists of lichens, mosses, certain grasses, herbs, and shrubs
Forests	Temperatures usually more moderate than surrounding areas; protection from light, heat, wind, and moisture evaporation; reduced wind action	Insects such as bees, beetles, crickets, certain grasshoppers and katydids; centipedes and millipedes; spiders; snakes; tree toads; numerous birds, mammals

A region that is included here with the Australian region is sometimes called the Polynesian region (Gr. *poly,* many; islands) and consists of the oceanic islands of the tropical Pacific, including the Hawaiian Islands, Samoa, and the Fiji Islands. They were formed by volcanic eruptions and are fringed with coral reefs. The vegetation is often large and herbaceous, such as palm and

Table 34-2 Geographic regions of the world

Region	Location	Some typical animals
Nearctic (Gr. *neo-*, new or late)	North America (down to edge of Mexican plateau), Greenland	Blue jays, rattlesnakes, raccoons, opossums, skunks, prairie dogs, American water dogs
Palaearctic (Gr. *palae-*, ancient)	Europe, Africa (north of Tropic of Cancer), Asia (north of Himalayas), Japan	Nightingale, Japanese water dog, camel, dromedary (of central Asia and northern Africa)
Neotropical (Gr. *neo-*, new)	Central America, Mexico, South America, West Indies	Tapir, sloth, armadillo, llama, flat-nosed monkeys, tree anteaters, tree porcupines
Ethiopian (Gr. *aithiops*, black face)	Africa (south of Sahara Desert), southern Arabia, Madagascar	African elephant, hippopotamus, rhinoceros, zebra, giraffe, lion, leopard, gorilla, chimpanzee
Oriental (L. *orientalis*, eastern)	India (south of Himalayas), southern China, Philippines, Borneo, Java, Sumatra	Indian elephant, rhinoceros, tiger, gibbons, cobra, jungle fowls (ancestors of domestic fowls)
Australian	Australia, New Zealand, New Guinea, Tasmania, Papua	Marsupial animals (with pouch for carrying young), such as kangaroo, duckbill platypus; certain wingless birds

banana trees. There are no land mammals, except bats, and no amphibia. The types of living animals are quite limited.

GEOGRAPHIC DISTRIBUTION

There are two general types of geographic distribution of organisms: (1) lateral (longitudinal) distribution, in which the organisms are spread over the surface of the earth in the various geographic regions, and (2) vertical distribution, in which organisms are distributed throughout the various altitudes (Fig. 34-5). The latter type emphasizes the differences in distributions on mountains, in valleys, in caves, and in the depths of the sea. There are regions of distribution as we ascend from the lowest depths of the ocean to the top of the highest mountain.

Principles of distribution
Principle of dispersion

This principle illustrates the natural tendency of some animals to migrate from their birthplace, because more offspring are produced than can be accommodated in that habitat. This reproductive pressure tends toward overpopulation, and dispersal is an attempt to remedy it. Offspring and parents compete with each other.

Principle of definite habitats

The habitat (home) of a particular species is determined by several factors. (1) The availability of usable foods is essential. Herbivorous animals, such as deer, must have suitable vegetation, whereas carnivorous animals, such as tigers and lions, must have suitable flesh foods. (2) The availability of water also affects the selection of a habitat. Some water is essential for all animals. Many species in dry climates prevent excessive evaporation by some type of covering. Some species found under rocks are there for moisture and protection. The depth, salinity, and pH of water influence the selection of a habitat. (3) Oxygen content of the air is important in the habitat of both terrestrial and aquatic types. (4) An optimum temperature must be present. Ani-

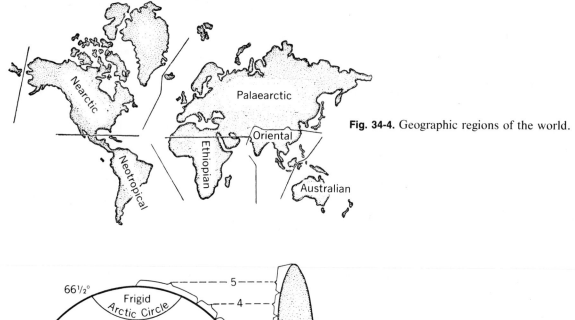

Fig. 34-4. Geographic regions of the world.

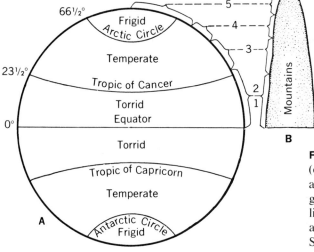

Fig. 34-5. Parallel distribution of organisms (diagrammatic). **A,** In longitude; **B,** in altitude, such as mountains. **1** and **2,** Tropical and subtropical organisms; **3,** deciduous trees; **4,** evergreen trees; **5,** limited varieties of such plants as mosses, lichens; a similar transition exists from the equator to the South Pole.

mals tend to seek the temperature for which they are best suited. Many animals in the tropics pass the summer in a condition of aestivation or semi-torpid condition. Certain animals living in colder climates pass the winter in various ways: (1) hibernation, or a period of inactivity in some protected location, (2) migration by species of birds that fly from the Arctic regions to the tropics, and (3) adaptation to the cold habitats by increasing their fat layers as well as their coats of hair.

Principle of barriers and highways

What is a barrier for one species may serve as a favorable highway for another. Some of the more common factors that interfere with dispersal are: (1) There may be lack of proper food along the migration route. (2) Water may be a

Fig. 34-6. Biomes of North America (boundaries of various regions are given in a general way).

barrier for terrestrial forms but may be used successfully by aquatic types. The size, depth, temperature, acidity, and pressure in bodies of water influence dispersal of aquatic organisms. The aridity and humidity of terrestrial environments act as barriers or highways, depending on the type of animal. Salt water is a barrier for freshwater forms, and fresh water serves as a barrier for marine organisms. (3) Various kinds of land serve as barriers or highways, depending on the animal. Forests act as barriers for open-country or prairie-inhabiting species, while deserts and open countries limit the spread of forest-inhabiting types. (4) Winds, especially if strong, carry certain animals in various directions, which results in their migration into more or less favorable habitats. (5) Temperature prevents the dispersal of many animals, either by its direct effect on the migrant or by its effect on the vegetation that influences food and shelter. (6) The lack of inherent adaptive ability may prevent sufficiently quick adaptation to the new conditions. This is frequently called a biologic barrier.

The following methods of dispersal are common in the animal world: (1) Driftwood may transport animals for great distances. William Beebe, on his *Arcturus* voyage, observed fifty-four species of marine fishes, worms, and crabs on one floating log. (2) Ships transport various types of organisms from port to port. Unknown

Table 34-3 Biomes of North America

Biomes	Plants typically present	General characteristics and location
Tundras	Certain mosses, lichens, grasses, sedges, herbs, low shrubs, dwarf trees	North of latitude 55° to 60°; long, cold winters, with low soil temperatures and limited amounts of cold water; subsoil usually frozen; short growing season; strong winds
Deserts	Sagebrush, cacti, yucca trees, bunch grasses, small herbs	Arizona, Nevada, parts of New Mexico, California, Texas, northern Mexico, and peninsula of southern California; small rainfall; intense heat and light; high loss of water from plants; fairly strong winds
Grasslands	Bunch grasses, cacti, various shrubs	Central Texas to Manitoba and along foothills of Rockies from New Mexico to Alberta; light rainfall; humus soil over sand and clay; few trees because of soil moisture and great evaporation from heat
Forests		
(a) Northern evergreen	Cone-bearing trees, such as spruce, balsam fir, white pine, red or Norway pine, Jack pine, hemlock, arborvitae; deciduous trees, such as balsam poplar, aspen, white birch	From Atlantic to Pacific oceans between tundra on north and Great Lakes on south, and extending northwestward to Alaska
(b) Southern evergreen	Longleaf pine, shortleaf pine, bald cypress, magnolia trees, water oaks, gum trees	Southeastern United States from Texas to Florida and Virginia; many low, rolling sandy plains; also high coastal plains farther from the ocean; many swamps
(c) Deciduous forests	White oaks, black oaks, hickories, chestnuts, walnuts, maples, ash, elm, birch, and certain cone-bearing trees	From central New York to Texas and Louisiana; from Wisconsin to Oklahoma
(d) Rocky Mountain forests	Western yellow pine; lodgepole pine, Douglas fir; western hemlock, western larch	Along Rockies from southern Mexico to Columbia; mountains of various elevations and climates; variety of trees exist, but none above 10,000 feet where low, tundra-like vegetation occurs
(e) Pacific Coast forests	California region—redwoods and sequoia trees Washington-Oregon region—with mild winters and good rainfall—Sitka spruce, Douglas fir, western hemlock, white pine; birch, maples, poplars; dense growths of ferns Canadian-Alaska region—Sitka spruce, Douglas fir, western hemlock	Extend along western slopes of mountains from California to Alaska

Table 34-3 Biomes of North America—cont'd

Biomes	Plants typically present	General characteristics and location
(f) Tropical forests	Various palms, tropical orchids, mangrove swamps, lianas (woody, climbing vines); dense jungles in certain places	West Indies, Central America, coasts of Mexico, southern tip of Florida

numbers of rats have had free transportation in this manner. (3) Water and floods transport organisms mechanically, drive them from their original habitats, or change the food supplies sufficiently to require dispersal. (4) Aquatic animals transport other animals on or within their bodies; the larvae of clams may be carried on the gills and fins of fishes. (5) Terrestrial animals transport other animals on or within their bodies. Birds may carry eggs, larvae, pupae, or adults of smaller animals. (6) Winds direct the course of certain animals or blow objects to which certain types are attached. (7) Glaciers cause animal migrations by actually transporting them or by changing the temperature or food supply. (8) Man, either knowingly or unknowingly, aids in animal dispersal by means of automobiles, trains, and airplanes.

Principle of discontinuous distribution

This is illustrated by the presence of the same species of animal in two widely separated regions. It is thought that the distribution of that species was continuous between the regions at some earlier time. In the Pliocene epoch of geologic history, tapirs were distributed over nearly all of North America, Europe, and northern Asia, but today they are present only in Central and South America, southern Asia, and the Malay archipelago.

Principle of vertical distribution

According to the principle of vertical distribution, the organisms of higher elevations of mountains tend to simulate those of the polar regions of the earth. As we proceed downward in altitude, the forms simulate those that might be found in traveling from the poles toward the equator. In general, the temperate zones extend not only laterally north and south of the equator, but also vertically in parallel succession from the somewhat mild conditions at sea level to the somewhat frigid conditions at the tops of higher mountains.

BIOMES

The major land masses of the earth are often studied in terms of the most characteristic plant life that occurs there. In these terms the areas are referred to as biomes. For example, the North American continent can be considered as a number of distinct biomes (Fig. 34-6). These are summarized in Table 34-3, in which the general environmental characteristics and the plants typically present in each region are given. The environmental characteristics differ in tundras, deserts, grasslands, and forests. If the distribution of forests is studied in detail, it is apparent that different types of trees are present in various parts of the continent, depending upon the environment peculiar to each area.

LIVING COMPONENTS OF THE ECOSYSTEM

In addition to the physical factors considered above, an ecosystem includes the living components. The pressures of survival for living organisms concern primarily food, shelter, mates, and the growth and dispersal of offspring. As indicated before, an ecosystem is relatively stable except for energy inputs. The necessary energy comes from the sun to the plants in the system.

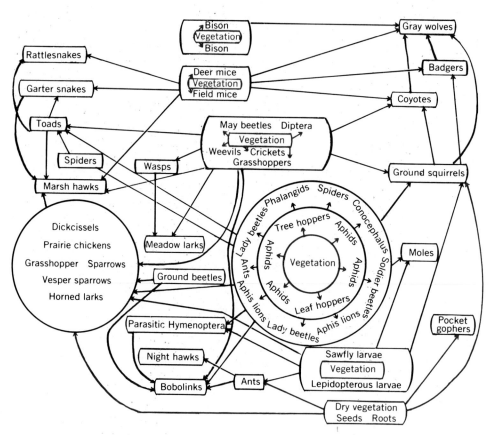

Fig. 34-7. Food interrelationships in a hypothetical prairie community. Arrows point toward the consumers.
(Adapted from Shelford; from Potter, G. E.: Textbook of zoology, ed. 2, St. Louis, The C. V. Mosby Co.)

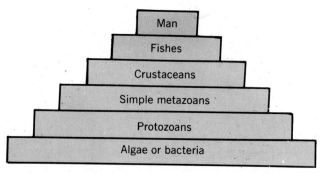

Fig. 34-8. Food pyramid. The sizes of the boxes are not an indication of volumes required at each level.

Trophic levels

The trophic (food related) activities of an ecosystem are considered in a fairly simple pattern. Organisms fill one of three basic functions. They are either (1) *producers,* the plants that manufacture foods by photosynthesis, (2) *consumers,* the animals that consume foods, or (3) the *decomposers,* those organisms such as bacteria and fungi that decompose or break down the organic residue of dead plants and animals, returning the components to the soil as minerals or fertilizers to be used by plants.

Modes of nutrition

The trophic levels of a biotic community may be further elaborated by considering modes of nutrition. The producers in the community show *autotrophic nutrition.* They are able to capture the energy of the sun by photosynthesis or in a small extent in some bacteria by a related process called chemosynthesis. These plant producers are eaten by animals called *heterotrophs.*

These are the consumers. The *primary consumers* are herbivores, that is, plant eaters. These in turn are eaten by *secondary consumers* called carnivores (flesh eaters). *Tertiary consumers* may include scavengers of various types or *omnivores,* which eat plant or animal matter. At the end of the line are the various types of decomposers, the various types of bacteria and fungi that assist in the decomposition of decaying plant and animal matter.

Food chains

A food chain consists of a connected group of producers, consumers, and decomposers. It begins with energy from the sun and nutrients from the soil. These pass through a plant and one or several consumers to a final consumer that is not fed upon by any other. Even this one may have parasites, and in time it will certainly die. A common example of a food chain is one involving the plants and grazing antelopes in Africa. The consumers, antelopes, are herbivores. They in turn are preyed upon by various large cats such as lions and cheetahs. These secondary consumers are carnivorous. They eat their fill but do not completely consume the antelope. Tertiary consumers, the scavenging vultures, then feed on the remains. After they finish, the bones may be crushed by jackals and hyenas. By this time it appears that not much is left of the antelope; however, there is still sufficient nutrient to attract a large number of insects that can fairly well clean up a skeleton. Finally, whatever is left, including the feces or dung of all of the consumers, may be decomposed by bacteria and fungi. The debris is then gradually returned to the soil where it provides the nutrients and natural fertilizer required for the growth of more vegetation. A variety of interrelated food chains that might be found in a prairie community is shown in Fig. 34-7.

Food pyramids

One of the consequences of a food chain is loss of energy at each step. The more consumers in the chain, the less energy is available at the end. If a food chain is constructed so that the number or energy content of the producers is at the bottom and the layers of consumers are placed on top of this, a pyramid is said to be constructed. A great number of producers is needed to support a lesser number of primary consumers (for example, herbivores), which in turn can support a still lesser number of secondary consumers, and even fewer tertiary consumers, and so on. Fig. 34-8 gives some idea of this in terms of relative loss of mass and energy at each level of a food chain.

CYCLES OF MATERIALS

A definition of an ecosystem was given early in this chapter. One factor discussed was the relative stability of the system. Only a constant supply of energy is needed to make the system self-sufficient. The other components of the system are constantly recycled. Some of the more important cycles will now be considered.

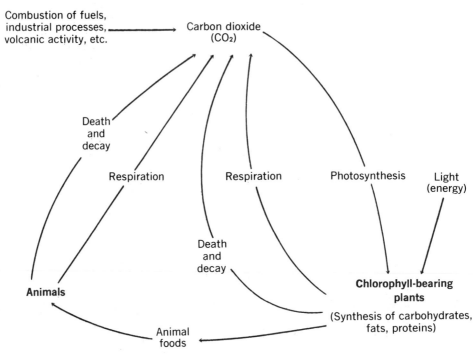

Fig. 34-9. Carbon cycle in nature.

Carbon cycle

Carbon can be considered an essential element in any ecosystem. Not only is it the basis of the organic compounds that make up living things, but it is involved in the primary energy-capturing activities of the green plant. About half of the dry weight of living organisms is carbon. It originates from the carbon dioxide of the atmosphere during the process of photosynthesis. The end result is some type of carbohydrate in the plant. These either die and decay or they are eaten by an animal. In either case the result is carbon dioxide returned to the atmosphere. This may result from plant or animal metabolism and respiration.

Carbon dioxide may also be returned to the atmosphere as the result of urine and feces decay by bacterial action. Even the natural death of the plant is followed by bacterial and fungal action and results in decay, which releases carbon dioxide to the atmosphere.

The bodies of plants and animals in the past were often covered with layers of earth or water. The pressure on these bodies over long periods of time produced what we now call the fossil fuels—peat, coal, petroleum, and their by-product, natural gas. When any of these is burned, one of the by-products is carbon dioxide. Most of the carbon retained in the earth is in the form of limestone, marble, and various carbonates. In time, these also will break down releasing carbon dioxide to the atmosphere. The carbon cycle is outlined in Fig. 34-9.

Nitrogen

Nitrogen is one of the basic elements of amino acids, which make up proteins. About 78% of the

Table 34-4 Nitrogen transformations as produced and used by living organisms

	Bacterial organisms	Results
Nitrogen fixation		
Symbiotic	*Rhizobium*	Nitrates produced from free nitrogen of atmosphere in root nodules of leguminous plants (clovers, alfalfas, peas)
Nonsymbiotic	*Azotobacter, Clostridium*	Nitrates produced from free nitrogen of atmosphere within soil
Nitrification	Ammonifying bacteria	Change nitrogenous compounds into ammonia (NH_3) by process of ammonification
	Nitrosomonas, Nitrosococcus	Change ammonia to nitrites (NO_2)
	Nitrobacter	Change nitrites to nitrates (NO_3), which are usable by plants
Denitrification	Certain bacteria	Convert nitrates to nitrites, oxides of nitrogen, and free nitrogen

atmosphere is in the form of nitrogen gas. It is a major component of the air we breathe. It is exhaled in essentially uncharged form. Neither plants nor animals are able to use nitrogen in its gaseous form. It must first be converted to soluble forms.

Nitrogen fixation. This is primarily the result of bacteria such as *Rhizobium* sp., which live in a symbiotic relationship in swellings on the roots of various plants, especially legumes. Other free living bacteria, blue-green algae and some fungi are also able to convert gaseous nitrogen into soluble nitrates or ammonia (Table 34-4). In either case they are absorbed by plants and used to make amino acids and more complex compounds.

Nitrification. Other bacteria such as *Nitrosomonas* work on ammonia compounds produced as a result of decay and putrefaction. They convert ammonia compounds into nitrite, compounds containing NO_2. These compounds are soluble. A second group of bacteria such as *Nitrobacter* then act to convert the nitrites into nitrates, containing NO_3. These too are absorbable by plant roots.

Denitrification. Another group of bacteria are capable of converting both nitrates and ammonium compounds into gaseous nitrogen (N_2). This happens after excretion of decay of or-

ganisms. In this last step the gaseous nitrogen is returned to the atmosphere. A diagram of a nitrogen cycle is indicated in Fig. 34-10.

Phosphorus

Unlike carbon and nitrogen, the reservoir for phosphorus is not in the atmosphere. It is found in the rocks and phosphate deposits in the soil. Estimates of its availability vary depending on the rate of use as fertilizer. If carefully conserved, it appears that reserves are good for 400 to 500 years. If the use continues to rise, it may be seriously depleted in 50 to 100 years.

Phosphorous is essential to plant and animal growth. It is obtained by plants in the form of inorganic phosphate obtained by weathering of soils and rocks or by application of fertilizer. Animals get phosphorous in drinking water or by eating plants or meat. In either case inorganic phosphorus is converted to many different organic compounds needed for normal metabolism.

The phosphorus cycle is being gradually depleted by runoff in rainwater. This eventually ends up in the oceans. Some of this may enter marine food chains in algae and plankton that end up as fish or bird food. When the fishes or birds are eaten, or when the birds produce guano or feces, the phosphorus may again be recycled.

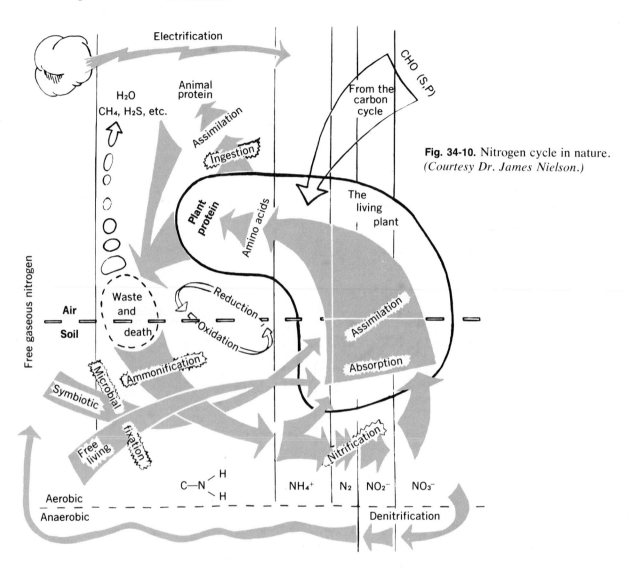

Electrification

H₂O
CH₄, H₂S, etc.

Animal
protein

Assimilation

Ingestion

From the
carbon
cycle

CHO (S,P)

Fig. 34-10. Nitrogen cycle in nature.
(Courtesy Dr. James Nielson.)

The
living
plant

Plant
protein

Amino acids

Free gaseous nitrogen

Waste
and
death

Reduction

Oxidation

Assimilation

Air

Soil

Absorption

Ammonification

Microbial

Symbiotic

Free
living

fixation

Nitrification

$$C-N\begin{smallmatrix}H\\\\H\end{smallmatrix}$$

NH₄⁺   N₂   NO₂⁻   NO₃⁻

Aerobic

Anaerobic

Denitrification

Water

The great natural reservoir of water is the ocean. The action of sun and wind cause it to evaporate and rise to the atmosphere where it forms clouds. The action of winds blows the clouds over land masses where the change in temperature causes the moisture to condense into rain or snow.

Some of the rain may simply reevaporate into the atmosphere. Some water flows over the soil as runoff. Some water soaks into the ground where it is absorbed by plant roots. Some continues to percolate through the soil into the water table.

The water in organisms is returned by transpiration in plants and by breathing in animals. At any rate, one way or another the water is recycled to the rivers, lakes, and oceans where it

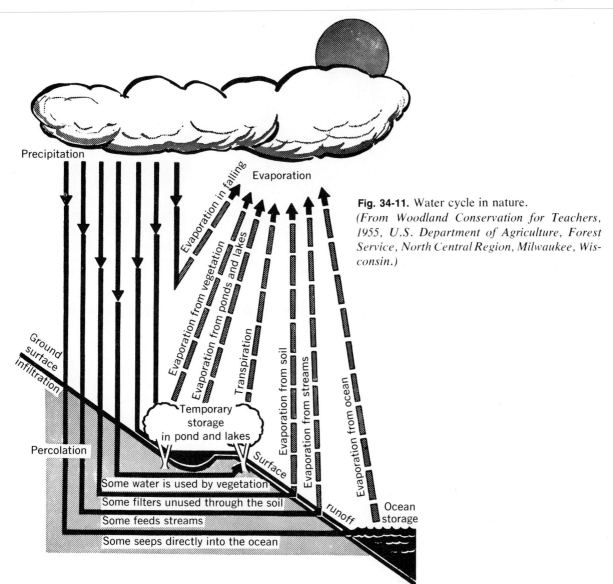

Precipitation

Evaporation in falling

Evaporation

Fig. 34-11. Water cycle in nature.
*(From Woodland Conservation for Teachers,
1955, U.S. Department of Agriculture, Forest
Service, North Central Region, Milwaukee, Wis-
consin.)*

Ground surface infiltration

Evaporation from vegetation

Evaporation from ponds and lakes

Transpiration

Evaporation from soil

Evaporation from streams

Evaporation from ocean

Temporary
storage
in pond and lakes

Percolation

Surface

Some water is used by vegetation

Some filters unused through the soil

Some feeds streams

runoff

Ocean
storage

Some seeps directly into the ocean

reenters the atmosphere. A diagram of the water
cycle is shown in Fig. 34-11.

SUCCESSION

In the discussion of ecosystems it was indicated that the association of living and nonliving

elements was relatively self-sufficient and relatively stable. As we have seen, the self-suf-
ficiency depends on the energy input through the
primary producers, that is, green plants.

Relative stability is indicated because the very
presence of living organisms causes a change in
the total environment. Also, changes in the

Fig. 34-12. Photograph of a pond to illustrate succession. Note areas of open water, emergent plants, shore vegetation, open field with grasses and shrubs, conifers, and deciduous forest.
(Photograph by F. S. DeMartino, University of Dayton.)

physical environment caused by erosion, floods, fires, earthquakes, and so on alter the relations of living and nonliving parts of the ecosystem. The result of these changes is that the area is less suited for the organisms and new species of plants and animals may be better adapted to the area and replace the earlier species.

The changes in plant and animal species in an ecosystem result in ecological *succession*. Each new community changes the environment and in its turn is followed by a different species complex. These changes in species composition continue until a fairly stable ecosystem is attained.

The last groups of vegetation form the *climax* (stable) *community*. The biomes discussed earlier are examples of climax communities on a large scale.

Old field succession

Succession can be observed on a small scale in an area that has been burned or in a field that is left unplanted for some years. An abandoned farm is quickly overgrown by annual weeds. Seedlings of crabgrass and herbaceous and perennial plants appear during the first year. These continue to grow during the next year as the perennials grow tall and produce flowers and seeds. Other windblown or bird-carried seed take hold during the second and third years.

These are replaced by bunch grass and small woody shrubs. About this time pine seedlings begin to appear. After about 10 years or so they form a young well spaced pine forest. As the pines get taller, they begin to shade the other plants and also their own seedlings. Pine seedlings require full sunlight in order to thrive. In effect the pine shades and kills its own offspring. The pine trees compete with seedlings and with each other for sunlight and for moisture in the soil.

Broadleaf plants can thrive in a semishaded environment. These begin to grow as the pines provide shade. This growth continues for long periods of time. After 50 to 100 years the pines begin to die off. They are not replaced by other pines but by more broadleaf trees. In time then the old field has produced a broadleaf climax forest.

Pond succession

The same sort of succession might happen on the shores of a lake or pond. This system begins as a body of open water with sandy shores. This may be an active ecosystem with various aquatic

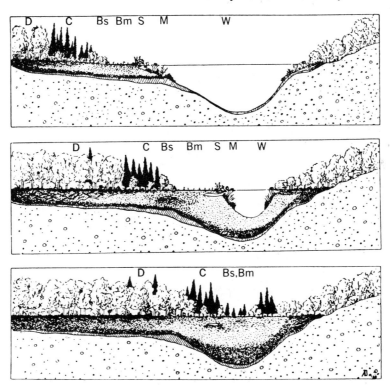

Fig. 34-13. Plant successions in a pond. In stages I, II, and III, a pond fills with deposits of soil runoff and dead vegetation, and the original pond is finally covered by a forest. Successive stages are open-water plants, *W;* marginal plants, *M;* shore plants, *S;* bog-meadow plants, *Bm;* bog shrubs, *Bs;* coniferous forest, *C;* and deciduous forest, *D.* Deciduous trees are the climax vegetation and will eventually cover the entire area.
(From Dachnowski, A.: Peat deposits of Ohio; courtesy Geological Survey of Ohio, Columbus, Ohio, Fourth Series, Bulletin 16.)

plants and animals. Runoff from rain carries mud and other debris to the bottom of the pond. Mud now covers a former sandy bottom and also provides a microenvironment for plants and animals. The mud and silt along the banks provide a suitable habitat for aquatic and semiaquatic vegetation such as algae, elodea, cattails, and bullrushes. This also provides a suitable habitat for many animals (Fig. 34-12).

As more runoff is provided and as the emergent vegetation dies and decays, the edges of the pond are gradually filled in, producing a marsh or bog. This provides a suitable area for grasses and shrubs and more area is filled in. The process of growth, death, and decay leads to a filling in of the open water area. If this continues long enough the area of open water becomes smaller and smaller and finally disappears. Eventually the shrubs will be replaced by trees, perhaps by conifers and eventually be a deciduous forest (Fig. 34-13).

Silting and eutrophication

The activities of humans may speed up the process of succession, often to the detriment of mankind. A large number of recreational lakes have been built up in this country in recent years by damming streams. Many of them have encountered problems caused by the accumulation of mud and silt from runoff. As this builds up, the lake is gradually filling in. In a few extreme cases, obvious silting has taken place within 5 to 10 years after the lake was established. The principles of succession might be a warning to developers of such areas.

On a more important level the runoff of farmlands often contains large amounts of fertilizers. When these are carried into a lake or stream there may be a surge of growth by aquatic plants. Probably the most important limiting factor to plant growth in lakes is a source of phosphorus. Fertilizers, sewage, detergents, and other compounds commonly found in runoff waters may contain large amounts of phosphorous. When this reaches the lake, a massive growth of algae may

result. This may cover the surface of the water with scum and rotting vegetation (Fig. 34-14).

As the plant life grows, dies, and decays, it sinks to the bottom of the lake. This provides the basis for massive bacterial growths and decomposition of organic matter. As a result the amount of oxygen in the lake is depleted and this affects the animal population. The total change that results from human interference with the natural succession in a lake is called *eutrophication*.

BIOTIC COMMUNITIES AND POPULATIONS

Each area or region is inhabited by a number of plants or animals, with many interrelationships. Together with environmental factors these interrelationships determine the number and type of species in the area. All of these species of organisms living in a given area of habitat constitute a *biotic community*, composed of intimately associated smaller groups called populations. A *population* is a group of plants or animals of the same species that lives in a given area. All of the different populations in one area form a community.

In general, communities may be widely separated from each other, but if their environmental factors are alike, similar kinds of living organisms will be found in them. Similar organisms will generally be found in somewhat similar environments, although there are exceptions. In the biotic community the idea of a natural association of organisms that are bound together by their requirement of the same environmental conditions is emphasized.

Communities do not always have definite boundaries and often overlap and grade into each other. Some animals shift from one community to another because of seasonal changes; others stay in one community during the day and spend the night in another.

Each community is organized and divided into definite horizontal and vertical strata. This principle of stratification is well illustrated by forest communities, terrestrial communities, and aquat-

Fig. 34-14. Overgrowth of a woodland pond. *(Photograph by F. S. DeMartino, University of Dayton.)*

ic communities. In a forest community there may be vertical stratification, shown by some organisms living on the forest floor, others only a few feet above the floor, and still others at greater heights above the floor. In aquatic communities some organisms also tend to live at different levels.

Plant communities contain only plants, but usually certain types of animals share that area with them. A community may consist of a mass of algae in water, some lichens on a rock, a swamp, the plants in a certain field, and a forest composed of certain types of trees.

Population density and growth curve

A population has characteristics that result from the group as a whole and not from the individuals. Among such characteristics are population density (number of individuals per unit area or volume) such as the number of trees per acre in a forest or the number of human beings per square mile. Population density is an index of a population's successful living in that region under the existing conditions; it is often difficult

to measure in terms of inidividuals but it can be estimated by careful countings. A population growth curve is a graph in which the number of organisms is plotted against time. Such curves are characteristic of populations rather than of single species, and they are similar for the populations of most organisms, from bacteria and protozoans to man (see Fig. 35-9). In 1925 Raymond Pearl estimated that the human population of 2.2 billion would reach 2.6 billion in the year 2100, after which it would stabilize unless certain contributory factors should arise. We know that this was a conservative estimate. Recent estimates by the United Nations show that the world population in 1978 was over 4.2 billion. At the present rate of growth, world population will have doubled in 35 years, passing 5 billion persons by 1985 (Table 34-5).

Birthrate and mortality rate

The birthrate of a population is the number of new individuals produced per unit of time. The maximum birthrate is the largest number of organisms that could be produced per unit of time under ideal, or optimal, conditions when no limiting factors are involved. This rate is somewhat constant for a species and is influenced by such physiologic factors as the proportion of females to the number of eggs produced. It is difficult to ascertain the maximum birthrate because of the difficulty in eliminating all the limiting factors, and it is always greater than the actual birthrate.

The mortality rate of a population is the number of individuals dying per unit of time. There is a minimum mortality, which is the number of deaths that would occur under ideal conditions, that is, the deaths simply resulting from physiologic changes of old age. This is rather constant for a given population. The actual mortality rate varies with physical factors and size and composition of the population. In some countries man has so improved his average life expectancy by modern health measures that the curve for human survival approaches the curve for minimum mortality.

Since the death rate is more variable and af-

fected more by environmental factors than is the birthrate, it plays a vital role in population control. Populations that differ in the relative numbers of young and old individuals, especially if past the reproductive age, will have quite different characteristics, birthrates, and death rates. Death rates usually vary with age, and birthrates are often proportional to the number of individuals of reproductive age. In general, rapidly increasing populations have a high proportion of young individuals (Table 34-5).

Biotic potential and environmental resistance

The biotic or reproductive potential expresses the inherent ability of a population to increase its numbers when the age ratio is stable and all environmental conditions are optimal. Under ordinary conditions when the environment is suboptimal, the rate of population growth is less, and the difference between the potential ability of a population to increase and the actual observed performance is a measure of what is called *en-*

Table 34-5 Population, birthrate, and death rates of selected countries*

Area	1973 population (in millions)	Birthrate per 1000 population	Death rate per 1000 population	Number of years to double	Projected population in 1985 (in millions)
World	3,860	33	13	35	4,933
Mexico	56.2	43	10	21	84.4
Canada	22.5	15.6	7.3	58	27.3
United States	210.3	15.6	9.4	87	235.7
Columbia	23.7	45	11	21	35.6
Brazil	101.3	38	10	25	142.6
Uruguay	3	23	9	50	3.4
Russia	250	17.8	8.2	70	287
United Kingdom	57	14.9	11.9	231	61.8
Austria	7.5	13.8	12.6	700	8.0
Morocco	17.4	50	16	21	26.2
Nigeria	56.9	50	25	27	84.7
Egypt	36.9	37	16	33	52.3
Pakistan	68.3	51	18	21	100.9
China	799.3	30	13	41	964.6
India	600.4	42	17	28	807.6

*Data from Population Reference Bureau, Washington, D.C.

vironmental resistance. When a population is expanding rapidly, each individual organism of reproductive age reproduces at the same rate as before; the increased numbers result from increased survival.

Environmental resistance consists of all the physical and biologic factors that prevent a species from reproducing at its maximal rate. When a species is first introduced into a new area, the environmental resistance is often low, resulting in a great increase in numbers, as illustrated by the introduction of the English sparrow into the United States or the rabbit into Australia. As a species increases in number, the environmental resistance may also increase in the form of other organisms that parasitize it or prey upon it and in the competition of members of the species for food, living space, and mates.

After a population has become established in a region and has reached an equilibrium level, the numbers may change from time to time, being affected by variations in environmental resistance, by factors inherent to the population, or by a combination of the two. The variations of some populations are very irregular, whereas others are regular and occur in cycles.

Balance in nature

When all organisms in an area are studied, much competition and many struggles are evident, even though there is also much cooperation among the inhabitants. The various organisms and their activities somehow seem to counteract each other, resulting in a balance in nature. If one group of organisms is eliminated in an area, another group may take its place, or the remaining organisms may fill the vacancy, thus producing a new balance. There is much truth to the statement that you cannot alter *one* factor in the balance.

Review questions and topics

1 How can knowledge of ecology be beneficial in the successful cultivation of flowers and vegetables?
2 How can knowledge of ecology be helpful in the proper operation and care of an outdoor pool? Of an aquarium? Of a terrarium?
3 List some of the factors that might influence migrations of certain animals or dispersals of certain plants.
4 Explain how the distribution of certain types of plants might influence the distribution of certain types of animals and vice versa.
5 Discuss biotic communities, including advantages and disadvantages of such biotic relationships. What are the benefits of efficient communal life?
6 What is meant by the term population?
7 Explain what is meant by a so-called food chain, including its importance in a biotic community.
8 Explain what is meant by plant successions.
9 Explain the relationships of the nitrogen, oxygen, carbon, and sulfur cycles in soil fertility.

Selected references

Allee, W. C., and others: Principles of animal ecology, Philadelphia, 1949, W. B. Saunders Company.
Carson, R.: The sea around us, New York, 1952, Oxford University Press, Inc.
Cole, L. C.: The ecosphere, Sci. Amer. April 1958.
Colinvaux, P. A.: Introduction to ecology, New York, 1973, John Wiley and Sons.
Dasmann, R. F.: Environmental conservation, ed. 3, New York, 1972, John Wiley & Sons, Inc.
Emlen, J. M.: Ecology: an evolutionary approach, Reading, Mass., 1973, Addison-Wesley Publishing Co., Inc.
Evans, F. C.: Ecosystem as the basic unit in ecology, Science **123:**1127-1128, 1956.
MacArthur, R. H.: Geographical ecology, patterns in the distribution of species, New York, 1970, Harper and Row, Publishers.
Odum, E. P.: Fundamentals of ecology, ed. 2, Philadelphia, 1959, W. B. Saunders Company.
Ricklefs, R. E.: Ecology, Newton, Mass., 1973, Chiron Press, Inc.
Waldbott, G. E.: Health effects of environmental pollutants, St. Louis, 1973, The C. V. Mosby Co.
Went, F. W.: The ecology of desert plants, Sci. Amer. April 1955.
Wessells, N. K., editor: Vertebrate adaptations. Readings from Scientific American, San Francisco, 1969, W. H. Freeman and Co. Publishers.
Woodwell, G. M.: Toxic substances and ecological cycles, Sci. Amer. **216:**24-31, 1967.

35 Behavior and relationships among organisms

Anyone who has stood before a display of tropical fish knows the fascination of watching these animals as they swim, feed, fight, and produce young. This same sense of wonder is shared by thousands of amateur and professional students of animals behavior who study insects, fishes, birds, and mammals.

In 1973, a Nobel Prize was awarded to Karl von Frisch and Konrad Lorenz of Austria and Nikolas Tinbergen of the Netherlands for "their discoveries concerning organization and elicita-

tion of individual and social behavior patterns." They worked with bees, fishes, and birds, but the basic principles have proved to be applicable to mammals and perhaps to man.

These men and their co-workers were among the pioneers and developers of the science of *ethology*, the comparative study of behavior. Details of their work and of related studies will be considered in the following accounts.

SIMPLE RELATIONSHIPS

Social behavior patterns involve organisms of the same species gathered for their mutual interests. Behavior implies that the members are able to communicate and therefore to respond in some fashion. Probably the simplest of these is a group of parents and offspring. There is at least a minimum of parental care, feeding, and training.

Working with greylag geese in the 1930s, Konrad Lorenz investigated the phenomenon known as *imprinting*. He divided a clutch of eggs into two groups. One of these was hatched by the goose and the other was hatched in an incubator. The goslings hatched by the actual mother, the goose, began to follow her as she walked around the area. The first thing that the incubator hatched birds saw was "mother" Lorenz, whom

Fig. 35-1. In Lorenz's experiment, the incubator-raised goslings were "imprinted" to think of him as their "mother."

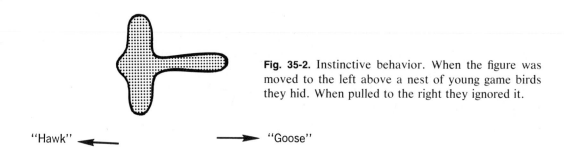

Fig. 35-2. Instinctive behavior. When the figure was moved to the left above a nest of young game birds they hid. When pulled to the right they ignored it.

"Hawk" ←⎯⎯⎯⎯⎯   ⎯⎯⎯⎯⎯→ "Goose"

they proceeded to follow as he walked away.

Lorenz next marked the goslings so that he could identify the members of each group. He placed all of them under a large box. After a short time he removed the box. The members of each group quickly separated and ran to their mothers—the goose-hatched young to the goose and the incubator-hatched forms to Lorenz (Fig. 35-1).

Lorenz called this type of behavior, in which the early experience of the goslings determines later behavior, *imprinting.* Other investigators have observed similar behavior in birds, sheep, goats, guinea pigs, and related animals.

Lorenz and Tinbergen also worked with what they called *instinctive* behavior. They discovered that different behaviors, such as fear or indifference, could be elicited by moving cardboard silouettes over pens of wild birds. The same gen-

eral shape moved in one direction produced hiding behavior, whereas moved in the opposite direction, it did not lead to the fear reaction. They reasoned that in one direction the shape seemed to be that of a long tailed hawk while in the other it appeared to be a long necked goose (Fig. 35-2).

Larger groups of a more gregarious type include the herds of mammals, flocks of birds, and schools of fishes. The advantage of numbers appears to be that of protection from predators. The animals at the edges of the group serve as sentries or lookouts. Because of their location they are the first to observe or come in contact with a potential predator. When a bird cries out or flies, or a deer flashes its white tail (Fig. 35-3), the other animals are alerted and may escape or band together making it difficult or dangerous for the predator to attack. The arctic musk oxen are good examples of this. When attacked by wolves

Fig. 35-3. Alarm reaction in white tailed deer. When the deer is startled it raises its tail, exhibiting the white hairs on the underside.

or other predators, they form a tight circle with their horns facing outwards and the young in the center.

COMPLEX BEHAVIOR

A higher level of behavior is exhibited in some animal societies that seem to have a leader. Close observation has shown this to be somewhat misleading in some cases. In fishes the leader is merely the fish or groups of fishes at the leading edge of the school. When the school changes direction, it also changes leaders. Migrating flocks of birds show the same change of leader with a change in direction.

In other cases, such as packs of wolves, there is a definite dominant leader. The same is true with primates such as baboons, chimpanzees, and gorillas.

Dominance

Probably the best known examples of dominance of the leader is in the pecking order of chickens. Generally one bird is the dominant individual and has first choice of food, nests, and mates and may peck others without reprisal. Sometimes this is a simple dominance with a leader bird and all others of minor rank. In other cases there is a chain of peck orders. Here bird one dominates all others, and bird two dominates all except bird one, bird three all except one and two, and so on.

Although pecking order terminology originated with chickens, it is also loosely applied to such groups as wolves, baboons, and chimpanzees. The relative status in a particular group is initially established by some sort of aggressive behavior or fighting. Further encounters usually do not result in injury; rather, they serve to reassert the dominance of the leader.

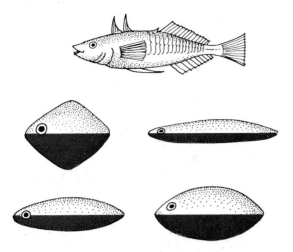

Fig. 35-4. Tinbergen's of the stickleback. In breeding season the male will attack the models with red bellies more often than the more lifelike model without the red color.

Fig. 35-5. Lack's models of the European robin. Male robins attacked the tuft of red feathers at the right far more than the more birdlike model on the left.

Territoriality

Many animals, either singly or in groups, occupy and defend a particular area. They attack and drive away other members of their species. Sometimes they define the territory by aggressive behavior, by sound as in bird songs, or by odor. In some species an individual bird or fish will aggressively defend a territory only during the breeding season. At other times it may be part of a larger group or simply live as a solitary wanderer.

In Tinbergen's work with the three-spined stickleback he observed this aggressive defending of the nesting territory. During the breeding season the underside of males turn a distinct red color. If another male enters the area, it will be driven away. Tinbergen found that the stimulus for the attack was more in the red underside than in any other factor, even the shape of the model "fish" (Fig. 35-4).

In the same manner a bird may attack its reflection in a window. It may also attack an object with a color stimulus similar to another of its species. David Lack observed that in the European robin the aggressive actions were stimulated by the color of the breast feathers of a potential competitor. When he placed a reasonably accurate model of the robin in its territory, it was ignored unless the breast feathers were red. Similarly, when he placed a tuft of red feathers, which did not resemble the bird in any way except the red color, in the area, it was immediately attacked by the territory holder (Fig. 35-5).

Mating behavior in the stickleback

Tinbergen began working with a small fish, the three-spined stickleback (*Gasterosteus aculeatus*) in 1932. His studies eventually led him to the discovery of a number of distinct behavior patterns in these little fish.

Throughout most of the year the fish swim in small groups or schools. In the spring individual males leave the school and begin a distinct pattern of behavior. They begin to defend a particular area of the bottom as their special territory and attack any other fish that enters it. After a short period of aggressiveness they swim to the bottom of the stream (or aquarium if confined) and begin to dig a hole. After this, they gather bits of vegetation and construct a tube-like nest. The male makes an entrance and exit by a burrowing or tunneling action. During this time, as well as

Fig. 35-6. Honeybees working in the hive.
(Photograph by F. S. DeMartino, University of Dayton.)

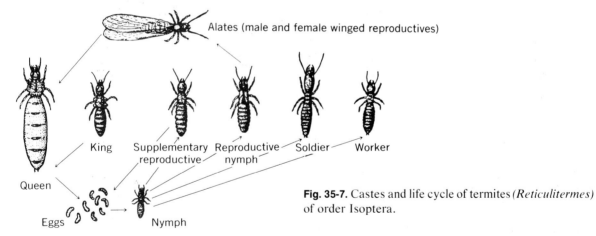

Alates (male and female winged reproductives)

King   Supplementary   Reproductive   Soldier   Worker
        reproductive   nymph

Queen

Eggs   Nymph

Fig. 35-7. Castes and life cycle of termites *(Reticulitermes)* of order Isoptera.

afterwards, he defends his nest-building territory by attacking other males or females if they come too close.

About the end of the nest building period the underside of the male turns a distinct red color. If a female in the breeding condition (abdomen distended with eggs) approaches, he swims toward her with a zig-zag motion. The female responds with a peculiar head up swimming motion. This leads to further zig-zag activity on the part of the male.

He now turns toward the nest, followed by the female. Upon reaching the nest, he turns on his side with the head pointing at the nest entrance. The female then enters the nest and because of its small size seems to stick out on both ends. The

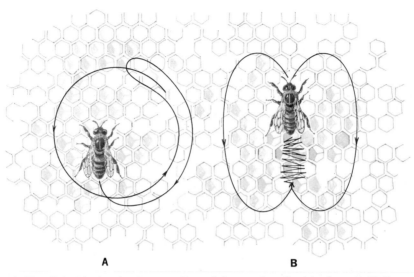

Fig. 35-8. Von Frisch's dancing bees. **A,** Round dance when food is close. **B,** Tail-waggling dance when food is farther away.

male then thrusts his head against her tail in a trembling motion. After this stimulus the female lays eggs in the nest and swims away.

The male now enters the nest and releases sperms over the eggs. The entire process takes only a few minutes at most. After the male leaves the nest, he stays near, fanning fresh water through the nest with his fins until the eggs hatch.

Tinbergen and his colleague J. van Iersel found that the entire sequence of behavior in both male and female was initiated by distinct stimuli. A male will attempt to drive off any fish or other object that has a red underside. This is true regardless of the general size or shape of the object. The color red serves as a distinct stimulus (Fig. 35-4).

Similarly he will court, that is, swim in a zig-zag motion toward, a fish with swollen sides whether it is an egg-filled female or a food-filled male. The female is attracted by the zig-zag "dance" of the male and follows it. She will attempt to enter a nest or a mound of debris if the fish or a red bellied model is poked into it.

Once inside, the female will not release eggs unless prodded at the base of the tail by the male. This prodding stimulus can also be induced by poking her with a glass rod. The touch stimulus is enough to cause her to release eggs.

Neither fish will respond to any of these stimuli except during the breeding season. Tinbergen suggests that all of these activities are initiated by the buildup of hormone levels during the spring.

Orientation in the digger wasp

Tinbergen studied the landmark orientation of the digger wasp *Philanthus triangulum*. After mating, these wasps dig a hole in the ground, stock it with insect larvae, lay eggs on them, and seal up the hole. Once the hole is dug the wasp circles the area and then flies away looking for food. Tinbergen discovered that the wasp oriented on nearby landmarks so that it could find the nest.

He made a circle of pine cones around a nest that was being constructed. When the wasp left, he moved the cones so that the nest was now outside of the circle. When the wasp returned, it headed for the center of the cones and did not succeed in finding the nest even though it was clearly visible (at least to Tinbergen).

In another study he moved the pine cones, changing their order from a circle to a triangle and surrounding the nest with a circle of stones. This time the wasp flew to the nest opening. It was the arrangement rather than the type of objects that gave the wasp its orientation.

SOCIAL INSECTS

Most insects live more or less solitary lives, coming together only for mating. Some species may aggregate in mating swarms, whereas others such as bees, ants, and termites, have developed rather complex societies in which the labor is divided and adults and young live in a cooperative community.

In the honeybee community, more than 50,000 bees may live in one hive. This society usually consists of one queen, a few hundred male drones, and thousands of infertile female workers. During a mating flight the queen is fertilized by one of the drones. Enough sperms are stored during this mating to last the reproductive lifetime of the queen, often five seasons. During this time she may produce hundreds of thousands of fertilized eggs.

The drones usually live for one season and are then killed or driven out of the hive by the workers. The workers move from job to job as they get older, acting as hive bees, cleaning and feeding the larvae, and tending the queen (Fig. 35-6). Some produce beeswax, which is made into the honeycomb. Others ventilate the hive by fanning their wings. Later they become forager or field bees, gathering pollen and nector for the hive. Most workers live just a few weeks.

Ant colonies contain one or more fertile females or queens, fertile males, and infertile females. The males provide sperms in mating and are not too numerous. Some of the females are food gatherers; others act as nurses, and others act as soldiers to guard the colony.

The complex colonies of termites contain two castes: fertile males and females, and infertile males and females (Fig. 35-7). The swarms of winged termites occurring in the spring are mating swarms. New colonies may be formed by each pair of mated termites.

A colony will have supplementary reproducers with small nonfunctional wings or no wings at all. In the event of death of the king and queen, these forms take over the egg laying function of the colony. The sterile, wingless caste includes both workers and soldiers. The workers maintain the colony by obtaining food, building tunnels, caring for eggs and young, and tending the queen.

Karl von Frisch and the dancing bees

Karl von Frisch began his investigations about 40 years ago and showed that bees are not totally color-blind. They have a definite color sense and can be trained to seek food on the background of a specific color that they distinguish from other colors. They are blind to the red end of the color spectrum.

When a dish of sugar water was left near a hive it was eventually visited by a field bee. Shortly after, von Frisch noted an increasing number of bees collecting on the food source. Using glass walled observation hives and paint spots to mark the bees, he saw that after visiting the sugar source and returning to the hive, the bees went through some distinct motions. After first feeding several of the hive bees, the forager bee began a peculiar round dance, circling first to the right, then to the left, then to the right and repeating this many times. After this, many bees left the hive and searched nearby for food.

If he placed four separate containers at an equal distance but in different directions from the hive, bees appeared at all four and not just at the original source. If a flower odor was placed in one dish, more bees found it than the others. He concluded that the *round dance* indicated that food was near the hive and that the addition of the scent indicated the particular kind of flower to look for (Fig. 35-8).

The vigor of the dance that guides the bees is determined by the ease with which nectar is secured. When the supply of nectar in a certain type of flower is giving out, the bees visiting it slow down or stop their dance.

When he moved the food source some distance away from the hive, the bees used a different type of dance. In this dance the forager bee moved along a straight line and wiggled her abdomen from side to side. At the end of a straight run she circled back to the starting point and started the *tail wiggling dance* again. This time she circled back by going in the opposite direction; at the end of the first run she circled right and at the end of the second she circled left. These alternate circles were used to get back to the start of the tail wiggling run.

Von Frisch trained two groups of bees from the same hive to feed at different places. One group marked with a blue stain was trained to visit a feeding place a few meters away, whereas a group marked red was fed at a distance of 300 meters. All of the red bees performed wagging dances; all the blue ones performed circling dances. By a series of steps he then moved the nearer feeding place farther and farther from the hive. At a distance between 50 and 100 meters, the blue bees changed from a circling dance to a wagging dance. Conversely, the red bees, when brought gradually closer to the hive, changed from wagging to circling in 50- to 100-meter interval. He also found that the frequency of the turns gave a fairly good indication of the distance. When the feeding site was 100 meters away, the bee made about 10 short turns in 15 seconds. To indicate 3,000 meters, only 3 long turns in 15 seconds were made.

He also found that the direction of the dance was oriented to the sun. For instance, if the food source was directly toward the sun, she ran directly upwards. If the flower source was 50 degrees right of a line straight to the sun, her wiggley dance was also in a line 50 degrees right of a vertical line. Although more recent workers have added some refinements to the interpretation of bee behavior, the pioneering work of von Frisch stands as a landmark in animal behavior.

COMPETITION

The need of a number of organisms of the same or different species for some limited resource is called competition. The resources in short supply may be such obvious things as food, water, and shelter. They may also be such things as nutrients in the soil and exposure to sunlight for plants and nesting sites and living space for both plants and animals.

A rapidly growing plant species may effectively prevent growth of other species by shading them from the sun or by absorbing some limited nutrient at a faster rate.

Competitive exclusion

The result of interspecific (two species) competition is frequently the extinction of one of them. This observation was formalized by G. F. Gause in the *Principle of Competitive Exclusion*. Where two species compete for the same factor, only one of them survives.

Gause worked with two species of *Paramecium*. When grown in the laboratory under controlled conditions, *Paramecium aurelia* reached a stable population size in about 8 to 10 days. A second species *P. caudatum* increases at a slower rate, reaching a population peak about half that of *P. aurelia*. If the two species are mixed in the same culture, competition for food and other resources begins. The result is typically that *P. aurelia* survives and *P. caudatum* dies out. *P. aurelia* eventually reaches its typical population size, but a longer time is required (Fig. 35-9).

The same patterns of growth were observed in studies of flour beetles by Thomas Park at the University of Chicago. Park worked with two species of the genus *Tribolium*, *T. confusum* and *T. casteneum*. When kept together in the same

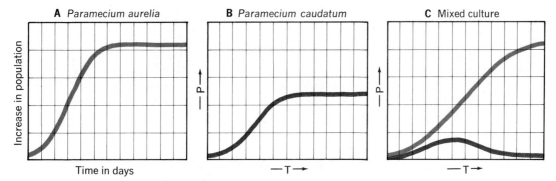

Fig. 35-9. Competitive exclusion. **A,** *Paramecium aurelia* when grown in a culture reached a stable population in 8 to 10 days. **B,** *Paramecium caudatum* increases its population at a slower rate reaching a peak about half that of **A. C,** When grown together competition favors the faster growing *P. aurelia* and the other species dies out.

Fig. 35-10. Lichens growing on a log.
(Photograph by F. S. DeMartino, University of Dayton.)

container of flour, one or the other species died out. The determining factor seemed to be the temperature and humidity of the flour. When grown in hot and humid conditions, *T. casteneum* was the survivor. In cool dry conditions *T. confusum* usually was the survivor. The limiting resource in the competition was space.

In his book *On Aggression,* Konrad Lorenz refers to the exclusion of the larger and stronger marsupials, the tasmanian devil and the marsupial wolf, by the introduction of the dog-like dingo. When brought into Australia by primitive man, the dingo was able to hunt better and faster than the other two. Since they were all predators, feeding on the same quarry, they competed for food. The dingo was a better hunter and in time effectively excluded the others from its range. Today they exist only in areas where the dingo is absent.

SYMBIOSIS

In true symbiosis (Gr. to live together) there is a prolonged and intimate relationship between two species. This often involves direct body contact and a physiologic relationship. Three basic forms are recognized: *mutualism* wherein both species are benefited, *commensalism* where one species benefits and the other is neither benefited or harmed, and parasitism where one species benefits at the expense of the other.

Mutualism

When two species both gain from their association and are unable to live separately, the phenomenon is called mutualism (L. *mutus,* exchange).

In lichens (Fig. 35-10), in which an alga and a fungus are associated, the alga photosynthesizes food, part of which the fungus uses. The fungus furnishes water, essential elements, and probably some protection against drying, so that both benefit. Certain green algae live with several kinds of animals, including certain protozoans, freshwater sponges, and *Hydra.* For example, a green paramecium *(Paramecium bursaria)* may contain numerous algae cells.

Certain flagellated protozoans that live in the intestines of termites obtain food by assisting in the digestion of the cellulose of the wood eaten by the termites. In return, the protozoans receive protection and are supplied with a stable environment and food. Plant aphids secrete foods that are used by ants, and the ants give protection and shelter to the aphids.

The association between certain fungi and the root tips of higher plants, known as mycorrhiza (Gr. *mykes,* fungus; *rhiza,* root), is another example. The fungi increase root absorption by the plant and receive some benefits in return. Another example is the symbiotic relationship between the nitrogen-fixing, root-tubercle bacteria and the leguminous plants in which they live.

Commensalism

When organisms of different species live together habitually in such a way that one species benefits and the other is neither harmed nor benefited, the phenomenon is called commensalism (L. *cum,* with; *mensa,* table), or literally "eating at a common table." Sea anemones may attach themselves to the shell of a hermit crab, thus obtaining food and giving some camouflage to the crab. The peculiar tropical fishes called remoras *(Echeneis remora)* have a modified dorsal fin that forms a sucker used to attach to sharks and other large aquatic animals. Their food consists of scraps of food left by the shark, who remains unharmed.

Parasitism

In parasitism (Gr. *para,* beside; *sitos,* food) an organism known as the parasite lives on, or within, and at the expense of another living organism known as the host. When a parasite lives on the outside of the body of the host, it is an ectoparasite (Gr. *ekto-,* outside); when it lives within the body of the host, it is an endoparasite (Gr. *endon,* within). When the effects of para-

Fig. 35-11. Starfish attacking an oyster: Note the tube feet on the underside of the starfish arms.
(Courtesy The American Museum of Natural History, New York, New York.)

sitism on the host result in discernible, abnormal symptoms, the condition is known as disease production or pathogenesis (Gr. *pathos,* disease or suffering; *genesis,* origin). The malaria-causing parasite *Plasmodium vivax* is a good example of this.

In evolving their dependent mode of living, parasites at times lose some structures and abilities that are no longer used, or they develop modifications in certain systems. The loss of a digestive system and the development of a more complex reproductive system by the tapeworm (see Fig. 20-8) are illustrative. In general, parasites do not use complex equipment for obtaining foods, because they are usually supplied with ready-made foods.

Parasites are usually restricted to one or a few hosts, and they are thought to have evolved from free-living types rather than vice versa. The relationship between parasite and host may be quite harmful at first, but in time the forces of natural selection usually result in a decrease of the harmful effects, and in development of a give-and-take in the parasite-host relationship. As a general rule, the most successful parasite is the one that causes the least harm.

PREDATION

Predation or predaciousness (L. *praeda,* prey or booty) is a condition in which one organism, the predator, captures another living organism, using it for food (Fig. 35-11). These predatory habits are exhibited by a great number of animals, whose methods of capturing and devouring their prey vary greatly. *Amoeba, Paramecium,* and other protozoans capture a variety of living organisms for food. *Hydra* and other coelenterates devour aquatic organisms. The Portuguese man-of-war preys on fishes and crustaceans. Planarians feed on mollusks and arthropods; squids capture fishes; starfishes capture oysters and other animals; dragonflies destroy flies and mosquitoes; certain insects, such as praying mantises (Fig. 35-12), ground beetles, ladybird beetles, and aphis lions, destroy other insects. Fishes devour worms, crustaceans, and insects; frogs capture worms and insects; snakes destroy frogs and birds; owls kill rabbits and mice. Predacious mammals include wolves and cougars, which kill sheep, cattle, and big game; dogs and cats are predacious when they kill animals such as rats and mice.

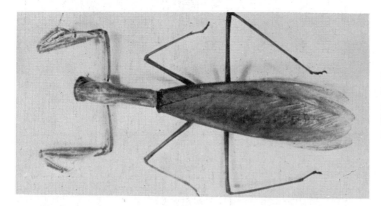

Fig. 35-12. Praying mantis, a predacious insect of order Orthoptera. This insect is so named because of the front legs folded as if in prayer. It is also called the preying mantis.

Fig. 35-13. Pitcher plant, an insectivorous plant, showing the pitcher-shaped leaves for capturing insects.
(Courtesy General Biological Supply House, Inc., Chicago, Illinois.)

It should be obvious that in a predator-prey situation if the number of predators increases, the chance of catching the quarry also increases. When this happens, fewer prey are left to breed and serve as food for the next season's predators. When there is less food, some of the predators starve or fail to grow. This results in fewer predators so there is less pressure on the prey and more of them survive to breed. The relative numbers in the population of predators follow the same relative numbers of the prey.

INSECTIVOROUS PLANTS

The so-called insectivorous or carnivorous plants possess special organs, usually modified leaves or parts of leaves, by which they are able to trap insects or other small animals for part of their food. The specialized structures secrete enzymes for the digestion of the food, which is then absorbed by the plant. The pitcher plants *(Sarracenia)* (Fig. 35-13), common in bogs, have a pitcherlike leaf that is filled with water in which insects drown. Escape is prevented by inwardly

Fig. 35-15. ''Trap'' (modified leaves) of Venus's-flytrap (much enlarged).
(Courtesy Carolina Biological Supply Company, Burlington, North Carolina.)

Fig. 35-14. Venus's-flytrap, an insectivorous plant, showing flowers and specialized leaves for capturing insects.
(Courtesy Carolina Biological Supply Company, Burlington, North Carolina.)

directed spines, and the digested insects are absorbed. In Venus's-flytrap *(Dionaea)* (Figs. 35-14 and 35-15) the specialized leaves possess a row of teeth on the outer margins of each half of the leaf blade. On the upper surface are sensitive hairs that when stimulated cause the two halves to spring together to entrap the insect. Digestion by enzymes somewhat resembles that in the pitcher plant. In the sundew *(Drosera)* the leaves are covered with long glandular hairs that secrete a sticky substance to capture and digest insects (Fig. 35-16). There are numerous species of chlorophyll-bearing plants that are insectivorous or carnivorous. In the bladderwort plants, present in ponds and lakes, there are tiny ''bladder traps'' on the submerged stems. Each bladder has a one-way trapdoor through which aquatic animals enter and in which they are digested.

Fig. 35-16. Sundew *(Drosera)*, an insectivorous plant.
(Courtesy Carolina Biological Supply Company, Burlington, North Carolina.)

Review questions and topics

1 Discuss symbiosis, mutualism, and commensalism, including differences, similarities, and examples of each that you may have observed.
2 Discuss gregariousness, including the advantages and disadvantages of such a biologic relationship.
3 Discuss the effects of predatism, including examples from your own observations.
4 Discuss the unique biotic relationships displayed by insectivorous plants, including several examples with their specific activities.
5 Discuss epiphytism in plants, including examples.
6 Discuss the nature and significance of saprophytes in the living world.

Selected references

Alcock, J.: Animal behavior: an evolutionary approach, Saunderland, Mass., 1975, Sinauer Associates, Inc.
Baer, J. G.: Ecology of animals parasites, Urbana, Ill., 1951, University of Illinois Press.
Barash, D. P.: Sociobiology and behavior, New York, 1977, Elsevier North-Holland.
Boughey, A. S., editor: Population and environmental biology, Belmont, Calif., 1967, Dickenson Pub. Co., Inc.
Brown, J. L.: The evaluation of behavior, New York, 1975, W. W. Norton Co., Inc.
Burnett, A. L., and Eisner, T.: Animal adaptation, New York, 1964, Holt, Rinehart and Winston, Inc.
Caullery, M.: Parasitism and symbiosis, New York, 1956, John Wiley & Sons, Inc.
Cheng, T. C.: The biology of animal parasites, Philadelphia, 1964, W. B. Saunders Company.
Davis, D. E.: Integral animal behavior, New York, 1966, The Macmillan Company.
Dethier, V. G., and Stellar, E.: Animal behavior, ed. 3, Englewoods Cliffs, N.J., 1970, Prentice-Hall, Inc.
Eibl-Eibesfeldt, I.: Ethology, the biology of behavior, translated by E. Klinghammer, New York, 1970, Holt, Rinehart and Winston, Inc.
von Frisch, Karl: Bees, their vision, chemical senses, and language, Ithaca, N.Y., 1950, Cornell University Press.
Goodall, J. van Lawick: In the shadow of man, New York, 1971, Dell Publishing Co., Inc.
Linbaugh, C.: Cleaning symbiosis, Sci. Amer. Aug. 1961.
Lorenz, Konrad; Studies in animal and human behavior, translated by R. Martin, Cambridge, Mass., 1970, Harvard University Press.
McGaugh, J. L., Weinberger, N. M., and Whalen, R. E., editors: Psychobiology; the biological bases of behavior. In Readings from Scientific Amercian, San Francisco, 1967, W. H. Freeman and Co. Publishers.
Petrunkevitch, A.: The spider and the wasp, Sci. Amer. Aug. 1952.
Platt, R. B., and Griffiths, J. F.: Environmental measurement and interpretation, New York, 1964, Reinhold Publishing Corp.
Rogers, W. P.: The nature of parasitism, New York, 1962, Academic Press, Inc.
Shaw, E.: The schooling of fishes, Sci. Amer. June 1962.
Social behavior in animals with special reference to vertebrates, New York, 1953, John Wiley & Sons, Inc.
Tinbergen, Nikolass: Animal behavior, New York, 1965, Time Inc.
Wilson, E. O.: Sociobiology: the new synthesis, Cambridge, Mass., 1975, Harvard University Press.

Glossary

PREFIXES, SUFFIXES, AND COMBINING FORMS USED IN BIOLOGY

The following list is by no means complete, but it may help in understanding some difficult terminology. The following abbreviations are used: Gr., from Greek; L., from Latin; A. S., from Anglo-Saxon; Sp., from Spanish.

A

a- or **an-** (Gr., without or absent) *asexual,* without sex; *anaerobe,* organism that lives without free oxygen.

ab- (L., away from or without), *aboral,* away from the mouth.

ad- (L., toward, upon, or equal), *adrenal,* relating to the kidney; *adduct,* to draw one part toward another.

-ae (L., plural ending of singular Latin nouns ending in a) *alga* and *algae* (pl.).

aer- (Gr., air), *aerobe,* organism that requires free air.

alb- (L., white), *albino,* organism exhibiting no pigment.

-algia (Gr., pain), *neuralgia,* pain in a nerve.

ambi- (L., both), *ambidextrous,* being able to use either hand.

amphi- (Gr., on both sides), *Amphibia,* class of vertebrate animals living in water and on land.

amyl- (L., starch), *amylase,* enzyme that changes starch to sugar.

ana- (Gr., back or again), *anabolism,* building-up process of metabolism.

angio- (Gr., enclosed), *angiosperm,* plant with enclosed or protected seeds.

ante- (L., before in time or space), *antedorsal,* placed before dorsal.

anti- (Gr., opposed or opposite), *antitoxin,* antibody opposed to or neutralizing a toxin.

antr- (L., cavity), *antrum,* cavity of a bone.

apo- (Gr., away or separate), *apodeme,* ingrowth from the exoskeleton of most arthropods.

aqua- (L., water), *aquatic,* living in water.

arch- (Gr., early or chief), *archenteron,* early digestive tract or enteron; *Archeozoic,* earliest era of geologic history.

areol- (L., space), *areolar,* containing minute spaces.

arthr- (Gr., joint), *Arthropoda,* phylum of invertebrate animals with jointed appendages or feet.

asco- (Gr., sac or bag), *Ascomycetes,* class of sac-bearing fungi.

-ase (suffix designating an enzyme), *protease* enzyme that acts on proteins.

aster- (Gr., star), *Asteroidea,* class of echinoderms resembling stars.

auto- (Gr., self), *autosynthesis,* self building up.

B

bacter- (Gr., *baktron,* a stick), *bacteria,* rod-shaped organisms.

basi- (Gr., base), *basidiospore,* spore formed at the base of a basidium

bi (Gr., base), *bilateral,* similar on both sides.

bio- (Gr., life), *biology,* science of life.

blast- (Gr., bud or young), *blastoderm*, primitive germ layer.

brachy- (Gr., short), *brachydactyly*, abnormal shortness of the digits.

brevis (L., short), *adductor brevis*, short adductor muscle.

bryo- (Gr., moss), *bryophyte*, plant of the phylum comprising the mosses.

C

caec- (L., blind), *cecum (caecum)*, blind pouch.

calci- (L., lime), *calcareous*, containing lime.

-carp (Gr., fruit), *pericarp*, wall around the plant ovary.

cata- (Gr., down), *catabolism*, breaking-down process of metabolism.

cauda- (L., tail), *caudal*, relating to a tail.

cav- (L., hollow), *vena cava*, hollow vein.

ceno- (Gr., recent), *Cenozoic*, recent era of geological history.

centr- (L., center), *centrosome*, center of activity during mitosis.

cephalo- (Gr., head), *cephalic*, relating to, or toward the head.

chlor- (Gr., green), *chlorophyll*, green coloring matter of plants.

-chondro (Gr., granular), *mitochondria*, small, granular parts of protoplasm.

chondro- (Gr., cartilage), *chondrocranium*, part of the cranium developing from cartilage.

chrom- (Gr., color), *chromatophore*, color-bearing cell.

-cide (L., kill), *insecticide*, agent that kills insects.

cili- (L., eyelash), *cilia*, minute, hairlike processes.

circum- (L., around, *circumesophageal*, around the esophagus.

cloaca (L., sewer), *cloaca*, outlet for excretions.

cnido- (Gr., nettle), *cnidoblast*, nettle cell of certain animals.

coel- (Gr., hollow), *coelom (celom)*, hollow body cavity.

coeno- (Gr., common), *coenosarc*, common tissue in certain animals.

coleo- (Gr., sheathed), *Coleoptera*, order of sheathed insects, such as beetles.

com- (L., together), *commensalism*, living together.

con- (L., cone), *conifer*, cone-bearing tree; or (L., with), *concretion*, something that has grown together.

cotyl- (Gr., cup shaped), *cotyledon*, cup-shaped seed leaf.

creta- (L., chalk), *Cretaceous*, chalk period of geologic times.

cyan- (Gr., blue), *Cyanophyta*, phylum of blue-green algae.

cyst (Gr., sac), *cyst*, pouch or sac.

cyt- (Gr., cell), *cytology*, branch of biology studying cell structure and function.

D

de- (L., off), *degenerate*, to lose generative ability.

dendr- (Gr., brush or tree), *dendrite*, treelike structure of a nerve cell.

derm- (Gr., skin), *dermis*, part of the skin.

di- (Gr., twice), *diploblastic*, possessing two germ layers; *dicotyledon*, plant possessing two cotyledons.

dis- (L., away), *distal*, away from the point of origin.

dors- (L., back), *dorsal*, pertaining to the back.

dura- (L., tough), *dura mater*, tough, outer covering of the brain and spinal cord.

E

e- (L., out of, without), *egest*, to pass outside.

ec- (Gr., house or environment), *ecology*, study of the habitats of an organism.

ecto- (Gr., outside), *ectoderm*, outer layer of cells.

-ectomy (Gr., cut), *appendectomy*, removal of the appendix.

-emia (Gr., blood), *anemia*, blood deficiency.

en- (Gr., in or within), *encyst*, to cover with a membranous cyst.

endo- or **ento-** (Gr., within), *endoderm*, inner layer of cells.

eo- (Gr., dawn, or early), *Eocene*, early geologic period.

epi- (Gr., upon), *epidermis*, epithelial layer upon the dermis.

equi- (L., horse), *Equisetineae*, class to which the horsetails belong.

eu- (Gr., good or well), *eugenic*, being fitted for the production of good offspring.

ex- (Gr., external), *exoskeleton*, external skeleton.

extra- (L., beyond), *extracellular*, beyond or outside the cell.

F

-fer (L., to bear), *Porifera*, phylum comprising pore-bearing sponges.

fil- (L., thread), *filiform*, threadlike.

flex- (L., bend), *flexor*, muscle that bends joints.

-form (L., shape), *uniform*, all one shape.

G

gam- (Gr., marriage), *gamete*, reproductive cell.

gastr- (Gr., stomach), *gastric*, pertaining to the stomach.

-gen (Gr., to produce), *pathogenic*, capable of causing disease.

geo- (Gr., earth), *geology*, science of the earth.

-gest (Gr., to bear or hold), *ingest*, to take in.

-glea (Gr., jelly), *mesoglea*, middle, jellylike layer in certain animals.

glyc- (Gr., sweet or carbohydrate), *glycogen*, animal starch.

gono- (Gr., seed or reproduction), *gonad*, organ of reproduction.

gymn- (Gr., naked), *gymnosperm*, class of seed plants whose seeds are not enclosed in an ovary.

H

haem- (Gr., blood), *hemoglobin (haemoglobin)*, substance in the blood.

hemi- (Gr., half), *hemisphere*, one half of a sphere.

hepat- (Gr., liver), *hepatic*, pertaining to the liver.

hetero- (Gr., other or different), *heterogeneous*, consisting of different constituents.

hex- (Gr., six), *hexagonal*, six sided.

homo- (Gr., same), *homogeneous*, of a similar kind.

hyal- (Gr., glass), *hyalin*, something that is transparent or glasslike.

hydr- (Gr., water), *dehydrate*, to remove water.

hymen- (Gr., membrane), *Hymenoptera*, order of insects with membranous wings.

hyper- (Gr., above), *hypersensitive*, especially sensitive.

hypo- (Gr., under), *hypoglossal*, situated under the tongue.

I

in- (L., in, into, not, without), *invaginate*, to infold one part within another.

infra- (L., below), *infraorbital*, below the orbit.

inter- (L., between), *intercellular*, between cells.

intra- (L., inside), *intracellular*, within a cell.

is- (Gr., equal), *isothermic*, having equal temperatures.

-itis (Gr., inflammation), *appendicitis*, inflammation of the appendix.

J

-juga (L., join), *conjugation*, a process of reproduction in which two animals are joined.

K

kata- or **cata** (Gr., down or destroy), *catabolism*, breaking-down process of metabolism.

kine- (Gr., move), *kinetic*, as in kinetic energy, which is energy of movement.

L

labi- (L., lip), *labium*, lip.

lac- (L., milk), *lactose*, milk sugar.

later- (L., side), *lateral*, relating to the side.

-lemma (Gr., covering), *neurilemma*, covering of a nerve.

lepi- (Gr., scale), *Lepidoptera*, order of insects with scale wings.

leuko- (Gr., white) *leukocyte*, white blood cell.

lip- (Gr., fatty), *lipoid*, fatty substance.

-log (Gr., study), *zoology*, study of animals.

luci- (L., light), *luciferin*, light-producing material.

lysis (Gr., destroy), *bacteriolysis*, destruction of bacteria.

M

macro- (Gr., large), *macronucleus*, large nucleus.

mal- (Gr., bad), *malnutrition*, bad nutrition.

mega- (Gr., large), *megaspore*, large spore.

mens- (L., table), *commensalism*, eating at a common source of food.

-mere (Gr., part), *micromere*, small part.

meso- (Gr., middle), *mesoderm*, middle cellular layer.

meta- (Gr., after), *metaphase*, later phase of mitosis.

micro- (Gr., small), *micronucleus*, small nucleus.

milli- (Gr., thousand), *millipede*, animal with a "thousand" legs.

mio- (Gr., less), *Miocene*, less recent period in geological history.

mito- (Gr., thread), *mitosis*, cell division with the formation of threadlike structures.

mono- (Gr., one), *monograph*, something written about one subject.

morph- (Gr., form), *morphology*, study of form.

multi- (L., many), *multicolored*, of many colors.

muta- (L., to change), *mutation*, abrupt hereditary change.

myco- (Gr., fungus), *mycology,* study of fungi.

myxo- (Gr., slime), *Myxomycophyta,* phylum comprising the slime molds.

N

nema- (Gr., thread), *nematocyst,* threadlike structure of coelenterates.

neo- (Gr., young or recent), *Neotropical,* constituting a recent biogeographic region in the tropics.

nephro- (Gr., kidney), *nephridium,* tubular excretory organ.

non- (L., not), *nonirritant,* not irritating.

nuc- (L., kernel or center), *nucleus,* central portion of a cell.

O

octo- (L., eight), *octopus,* animal with eight appendages.

oedo- (Gr., swollen), *edema (oedema),* swollen condition.

-oid (Gr., like), *ameboid (amoeboid),* like an *Amoeba.*

oligo- (Gr., few or little), *oligotrichous,* having few cilia.

-oma (Gr., swelling or tumor), *carcinoma,* malignant growth (cancer).

oo- (Gr., egg), *oogenesis,* formation and development of an egg.

or- (L., mouth), *oral,* pertaining to the mouth.

ortho- (Gr., straight), *Orthoptera,* order of insects with straight wings.

os- (Gr., bone), *osseous,* pertaining to bone.

ovi- (L., egg), *ovum,* egg.

P

palaio- (Gr., ancient), *paleontology,* study of ancient life.

para- (Gr., beside), *parapodia,* appendages beside others.

path- (Gr., disease), *pathogenic,* capable of causing disease.

ped- (L., feet), *pedal,* pertaining to the foot.

peri- (Gr., around), *peristome,* region around an opening or mouth.

phaeo- (Gr., dark or brown), *Phaeophyta,* phylum of brown algae.

phago- (Gr., to eat), *phagocyte,* cell that eats or destroys.

phor- (Gr., to bear), *sporophore,* part of a sporophyte that bears spores.

photo- (Gr., light), *photosynthesis,* formation of carbohydrates in the presence of light.

-phil (Gr., loving), *thermophile,* heat-loving organism.

phyco- (Gr., alga, or seaweed), *Phycomycetes,* algalike fungus.

-phyll (Gr., leaf), *mesophyll,* middle part of a leaf.

physio- (Gr., nature), *physiology,* study of the nature or function of living matter.

-phyte (Gr., plant), *sporophyte,* spore-bearing plant.

-plasm (Gr., formed), *ectoplasm,* outer region of the cell cytoplasm.

-plast (Gr., living), *chloroplast,* green body in certain living plants.

platy- (Gr., flat), *Platyhelminthes,* phylum of flatworms.

plio- (Gr., more), *Pliocene,* more recent geologic period.

poly- (Gr., many), *polymorphous,* having many forms.

post- (L., after), *postnatal,* after birth.

-pous (Gr., foot), *octopus,* animal with eight feet or appendages.

pre- (L., before), *prenatal,* before birth.

pro- (Gr., before), *prostomium,* portion of the head situated before the mouth of certain worms and mollusks.

proto- (Gr., first or essential), *protoplasm,* essential material of all plant and animal cells.

prox- (L., nearest), *proximal,* nearest.

pseudo- (Gr., false), *pseudopodia,* false feet.

-ptero (Gr., wing), *Diptera,* order of insects with two wings.

R

re- (L., again or back), *regenerate,* to form again.

ren- (L., kidney), *renal,* pertaining to the kidney.

rept- (L., creeping), *reptile,* creeping animal.

retro- (L., backward), *retrolingual,* backward from the tongue.

rhizo- (Gr., root), *Rhizopoda,* subclass of animals with rootlike appendages.

rhodo- (Gr., red), *Rhodophyta,* phylum of red algae.

roti- (L., wheel), *rotifer,* animal with a wheel-like structure on its head.

S

-sarc (Gr., flesh), *ectosarc,* outer flesh or layer of protoplasm.

schizo- (Gr., to divide), *Schizomycophyta,* phylum of fission fungi (bacteria).

scler- (Gr., hard), *sclerotic,* hard.

-scope (Gr., see), *microscope,* instrument enabling one to see minute objects.

-sect (L., to cut), *dissect,* to cut.

semi- (L., half), *semicircle,* half of a circle.

sept- (L., wall), *septum,* partition.

set- (L., bristle), *seta,* bristlelike structure.

sinu- (L., hollow), *sinus,* hollow cavity.

soma- (Gr., body), *somatoplasm,* protoplasm of the body.

spor- (Gr., seed), *spore,* reproductive structure.

stoma- (Gr., opening), *stoma,* opening, such as is found in leaves.

sub- (Gr., under), *submaxillary,* under the maxilla.

super- (L., over or above), *superior,* higher, upper, or above.

supra- (L., above) *suprarenal,* above the kidney.

sym- (Gr., together), *symbiosis,* living together.

syn- (Gr., together), *synapsis,* association or union.

T

telo- (Gr., complete or end), *telophase,* end stage of cell division.

terato- (Gr., marvel, or monster), *teratology,* study of monstrosities or deviations from the normal.

tetra- (Gr., four), *tetrapod,* something that has four feet.

-thec (Gr., case), *spermatheca,* sperm case.

thermo- (Gr., heat), *thermotropism,* reaction to heat.

thigmo- (Gr., contact), *thigmotropism,* reaction to contact.

-tom (Gr., to cut), *microtome,* instrument to cut small sections.

toxi- (Gr., poison), *toxin,* poison.

trans- (Gr., across), *transfer,* to carry across.

tri- (Gr., three), *trilobed,* having three lobes.

tricho- (Gr., hair), *trichocyst,* hairlike structure.

trop- (Gr., reaction), *tropism,* reaction to stimuli.

U

ultra- (L., beyond), *ultramicroscopic,* so small that it is beyond the microscope.

uni- (L., one), *unilateral,* on one side.

-ur (Gr., tail), *Anura,* animals without tails.

V

vas- (L., vessel), *vas deferens,* vessel to transmit male sex cells.

ventr- (Gr., belly), *ventral,* pertaining to the lower or belly side.

vit- (L., life), *vital,* essential to life.

vorti- (L., to turn), *Vorticella,* animal that turns as it moves.

Z

zoo- (Gr., life or animal), *zoology,* study of animals.

zyg- (Gr., union), *zygote,* cell produced by the uniting of male and female sex cells.

zym- (Gr., a ferment), *zymase,* enzyme that acts on a certain carbohydrate to produce carbon dioxide and water, or alcohol and carbon dioxide.

TERMS

A

abductor (ab-duk′ tor) (L. *ab,* away; *ducere,* to lead) leading away from the center or median line; contrast to **adductor.**

aboral (ab-or′ al) (L. *ab,* from; *os,* mouth) opposite to or away from the mouth.

abscission (ab-sish′ un) (L. *abscissus,* cut off) separation of leaves and other plant parts, usually by dissolution of certain cell layers.

absorption (ab-sorp′ shun) (L. *ab,* away; *sorbere,* to move) the taking up of substances or their passage through the walls of cells.

acetabulum (as e-tab′ u lum) (L. *acetabulum,* vinegar cup) cup-shaped cavity on either side of the pelvic bone into which the femur fits.

acetylcholine (as et il-ko′ lin) a substance (choline ester) released from the axon terminal of a cholinergic nerve fiber when impulses are transmitted across synapses.

acoelomate (a-se′lo mat) (Gr. *a-,* without; *koilos,* hollow) without a hollow, true body cavity, or coelom.

acromegaly (ak ro -meg′a li) (Gr. *akron,* point; *megas,* large) a condition in which the head, hands, and feet become enlarged, caused by overactivity of the pituitary gland.

acrosome (ak′ ro som) (Gr. *akros,* outermost; *soma,* body) structure at tip of sperm head which contacts the egg during fertilization.

adaptation (ad ap-ta′ shun) (L. *ad,* to; *aptare,* to fit) becoming fitted to an environment, or the mutual fitness of an organism and its environments.

adductor (a duk′ tor) (L. *ad*, toward; *ducere*, to lead) leading toward the center or median line; contrast to **abductor.**

adenine (ad′e nin) a pyrimidine component of nucleotides and nucleic acids.

adenosine (di-, tri-) **phosphate** (ADP, ATP) certain phosphorylated organic compounds through which respiratory energy is distributed into other metabolic processes in cells.

adipose (ad′i pos) (L. *adeps*, fat) fatty or greasy material.

adrenal (ad-re′nal) (L. *ad*, near; *renes*, kidneys) an endocrine (ductless) gland near the kidney.

adrenaline (ad-ren′a lin) epinephrine; hormone secreted by the inner or medullary part of the adrenals.

adrenergic (ad ren-ur′ jik) (Gr. *ad*, on; *renes*, kidneys; *ergon*, work) applied to nerve fibers that release adrenaline-like substances from their axon terminals when impulses are transmitted across synapses; contrast to **cholinergic.**

adventitious (ad ven-tish′us) (L. *adventicius*, foreign) arising from an unusual place.

aecium (e′si um) (Gr. *aikia*, injury, or *oikidion*, little house) in rust fungi, a cup at the surface of the host in which dikaryotic spores called aeciospores are produced (plural, aecia).

aerobe (a′er ob) (Gr. *aer*, air; *bios*, life) an organism requiring free oxygen for living; contrast with **anaerobe.**

afferent (af′er ent) (L. *ad*, to; *ferre*, to bear) conveying toward a center.

agglutination (a gloo ti-na′shun) (L. *ad*, to, together; *glutinare*, to glue) clumping of cells in a liquid due to specific substances known as agglutinins.

albinism (al′bi nizm) (L. *albus*, white) the absence of normal pigments in the hair, skin, and eyes of animals, or the absence of chlorophyll in plants that normally possess it.

alga (al′ ja) (L. *alga*, seaweed) simple, green chlorophyll-bearing; plants; plural, algae.

alimentary (al i-men′ta ri) (L. *alimentum*, food) pertaining to foods and digestion.

allantois (a-lan′to is) (Gr. *allantoes*, sausage-shaped), an embryonic membrane of higher vertebrates.

allele (a-lele′) (Gr. *alleon*, of one another) one of a pair, or series, of genes similarly located on homologous chromosomes.

allergy (al′er ji) (Gr. *allos*, others; *ergon*, activity) reaction to a foreign substance, especially protein.

alternation of generations the succession of sexual and asexual generations in the life cycles of many plants and such animals as *Obelia.*

alveola (al-ve′o la) (L. *alveolus*, small cavity) small, foamlike cavity.

ambulacrum (am bu-lak′rum) (L. *ambulare*, to walk) region in echinoderm animals in which are located the ambulacral tube feet for locomotion.

amino acid (a-me′no) (amine, an organic compound) an organic acid containing the amine group (NH_2) and serving as building material for proteins.

amnion (am′ni on) (Gr. *amnos*, lamb) thin membranous sac enclosing the embryos of reptiles, birds, and mammals.

amoeboid (a-me′boid) (Gr. *amoibe*, change) resembling an ameba.

amphibious (am-fib′ i us) (Gr. *amphi*, both; *bios*, life) living both in water and on land.

amphiblastula (am fi-blas′tu la) (Gr. *amphi*, both; *blastos*, bud or origin) free-swimming larval stage of sponges.

ampulla (am-pul′a) (L. *ampulla*, flask) saclike structure of the ambulacral system of starfishes.

amylase (am′i lase) (L. *amylum*, starch; -*ase*, enzyme) carbohydrate-splitting enzyme.

anaerobe (an′ a er ob) (Gr. *an-*, without; *aer*, air; *bios*, life) living without free oxygen; contrast to **aerobe.**

anal (a′ nal) (Gr. *anus*, anus) pertaining to the anus.

anaphase (an′a faz) (Gr. *ana*, up; *phasis*, appearance) stage in cell division in which chromosomes move toward the poles of the cell.

anastomose (a-nas′to moz) (Gr. *a* **amphiblastula** (am blas′tu la) (Gr. *amphi*, both; in) free-swimming larval stage of sponges.
ee-swimming larval stage of sponges.
structure of organs, especially as revealed by dissection.

androgen (an′dro jen) (Gr. *andro*, man; *genes*, origin) any of a group of male sex hormones.

angiosperm (an′ji o spurm) (Gr. *angio*, covered; *sperma*, seed) plants with seeds enclosed by carpels, i.e., ovary; contrast to **gymnosperm.**

anhydrase (an-hi′ dras) (Gr. *an*, not; *hydro*, water) an enzyme that removes water from a compound. Carbonic anhydrase specifically catalyzes the conversion of carbonic acid into carbon dioxide and water, or the reverse.

annular ring ringlike structures of xylem (wood) in stems of higher plants that show seasonal growth.

anterior (an-te′ ri or) (L. *ante*, front) head end.

anther (an′ther) (Gr. *anthos*, flower) pollen producing part of a plant stamen.

antheridium (an ther-id′ i um) (Gr. *anthos*, flower; *idion*, diminutive) the plant sexual organ that produces sperms.

antheridiophore (an ther -id′ i o for) (Gr. *anthos*, flower; *phoreo*, to bear) antheridium-bearing structure.

anthocyanin (an tho-cy′ a nin) (Gr. *anthos*, flower; *kyanos*, blue) a coloring matter of certain higher plants, which imparts a bluish or reddish color.

anthropology (an thro-pol′ o ji) (Gr. *anthropos*, human; *logos*, science) the science of ancient man and his development.

antibiotic (an ti bi -ot′ ik) (Gr. *anti*, against; *bios*, life) antagonism of one organism toward another; a drug, derived chiefly from fungi and bacteria.

antibody (an′ ti bod i) (Gr. *anti*, against; A.S. *bodig*, body) a substance engendered in an organism by the presence of a foreign material, especially bacterial proteins; an antibody is specifically antagonistic to the antigen or substance under whose influences it was formed.

antigen (an′ ti jen) (Gr. *anti*, against; *gen*, to form) a substance causing the formation of an antibody.

antitoxin (an ti-tok′ sin) (Gr. *anti*, against; *toxikon*, poison) a specific defensive substance in a body, either existing naturally or produced as a result of the presence of a specific toxin, which it tends to neutralize.

aorta (a-or′ ta) (Gr. *aorte*, lift, raise) main artery arising from the heart.

apical (ap′i kel) (L. *apex*, summit) pertaining to the apex, tip, or summit.

appendix (ap-pen′dix) (L. *ad*, to; *pendere*, to hang) an outgrowth, such as the vermiform appendix.

archegoniophore (ar ke-go′ ni o for) (Gr. *archegonos*, originator; *phoreo*, to bear) basal structure that bears the archegonium.

archegonium (ar ke -go′ ni um) (Gr. *archegonos*, originator) female, multicellular sex organ in certain plants.

archenteron (ar-ken′ ter on) (Gr. *arche*, beginning; *enteron*, gut, intestine) the future digestive tract.

artery (ar′ ter i) (Gr. *arteria*, artery) vessel conducting blood away from the heart.

ascospore (as′ co spor) (Gr. *askos*, sac; *spores*, spore) a spore contained in a saclike ascus.

ascus (as′ kus) (Gr. *askos*, sac) a spore sac of a class of fungi that typically contains eight ascospores.

asexual (a-sek′ shu al) (Gr. *a*, without; *sexus*, sex) pertaining to reproduction without sex cells.

assimilation (as sim i -la′ shun) (L. *assimilatio*, conversion) conversion of digested food into living protoplasm.

aster (as′ ter) (Gr. *aster*, star) starlike figure of radiating lines about the centrosome during certain stages of animal mitosis.

atlas (at′ las) (Gr. *Atlas*, name of a god whose pillars upheld the heavens) the first or anterior vertebra of the neck.

atrophy (at′ ro fi) (Gr. *a*, not; *trophe*, nourishment) wasting away of an organ or a part of it contrast to **hypertrophy.**

auditory (o′ di to ri) (L. *audire*, to hear) pertaining to sound reception and interpretation.

auricle (o′ ri kl) (L. diminutive of *auris*, ear) a heart chamber receiving blood from the circulation and pumping it into a ventricle.

autogamy (o- tog′ a mi) (Gr. *autos*, self; *gamos*, marriage) nuclear reorganization and self-fertilization within the same individual, as in *Paramecium.*

autonomic (ot o-nom′ ik) (Gr. *autos*, self; *nomos*, control) pertaining to a system of nerves and ganglia that regulate involuntary muscles, blood vessels and so forth, and connect with the central nervous system by the cranial and spinal nerves; movements in plants initiated by internal stimuli.

autotomy (o-tot′ o mi) (Gr. *autos*, self; *tome*, cut) self-mutilation of an organism, such as the loss of an appendage.

autotrophic (o to-trof′ ik) (Gr. *autos*, self; *trophe*, nourishment) capable of self-nourishment by using chemical elements for food, i.e., green plants; contrast to **heterotrophic.**

auxin (ok′ sin) (Gr. *auxein*, to increase) plant hormone that influences growth.

auxospore (ok′ so spore) (Gr. *auxein*, to grow; *spora*, spore) a reproductive cell in diatoms, usually resulting from the fusion of two diatoms.

axial skeleton (L. *axis*, axis) the main axis of a skeleton to which appendages are attached.

axil (ax′ il) (L. *axilla*, armpit) the angle formed by a leaf or petiole with the upward internode of a stem.

axon (ak′ son) (Gr. *axon*, axis) elongated process of a nerve cell for conduction of impulses away from the cell body; contrast to **dendrite.**

B

bacillus (ba-sil′ us) (L. diminutive of *baculum*, rod) rod-shaped bacteria.

backcross a cross between an individual of the first filial generation (F₁) and one of the parental types.

bacteriophage (bak-ter′ i o fage) (Gr. *baktron*, bacteria; *phagein*, to destroy) a type of virus that destroys bacteria; also called **phage**.

basidium (ba -sid′ i um) (Gr. diminutive of *basis*, base) a paddle-shaped, spore-bearing organ of a class of fungi; typically bears four spores.

bicuspid (bi-kus′ pid) (L. *bis*, two; *cuspis*, point) ending in two points such as bicuspid heart valve, or bicuspid tooth.

biennial (bi-en′ i al) (L. *bis*, two; *annus*, years) a plant that lasts 2 years (seasons), producing only leaves the first season, and flowers and seeds the second.

binomial nomenclature (bi-no′mi al; no′ men kla tur) (L. *bis*, two; *nomen*, name) (L. *nomen*, name; *clacare*, to call) the scientific method of naming organisms by two Latin (or latinized) words, the first the genus, the second the species.

biogeography (bi o je-og′ ra fi) (Gr. *bios*, life; *geo*, earth; *graphein*, to write) study of geographic distributions of organisms in space.

biology (bi -ol′ o ji) (Gr. *bios*, life; *logos*, science) study of living organisms.

bioluminescence (bi o lu mi -nes′ ens) (L. *luminare*, to illuminate) light emission by living organisms not directly attributable to the heat that produces incandescence.

biome (bi′ om) (Gr. *bios*, life) a large, natural assemblage of animals and plants, extending over large regions of the earth's surface.

blastocoele (blas′ to seal) (Gr. *blastos*, bud; *koilos*, hollow) hollow segmetation cavity of an embryo.

blastopore (blas′ to pore) (Gr. *blastos*, young; *poros*, pore) a pore of the blastula stage that opens into the archenteron.

blastula (blas′ tu la) (Gr. *blastos*, young) spherical, hollow mass of cells resulting from the division of the egg.

Bowman's capsule (after *Bowman*, English histologist) the enlarged end of a kidney tubule in which is found a mass of thin-walled capillaries, known as a glomerulus.

brachial (brak′ i al) (L. *brachium*, arm) pertaining to the arm.

branchial (brang′ ki al) (Gr. *branchia*, gills) pertaining to gills.

bronchus (brong′ kus) (Gr. *bronchos*, windpipe) tube leading from trachea to lungs.

buccal (buk′ al) (L. *bucca*, mouth) pertaining to the mouth.

C

caecum (cecum) (se′kum) (L. *caecus*, blind) a blind pouch, open at one end.

calcareous (kal-kar′ e us) (L. *calcarius*, lime) pertaining to a limy composition.

calciferous glands (kal-sif′ er us) (L. *calcarius*, lime) (Fr. *ferre*, to bear) glands carrying lime, as in earthworms.

calorie (kal′ o ri) (L. *calor*, heat) a unit of heat measurement, usually the amount of heat required to raise the temperature of 1 gram (ml.) of water 1° C. A kilocalorie is the amount of heat required to raise the temperature of 1 kilogram of water by 1° C.

calyptra (ka-lip′ tra) (Gr. *kalyptra*, covering) the archegonium of a moss or liver-wort distended or modified by the growth of the sporophyte. In certain mosses it is carried to the top of the capsule to form a sheath-like hood.

calyx (ka′liks) (Gr. *kalix*, husk or cup) the outer whorl of floral leaves known individually as sepals.

cambium (kam′ bi um) (L. *cambiare*, change) the growing meristematic tissue from which the secondary phloem and xylem tissues arise in roots and stems; located between wood and bark.

Cambrian (from Cambria, Wales) earliest geologic period in which fossils are found abundantly.

capillary (kap′ i lar i) (*capillaris*, hair) minute blood vessels whose walls are one cell thick and which connect arteries and veins.

capillitium (kap il -ish′ i um) (L. *capillaris*, hair) delicate network in the sporangia of slime fungi.

capsule (kap′ sul) (L. *capsula*, little box) the spore case of bryophytes; in angiosperms, a dry dehiscent fruit, composed of more than one carpel.

carapace (kar′ a pas) (Sp. *carapacho*, hard shield) a covering on the back of certain animals, such as lobsters and turtles.

carbohydrate (kar bo-hi′ drate) (L. *carbo*, carbon; *hydro*, water) substances (sugars, starches, etc.) composed of carbon, hydrogen, and oxygen, with the latter two usually in the same ratio as found in water.

cardiac (kar′ di ak) (Gr. *kardia*, heart) pertaining to the heart.

carnivorous (kar-niv′ or us) (L. *carnis*, flesh; *varare*, to devour) flesh eating.

carotene (kar′ o tene) (L. *carota*, carrot) reddish-orange pigments of certain plants.

cartilage (kar′ ti lage) (L. *cartilago*, gristle) elastic, flexible connective tissue.

catalyst (kat′ a list) (Gr. *kata*, down; *lysis*, loosening) a substance that accelerates a chemical reaction but does not become a part of the product produced.

caudal (ko′ dal) (L. *cauda*, tail) pertaining to the tail.

cell (L. *cella*, small space) a small mass of protoplasm containing nuclear materials and enclosed by an outer covering.

cellulose (sel′ u los) (L. *cellua;* little cell) a complex carbohydrate in plant cell walls.

Cenozoic era (sen o -zo′ ik) (Gr. *kenos*, recent; life) the most recent geologic era, characterized by mammals, birds, modern insects, and so forth.

centriole (sen′ tri ol) (L. *centrum*, center) small central granule of most centrosomes.

centromere (sen′ tro mere) (Gr. *kentron*, center; *meros*, part) a region of a chromosome where it attaches to a spindle fiber during mitosis and meiosis.

centrosome (sen′ tro some) (Gr. *kentron*, center; *soma*, body) central body; a body enclosing a centriole and located in the center of the aster during mitosis.

cephalic (se-fal′ ik) (Gr. *kephale*, head), pertaining to the head.

cephalothorax (sef′ a lo-thor′ aks) (Gr. *kephale*, head; *thorax*, chest) head fused with the thorax.

cercaria (ser-ka′ ri a) (Gr. *kerkos*, tail) tailed larva of a fluke, produced by a redia.

cerebellum (ser e-bel′ um) (L. diminutive of *cerebrum*, brain) anterior hemispheric part of the vertebrate brain dorsal to and anterior to the medulla oblongata.

cerebrum (ser′ e brum) (L. *cerebrum*, brain) anterior hemispheric part of a vertebrate brain.

cervical (sur′ vi cal) (L. *cervix*, neck) pertaining to the neck.

chelicera (ke -lis′ er a) (Gr. *chele*, claw; *keros*, horn) the most anterior pair of pincer-like appendages of spiders, scorpions, king crabs, and so forth.

chemotaxis (kem o tax′ is) (Gr. *chemo*, chemical or juice; *taxis*, locomotion) simple locomotor response, either positive or negative, to chemical stimuli.

chemotropism (kem- ot′ ro pizm) (Gr. *chema*, chemical; *trophe*, turn) response to chemicals, frequently by growth.

chitin (ki′ tin) (Gr. *chiton*, covering) outer, horny covering of insects, crustacea, and so forth.

chlorenchyma (klor -engk′ i ma) (Gr. *chloros*, green; *enchyma*, infusion) a tissue that possesses chloroplasts, usually a parenchyma tissue.

chlorogogen (klo- rog′ e jen) (Gr. *chloros*, green; *gogne*, a leading) cells on the outer surface of the earthworm intestine.

chlorophyll (klo′ ro fil) (Gr. *chloros*, green; *phyllon*, leaf) the green pigment of many plants used for photosynthetic purposes.

chloroplast (chloroplastid) (klo′ ro plast) (Gr. *chloros*, green; *plastos*, moulded) a body that contains chlorophyll.

choanocyte (ko′ an o site) (Gr. *choana*, funnel; *kytos*, cell) see **collar cell.**

cholinergic (ko lin-er′ jik) a type of nerve fiber that releases acetylcholine from the axon terminal when impulses are transmitted across synapses; contrast to **adrenergic.**

chordata (kor-da′ ta) (Gr. *chorde*, chord or string) animals having a temporary or permanent dorsal skeletal notochord.

chorion (ko′ ri on) (Gr. *chorion*, membrane) outer membrane enveloping the mammalian fetus and enclosing the amnion membrane.

choroid (ko′ roid) (Gr. *chorion*, membrane; *eidos*, form) a vascular layer between the retina and the sclerotic layer (sclera) of the eye.

chromatid (kro′ ma tid) (Gr. *chroma*, color; *id*, daughter) one of two threads and its matrix in a chromosome.

chromatin (kro′ ma tin) (Gr. *chroma*, color) the part of the nucleus that stains well.

chromatophore (kro′ ma to for) (Gr. *chroma*, color; *phoreo*, to bear) a colored plastid or cell, such as a chloroplast.

chromonema (kro mo-nem′ a) (Gr. *chroma*, color; *nema*, thread) thread-like structures within the chromosome; plural, chromonemata.

chromoplast (kro′ mo plast) (Gr. *chroma*, color; *plastos*, formed or moulded) a colored plastid, other than a chloroplastid.

chromosome (kro′ mo som) (Gr. *chroma*, color; *soma*, body) deeply stained bodies formed in the cell nucleus during mitosis and meiosis; they carry the materials of heredity.

chyme (kime) (L. *chymus*, juice) semiliquid, partially digested food in the stomach.

cilium (sil′ i um) (L. *cilium*, eyelash) hairlike, vibratile structure on certain cells and certain protozoans.

cleavage (kle′ vij) (A.S. *cleofan*, to separate) division of a zygote into cells known as blastomeres.

clitellum (kli-tel′ um) (L. *clitellae*, pack saddle) thickened area on certain annelids that assists in reproduction.

cloaca (klo-a′ ka) (L. *cloaca*, sewer) common organ into which the intestine, kidneys, and sex organs discharge their products.

clone (klone) (Gr. *klon*, twig) all the asexual offspring of an individual that are identical in regard to their gene content.

cnidoblast (ni′ do blast) (G. *knide*, nettle; *blastos*, bud) sac-shaped stinging or nettle cell with a permanent long, barbed thread and poisonous fluid, such as found in certain coelenterates.

cnidocil (ni′ do sil) (Gr. *knide*, nettle; L. *cilium*, eyelash) small, trigger-like process for ejecting the thread from a cnidoblast.

coagulation (ko ag u -la′ shun) (L. *coagulare*, change) change in the physical state of proteins due to the destruction of the internal structure of the protein molecule; gross result is a solidification of previously liquid or jellylike proteins; contrast to **denaturation.**

coccus (kok′ us) (Gr. *kokkus*, berry) spherical unicellular organism.

cocoon (ko koon′) (F. *cocon*, cocoon) the enclosed stage of certain insects in which the pupa enters the cocoon and the adult emerges from it; egg case of earthworms and spiders.

coelom (se′ lom) (Gr. *koilos*, hollow) a hollow, true body cavity containing organs, and lined with mesoderm tissue.

coenocyte (se′ no site) (Gr. *koinis*, shared; *kytos*, cell) a cell with several to many nuclei.

coenzyme (ko -en′ sim) (L. *co*, together) a substance, usually organic, required to activate a given enzyme.

cochlea (kok′ le a) (Gr. *kochlias*, snail) spirally coiled part of the inner ear, containing receptors for hearing.

collar cell choanocyte; cell with a collar-like structure on the surface such as sponges have.

collenchyma (kol-engk′ i ma) (Gr. *kolla*, glue; *enchein*, to pour in) a plant tissue of living cells with thick walls that consist largely of cellulose for strengthening purposes.

colon (ko′ lon) (L. *kolos*, colon) portion of large intestine between caecum and rectum.

commensalism (kom-men′ sal izm) (L. *cum*, with; *mensa*, table) an association of members of two or more species (not truly parasites) that live in, on, or with each other, and usually partake of the same food.

commissure (kom′ i shoor) (L. *commissura*, join together) a circle of nervous tissue that connects various regions in earthworms, snails, insects, and so forth.

companion cell one that usually accompanies sieve tubes in phloem tissues of plants.

compound eye one made of numerous units, called ommatidia, such as those found in certain arthropods.

congenital (kon-jen′ i tal) (L. *con*, together; *gigno*, to bear) present at birth.

conidiophore (ko-nid′ i o for) (Gr. *konis*, dust; *idion*, diminutive; *phoreo*, to bear) structure that bears conidiospores.

conidiospore (ko-nid′ i o spor) (Gr. *konis*, dust; *idion*, diminutive; *sporos*, spore) spore formed by constricting the tip of a hypha in certain fungi (molds).

connective tissues (kon-nek′ tiv) (L. *cum*, together; *nectere*, to bind) similar cells for support and binding.

contractile vacuole (kon-trak′ til; vak′ u ol) (L. *con*, together; *trahere*, to draw) (L. *vaccus*, empty), a hollow structure that alternately contracts and expands, as found in amebae, paramecia, and so forth.

conus arteriosus (ko′ nus; ar ter ri-o′ sus) (Gr. *konus*, cone shaped) cone-shaped structure between the ventricle and arteries of certain animals.

copulation (kop u -la′ shun) (L. *copulare*, to couple) sexual union.

cork cambium a ring of dividing cells beneath the epidermis in woody plants, originating parenchyma tissue on the inside and cork on the outside.

cornea (kor′ ne a) (L. *corneus*, horny) transparent part of the sclerotic coat (sclera) of the eye that covers the iris and pupil.

corolla (ko-rol′ a) (L. diminutive of L. *corona*, crown) all the flower petals considered together.

corpus luteum (kor′ pus; lu′ te um) (L. *corpus*, body; *luteum*, yellow), a yellowish body developed from a graafian follicle after extrusion of the ovum (egg).

cortex (kor′ tecks) (L. *cortex*, covering or bark) the outer covering.

cotyledon (kot i -le′ don) (Gr. *kotyledon*, cup-shaped embryonic leaf) embryonic seed leaf, usually containing stored food.

cranium (kra′ ni um) (Gr. *kranion*, skull or brain case) the brain case.

cretinism (kre′ tin izm) (L. *christianus*, human being) one who is physically and mentally deficient due to a deficient thyroid gland during youth.

cross fertilization union of gametes (sex cells) produced by different individuals, either plants or animals.

crossing over rearrangement and crossing over of linked characters as a result of exchange of genes between homologous chromosomes during synapsis.

cutaneous (ku-ta′ ne us) (L. *cutis*, skin) pertaining to the skin.

cuticle (ku′ tik l) (L. *cutis*, covering) transparent covering.

cutin (ku′ tin) (L. *cutis*, covering) waxy substance covering leaves, making them impervious to water.

cyclosis (si- klo′ sis) (Gr. *kyklosis*, circulate) circulatory movement of protoplasm in a cell.

cyst (sist) (Gr. *kystis*, sac) protective covering around an object.

cysticercus (sis ti-sur′ kus) (Gr. *kystis*, sac; *kerkos*, tail) larval bladderworm stage of certain tapeworms.

cytochrome (si′ to krom) (Gr. *kytos*, cell; *chroma*, color) a substance that contains iron and acts as a hydrogen carrier for the eventual release of energy in aerobic respiration. Blood hemoglobin also contains iron but it carries oxygen for the eventual release of energy also.

cytokinesis (si to ki -ne′ sis) (Gr. *kytos*, cell; *kinesis*, movement), cytoplasmic division by cell wall formation, usually following mitosis or nuclear division.

cytology (si-tol′ o ji) (Gr. *kytos*, cell; *logos*, science) study of cells.

cytoplasm (si′ to plazm) (Gr. *kytos*, cell; *plasma*, liquid) the portion of the protoplasm outside the nucleus.

D

deciduous (de-sid′ u us) (L. *de*, away; *cadere*, to fall) falling off at the end of a period of growth.

dehydration (de hi-dra′ shun) (Gr. *de*, from; *hydro*, water) extraction of water.

deltoid (del′ toid) (L. *delta*, triangle) triangular, like the deltoid muscle.

denaturation (de na tur -a′ shun) partial physical disruption of the internal structure of a protein molecule. Denaturation is usually reversible, whereas coagulation is not.

dendrite (den′drite) (Gr. *dendron*, tree) dendron; branched process that carries impulses toward a nerve cell (neuron); contrast to **axon.**

denitrifying (de -ni′ tri fy ing) (Gr. *de*, from; *nitrogen*, nitrogen) converting nitrogenous substances, such as nitrates, to ammonia and free (molecular) nitrogen.

dephosphorylation (de fos fo ri-la′ shun) removal of organic phosphorus from a molecule.

dermis (der′ mis) (Gr. *derma*, skin) derma; true skin underlying the epidermis; same as **corium.**

desoxyribonucleic acid (DNA) a nucleic acid containing pentose, a 5-carbon sugar; see **desoxyribose.**

desoxyribose (des-oks′ i ri′ bos) (Gr. *desoxy*, loss of oxygen; *ribose*, pentose sugar) a 5-carbon sugar having one oxygen atom less than ribose sugar; a component of desoxyribonucleic acid (DNA).

diabetes (di a -be′ tez) (Gr. *diabainein*, to pass through) an abnormal condition marked by sugar excretion in urine and low blood glucose levels caused by an insufficiency of insulin.

diaphragm (di′ a fram) (Gr. *diaphragma*, partition) muscle that separates the thoracic and abdominal cavities.

diastole (di-as′ tole) (Gr. *diastole*, moved apart) the relaxation of auricles or ventricles, during which time they fill with blood; contrast with **systole.**

dichotomous (di-kot′ o mus) (Gr. *dicha*, in two; *tomein*, to divide) pertaining to repeated forking into two parts.

dicotyledon (di kot i -le′ don) (Gr. *dis*, twice; *kotyledon*, cup-shaped), a plant having two seed leaves or cotyledons; contrast with **monocotyledon.**

diecious (di -e′ shus) (Gr. *di*, two; *oikos*, house) dioecious; having male and female sex organs in separate individuals; contrast with **monecious.**

diffusion (di-fu′zhun) (L. *diffusia*, spread) passage of molecules of one substance among those of another, from a region of greater concentration to one of lower concentration.

digestion (di-jes′ chun) (L. *digestio*, arrange food) preparing food, through the action of enzymes, for absorption.

dimorphism (di -mor′ fizm) (Gr. *di*, two; *morphe*, form) two forms or types belonging to the same species, such as males and females of the same species, but differing from each other.

diploblastic (dip lo- blas′ tik) (Gr. *diplos*, two; *blastos*, germ) having two primary germ layers (ectoderm and endoderm).

diploid (dip′ loid) (Gr. *diploos*; *eidos*, like) double num-

ber of chromosomes found in the sporophyte generation of plants and in the body cells of animals, as
contrasted with the single (monoploid) number in
germ cells; contrast to **monoploid, haploid.**

disaccharide (di-sak' a rid) (Gr. *dis*, twice; *sakcharon*,
sugar) a sugar composed of two monosaccharides;
usually refers to 12-carbon sugars.

distal (dis' tal) (L. *dis*, apart; *stare*, to stand) farthest
from median line; or farther from a fixed point.

DNA abbreviation for desoxyribonucleic acid.

dominance dominant character; one of a pair of alternative characters that is always expressed when
its gene is present and that appears to exclude the
other (recessive) character.

dorsal (dor' sal) (L. *dorsum*, back) the back side of
animals.

dorsal aorta chief artery that arises from the heart to
distribute blood to the body.

dorsal horn (root) sensory root of a spinal nerve that
carries impulses into the spinal cord; ventral root
carries impulses from the cord.

ductus arteriosus (duk' tus; ar te ri-o' sus) an artery,
present in the embryo and fetus, that conducts blood
from the pulmonary artery to the aorta; it disappears
at birth when lungs begin to function.

duodenum (du -od' e num) (L. *duodeni*, twelve) anterior part of the small intestine, 12 finger widths
long.

E

ecdysis (ek' di sis) (Gr. *ekdyein*, to shed) losing or molting of an outer structure, as in crayfishes, insects,
and so forth.

ecology (e-kol' o ji) (Gr. *oikos*, home; *logos*, study)
scientific study of living organisms and their living
and nonliving environments.

ectoderm (ek' to durm) (Gr. *ektos*, outside; *derma*,
skin) outer layer of embryonic germ layers.

ectoplasm (ek' to plazm) (Gr. *ektos*, outer; *plasma*,
liquid form) ectosarc; outer layer of cell cytoplasm;
contrast to **endoplasm.**

efferent (ef' er ent) (L. *ex*, out; *ferro*, to carry) conveying away from.

egest (e-jest') (L. *ex*, out; *gerere*, to carry) to throw
out, usually indigestible, material.

egg ovum; the mature female sex cell of an animal or
plant.

elater (el' a ter) (Gr. *elater*, driver) a springlike structure of various plants that disperses spores, as found
in horsetails.

elytra (el' i tra) (Gr. *elytron*, sheath) sheath-like wings
of beetles.

embryo (em' bri o) (Gr. *embryon*, embryo) early stage
of development of an organism.

embryology (em bri- ol' o ji) (Gr. *embryon*, embryo;
logos, study) study of early development of organisms.

embryo sac the female gametophyte of angiospermous
plants, within which the embryo begins development.

embryophyte (em' bri o fit) (Gr. *embryon*, embryo;
phyta, plants) a plant producing an embryo and developing vascular tissues.

endocardium (en do-kar' di um) (Gr. *endo*, within;
kardium, heart) inner lining of heart.

endocrine glands (en' do krin) (Gr. *endo*, within;
krinein, to separate) ductless glands that produce
internal secretions from materials brought to them by
blood, and whose secretions are carried away by the
blood.

endoderm (en' do durm) (GR. *entos*, within; *derma*,
skin) inner germ layer; contrast to **ectoderm.**

endoplasm (en' do plazm) (Gr. *endo*, within; *plasma*,
liquid) inner layer of cytoplasm of a cell; contrast to
ectoplasm.

endoskeleton internal skeleton.

endosperm (en' do spurm) (Gr. *endo*, within; *sperma*,
seed) nutritive substances within the seed coats, but
not part of the embryo proper.

energy (en' er ji) (Gr. *energos*, action) capacity to do
work, with the time rate of doing work being called
power.

enteric (en-ter' ik) (Gr. *enteron*, intestine) pertaining to
digestion or digestive tract.

enterocoele (en' ter o -seal) (Gr. *enteron*, gut; *koilos*,
hollow) a type of coelom formed by the outpouching
of mesodermal sac from the entoderm of the primitive gut.

enterokinase (en ter o -ki' nas) (Gr. *enteron*, gut;
kinetos, moving) an enzyme in the intestinal juice
which converts trypsinogen into trypsin.

enzyme (en' sim) (Gr. *en*, in; *zyme*, leaven) organic
(biological) catalyst that is secreted by protoplasm to
bring about or hasten a reaction, but that is not consumed in the process.

epicotyl (ep' i kot il) (Gr. *epi*, upon, *kotyle*, seed leaf,
cotyledon) portion of the embryo's axis, above the
attachment of the cotyledon, that forms a young
stem.

epidermis (ep i -dur′ mis) (Gr. *epi*, upon; *derma*, skin) outer layer of skin cells.

epiglottis (ep i -glot′ is) (Gr. *epi*, upon; *glotta*, tongue) tongue-like covering of the glottis during swallowing.

epinephrine (ep i -nef′ rin) (Gr. *epi*, upon; *nephros*, kidney) adrenaline; hormone of the inner medulla of the adrenals, which are located on the kidneys.

epithelium (ep i -the′ li um) (Gr. *epi*, upon; *thele*, teat or nipple) membranes lining or covering a surface, including secretory glands.

equatorial plate middle or equator of the spindle during cell division.

erythrocyte (e -rith′ ro sit) (Gr. *erythros*, red; *kytos*, cell) red blood corpuscle.

estrogen (es′ tro jen) (Gr. *oistros*, frenzy; *genes*, origin) one of a group of female sex hormones; an estrus-producing hormone.

eustachian tube (u -sta′ shun) (*Eustachio*, an Italian anatomist) tube connecting the pharynx and middle ear.

evagination (e vaj i -na′ shun) (L. *evageri*, to go forth) outgrowing of a layer of cells from a cavity.

evolution (ev o -lu′ shun) (L. *evolvere*, to unroll) theory that all living things have undergone, and do undergo, gradual changes through successive generations; that all living organisms are constantly changing (evolving).

excretion (eks -kre′ shun) (Gr. *ex*, out; *cernere*, to separate) elimination of wastes.

excurrent (eks-kur′ ent) (Gr. *ex*, out; *currens*, to run) conducting away from a cavity organ.

exoskeleton (eks o -skel′ e ton) (Gr. *ex*, outside; *skeleton*, mummy, dried) outer skeleton.

extensor (eks-ten′ sor) (Gr. *ex*, out; *tendere*, to stretch) muscle that extends a limb or part.

F

F₁, F₂, F₃ abbreviations for first, second, and third filial generations in heredity.

fallopian tube (fa-lo′ pi an) (after *Fallopius*, a physician, who died in 1562) the oviduct of mammals.

fascia (fash′ i a) (L. *fascia*, band) band-like covering of connective tissue.

fatty acid one of a group of organic acids, such as acetic, butyric, oleic, stearic, etc., that contain only one carboxyl group (COOH).

feces (fe′ sez) (L. *faeces*, dregs) wastes or excrements.

feedback information passing from an effector to a re-ceptor, as a signal indicating the action performed by an effector.

femur (fe′ mur) (L. *femur*, thigh) thigh bone or the third segment of an insect leg from the proximal (near) end.

fermentation (fur men-ta′ shun) (L. *fermentum*, ferment or leaven) change in an organic substance caused by a ferment, such as souring of milk, with little or no oxygen involved, ie, anaerobic respiration.

fertilization (fer til i -za′ shun) (L. *ferre*, to produce) union of sperm and egg in sexual reproduction.

fetus (fe′ tus) (L. *fetus*, offspring) foetus; the later embryo of a vertebrate (after the third month in a human being).

fibrinogen (fi-brin′ o jen) (L. *fibra*, thread; Gr. *gignesthai*, to form) a protein constituent of blood that assists in fibrin formation during clotting.

fibula (fib′ u la) (L. *fibula;* buckle) outer, smaller bone of lower leg.

filial (fil′ i al) (L. *filia*, daughter; *filius*, son) one or more successive generations after the parents' generation.

fission (fish′ un) (L. *fissus*, cleft) asexual division into two or more parts.

flagellum (fla-jel′ um) (L. *flagellum*, whip) whip-like protoplasmic process for locomotion.

flame cell an excretory cell with a bunch of cilia that expel wastes to the outside; the actions of the cilia somewhat resemble a flickering flame, as in certain flatworms.

flexor (flek′ ser) (L. *flexus*, bend) muscle to bend a joint.

foramen (fo-ra′ men) (L. *foramen*, opening) an opening in a structure.

fossil (fos′ il) (L. *fossilis*, dug up) preserved record of an ancient organism.

fovea centralis (fo′ ve a; sen-tra′ lis) (L. *fovea*, pit) a small, pit-like area in the optic center of the eye's retina; only cone cells are present and their stimulation leads to acute vision.

fraternal twins those produced by fertilization of two separate eggs, which usually have different hereditary traits; sometimes called nonidentical or dizygotic twins; contrast to **identical** or **monozygotic twins.**

frond (L. *frond*, leafy branch) fern leaf.

fruit (L. *fructus*, fruit) a mature ovary, together with any other structures that ripen and form a unit with it.

fucoxanthin (fu ko -zan′ thin) (L. *fucus*, seaweed, rock lichen; *xanthos*, yellow) yellowish-brown pigment of algae, diatoms, and certain flagellates.

G

gall bladder sac near the liver in which bile is stored.

gametangium (ga me -tan' ji um) (Gr. *gamos*, gametes; *angios*, case) a gamete-producing structure.

gamete (gam' et) (Gr. *games*, gamete) mature male or female sex cell.

gametogenesis (gam e to-jen' e sis) (Gr. *gamos*, gamete; *genesis*, origin) production and maturation of gametes (sex cells).

gametophyte (gam' e to fit) (Gr. *gamos*, gamete; *phyton*, plant) plant producing gametes, i. e., sex cells; contrast to **sporophyte.**

ganglion (gang' li on) (Gr. *ganglion*, a swelling) an enlargement of a nerve that contains nerve cells and acts as a center of influence; plural, ganglia.

gastric (gas' trik) (Gr. *gastros*, digestive, stomach) pertaining to the stomach or to digestion.

gastrin (gas' trin) (Gr. *gastros*, stomach) a hormone that is produced by the stomach wall when food contacts it and that stimulates other parts of the stomach to secrete gastric juices.

gastrovascular (gas tro-vas' ku lar) (Gr. *gastros*, stomach, digestive; L. *vasculum*, vessel or circulation), pertaining to digestive-circulatory cavity of *Hydra* and so forth.

gastrula (gas' troo la) (Gr. *gastros*, stomach) a two-layered or, usually, three-layered embryonic stage of animals that forms a hollow invaginated sac with the blastopore as the opening.

gastrulation (gas troo-la' shun) (Gr. *gastros*, digestive) the formation of the gastrula stage in embryonic development by an invagination by which the future digestive tract is formed.

gemma (jem' a) (L. *gemma*, bud) small, asexual reproductive bodies found in certain organisms, such as *Marchantia* (liverwort).

gemmule (jem' ul) (L. *gemma*, bud) asexual reproductive body of several cells found in certain sponges.

gene (Gr. *jen*, to form) factor (determiner) on a chromosome that influences the development of a hereditary trait.

genetics (je-net' iks) (Gr. *genesis*, origin) science of trait transmissions from parents or other ancestors to offspring.

genital (jen' i tal) (L. *gignere*, to beget) pertaining to reproduction.

genome (jen' om) (Gr. *genesis*, origin) the total of the genes in a monoploid (haploid) set of chromosomes, and therefore the total of all the different genes in a cell that make one complete set.

genotype (jen' o tipe) (Gr. *genos*, race; *typos*, model) hereditary constituent of an organism or a group of organisms based upon gene content; contrast to **phenotype.**

genus (Gr. *genos*, race) somewhat similar organisms having one or more species which are structurally or physiologically related; plural, genera.

geotaxis (je o-tak' sis) (Gr. *geo*, earth; *taxis*, arrangement or locomotion), locomotor response due to gravity.

geotropism (je -ot' ro pizm) (Gr. *geo*, earth; *trope*, turning) reaction of organisms to gravity.

germ cell male or female reproductive cell.

germ layer three embryonic cellular layers (ectoderm, mesoderm, and entoderm) from which future adult tissues and organs may arise.

germination (L. *germinare*, to sprout) the resumption of growth by a spore or seed after a period of dormancy.

gestation (jes' ta-shun) (L. *gestatio*, to bear) carrying of young, normally in the uterus, from conception to delivery (birth).

gibberellin (jib ber el'-in) (after a sac fungus, *Gibberella*) a growth-regulating plant hormone.

gill filamentous or platelike structure with blood vessels for respiration in water.

gill slits paired openings in vertebrates that connect the pharynx with the exterior and permit the exit of water (same as pharyngeal cleft).

gizzard (giz' ard) (F. *giser*, gizzard) muscular grinding organ for the purpose of digestion.

gland (L. *glans*, nut) a cell or group of cells for the purpose of secretion.

glochidium (glo-kid' i um) (Gr. *glochis*, point; *idion*, diminutive) bivalved larva of mollusks living parasitically on a fish for a period of time.

glomerulus (glo-mer' u lus) (L. *glomus*, ball) ball-like mass of capillaries at the enlarged end of a kidney tubule of higher vertebrates (malpighian body).

glottis (glot' is) (Gr. *glotta*, tongue) slitlike opening in the pharynx leading to the trachea (windpipe).

glycogen (gli' ko jen) (Gr. *glykys*, sweet) starchlike carbohydrate stored in the liver and other tissues and in certain algae and fungi.

glycerin (glis' er in) (Gr. *glyceros*, sweet) an organic compound that has a 3-carbon skeleton and that may unite with fatty acids to form a fat.

goiter (goi′ ter) (L. *guttur*, throat) pathologic condition of the thyroid gland.

Golgi bodies (apparatus) (gol′ je) (after *Golgi*, Italian histologist) special bodies in cytoplasm of certain cells that may assist in production of secretion.

gonad (gon′ ad) (Gr. *gone*, generate) male or female reproductive organ.

graafian follicle (graf′ i an) (fol′ i kl) (after *de Graaf*, Dutch physician; L. *folliculus*, small bag) small cavity in the ovary, especially in mammals, in which an egg develops.

grana (gran′ a) (L. *granum*, small grain) small particles of chlorophyll in chloroplasts.

green gland excretory gland of a crayfish.

growth hormone specific chemical substance in plants and animals, especially higher animals, that influences, regulates, or controls growth or other activities.

guard cell specialized cell of leaf epidermis that regulates the size of a leaf stoma.

gustatory (gus′ta to ri) (L. *gustare*, to taste) sense of taste.

guttation (gu-ta′ shun) (L. *gutta*, a drop) exudation of water drops from plants, especially leaves.

gymnosperm (jim′ no spurm) (Gr. *gymnos*, naked; *sperma*, seed) plant whose seeds are not enclosed by carpels, i.e., ovary; contrast to **angiosperm.**

H

habitat (hab′ i tat) (L. *habitare*, to dwell) usual or natural dwelling place.

halteres (hal-te′ rez) (Gr. *halter*, weight used in jumping) pair of capitate bodies used as balancers during flight of insects of the order Diptera. They represent the rudimentary posterior wings of these insects.

haploid (hap′loid) (Gr. *haploos*, single; *eidos*, form) single or reduced number of chromosomes in mature germ cells (gametes), or in the gametophyte generation of plants, in contrast to the diploid number; same as **monoploid.**

haustorium (how-sto′ ri um) (L. *haustum*, to draw) specialized organelle or organ through which a parasite secures nourishment from its host.

haversian canal (ha-vur′si an) (after *Havers*, an English physician of seventeenth century) small canal in bone that conducts blood and other substances.

hemocoel (he′mo sel) (Gr. *haima*, blood; *koilos*, hollow) special portion of the coelom used for blood circulation in certain animals.

hemoglobin (he mo-glo′ bin) (Gr. *haima*, blood; *globos*, globe) reddish, oxygen-carrying substance of red blood corpuscles.

hemophilia (he mo-fil′ i a) (Gr. *haima*, blood; *philos*, loving) a disease, usually hereditary, that produces a tendency toward excessive bleeding, from even slight wounds.

hemorrhage (hem′ o raj) (Gr. *haima*, blood; *rhegnymi*, to break) loss of blood from broken blood vessel.

hepatic (he-pat′ ik) (L. *hepaticus*, liver) pertaining to the liver.

hepatic portal system the double blood supply of the liver of vertebrates.

herbaceous (hur-ba′ shus) (L. *herbaccous*, grassy) pertaining to plants without woody stems.

herbivorous (hur-biv′ or us) (L. *herba*, plant; *vorare*, to devour) pertaining to plant-eating organisms; contrast to **carnivorous.**

heredity (he-red′ i ti) (L. *hereditas*, heir) transmission of physical and mental traits from parent or other ancestor to offspring; see **genetics.**

hermaphrodite (her-maf′ ro dit) (Gr. *hermaphrodites*, containing both sexes), having both male and female reproductive organs in one individual; also known as monecious; contrast to **diecious.**

heterocyst (het′ er o sist) (Gr. *heteros*, different; *kystis*, sac) clear cell in an alga that separates a filament into hormogonia.

heterogamy (het er-og′ a mi) (Gr. *heteros*, different; *gamos*, marriage) anisogamy; union of unlike gametes, i.e., sex cells; contrast to **isogamy.**

heterosis (het er-o′ sis) (Gr. *heteros*, other) increased vigor which often occurs in offspring of parents of different inbred lines, species or varieties.

heterospory (het er-os′ po ri) (Gr. *heteros*, different; *spora*, spore) production of unlike spores; contrast with **homospory.**

heterotrophic (het er o-trof′ ik) (Gr. *heteros*, different; *trophe*, food) pertaining to organisms that are unable to manufacture their food and hence are parasites or saprophytes; contrast to **autotrophic.**

heterozygote (het er o -zi′ got) (Gr. *heteros*, unlike; *zygon*, yoke) an organism formed by union of gametes that are unlike in their genetic content; contrast with **homozygote.**

histogenesis (his to-gen′ e sis) (Gr. *histos*, tissue; *gen*, to form) tissue formation and development.

histology (his-tol′ o ji) (Gr. histos, tissue; logos, study) microscopic study of tissues.

holozoic (hol o -zo' ik) (Gr. *holos*, whole; *zoon*, animal) pertaining to an animal's securing nourishment by ingesting and digesting organic materials.

homeostasis (ho me-o-sta' sis) (Gr. *homos*, same; *stasis*, standing), a tendency to stability.

homeothermal (ho me o-ther' mal) (Gr. *homos*, same; *therma*, heat) maintaining a uniform temperature, such as warmblooded animals do; same as **homothermal, homoiothermal.**

homologous chromosomes (ho -mol' o gus) (Gr. *homos*, same, *logos*, speech), a pair of chromosomes, one from each parent, that have relatively similar structures and gene values.

homologous genes genes similarly located on homologous chromosomes, contributing to the same or different expression of a trait.

homozygote (ho mo-zi' got) (Gr. *homos*, same; *zygon*, yoke) union of gametes that are alike in their genetic content; contrast with **heterozygote.**

hormogonia (hor mo-go' ni a) (Gr. *hormos*, chain; *gonos*, offspring) ability of portions of bluegreen algal filaments to form new individuals.

hormone (hor' mon) (Gr. *hormaein*, to excite) a chemical substance that is secreted by one organ and produces specific effects elsewhere.

host (L. *hostis*, stranger) an organism in or on which a parasite lives.

humerus (hu' mer us) (L. *humerus*, shoulder) upper arm bone.

humoral (hu' mor al) (L. *humor*, fluid) pertaining to body fluids; as in humoral agent, a substance that is transported in body fluids and that has specific effects on specific tissues.

humus (hu mus) (L. *humus*, soil) the organic part of soil.

hyaline (hi' al in) (Gr. *hyalos*, glassy) clear or transparent, as in hyaline cartilage.

hybrid (hi' brid) (L. *hybrida*, mongrel) a crossbreed animal or plant; the offspring of two parents who differ in at least one trait.

hydroid (hi' droid) (Gr. *hydro*, water; *eidos*, like) resembling a hydra.

hydrolysis (hi-drol' i sis) (Gr. *hydro*, water; *lysis*, a loosening) destruction of a chemical substance by the addition of the elements of water.

hydrophyte (hi' dro fite) (Gr. *hydro*, water; *phyton*, plant) plant growing in water.

hydroponics (hi dro-pon' iks) (Gr. *hydro*, water; *ponos*, exertion) growth of plants in liquid culture media (soilless cultivation).

hydrotropism (hi-drot - ro pizm) (Gr. *hydro*, water; *trops*, a turning) response of organisms to water.

hyoid (hi-oid) (Gr. *hyoides*, Y-shaped) bone or cartilage at base of tongue.

hypertonic (hi per-ton' ik) (*hyper*, above; *tonos*, tension) possessing greater osmotic pressure than some related substance; contrast to **hypotonic.**

hypertrophy (hi-per' tro fi) (Gr. *hyper*, above; *trophe*, growth) excessive growth or development; contrast to **atrophy.**

hypha (hi' fa) (Gr. *hyphe*, web) threadlike element of the mycelium of a fungus; plural, hyphae.

hypocotyl (hi po-kot' il) (Gr. *hypo*, below; *kotyle*, seed leaf, cotyledon) portion of the embryo axis below the attachment of cotyledons and forming the primary root of a seedling.

hypothalamus (hi po-thal' am us) (Gr. *hypo*, under; L. *thalamos*, chamber) the part of the forebrain containing centers of the autonomic nervous system.

hypothesis (hi -poth' e sis) (Gr. *hypo*, under; *tithenai*, to put) a suggested solution of a problem that must be proved or validated by securing additional information and data, through experimentation if possible.

I

identical twins those produced by the division of a single, fertilized egg that results in two separate individuals with identical hereditary traits; same as monozygotic twins; contrast to **fraternal** or nonidentical **twins.**

ileum (il' e um) (L. *ileum*, groin) posterior or lower part of small intestine.

ilium (il' i um) (L. *ilium*, flank) dorsal part of hip or pelvic bone.

imbibition (im bi -bish' un) (L. *imbibere*, to drink in) the process by which solid particles (chiefly colloidal) absorb liquids and swell.

immunity (im -mu' ni ti) (L. *im*, not; *munia*, obligation) ability of an organism to resist disease.

inbreeding mating or crossing closely related types of animals or plants.

incisor (in-size' or) (L. *incisis*, cut) adapted for cutting, as are certain teeth.

incomplete dominance a situation in which neither of two genic factors completely dominates the other.

incus (ing'kus) (L. *incus*, anvil) middle or anvil bone of the ear of certain vertebrates.

independent assortment a mendelian principle that

states that genes representing contrasting pairs of traits are segregated to gametes independently of each other in meiosis.

inductor (in- duk′ tor) (L. *inducere*, to introduce), an embryonic tissue or organ that causes differentiation of another tissue or organ.

indusium (in-du′ si um) (L. *indusium*, covering) membranous cover of a fern sorus.

infection (in-fek′ shun) (L. *in*, in; *facere*, to make) invasion of tissues by pathogenic organisms, or their products, with a resultant pathologic condition.

infundibulum (in fun-dib′ u lum) (L. *infundere*, pour into) funnel-like growth such as that from the ventral part of the diencephalon of the brain; see **pituitary gland.**

ingest (in-jest′) (L. *ingestus*, take in) to take in food.

inheritance transmission of traits from one generation to another.

inhibitor (in-hib′ i tor) (L. *in*, in; *habeo*, to have) something that restrains or checks.

innominate (in-nom′ i nate) (L. *in*, not; *nomen*, name) nameless.

inorganic (in or-gan′ ik) (L. *in*, not; organic) not organic but pertaining to nonliving; a substance that is not a hydrocarbon.

insectivorous (in sek-tiv′ or us) (L. *insectus*, insect; *vorare*, to eat) insect-eating.

insertion (in-sur′shun) (L. *insertus*, join) place of attachment, such as the more movable end of a muscle (contrast to origin of muscle).

instinct (in′ stingt) (L. *instinguere*, to incite) subconscious act due to an arrangement of an inherited pattern of nervous tissue, or protoplasm.

insulin (in′ su lin) (L. *insula*, island) hormone secreted by the isles of Langerhans of the pancreas.

integument (in-teg′ u ment) (L. *integumentum*, covering) covering or investing layer.

intercellular (in ter-sel′ u lar) (L. *inter*, between; *cellula*, diminutive of *cella*, cells) between cells.

intermedin (in ter-me′ din) (L. *inter*, between; *mediatus*, middle) hormone secreted by the midportion of the pituitary gland; adjusts expansion of the pigment cells (melanophores) of the skin of certain vertebrates, e.g., frogs.

internode (in′ ter node) (L. *inter*, between; *nodus*, joint) space between two successive joints or nodes on a plant stem.

intestine (in-tes′tine) (L. *intestinus*, internal) part of the digestive tract beyond the stomach.

intracellular (in tra-sel′ u lar) (L. *intra*, within; *cellula*, cells) within cells; contrast to **intercellular.**

invaginate (in-vaj′ i nate) (L. *in*, in; *vagina*, sheath) to fold in, as in a gastrula stage.

invertebrate (in ver′ te brate) (L. *in*, not; *vertebratus*, vertebra) lower animal without vertebrae or notochord; contrast to **vertebrate.**

iris (L. *iris*, rainbow) colored part of the eye.

irritability (ir i ta -bil′ i ti) (L. *irrito*, to excite) ability to receive and respond to external or internal stimuli.

ischium (is′ ki um) (Gr. *ischion*, hip) posterior and dorsal bone of the pelvic girdle.

isogametes (i so ga-metes′) (Gr. *isos*, equal; *gamos*, gamete) similar gametes (sex cells).

isogamy (i -sog′ a mi) (Gr. *isos*, equal; *gamos*, marriage) union of two similar gametes; contrast to **heterogamy.**

isotonic solution (i so-ton′ ik) (Gr. *isos*, equal; *tonos*, tension) one with osmotic pressure equal to that of protoplasm.

J

jejunum (je-joo′num) (L. *jajunus*, empty) middle or second part of the small intestine, between the duodenum and the ileum.

jugular (jug′ u lar) (L. *jugulum*, collar bone) pertaining to the neck, such as the jugular vein in the neck.

K

kappa particle a "killer" particle in a paramecium.

karyogamy (kary-og′ a mi) (Gr. *karyon*, nucleus or nut; *gamos*, union) fusion of two nuclei.

keratin (ker′ a tin) (Gr. *keras*, horny) insoluble substance, similar to chitin, forming the basis of horns, hoofs, and so forth.

Krebs cycle tricarboxylic acid cycle; one of the most basic processes in respiration, in which starch (or glycogen), by a series of complex processes, is converted successively to PGAL, PGA, pyruvic acid, and other substances, and eventually to CO_2.

L

labial (la′ bi al) (L. *labium*, lip) pertaining to the lip.

lactase (lak′ tase) (L. *lac*, milk; *ase*, enzyme) enzyme that changes lactose (milk sugar) into dextrose and galactose.

lacteals (lak′te als) (L. *lacteus*, milky) lymphatic vessels of small intestine that convey the milky chyle from the intestine through the mesenteric glands to the thoracic duct.

lamella (la-mel′ a) (L. *lamella*, small plate) structure made of small plates, as lamella of bone; or layers of a cell wall.

larva (lar′va) (L. *larva*, mask) active, immature stage of development; contrast with **pupa.**

larynx (lar′inks) (Gr. *larynx*, larynx) enlarged anterior end of trachea (windpipe) that contains vocal folds; present in all vertebrates except birds.

legume (leg′-yum) (L. *legumen* from *legere*, to gather) family of plants in which the seed vessel is two-valved and contains a linear arrangement of seeds, such as beans, peas, alfalfa, etc.

lethal factor (le′ thal) (L. *lethum*, death) genetic factor that brings premature death to an organism.

leucocyte (lu′ ko sit) (Gr. *leukos*, white; *kytos*, cell) colorless blood corpuscle.

leucoplast (lu′ ko plast) (Gr. *leukos*, white; *plastos*, formed) colorless plastid that forms starch in many types of plant cells.

leukemia (lu-ke′ mi a) (Gr. *leukos*, white; *aima*, blood) a cancerous condition of blood characterized by overproduction of leucocytes.

levator (le-va′tor) (L. *lavare*, to rise) a muscle that elevates a structure.

lichen (li′ ken) (Gr. *leichen*, "liverwort") plant composed of a chlorophyll-bearing alga and a fungus living together symbiotically.

life cycle various stages from development to maturity.

ligament (lig′ a ment) (L. *ligare*, to bind) band of connective tissue that binds one bone to another; or supports an organ.

lignin (lig′nin) (L. *lignum*, wood) chemical substance that is related to cellulose and constitutes the essential part of woody tissue (xylem).

linkage inheritance of traits in groups because their genes are near each other (linked) on the same chromosome.

lipase (li′pase) (Gr. *lipos*, fat) a fat-splitting enzyme that converts fat into fatty acids and glycerin.

lipid (li′pid) (Gr. *lipos*, fat) fats and similar fatlike chemical compounds, that are insoluble in water but soluble in certain organic solvents.

luciferase (lu-sif′ er ase) (L. *lux*, light; *ferre*, to bring) an enzyme involved in light production by its action on luciferin in organisms.

lumbar (lum′ bar) (L. *lumbus*, loin) pertaining to the loins that are posterior to the ribs.

lumen (lu′men) (L. *lumen*, cavity) space within an organ or tube.

lymph (limf) (L. *lympha*, liquid) the blood plasma and white blood corpuscles that have passed from the circulatory vessels and that surround tissues and cells.

lymphocyte (lim′ fo site) a type of white blood corpuscle with a rounded or kidney-shaped nucleus.

M

macrogamete (mak ro ga-met′) (Gr. *makros*, large; *gamos*, gamete) large female gamete (sex cell) produced by an organism that exhibits heterogamy.

macronucleus (mak ro -nu′ kle us) (Gr. *makros*, large; L. *nucleus*, nucleus) larger nucleus of certain protozoans, as distinguished from the smaller micronucleus.

madreporite (mad′ re po rit) (L. *mater*, mother; Gr. *poros*, pore; *ite*, nature of) porous plate leading to the water-vascular system of certain echinoderms, such as starfish.

malaria (ma-la′ ri a) (L. *mal*, bad; *aria*, air) fever produced by a protozoan (class Sporozoa); it was formerly thought to be due to "bad" air.

malpighian corpuscle (mal-pig′ i an) (after *Malpighi* of Pisa, Italy) a body in the vertebrate kidney; see **Bowman's capsule.**

maltose (mol′toz) malt sugar; a 12-carbon sugar formed by the union of two glucose units.

mammal (mam′al) (L. *mamma*, breast) vertebrate that has milk-giving breasts.

mandible (man′ di bl) (L. *mandibula*, to chew) a chewing jaw.

mantle (man′tl) (L. *mantellum*, cloak) sheetlike tissue in clams, oysters, and snails that secretes shell.

marsupial (mar-sup′ i al) (L. *marsupium*, pouch) mammal that carries young in an abdominal pouch, such as opossum or kangaroo.

matrix (ma′triks) (L. *mater*, mother) noncellular material in which cells are embedded, such as cartilage or bone.

maturation (mat u -ra′ shun) (L. *maturus*, mature) maturing of structures, such as sperms or eggs.

maxilla (maks -il′a) (L. *maxilla*, jaw) a jaw, especially the upper one in higher animals.

medulla (me-dul′ la) (L. *medulla*, inner, marrow) inner portion of an organ such as the medulla of a kidney; **medulla oblongata** is the posterior part of the brain (hindbrain).

medullary ray pith ray that separates the vascular bundles in certain higher plants.

medullary sheath the covering of a medullated nerve fiber.

medusa (me-du' sa) (Gr. *medousa,* one who rules) free-swimming stage of certain coelenterates, e.g., jellyfish.

megagametophyte (meg a game' to fit) (Gr. *megas,* large; *gamos,* gamete; *phyton,* plant) female gametophyte that results from the development of a megaspore and produces female gametes (eggs).

megasporangium (meg a spo-ran' ji um) (Gr. *megas,* large; *spora,* spore; *angios,* case) a spore-bearing structure that bears megaspores that develop into megagametophytes.

megaspore (meg' a spor) (Gr. *mega,* large; *spora,* spore) a large spore produced in a megasporangium.

meiosis (mi-o' sis) (Gr. *meioun,* to make less) the preparation and maturation of a cell, whereby the chromosome number is reduced one-half (reduction division).

melanin (mel' a nin) (Gr. *melas,* black) blackish pigment.

melanophore (mel' an o for) (Gr. *melas,* black; *phorein,* to bear) chromatophore that bears a blackish pigment.

mendelism Mendel's laws; the proposition that characteristics are inherited as units independently of each other; genes (factors) separate (segregate) from one another and later recombine in various ways in germ cells; characteristics are in pairs, one of which is dominant over the other, or recessive, one. These laws were formulated by Gregor Mendel.

meninges (me-nin' jez) (Gr. *meninx,* membrane) three membranous coverings of the brain and spinal cord: outer dura mater, arachnoid, and inner pia mater.

menopause (men' o poz) (Gr. *menos,* month; *pauein,* to cease) cessation of the female menstruation cycle and of reproduction.

menstruation (men stroo-a' shun) (L. *mensis,* month) the discharge of blood and uterine tissue from the vagina at the end of a menstrual cycle.

meristem (mer' i stem) (Gr. *merizen,* to divide) undifferentiated tissue of plants composed of cells actively dividing.

mesencephalon (mes en-sef' a lon) (Gr. *mesos,* middle; *kephals,* head) midbrain; third region of vertebrate brain.

mesenchyme (mes' eng kime) (Gr. *mesos,* middle; *en,* in; *chein,* to pour) middle layer of embryos that forms connective tissues and so forth.

mesentery (mes' en ter i) (Gr. *mesos,* middle; *enteron,* intestine) membrane that invests and suspends internal organs such as the intestine.

mesoderm (mes' o derm) (Gr. *mesos,* middle; *derma,* covering) middle germ layer of cells that gives rise to certain tissues and organs.

mesoglea (mes o -gle' a) (Gr. *mesos,* middle; *gloios,* glue) noncellular gelatinous layer between the ectoderm and the entoderm of animals such as sponges and coelenterates.

mesophil (mes' o fil) (Gr. *mesos,* middle; *philain,* to love) an organism that grows at temperatures between 20° C. and 40° C.; contrast to **psychrophil** and **thermophil.**

mesophyll (mes' o fil) (Gr. *mesos,* middle; *phyllon,* leaf) plant leaf tissue located between the upper and the lower epidermis.

mesophyte (mes' o fite) (Gr. *mesos,* middle; *phyton,* plant) plant that requires only medium moisture.

mesothelium (mes o -the' li um) (Gr. *mesos,* middle; *thelium,* lining) lining of the peritoneal cavity.

mesothorax (mes o -thor' aks) (Gr. *mesos,* middle; *thorax,* chest) middle of the three thoracic segments of certain arthropods.

metabolism (me-tab' o lizm) (gr. *metabole,* to change) sum of constructive (anabolic) and destructive (catabolic) phases of metabolism in protoplasm.

metamorphosis (met a -mor' fo sis) (Gr. *meta,* after; *morphe,* form) abrupt change from one stage of embryonic development to another, such as from larval stage to pupal stage in insects.

metaphase (met' a -faz) (Gr. *meta,* after; *phasis,* appearance) a period in cell division that occurs between the prophase and the anaphase stages.

metathorax (met a-thor' aks) (Gr. *meta,* after; *thorax,* chest) posterior part of the thorax of certain arthropods, e.g., insects.

metazoa (met a-zo' a) (Gr. *meta,* after; *zoa,* animals) higher, multicellular animals; contrast to **protozoa.**

microgamete (mi kro-gam' et) (Gr. *mikros,* small; *gamos,* gamete) the smaller of two gametes formed by a heterogamous organism; male gametes.

microgametophyte (mik ro ga -me' to fite) (Gr. *mikros,* small; *gamos,* gamete; *phyton,* plant) male gametophyte resulting from the development of a microspore.

micrometer (mi′ kro met ur) (Gr. *mikros*, small; *metron*, measure) one-thousandth part of a millimeter, or one twenty-five thousandth of an inch; also called micron.

micronucleus (mi kro -nu′ cle us) (Gr. *mikros*, small; nucleus) the smaller, reproductive nucleus of certain protozoans, in contrast to the larger nutritive macronucleus.

microorganism (mi kro -or′ gan izm) (Gr. *mikros*, small; organism) a microscopic organism such as a bacterium or a protozoan.

micropyle (mi′ kro pile) (Gr. *mikros*, small; *pyle*, gate) small opening.

microsporangium (mi kro spor- an′ ji um) (Gr. *mikros*, small; *spora*, spore; *angios*, case) a spore case containing microspores.

microspore (mi′ kro spor) (Gr. *mikros*, small; *spora*, spore) minute spore that grows into a male gametophyte; in seed plants the microspore is the young pollen grain.

microsporophyll (mi kro -spor′ o fil) (Gr. *mikros*, small; *spora*, spore; *phyllon*, leaf) a leaf that bears a microsporangium in order to produce microspores.

middle lamella a thin layer of pectic substance that joins adjacent cells.

miracidium (mi ra-sid′ i um) (Gr. *meirakidion*, young) ciliated larval stage of flukes.

mitochondria (mi to-kon′ dri a) (Gr. *mitos*, thread; *chondros*, grain or grit) somewhat regularly shaped bodies in cytoplasm for producing cellular enzymes; singular, mitochondrion.

mitosis (mi-to′ sis) (Gr. *mitos*, thread) nuclear division characterized by the splitting of chromosomes, by spindle formation and so forth.

mitral (mi′ tral) (L. *mitra*, shaped like a miter) a membranous valve between the left auricle and left ventricle of the heart; it is also known as the bicuspid valve.

mold (A.S. *molde*, earthy) saprophytic fungus; a type of fossil.

molecule (mol′ e kul) (L. *molecula*, little) two or more atoms combined chemically.

monecious (mo -ne′ shus) (Gr. *mono*, one; *oikos*, household) both male and female reproductive organs in the same individual; same as **hermaphroditic;** contrast to **diecious.**

monocotyledon (mon o kot i -le′ don) (Gr. *mono*, one; *kotyle*, cup-shaped embryonic seed leaf) a plant that has one embryonic seed leaf or cotyledon; contrast to **dicotyledon.**

monohybrid (mon o -hi′ brid) (Gr. *mono*, one; L. *hybrida*, mongrel) offspring from parents who differ in one trait.

monosaccharide (mon o -sak′ a rid) (Gr. *mono*, one; *sakcharon*, sugar) a simple sugar composed of five or six carbons.

morphology (mor-fol′ o ji) (Gr. *morphe*, form; *logos*, study) deals with the form and the structure of an organism.

morula (mor′ u la) (L. *morum*, berry) mass of cells, called blastomeres, formed by cleavage of the egg in the early development of many animals.

mother cell one that gives rise to new (daughter) cells.

motor fiber (L. *moveo*, move) nerve fibers whose impulses cause movment in muscles.

mucous membrane (mu′ kus) (L. *mucus*, slime) lining of certain internal cavities, such as the digestive tract and the respiratory system.

multiple factors two or more pairs of genes that have a similar or cumulative effect on a trait.

mutation (mu-ta′ shun) (L. *mutare*, to change) an abrupt heritable variation, due to an alteration of a gene or a chromosome.

mycelium (mi-se′ li um) (Gr. *mykes*, fungus) mass of filamentous hyphae of true (higher) fungi.

mycorrhiza (mi ko-ri′ za) (Gr. *mykes*, fungus; *rhiza*, root) a symbiotic association of a fungus with the roots of higher plants.

myelin (mi′ e lin) (Gr. *myceles*, marrow) a fatty substance surrounding the axons of nerve cells in the central nervous system.

myofibril (mi o -fi′ bril) (Gr. *myos*, muscle; fibril) a contractile filament in a muscle or muscle fiber.

myosin (mi′ o sin) (Gr. *myos*, muscle) a protein of muscle.

myxamoeba (miks a -me′ ba) (Gr. *myxos*, slime; *amoeba*, change) swarm cell of slime fungi.

myxedema (mix se-de′ ma) (Gr. *myxos*, slime; *eidema*, a swelling) a disease resulting from thyroid deficiency in adults and characterized by swellings under the skin.

N

nacreous (na′ kre us) (L. *nacre*, mother-of-pearl) pearly.

naiad (ni′ ad) (Gr. *naias*, water nymph) a larval aquatic stage of certain insects.

nares (na′ res) (L. *nare*, nostril) nostrils.

nastic movements (Gr. *nasso*, close up) movements of flat organs (such as leaves and petals) caused by either growth differences or turgor of cells.

nematocyst (nem′ a to sist) (Gr. *nema*, thread; *kystis*, sac) a structure from which a stinging fiber, called a nematocyst fiber, is discharged, such as occurs in *Hydra*.

nematode (nem′ a tod) (Gr. *nematos*, thread) a roundworm.

nephridium (ne-frid′ i um) (Gr. *nephros*, kidney) tubular excretory organ of lower animals such as earthworms.

nephron (nef′ ron) (Gr. *nephros*, kidney) functional unit of a vertebrate's kidney, consisting of the glomerulus, capsule, convoluted tubules, Henle's loop, and the collecting tubule.

nephrostome (nef′ro stom) (Gr. *nephros*, kidney; *stoma*, opening), ciliated opening of the inner end of a nephridium.

nerve (L. *nervus*, sinew) group of nerve fibers, end to end and side by side, held together by special connective tissue called neuroglia.

neurolemma (nu ro-lem′ a) (Gr. *neuron*, nerve; *lemma*, covering) membranous covering of a nerve.

neuron (nu′ ron) (Gr. *neuron*, nerve) unit of the nervous system composed of dendrite, cyton, and axon.

nitrification (ni tri fi-ka′ shun) (L. *nitrum*, niter) preparation of nitrogenous materials for use by organisms.

nitrifying bacteria those capable of changing ammonia into nitrites, or nitrites into usable nitrates.

nitrogen-fixing bacteria those capable of combining free nitrogen of the air with oxygen, either in the nodules of the roots of leguminous plants (such as clover, peas, alfalfa, etc.), or by other species of bacteria that live freely in the soil (nonsymbiotic nitrogen-fixation).

node (L. *nodus*, knot or joint) the joint of a plant stem where branches and leaves join the stem.

node of Ranvier (ran-vya′) places on a nerve fiber where the membranous covering (medullary sheath) is interrupted.

notochord (no′ to kord) (Gr. *noto*, back; *chord*, rod) rodlike structure in the dorsal (back) side that is the forerunner of the backbone.

nucellus (nu-sel′us) (L. *nux*, nut) the megasporangium of an ovule, located inside the integument and enclosing the megagametophyte.

nucleic acid (nu-kle′ik) (L. *nucleus*, nut) a class of molecules composed of joined nucleotides; chief types are desoxyribonucleic acid (DNA) found in cell nuclei (chromosomes), and ribonucleic acid (RNA) found in cytoplasm (ribosomes), nucleoli, etc.

nucleolus (ne-kle′ o lus) (L. diminutive of *nucleus*) the somewhat spherical body within a nucleus; site of RNA synthesis.

nucleoplasm (nu′ kle o plazm) (L. *nux*, nut; *plasma*, liquid) liquid part of the nucleus.

nucleoprotein (nu′ kle o-pro′ te in) a molecule composed of nucleic acid and protein; there are two types, dependent on whether the nucleic acid portion is DNA or RNA.

nucleus (nu′ kle us) (L. *nux*, nut) specialized, central, and organized structure found in most cells; it contains the chromosomes.

nymph (nimf) (Gr. *nymphe*, young), a stage in the gradual metamorphosis of insects such as grasshoppers.

O

obligate (ob′ li gate) (L. *ob*, about; *ligo*, bind) unable to change life habits to suit varying conditions.

occipital (ok-sip′ i tal) (L. *occiput*, back of head) base of the skull.

ocellus (o -sel′ us) (L. *ocellus*, little eye) simple eye; contrast to **compound eye** (plural, ocelli).

oculomotor (ok u lo-mo′ tor) (L. *oculus*, eye; *movere*, to move) movement of the eye.

olfactory (ol-fak′ to ri) (L. *olfacere*, to smell) pertaining to odors or to sense of smell.

ommatidium (om a -tid′ i um) (Gr. *ommation*, little eye; *idion*, diminutive) unit of which compound eyes are made, such as in crayfish, and in certain insects.

omnivorous (om-niv′ or us) (L. *omnis*, all; *vovare*, to eat) eating both plant and animal tissue.

oocyte (o′ o site) (Gr. *oon*, egg; *kytos*, cell) female egg before fertilization.

oogamy (o -og′ a mi) (Gr. *oon*, egg; *gamos*, marriage, union) union of nonmotile egg and male gamete.

oogenesis (o o -jen′ e sis) (Gr. *oon*, egg; *genesis*, origin) formation of an egg and its preparation for fertilization and development.

oogonium (o o -go′ ni um) (Gr. *oon*, egg; *gonos*, offspring) egg cell before maturation; the one-celled female structure in certain thallophytes that produces one or more eggs.

oospore (o o-spor′) (Gr. *oon*, egg; *spora*, spore) a

thick-walled resting cell formed by the union of an egg and a sperm, as found in certain thallophytes.

optic (op'tik) (Gr. *optikos*, sight) pertaining to sight or to the eye.

organ (Gr. *organon*, an implement) a group of different tissues all performing a common function.

organelle (or gan-el') noncellular structure in a cell that serves a specific function.

organic (or-gan' ik) a term applied to molecules containing carbon, except those that are derivatives of carbon dioxide; practically all organic molecules contain carbon atoms linked together.

organism (or' gan izm) (Gr. *organon*, implement) an independent living being.

osmosis (os-mo' sis) (Gr. *osmos*, push) diffusion of substances through a semipermeable membrane.

osmotic pressure pressure exerted by substances in a solution, caused by molecular activity.

osseus (os' e us) (L. *os*, bone) pertaining to bone.

osteology (os te -ol' o ji) study of bones.

ostium (os' ti um) (L. *ostium*, pore or opening) mouth-like opening.

ovary (o' va ri) (L. *ovarium*, ovary) female reproductive organ in which the egg develops; the enlarged, basal part of a pistil (female) within which seeds develop.

oviduct (o' vi duct) (L. *ovum*, egg; *ducere*, to lead) tube to carry eggs from the ovary.

oviparous (o-vip' a rus) (L. *ovum*, egg; *parere*, to bear) producing eggs that hatch after they have been excluded from the body.

ovoviviparous (o vo vi -vip' a rus) (Gr. *ovum*, egg; F. *vivipare*, produce) forming eggs that develop within the body of the parent, but without nutritive or other metabolic aid by the parent.

ovulation (o vu- la' shun) discharging mature eggs from the ovary.

ovule (ov' ule) (L. *ovum*, egg) structure consisting of a female gametophyte, nucellus, and integuments, that, after fertilization, develop into a seed.

P

Paleozoic (pa le o-so' ik) (Gr. *palaios*, old: *zoe*, life) the geologic era between the pre-Cambrian and Mesozoic, approximately 200 to 550 millions years ago.

palisade layer (pal' i sade) (L. *palus*, column) columnar cells with chloroplasts in the mesophyll tissues of leaves, just below the upper epidermis.

pancreas (pan' kre as) (Gr. *pan*, all; *kreas*, flesh) an accessory digestive gland.

paramecin a "killer" particle in a paramecium.

paraphyses (pa-raf' i sez) (Gr. *para*, beside; *physis*, growth or nature) sterile, hair-like structures associated with sex structures in certain algae, fungi, mosses, and so forth.

parasite (par' i sit) (Gr. *para*, beside; *sitos*, food) a plant or animal living in or on another living organism at the organism's expense.

parasympathetic (par a sim pa-thet' ik) (Gr. *para*, beside; *sympathos*, sympathetic) a subdivision of the autonomic nervous system; centers are located in the brain and spinal cord.

parathyroid (par a -thi' roid) (Gr. *para*, beside; thyroid) four small endocrine glands adjacent to the thyroid.

parenchyma (par- eng' ki ma) (Gr. *para*, beside; *en*, in; *chein*, to pour) spongy, mesodermal tissues of lower animals; they are fundamental plant tissues as opposed to more highly differentiated plant tissues.

parotid (pa-rot' id) (Gr. *para*, beside; *otos*, ear) salivary gland below the ear.

parthenogenesis (par the no -jen' e sis) (Gr. *parthenos*, virgin; *genesis*, origin) development of an egg without fertilization by a male sperm.

pasteurization (pas tyur i -za' shun) (after *Pasteur*) killing of certain organisms by heating a liquid to between 142° F and 145/ F for 30 minutes (212° F is boiling).

patella (pa-tel' a) (L. *patena*, pan) kneecap.

pathogenic (path o -jen' ik) (Gr. *pathos*, suffering; *genesis*, origin) disease-producing.

pathology (pa -thol' o ji) (Gr. *pathos*, suffering or disease; *logos*, study) study of diseased or abnormal conditions.

pectoral (pek' to ral) (L. *pectus*, breast) pertaining to the chest or breast.

pedal (ped' al) (L. *pes*, foot) pertaining to the foot.

pellicle (pel' i kl) (L. *pellicula*, small skin) thin outer layer on certain cells.

pelvis (pel' vis) (L. *pelvis*, basin) group of bones to support abdominal organs and for attachment of lower (hind) limbs.

penis (pe' nis) (L. *penis*, penis) male organ of copulation.

pepsin (pep'sin) (Gr. *pepsis*, digest) protein-digesting enzyme of the stomach.

peptidase (pep' ti das) (Gr. *peptein*, to digest) enzyme that breaks down a peptide into amino acids.

perennial (per -en' i al) (L. *per*, through; *annus*, year) living more than two years (contrast to annual and **biennial**).

perianth (per' i anth) (Gr. *peri*, around; *anthos*, flower) all the petals and sepals of a flower taken collectively.

pericardium (per i -kar' di um) (Gr. *peri*, around; *kardia*, heart) serous membrane that encloses the heart.

pericycle (per' i si kl) (Gr. *peri*, around; *kyklos*, circle) circle of plant tissue of stems and roots between the cortex and stele.

periosteum (per i -os' te um) (Gr. *peri*, around; *os*, bone) membranous connective tissue covering a bone.

peripheral nervous system that part of the nervous system composed of cranial and spinal nerves; contrast to central nervous system.

peristalsis (per i -stal' sis) (Gr. *peri*, around; *stallein*, to arrange) wave-like constriction passing along a tube, caused by muscular contraction, as found in the esophagus, intestine and so forth.

peritoneum (per i to -ne' um) (Gr. *peri*, around; *teinein*, to stretch) membrane that lines the coelom of vertebrates and covers the viscera of the coelom.

permeable membrane one that permits substances to pass.

petal (pet' al) (Gr. *petalon*, leaf) one of the inner whorls of a flower, usually colored; all petals taken collectively form a corolla.

petiole (pe' ti ol) (L. *petiolus*, little stalk) slender support for the blade of a leaf.

petrifaction (pet ri-fak' shun) (L. *petra*, rock) method of fossil formation in which mineral matter takes the place of the original organic or living material during the disintegration of the organism.

pH a symbol denoting the relative concentration of hydrogen ions in a solution; pH values range from 0 to 14; the lower the value the more acid, or hydrogen ions, in the solution, with pH 7 being neutral.

phage (fage) (Gr. *phagein*, to destroy) a type of virus.

phagocyte (fag' o sit) (Gr. *phagein*, to destroy; *kytos*, cell) type of leukocyte that engulfs foreign materials.

phagocytosis (fag o si-to' sis) destruction of foreign materials by the action of phagocytes (white blood corpuscles).

pharynx (far' inks) (Gr. *pharynx*, pharynx) tube connecting mouth to the esophagus and also to the larynx.

phenotype (fe' no tip) (Gr. *phaino*, show; *typos*, impression) a type or kind that is determined on the basis of visible traits, as distinguished from genotype, which is based on gene content.

phloem (flo' em) (Gr. *phloos*, bark) food-conducting tissue of plants; phloem and xylem together form a vascular bundle.

phosphate (fos' fate) a salt of phosphoric acid.

phosphoglyceraldehyde (PGAL) a compound produced during photosynthesis (CO_2 fixation), in which there is a loss of an atom of oxygen from a molecule of phosphoglyceric acid (PGA). PGAL has a formula $C_3H_5O_3$ to which a phosphate group (P) is attached as PO_3H_2.

phosphoglyceric acid (PGS) a product formed during CO_2 fixation (photosynthesis [hydrogen joined to CO_2]); each molecule has a formula $C_3H_5O_4$ attached to a phosphate group, such as PO_3H_2, (same as glycerophosphoric acid).

phosphorylation (fos for i -la' shun) the addition of a phosphate group, such as PO_3H_2, to a compound.

photolysis (fo-tol' i sis) (Gr. *phos*, light; *lysis*, loosening) a process whereby molecules of water may have their hydrogen split from the oxygen, because of light energy present in chlorophyll.

photoperiodism (fo to-peer' i o dizm) (Gr. *phos*, light; *peri*, around; *odos*, way) growth and developmental responses of plants to relative lengths of day and night, or to differing lengths of exposure to light.

photosynthesis (fo to-sin' the sis) (Gr. *phos*, light; *syntithenai*, to build) production of carbohydrates from water and carbondioxide by means of chlorophyll in presence of energy-supplying light.

phototaxis (fo to-tax' sis) (Gr. *phos*, light; *taxis*, arrangement or locomotion) locomotor response to light.

phototropism (fo -tot' ro pizm) (Gr. *phos*, light; *trope*, movement) growth movement toward light in an organism.

phycobilin (fi -co' bil in) (Gr. *phykos*, seaweed; L. *bilis*, bile) protein-like, water-soluble, accessory photosynthetic pigments (red or blue).

phycocyanin (fi co -si' a nin) (Gr. *phykos*, seaweed or alga; *kyanos*, blue), blue-green pigment of certain algae.

phycoerythrin (fi ko e -ryth' rin) (Gr. *phykos*, alga; *erythros*, red) red pigment of red algae.

phylum (fi' lum) (Gr. *phylon*, tribe), one of the main groups into which the animal and plant kingdoms are divided; plural, phyla.

physiology (fiz i -ol′ o ji) (Gr. *physis*, nature or function; *logos*, study) study of function.

pia mater (pi′ ah; ma′ ter) (L. *pia*, tender; *mater*, mother) innermost of the three membranes covering the brain and spinal cord.

pineal (pin′ e al) (L. *pineus*, pine cone) small gland between the two cerebral hemispheres.

pistil (pis′til) (L. *pistillum*, a pestle) ovule-producing part of a flower, composed of one or more carpels.

pith (A.S. *pitha*, pith) soft, spongy tissue in the center of stems of certain plants.

pituitary (pi -tu′ i ta ri) (L. *pituita*, phlegm) small, oval endocrine gland attached to the infundibulum of the brain, and whose two lobes have entirely different hormones; same as hypophysis.

placenta (pla-sen′ ta) (L. *placenta*, flat cake) flat vascular organ that aids in nourishing the fetus in the uterus; or attachment for plant seeds.

plasma (plaz′ ma) (Gr. *plasma*, liquid) liquid part of the blood, lymph, or milk.

plasmagene (plaz′ ma jen) (Gr. *plasma*, something formed; *genos*, descent), a gene within the cytoplasm, in contrast to a nuclear gene.

plasma membrane living, semipermeable membrane covering certain cells; see **cell membrane.**

plasmodesma (plaz mo-dez′ me) (Gr. *plasma*, something formed; *desma*, bond) protoplasmic connection between cells; plural, plasmodesmata.

plasmodium (plaz -mo′ di um) (Gr. *plasma*, liquid; *eidos*, form) naked, protoplasmic mass, such as found in slime fungi.

plasmolysis (plaz -mol′ i sis) (Gr. *plasma*, form; *lysis*, loosening) shrinking of cytoplasm in a living cell caused by loss of water.

plastid (plas′ tid) (Gr. *plassein*, to form) specialized protein body in a cell that is concerned with production of a certain substance.

pleura (ploor′ a) (Gr. *pleura*, rib or side) membranous lining of the thoracic cavity of mammals and covering the lungs in the cavity.

ploidy (ploi′ di) (Gr. *ploos*, fold) the number of chromosome sets per cell, eg, monoploidy (haploidy), diploidy, triploidy, etc.

pollen (pol′ en) (L. *pollen*, fine flour) dust-like grains of material produced by the male anthers of flowers.

pollen tube tube formed by a pollen grain; transports sperms to the eggs in an ovary.

pollination (pol i -na′ shun) application of male pollen to the female stigma, or ovule, of a plant.

polygamy (po-lig′ a mi) (Gr. *poly*, many; *gamos*, marriage) having more than one mate at a time.

polymorphism (pol i mor′ fizm) (Gr. *poly*, many; *morphe*, form) more than two types of castes of individuals in a colony or community that belong to the same species and are derived from the same parents. The various castes of honeybees, ants, termites, and so forth are typical.

polyp (pol′ ip) (L. *polypus*, many footed) sessile phase of the life cycle of certain coelenterate animals.

polypeptide (pol i -pep′ tid) (Gr. *poly*, many; *peptein*, to digest) a molecule consisting of many joined amino acids, but not as complex as a protein.

polyploid (pol′ i -ploid) (Gr. *poly*, many) having three or more sets of chromosomes.

polysaccharide (pol i -sak′ i rid) (Gr. *poly*, many; *saccharon*, sugar) a carbohydrate composed of many monosaccharide units, for example—starch, glycogen, and cellulose.

portal vein (port′ al) (L. *porta*, gate) blood vessel carrying blood to the liver from the spleen, pancreas, digestive tract, and so forth.

postcaval vein (post-ka′ val) (L. *post*, after; *cavus*, cavity) inferior (posterior or ascending) vena cava carrying blood to the heart from posterior parts of body; contrast to **precaval vein.**

posterior (pos-te′ ri or) (L. *posterior*, following) behind or opposite anterior (head).

precaval vein (pre -ka′ val) (L. *prae*, before; *cavus*, cavity) superior (anterior or descending) vena cava carrying blood to the heart from anterior parts of body; contrast to **postcaval vein.**

predaceous (pre-da′ shus) (L. *praeda*, prey) characterized by killing of animals; for example, owls killing (preying on) mice.

prenatal (pre-na′ tal) (L. *prae*, before; *natalis*, birth) before birth.

primate (pri′ mate) (L. *primus*, first) highest animals, such as man, apes, monkeys, etc.

procambium (pro-kam′ bi um) (Gr. *pro*, before; *cambiare*, to exchange) young tissue of a root or shoot that develops into vascular tissue or, more specifically, the future cambium between phloem and xylem of a vascular bundle.

progesterone (pro-jes′ ter on) (L. *pro*, before; *gestare*, to carry) hormone secreted by the corpus luteum and placenta that prepares the uterus for a fertilized egg and maintains the capacity of the uterus to retain the embryo and fetus.

proglottid (pro-glot' id) (Gr. *pro*, before; *glotta*, tongue) one of the individuals of a chain making up a cestode worm, such as tapeworm.

prophase (pro' faz) (Gr. *pro*, before or first; *phasis*, appearance) preparatory stage of cell division preceding metaphase.

prostate (pros' tat) (Gr. *pro*, before; *stare*, stand) an accessory male reproductive gland near the urethra.

protein (pro' te in) (Gr. *protos*, first) compound of carbon, hydrogen, oxygen, and traces of phosphorus or sulfur.

prothallus (prothallium) (pro-thal' us) (Gr. *pro*, before; *thallos*, young part) the reduced prethallus gametophyte of plants such as ferns and their allies.

prothorax (pro-thor' aks) (Gr. *pro*, before; *thorax*, chest), anterior segment of insect thorax that bears the first pair of legs.

prothrombin (pro-throm' bin) (Gr. *pro*, before; *thrombus*, clot) a constituent of blood plasma that is changed to thrombin by thrombokinase in the presence of calcium ions during blood clotting.

protonema (pro to-ne' ma) (Gr. *protos*, first; *nema*, thread) first thread-like growth from a germinating spore, such as found in mosses.

protoplasm (pro' to plazm) (Gr. *protos*, first or essential; *plasma*, liquid) substance of which all living organisms are composed.

proximal (prox' i mal) (L. *proximus*, near) nearest the main axis; opposed to distal.

pseudocoel (su' do seal) (Gr. *pseudo*, false; *koilos*, hollow) false internal cavity not lined by mesoderm but by ectoderm and entoderm, and hence not a true coelom.

pseudopodium (su do -po' di um) (Gr. *pseudo*, false; *pois*, foot) temporary protrusion of protoplasm from a cell, especially in certain protozoans, such as amoebae, that serves various purposes (plural, pseudopodia).

psychrophil (si' kro fil) (Gr. *psychros*, cold; *philein*, to love) growing at temperatures of about 20° C or below.

pubis (pu' bis) (L. *pubes*, adult) anterior part of hip (pelvic) girdle.

pulmonary (pul' mon a ri) (L. *pulmo*, lung) pertaining to the lung.

punnett square a checkerboard-like arrangement for determining the results of a cross in heredity; named after Punnett.

pupa (pu' pa) (L. *pupa*, baby, puppet) the quiet stage in the development of certain insects that occurs between the larval and adult stages; known as a cocoon in moths and as a chrysalis in butterflies.

purine (pu'rin) (L. *purum*, pus; urine) **purine base;** basic compounds related in structure to uric acid; examples include adenine and guanine, which are bases and constituents of nucleic acids.

pylorus (pi -lo' rus) (Gr. *pylorus*, gate keeper) opening between stomach and small intestine.

pyrimidine (pi -rim' id in) important organic compounds (bases), such as cytosine, thymine, and uracil, which are constituents of nucleic acids.

R

radial canal (L. *radius*, ray) canal radiating from a center, such as in starfish.

radicle (rad' i cl) (L. *radix*, root) lower part of a hypocotyl, which grows into a primary root of seedlings.

radula (rad'u la) (L. *radere*, to scrape) scraping organ for mastication in certain mollusks such as snails.

receptacle (re-cep' ta kl) (L. *receptaculum*, reservoid) a structure that contains or bears other parts; in flowers it is the end of the stem to which the other flower parts are attached.

recessive characters (L. *recedere*, to recede) those traits that are not expressed, even though their genes are present together with the gene for the opposite dominant allele.

rectum (rek' tum) (L. *rectus*, straight) terminal part of the intestine.

reduction division the division of chromosomes previous to sexual reproduction in plants and animals in which the normal diploid, somatic number of chromosomes is reduced to the monoploid (haploid), single number; same as **meiosis.**

reflex action (L. *re*, back; *flectere*, to return) automatic, involuntary response of nervous and motor mechanisms to stimuli.

regeneration (re gen er -a' shun) (L. *re*, again; *generare*, to beget) ability to replace a lost part or develop a new individual from a lost part.

renal (re' nal) (L. *renes*, kidneys) pertaining to kidneys.

renal portal system (L. *renes*, kidneys; *porta*, gate) blood vessels (veins) carrying blood from posterior part of the body to kidneys. Oxygenated blood is carried to the kidneys by renal arteries.

renin (re' nin) (L. *renes*, kidneys) a hormone produced by the kidneys that probably influences blood-pressure control.

reproduction (re pro-duk′ shun) (L. *re*, again; *pro*, forth; *duco*, to lead) production of offspring.

response reaction to a stimulus, external or internal.

reticulum (re-tik′ u lum) (L. *reticulum*, little net) network of fibers.

retina (ret′ ina) (L. *rete*, net) light-sensitive membrane of the eye that receives images.

rhizoid (ri′ zoid) (Gr. *rhiza*, root; *eidos*, like) root-like filaments in certain lower plants that function somewhat like roots.

rhizome (ri′ zom) (Gr. *rhizoma*, root) horizontal underground stem that may have the general appearance of a root.

ribonucleic acid (RNA) found chiefly in cytoplasm and nucleoli; may be involved in transmitting hereditary information from the nucleus to cytoplasm.

ribulose diphosphate (RDP) a product produced during photosynthesis when not all the PGAL molecules go into the formation of glucose, but some of the former molecules, by a complex process, are converted into RDP, which unites with more CO_2 to form PGA.

root cap the extreme, protective tip of a root.

root hair fine hair-like extension of the epidermis of plant roots for absorption.

S

saliva (sa-li′va) (L. *saliva*, spittle) secretion of salivary glands.

saprophyte (sap′ ro fit) (Gr. *sapros*, rotten; *phyton*, plant) an organism living on dead or decaying organic matter, particularly on plants.

scapula (skap′ u la) (L. *scapula*, shoulder blade) shoulder blade or dorsal part of pectoral girdle.

schizocoel (skiz′ o seal) (Gr. *schizein*, to split; *koilos*, hollow) a coelom formed by a splitting of the embryonic mesoderm; examples are annelids, mollusks, and arthropods; an organism having a schizocoel is called a schizocoelomate.

sclerenchyma (skle-reng′ ki ma) (Gr. *skler*, hard; *en*, in; *chein*, to pour) plant tissue whose cell walls are thickened for support and protection.

sclerotic (skle-rot′ik) (Gr. *skleros*, hard) tough, outer coat of the eyeball.

scolex (sko′leks) (Gr. *skolex*, worm) enlarged anterior end of a tapeworm.

scrotum (skro′tum) (L. *scrotum*, bag) sac containing testes and accessory organs in most mammals.

secondary sexual characters structural, functional, or behavioral differences between two sexes other than those pertaining to the different sex organs themselves.

secondary tissue derived from the cambium or other lateral meristem.

secretion (se-kre′ shun) (L. *secretus*, to separate) producing a substance by the action of a gland or cell.

sedentary (sed′ en ta ri) (L. *sedere*, to sit) temporarily attached and not entirely free moving.

seed a characteristic reproductive structure of seed plants, consists of an embryo, enclosed by a seed coat, and a food-storage tissue; see **endosperm.**

segregation law passage of one member of each pair of allelic genes to different germ cells during reduction division.

self-fertilization fertilization (fusion) of an egg by a sperm from the same individual.

semicircular canals (L. *semi*, half; *circulus*, circle) ear canals of vertebrates that maintain the sense of equilibrium.

seminal receptacle (sem′ i nal) (L. *semen*, seed fluid) organ for storing sperms from another animal until needed for fertilization.

seminal vesicle (sem′ i nal) (L. *semen*, seed fluid; *vesica*, bladder) sac-like organ for storing sperms during spermatogenesis, as in earthworms.

seminiferous tubules (sem i -nif′ er us) (L. *semen*, seed fluid; *ferre*, to bear; *tubule*, small tube) tube to conduct seminal fluid of male.

semipermeable permitting passage of certain molecules, but not others.

sepal (se′pal) (Gr. *skepe*, covering) one of the outer whorls of floral leaves that, taken as a group, is called the calyx.

septum (sep′tum) (L. *septum*, partition) partition separating two cavities (plural, septa).

serum (se′rum) (L. *serum*, whey) liquid (plasma) of blood that separates on clotting.

sessile (ses′il) (L. *sedere*, to sit) permanently attached and not free moving.

seta (se′ta) (L. *seta*, bristle) bristle-like structure (plural, setai).

sex-influenced traits those traits influenced or modified by the presence of a particular sex organ (and its secretions), such as the beard and voice of the male caused by hormones from the male testes.

sex-linked traits those traits whose genes are located on the sex chromosome.

sexual dimorphism (di-mor′ fizm) (Gr. *di*, two; *morphe*, for) two forms or types of a plant or animal due to its sex.

shoot the stem and leaves of a plant.

sieve element a fundamental kind of cell in mature phloem tissue, that is long, slender, thin-walled and has no nucleus in the cytoplasm.

sieve plate a perforated wall connecting two sieve elements.

sieve tube elongated, fused, conducting cells of plant phloem that have perforated sieve plates at their ends, and are composed of several sieve elements end to end.

sinus (si'nus) (L. *sinus*, cavity) a cavity, or depression, especially in bone.

siphon (si'fon) (Gr. *siphon*, a pipe) tubular structure for drawing in or forcing out.

smooth muscle one whose cells are not striated.

solute a component of a liquid solution whose particles are dispersed separately from each other; see **solvent.**

solution a liquid or gaseous mixture in which the dispersed particles are of ordinary molecular or ionic size.

solvent the continuous component of a liquid solution; see **solute.**

somatic (so-mat'ik) (Gr. *soma*, body) pertaining to the body, e.g., somatic mutation in which a stable gene change occurs in a body cell rather than in a germ (reproductive) cell.

sorus (sor'us) (Gr. *soros*, heap) group of sporangia as found on fern leaves.

specificity (spes i-fis' i ty) uniqueness, especially of enzymes in given reactions and of proteins in a given organism.

sperm (Gr. *sperma*, seed) male gamete (sex cell).

spermatid (sper ma tid') (Gr. *sperma*, seed) male cell that arises by division from a secondary spermatocyte and that later gives rise to a sperm.

spermatocyte (sper -mat' o site) (Gr. *sperma*, sperm, or seed; *kytos*, cell) male germ cell (arising from the spermatogonium) before it is mature.

spermatogenesis (sper ma to -jen' e sis) (Gr. *sperma*, sperm; *genesis*, origin) formation of mature sperms.

sphincter (sfingk' ter) (Gr. *sphinggein*, to bind tightly) circular muscle to close an opening, such as stomach, bladder, or anus.

spicule (spik' al) (L. *spiculum*, little dart) a slender, pointed structure, ie, the calcareous (limy), or siliceous skeletal secretions of certain sponge cells.

spindle (A.S. *spinnan*, to spin) fibrous structure of nucleus associated with chromosomes during cell division.

spirillum (spi-ril' um) (L. *spirilla*, little coil) an organism possessing a wavy, coiled, or spiral body, ie, certain types of bacteria.

splanchnic (splangk' nik) (Gr. *splanchnon*, entrail) pertaining to internal, visceral organs.

spleen (Gr. *splen*, spleen) ductless, vascular organ near the stomach.

spongy tissue plant mesophyll tissue with cells loosely arranged, with many intercellular (air) spaces, and located beneath the lower epidermis.

sporangium (spor-an' jium) (Gr. *spora*, spore; *angios*, case) structure containing spores.

spore (Gr. *spore*, spore) cell for reproductive purposes with resistant covering; one or several may be produced at one time, depending on the species.

spore mother cell a cell, which by cell divisions, produces usually four spores.

sporophyll (spor' o fil) (Gr. *spora*, spore; *phyllon*, leaf) a leaf that bears sporangia.

sporophyte (spor' o fite) (Gr. *spora*, spore; *phyta*, plants) spore-bearing (asexual) generation in plants that exhibit alternation of generations.

squamous (skwa' mus) (L. *squama*, scale) flat, scale-like.

stamen (sta' men) (L. *sta*, stand) pollen-bearing structure of a flower.

statolith (stat' o lith) (Gr. *statos*, stationary or standing; *lithos*, stone) a hard structure in a statocyst that assists in orientation and in sense of equilibrium.

steady state innate ability of protoplasm for continuous and almost instantaneous adjustments to its changing requirements, including its external and internal environmental fluctuations.

stele (ste' l) (Gr. *stele*, post or pillar) central cylinder of united vascular bundles in the roots and stems of dicotyledonous seed plants.

sterigma (ste-rig' ma) (Gr. *sterigma*, support) stalk for bearing basidiospores.

sterile (L. *sterilis*, barren) infertile; free from all types of organisms.

sternum (stur' num) (L. *sternum*, breast bone) breast bone.

steroid (ster' oid) (Gr. *steros*, solid; L. *oleum*, oil) an organic compound containing four fused carbon rings, e.g., cholesterol, sex hormones, vitamin D, and adrenocortical hormone (ACTH).

stigma (stig'ma) (Gr. *stigma*, a mark) upper part of pistil that receives pollen; same as light-sensitive eyespot in certain organisms.

stimulus (stim′ u lus) (L. *stimulare*, to excite) condition or substance that induces a response.

stoma (sto′ ma) (Gr. *stoma*, opening or mouth) small opening such as found in leaves (plural, stomata).

striated (stri′ a ted) (L. *stria*, channel) marked by channels, usually parallel.

style (Gr. *stylos*, pillar) stalk to support the stigma, and through which pollen tubes grow.

subcutaneous (sub-ku-ta′ ne us) (L. *sub*, under; *kutis*, skin) beneath the outer skin.

suberin (su′ ber in) (L. *substratus*, strewn under) a substance that is acted upon by an enzyme.

suture (su′-tur) (L. *sue*, to sew) junction of two bones, usually in an irregular serrated line.

symbiosis (sim bi-o′ sis) (Gr. *symbioun*, to live together) association of two different species of organisms for mutual benefit.

sympathetic nervous system see **autonomic** system.

sympathin (sim′ path in) (Gr. *sympatetikos*, sympathetic) a substance, released in the body by stimulation of the sympathetic nerve chain, that causes sympathetic stimulation of certain systems.

synapse (si naps′) (Gr. *syn*, together; *hapto*, unite) space between axon brush of one nerve cell and the dendrite of the next nerve cell.

synapsis (sin-ap′ sis) (Gr. *synapsis*, union) temporary conjunction of the pairs of homologous chromosomes (from male and female parent) previous to the maturation of germ cells.

syncytium (sin-sish′ i um) (Gr. *syn*, together; *kytos*, cell) undivided mass of protoplasm with several nuclei, as in certain muscles, fungi, and so forth.

synergid (si -nur′ jid) (Gr. *synergos*, working together), small cells near the egg at the micropylar end of the embryo sac in the ovule.

synergistic (sin er-jis′ tik) (Gr. *syn*, together, *ergon*, work) working together; e.g., metabolic agents (such as hormones) working together to reinforce each other's activities.

systole (sis′ tole) (Gr. *systole*, contraction) the phase of contraction of auricles or ventricles during which blood is pumped forward; contrast to **diastole.**

T

tactile (tack′til) (L. *tangers*, touch) concerning stimulation by contact.

taxis (Gr. *taxis*, arrangement or locomotion) response involving movement of an organism as a whole.

taxonomy (taks -on′ o mi) (Gr. *taxis*, arrangement; *nomos*, law) scientific classification of organisms.

telencephalon (tel on-sef′ a lon) (Gr. *telos*, end; *encephalon*, brain) the most anterior vesicle of the brain.

telophase (tel′o faz) (Gr. *telos*, end; *phasis*, appearance) final stage in cell division when daughter cells are formed.

terrestrial (ter-res′ tri al) (L. *terra*, earth) pertaining to land.

testes (tes′ tes) (L. *testis*, male gonad) male reproductive organs that produce sperms; contrast to **ovary;** singular, testis.

thallophyte (thal′ o-fite) (Gr. *thallos*, sheet-like; *phyta*, plants) simple thallus plants without true leaves, stems, or roots; contrast to **embryophyte.**

thermophil (ther′ mo fil) (Gr. *therme*, heat; *philein*, to love) growing at or above a temperature of 45° C.

thermotaxis (ther mo-tax′ sis) (Gr. *therme*, heat; *taxis*, locomotion or arrangement) locomotor response to heat or cold.

thermotropism (ther -mot′ ro pizm) (Gr. *therme*, heat; *trope*, turn) movement in response to heat.

thigmotaxis (thig mo-tax′ sis) (Gr. *thigema*, touch; *taxis*, locomotion or arrangement) locomotor response to contact or touch.

thigmotropism (thig -mot′ ro pizm) (Gr. *thigema*, touch; *trope*, turn) movement in response to contact or touch.

thorax (tho′ raks) (L. *thorax*, chest) part of the body between the head (or neck) and abdomen.

threshold (thresh′ old) (A.S. *therscold*, starting point) minimum amount of a stimulus that obtains a response.

thrombin (thromb′ in) (Gr. *thrombos*, clot) substance to aid in blood clot formation.

thrombokinase (throm bo-ki′ nase) (Gr. *thrombos*, lump; *kinein*, to move) an enzyme that initiates blood clotting, and transforms prothrombin into thrombin in the presence of calcium ions; thromboplastin.

thrombus (throm′ bus) (Gr. *thrombos*, clot) blood clot circulating in the blood.

thymus (thy′ mus) (Gr. *thymos*, thymus) ductless gland in the pharyngeal region of vertebrates.

thyroid (thy′ roid) (Gr. *thyros*, shield; *eides*, resemble) ductless gland in the neck of vertebrates that regulates metabolism, growth, and so forth.

thyroxin (thi-rok′ sin) hormone produced by the thyroid gland.

tibia (tib′ i a) (L. *tibia*, shin), larger inner bone of lower leg of vertebrates; in an insect leg, the part between the femur and tarsus.

tissue (tish′ u) (F. *tissu*, woven) group of similar cells performing a specific function.

trachea (tra′ ke a) (Gr. *tracheia*, tube) tube that carries air.

tracheal tube (tra′ ke al) (Gr. *tracheia*, tube) rather long tube of the plant xylem, composed of several hollow, elongated plant cells fused end to end.

tracheid (tra′ ke id) hollow, single, elongated plant cell with pitted walls in the xylem that conduct materials.

tracheophyta (tra ke -of′ i ta) (Gr. *trachoia*, vesssel or tube; *phyta*, plants) phylum of higher plants possessing vascular (conducting) tissues.

transduction (trans-duk′ shun) (L. *trans*, across; *ducere*, to lead) transfer of genetic materials from one bacterial cell to another by a virus (phage).

translocation (trans lo- ka′ shun) (L. *trans*, beyond; *locus*, place) transfer of soluble materials through the sieve tubes of the phloem of vascular plants; the exchange of parts of chromosomes.

transpiration (trans pi-ra′ shun) (L. *trans*, through; *spirare*, to breathe) loss of water from plants, especially from leaves.

tricarboxylic acid cycle see **Krebs cycle.**

trichocyst (trik′ o sist) (Gr. *thrix*, thread; *kystis*, sac) organelle producing hair-like fibers for offensive and defensive purposes in animals such as paramecium.

trihybrid (tri-hy′- brid) (L. *tres*, three; *hybridos*, mongrel), offspring of parents who differ in regard to three different traits.

triphosphepyridine nucleotide (TEN) a constituent of protoplasm that acts as the final acceptor of hydrogen; see **diphosphopyridine nucleotide (DPN).**

triploblastic (trip lo-blas′ tik) (Cr. *triplax*, triple; *blastos*, bud or layer) three primary germ layers (ectoderm, mesoderm, and entoderm) from which all organisms arise in higher and intermediate animals.

trochanter (tro-kan′ ter) (Gr. *trochanter*, runner) second segment of an insect's leg.

trophic (trof′-ik) (Gr. trophos, feeder) pertaining to nutrition; see **autotrophic, heterotrophic.**

tropism (tro′ pizm) (Gr. *trope*, a turning) response of living organism, or some part of it, to a stimulus, usually through differential growth, etc.

trypsin (trip′sin) (Gr. *truein*, rub down; *pepsis*, digest) protein-splitting enzyme of the pancreas.

tube nucleus one of the nuclei in a pollen tube that affects the growth and behavior of the tube (not involved in the sexual fertilization process).

turgor (tur′ gor) (L. *turgero*, to swell) pressure within a cell because of absorption of water.

tympanum (tim′ pan um) (Gr. *tympanon*, drum) eardrum.

typhlosole (tif′ lo sole) (Gr. *typhlos*, blind; *solen*, channel) dorsal furrow of earthworm intestine that increases absorption.

U

ulna (L. *ulna*, elbow) bone that together with the radius forms the forearm.

umbilical cord (um -bil′ i kl) (L. *umbilicus*, navel) cord composed of blood vessels and connective tissues that connects the fetus with the mother.

umbilicus (um -bil′ i kus) (L. *umbilicus*, navel) scar on abdomen where umbilical cord was attached.

ureter (u-re′ ter) (Gr. *oureter*, ureter) tube carrying urine from kidneys to bladder or to cloaca.

urethra (u-re′ thra) (Gr. *ourethra*, urine tube) tube carrying urine from bladder to outside.

urogenital (u ro-jen′ i tal) (L. *urina*, urine; *genitalis*, reproduction) urinogenital; pertaining to the organs of both the urinary and the reproductive systems collectively.

uterus (u ′ ter us) (L. *uterus*, womb) enlarged part of an oviduct; in female mammals, an organ for containing and nourishing the young before birth.

utricle (u′ tri kl) (L. *utriculus*, little bag) part of the inner ear containing the receptors for body balance. The semicircular canals lead to and from the utricle.

V

vaccine (vak′ seen) (L. *vacca*, cow) a substance composed of dead or weakened bacteria (or other pathogens), or their toxins, that induces immunity when introduced into a body.

vacuole (vak′ u ol) (L. *vacuum*, empty) space for receiving something.

vagina (va-ji′ - na) (L. *vagina*, sheath) in female mammals, a tube leading from the uterus to the exterior; or a place in which reproductive cells are stored.

vagus (va′ - gus) (L. *vagus*, wandering) the tenth cranial nerve, innervating organs in the chest and abdomen.

vas deferens (vaz; def′ er enz) (L. *vas*, vessel; *deferens*, carrying down) tube for carrying sperms to the exterior; plural, vasa deferentia.

vascular (vas′ ku lar) (L. *vasculum*, little vessel) pertaining to vessels, usually for conduction.

vascular bundle structure composed of tissues (xylem and phloem) for conducting liquids in higher plants, and for giving strength.

vasomotor nerves (vas o -mot′ or) (L. *vasa*, vessel; *movere*, to move) nerves controlling the caliber of the arteries by the contraction and expansion of muscles in their walls.

vegetative reproduction asexual reproduction by methods such as grafting, cuttings, and fragmentation.

vein (L. *vena*, vein) vessel carrying blood toward the heart; also a vascular bundle of a leaf.

venous (ve′ nus) (L. *vena*, vein), pertaining to veins.

ventral (ven′ tral) (L. *venter*, belly) lower or belly side.

ventricle (ven′ tri kal) (L. *ventriculus*, little belly or chamber) lower, heavier chamber of a heart from which blood is pumped out.

vertebrate (ver′ te brate) (L. *vertebratus*, backbone) animal having a vertebral column; contrast to **invertebrate.**

vesicle (ves′ i kl) (L. *vesica*, sac) saclike structure.

vessel a xylem tube formed from several vessel segments (modified trachoids with imperfect or no end walls), set end to end.

vestigial (ves-tij′ i al) (L. *vestigium*, trace) rudimentary part of an organism no longer functionally useful, but probably important for functions in organisms of the past.

villus (vil′ us) (L. *villus*, hair) minute projection of small intestine that increases absorptive surface area; plural, villi.

virus (vi′ rus) (L. *virus*, poison) ultramicroscopic, virulent organism composed of a nucleoprotein core and a protein shell, which causes certain plant or animal diseases.

viscera (vis′ er a) (L. *viscera*, internal organs) organs within a body cavity.

vitamin (vi′ ta min) (L. *vita*, life; *amine*, a chemical radicle NH_2) substance that is essential for the proper metabolism and regulation of body processes; vitamins were so named because they were thought originally to contain an amine radicle, which is incorrect.

vitreous (vit′ re us) (L. *vitrum*, glass) glassy; as in vitreous humor, the transparent jellylike material filling the anterior part of the eyeball.

W

warm blooded pertaining to animals whose blood retains a constant temperature regardless of external temperatures, such as in birds and mammals; contrast to **cold blooded.**

white blood corpuscle colorless blood cell, also called a leukocyte.

X

X chromosome a chromosome associated with the sex of many organisms.

xanthophyll (zan′ tho fil) (Gr. *xanthos*, yellow; *phyllon*, leaf) yellow-orange carotenoid pigment of certain higher plants, especially leaves; also present in animals.

xerophyte (ze′ ro fite) (Gr. *zeros*, dry; *phyton*, plant) plant adapted to dry conditions.

xylem (zi′ lem) (Gr. *xylon*, wood) wood, water-conducting portion of a vascular bundle.

Y

Y chromosome a special chromosome associated with the sex of many organisms; in human beings this chromosome is present only in males.

yolk (Gr. *yolke*, yellow) stored food in the egg's cytoplasm.

Z

zoogeography (zo o je-og′ r fi) (Gr. *zoon*, animal; *ge*, earth; *graphein*, to write) study of geographic distribution of animals.

zoology (zo -ol′ o ji) (Gr. *zoon*, animal; *logos*, study) study of animals.

zooplankton (zo o -plangk′ton) (Gr. *zoon*, animal; *planktos*, wandering) collective term for animals present in plankton; contrast to **phytoplankton.**

zoosporangia (zo o spor - an′ ji a) (Gr. *zoon*, animal; *spora*, spore; *angios*, case) a structure in which motile zoospores develop.

zoospore (zo′ o spor) (Gr. *zoon*, animal, *spora*, spore) motile spore.

zygospore (zy′ go spor) (Gr. *zygotos*, united; *spora*, spore) spore formed by the union of two gametes (male and female sex cells).

zygote (zy′ - gote) (Gr. *zygotos*, united) fertilized egg cell after fusion with male gamete.

A · Basic chemistry

Some aspects of biology cannot be distinguished from similar studies in chemistry or physics. It is not the intention here to emphasize these areas. And, in all probability, instructors using this text do not choose to emphasize biochemistry or biophysics. If you happen to be taking a course in chemistry, then ignore this section. Your chemistry course will give a better and more comprehensive treatment of these topics.

For the students whose only exposure to chemistry will be by way of this text, what follows is a brief introduction to some aspects of chemistry. They will help students to understand some of the biochemical aspects of living things.

CHEMICAL MATTER

The basic chemical material in nature is the *element*. There are 92 naturally occuring chemical elements and another dozen or so that have been produced only in laboratories. We recognize these elements by their names such as *oxygen, hydrogen, carbon, chlorine, iron, gold,* and so on. Each of the elements is further designated by a chemical symbol that is usually an abbreviation of its name. Thus, the symbols for the previously mentioned elements are O, H, C, Cl, Fe, and Au, respectively. The symbols of iron and gold are taken from their Latin names *ferrum* and *aurum*. The same is true for others such as sodium with the symbol Na *(natrium)* and potassium, K *(kalium)*.

The smallest amount of an element still identifiable as that element, is called an *atom* (Gr. *atomos,* indivisible). There are as many different atoms as there are elements. When atoms of elements combine, they form *compounds*. These may be simple with two or three atoms such as in salt—NaCl, carbon dioxide—CO_2, or water—H_2O (Fig. A-1). They may also be extremely complex, containing thousands of atoms such as in hemoglobin, starch, or various proteins.

Atoms

This smallest unit of an element is composed of three types of *elementary particles* arranged in a special way. The central atomic nucleus contains *protons,* which have a positive (+) electric charge, and *neutrons,* which are electrically neutral. The number of protons is different for each atom and determines what is called the *atomic number* of the element.

The third type of elementary particles, called *electrons,* have a negative (−) electric charge. They occur in "shells" or "clouds" surrounding the atomic nucleus. There are usually as many electrons (−) as there are protons (+) in the nucleus. The entire atom, therefore, is electrically neutral.

Electrons move around the nucleus with great speed. The greater the amount of energy they possess, the farther they are from the nucleus. The path around the nucleus is called the *orbital* of the electron, and its average path forms an atomic shell. Note that this shell is not a physical structure like an egg shell; rather it represents all of the possible sites that an electron may occupy. In a sense it might be likened to a road on a map showing where a particular car might be if the driver was on a round trip.

It has been shown that the number of electrons in a path is limited and that several orbitals may be formed by the total number of electrons. The number of shells, then, varies with the different number of electrons in different atoms (Fig. A-2). The maximum number of electrons in the nearest or first shell is 2, in the second 8, in the third 18, and then 32, 18, 12, and 2. Regardless of the total number of electrons or shells, the outermost shell never has more than 8 electrons in it (Table A-1).

Hydrogen, the simplest atom, has 1 proton in its nucleus and 1 electron in its nearest (and only) shell (Fig. A-3). As we shall see, it combines easily with other atoms. Helium has 2 protons in the nucleus and 2 electrons in its only shell. Since 2 electrons in the outer shell are the maximum for this level, helium is relatively stable and does not readily react with other elements.

Similarly, oxygen, with 8 protons and 8 electrons in shells of 2 and 6, is an active atom. Neon with 10 protons and shells of 2 and 8 electrons, and therefore a complete outer shell, is relatively stable. Stable atoms such as helium and neon and a few others with complete outer rings (usually 8 electrons) are called the inert gases.

In some cases some atoms of an element may have additional neutrons in the nucleus. These are isotopes (gr. *isos*, equal; *topos*, place) of the element. Examples are hydrogen and its isotopes deuterium (^2H) and tritium (^3H), carbon 12 and its isotope carbon 14, and oxygen 16 and its isotopes ^{11}O and ^{12}O.

Artificially produced radioactive isotopes are extremely valuable in the study of certain biologic problems. The use of such radioactive isotopes as "tracers" is possible, because they emit certain radiations whose presence can be detected by sensitive counters. Hence, the rate of absorption of iodine by the thyroid gland can be determined by the use of radioactive iodine, and this has assisted in the treatment of goiter. Radioactive phosphorus has been traced to the stems and certain parts of the leaves of tomato plants, whereas radioactive zinc concentrates in tomato seeds. The use of radioactive isotopes may be of great value in the study of animal and plant metabolism and in the diagnosis and treatment of certain diseases (Table A-2).

The chemical and physical behaviors of atoms are determined largely by the number and arrangements of the orbital electrons. Proper bombardment of certain atoms (by neutrons and protons) results in the release of tremendous amounts of atomic energy by the process of nuclear fission. For example, the energy released by the fission of one pound of ^{235}U (a fissionable isotope of uranium) is roughly equivalent to that secured from burning 10,000 tons of coal.

Chemical bonds

The tendency of atoms to fill the outer electron shell and become stable leads to combinations of atoms. In filling the outer shell, electrons may be *accepted* from another atom, thus completing a shell, *donated* to another atom, resulting in the filling of a complete inner shell, or *shared* with another atom to complete the outer shell of both atoms. The type of electron activity leads to the formation of different kinds of chemical bonds.

IONIC (ELECTROVALENT) BONDS

When one atom gives up electrons to another atom, both tend to complete an outer shell and become stable. The positive charge of the pro-

Element:	Sodium	Chlorine	Carbon	Oxygen	Hydrogen	Nitrogen	Sulfur	Iron
Symbol:	Na	Cl	C	O	H	N	S	Fe

Compound:		
NaCl	Sodium chloride (table salt)
CO_2	Carbon dioxide (respiratory gas in breath)
H_2O	Water
$C_{3032}H_{4816}O_{872}N_{780}S_8Fe_4$	Hemoglobin, oxygen carrier in blood

Fig. A-1. Elements, symbols, and compounds.

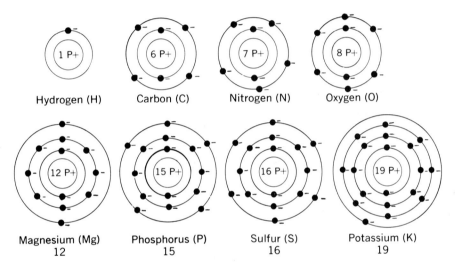

Hydrogen (H)   Carbon (C)   Nitrogen (N)   Oxygen (O)

Magnesium (Mg)   Phosphorus (P)   Sulfur (S)   Potassium (K)
12   15   16   19

Fig. A-2. Atomic structure of some elements that may be present in protoplasm (diagrammatic). The symbols in parentheses follow the name. **P,** Proton (positive electrical charge); black circle with dash, electron (negative electrical charge). The inner circle represents the nucleus of the atom; outer rings represent one or more orbits (''shells'') of electrons.

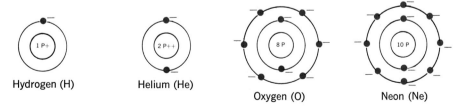

Hydrogen (H)   Helium (He)   Oxygen (O)   Neon (Ne)

Fig. A-3. Active and inert gases.

Table A-1 Maximum number of electrons in atomic orbits

Orbit	Number of electrons (maximum)
First	2
Second	8
Third	18
Fourth	32
Fifth	18
Sixth	12
Seventh	2

Table A-2 Some uses of radioactive isotopes

Element	Isotope	Use
Carbon	^{14}C	Fate of labeled nutrients
Carbon	^{14}C	Fate of injected drugs
Sulfur	^{35}S	Use of amino acids
Iron	^{59}Fe	Iron metabolism
Sodium	^{24}Na	Rate of sodium transfer
Phosphorus	^{32}P	Blood cell production
Potassium	^{42}K	Permeability of cell membranes
Cobalt	^{60}Co	Gamma ray treatment
Strontium	^{90}Sr	Radiation accumulation
Iodine	^{131}I	Thyroid metabolism
Iodine	^{131}I	Location of brain tumors
Gold	^{198}Au	Cancer treatments

tons attracts the negative charge of the electrons. Because of this attraction, an atom with a small number of electrons in the outer shell is more likely to give up electrons than one with a nearly complete shell.

An example of this type of bonding is the combination of sodium, which has 1 electron in the outer shell, and chlorine with 7 electrons in the outer shell. Sodium now has 1 more proton than electron and, therefore, has a *net* charge of 1^+, that is, before the transfer there were 11 protons (11^+) in the nucleus and 11 electrons, 2 in the first shell, 8 in the second, and 1 in the third (Fig. A-4). After the single electron in the outer shell is given up, two changes occur. First, the total number of electrons is now 10, which is 1 less than the number of protons. The net charge is now 1^+. Sec-

ondly, since there is no longer a third shell, the second shell becomes the outer. It now has 8 electrons and is relatively stable. Chlroine has gained 1 electron and now has a complete outer shell and a net charge of 1^-. Sodium with its excess positive charge tends to attract chlorine with its excess negative charge, and the two are held together by this electrical attraction. Atoms with an excess of positive or negative charges are called *ions*. The new combination, sodium chloride, is an *ionic compound*. When ionic compounds are placed in water or in living cytoplasm, which is largely water, they readily *ionize* or dissociate into separate positive (+) and negative (−) ions, for example, NaCl in $H_2O \rightarrow Na^+ + Cl^-$.

COVALENT BONDS

Many atoms, when in contact, do not transfer electrons; rather, they share them. These atoms rearrange the electrons in the outer shell so that neither atom loses electrons.

Hydrogen, with one electron in its shell may combine with another hydrogen atom, forming a *molecule* of hydrogen (H_2) held together by the energy attraction of a pair of shared electrons. The combination is called a *covalent* bond. A similar covalent bond is formed by two oxygen atoms (Fig. A-5). Many other atoms also form covalent bonds such as in the combining of hydrogen and carbon.

POLAR COVALENT BONDS

A polar covalent bond is formed whenever the atoms of a covalent molecule do not share the electrons equally. Recall that the ''shells'' around an atomic nucleus are really probable areas where an electron may occur. Since the electrons are moving rapidly, they are at different places in these orbital paths at different times.

In a *polar covalent bond* electrons are more likely to spend more time closer to one nucleus than to another. This means that there may be more electrons near one nucleus, making it more negative than the other. The entire molecule

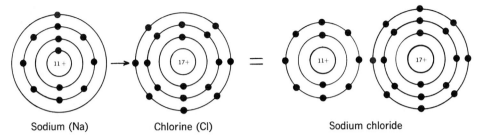

Sodium (Na)   Chlorine (Cl)   Sodium chloride

Fig. A-4. Formation of an ionic bond. When one atom, such as sodium, gives up an electron to another, they both tend to complete an outer shell and become stable.

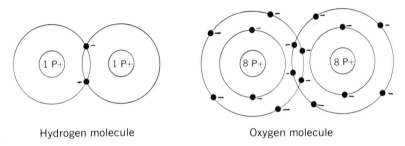

Hydrogen molecule   Oxygen molecule

Fig. A-5. Covalent bonds in hydrogen and oxygen (diagrammatic). In hydrogen sharing of one electron from each atom (total: two electrons) is involved in the combination. In oxygen two electrons from each atom (total: four electrons) are shared.

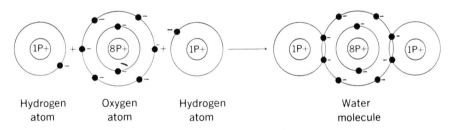

Hydrogen   Oxygen   Hydrogen   Water
atom   atom   atom   molecule

Fig. A-6. Formation of polar covalent bonds in a molecule of water. The shared electons are more likely to be near the oxygen nucleus than the hydrogen nucleus.

then may act as if it were an ion (Fig. A-6). In the water molecule, H_2O, the oxygen nucleus has a stronger attractive force, that is, a more positive force, than the hydrogen nucleus.

Therefore, the shared electrons are more likely to be nearer the oxygen nucleus. The water molecule then behaves as if it had a negative (−) oxygen end and a positive (+) hydrogen end.

As we will see later, polar compounds are important in such substances as the phospholipids of the cell membrane.

Valence or combining capacity

The electrons in the outer shell determine the *valance* or tendency to combine with other atoms. As we have seen, an atom of sodium with 1 electron in its outer shell forms a combination with 1 atom of chlorine, which has 7 electrons in the outer shell. Once a sodium atom has donated its sole *outer* electron, there are no others to donate to other atoms. The same situation exists with hydrogen, which may share its electron with 1 atom of chlorine, with another atom of hydrogen, or with any *one* other atom. Thus,

$$H- \; + \; -H \; \rightarrow \; H-H \quad \text{or} \quad H_2$$
$$H- \; + \; -Cl \; \rightarrow \; H-Cl \quad \text{or} \quad HCl$$

Oxygen may fill its two outer ring spaces by sharing 2 electrons with another oxygen atom, with 2 hydrogen atoms, or another combination of 2 electrons.

$$O= \; + \; =O \; \rightarrow \; O=O \quad \text{or} \quad O_2$$
$$O= \; + \; -H, H \; \rightarrow \; H-O-H \quad \text{or} \quad H_2O$$

Similarly, nitrogen and any other atom with a similar outer ring, may combine with 3 other atoms, $-N\big\langle$.

Carbon and similar elements will combine with four atoms, $-\overset{|}{\underset{|}{C}}-$.

$$-N \; + \; -H, -H, -H \; \rightarrow \; \overset{H}{\underset{H}{\diagdown}} N-H \quad \text{or} \quad NH_3$$

$$-\overset{|}{\underset{|}{C}}- \; + \; O=, O= \; \rightarrow \; O=C=O \quad \text{or} \quad CO_2$$

Several atoms of different kinds, such as C, H, O,

may combine in various compounds, for example,

$$\overset{\displaystyle H \quad H}{\underset{\displaystyle H \quad H}{H\overset{|}{C}-\overset{|}{C}-O-H}}, C_2H_5OH, \text{ ethyl alcohol}$$

The chemical properties of compounds are determined by several factors: (1) the *kinds of elements* in the compound, such as H, C, N, O, Na, Cl, (2) the *number of each kind* of atom involved, as in H_2, O_2, H_2O, $C_6H_{12}O_6$, and huge molecules such as $C_{708}H_{1130}O_{224}N_{180}S_4P_4$ (milk casein), (3) the *kinds of bonds* involved, whether ionic, covalent, polar, and double or single, and (4) the spatial arrangement of the atoms, that is, whether the atoms are arranged in straight lines

$C-C-C-C$; rings $\overset{\displaystyle C-C}{C \underline{\qquad} C \underline{\qquad} C}$ or other arrangements such as $-\overset{|}{\underset{|}{C}}-\overset{|}{\underset{|}{C}}-\overset{|}{\underset{|}{C}}-$ (Fig. A-7).

Chemical changes

When molecules of various chemicals are in contact, they may undergo chemical reactions. In *synthesis* or combination reactions 2 molecules combine to form a third, as in A + B → AB. Examples of this are carbon dioxide (CO_2) combining with water (H_2O) to form carbonic acid (H_2CO_3)

$$CO_2 + H_2O \; \rightarrow \; H_2CO_3$$

or two molecules of glucose combining to form the plant sugar sucrose.

$$C_6H_{12}O_6 + C_6H_{12}O_6 \; \rightarrow \; C_{12}H_{22}O_{11} + H_2O$$

In this last case a molecule of water was split out during combination.

In *decomposition reactions,* such as those that take place in the intestine during food digestion, a larger molecule is broken down, usually by

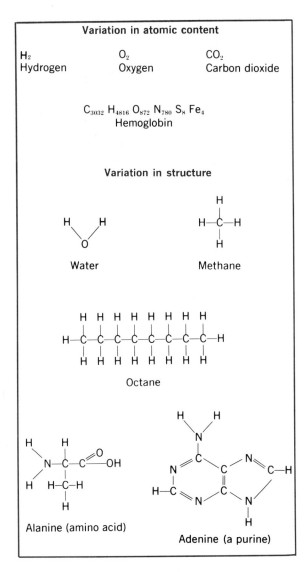

Fig. A-7. Molecular variation.

NaCl	H_2O	Na^+	+	Cl^-
HCl	H_2O	H^+	+	Cl^-
H_2CO_3	H_2O	H^+	+	HCO_3^-
NaOH	H_2O	Na^+	+	OH^-
NH_4OH	H_2O	NH_4^+	+	OH^-

Fig. A-8. Ion formation.

molecules exchange some atoms, for example, $AB + CD \rightarrow AD + BC$. In the gas exchange in the lungs during breathing, hemoglobin (Hb) gives up carbon dioxide in exchange for oxygen:

$$HbCO_2 + O_2 \rightarrow HbO_2 + CO_2 \uparrow$$

Rearrangement reactions occur when the position of atoms in a molecule is changed during a reaction such as $A \rightarrow B$. An example of this occurs during cell respiration when glucose—6—phosphate is rearranged to glucose —1— phosphate. The attachment of the phosphate group is changed from the sixth carbon in the molecule to the first.

IONIZATION OR IONIC DISSOCIATION

Compounds that are formed with ionic (electrovalent) bonds are not stable in water. The compound dissociated into its component charged particles or ions (Fig. A-8). Because this solution of ions in water can conduct an electric current, the compounds are often called *electrolytes*.

Ionization of different compounds results in different amounts of positive and negative charges. For example,

(a) NaCl in $H_2O \rightarrow Na^+ + Cl^-$
(b) H_2SO_4 in $H_2O \rightarrow H^+ + H^+ + SO_4^=$
(c) CH_3COOH in $H_2O \rightarrow CH_3COO^- + H^+$
(d) NH_4OH in $H_2O \rightarrow NH_4^+ + OH^-$
(e) $Fe(OH)_3$ in $H_2O \rightarrow Fe^{+++} + OH^- + OH^- + OH^-$

Because the ionization of (b) and (c) releases H^+ ions, the compounds are called *acids;* (b) is sul-

combining with water to form two smaller ones. Thus, $AB \rightarrow A + B$, as in the reverse of $A + B \rightarrow AB$. During digestion sucrose is broken down to form glucose, which is easily absorbed in the intestine.

$$C_{12}H_{22}O_{11} \xrightarrow{+H_2O} C_6H_{12}O_6 + C_6H_{12}O_6$$

In *exchange,* or displacement reactions, two

furic acid, and (c) is acetic acid. Similarly, both (d) and (e) release OH^- ions and are called bases.

Acids and bases are important concepts in chemistry and especially in the chemistry of living things. The measure of the amount of acidity or alkalinity (bases) is called *pH*.

If a solution has the same number of H^+ ions as OH^- ions, that is, the same ratio as in pure water, the solution is *neutral* and has a pH of 7. Acids have a pH range of 0 to 7, and bases have a pH of 7 to 14. Each unit more or less than 7, that is toward 0 or toward 14, indicates an amount of ion more than 7. This amount increases exponentially as a value of 10 ($10^2 = 100$, $10^3 = 1000$, and so on). Thus, a solution with a pH of 6 is 10 times more acid than pure water, pH 5 is 100 times more acid, and so on.

Human cytoplasm is slightly basic with a pH of about 7.4. This is important for the chemical reactions that characterize life. A change of only 0.2 or so in pH, for example, from pH 7.4 to 7.2, may result in serious harm or death to the individual.

Buffers are chemical systems that protect the organism from changes in pH. Buffers are normal components of living organisms. If there is an increase in the amount of H^+ ion in a cell, making it more acid, a buffer system combines with the H^+ ions to neutralize them and restore the original pH. The *bicarbonate* ($NaHCO_3$) buffer system works as follows:

1. $NaHCO_3$ ionizes in cells $\rightarrow$ $Na^+ + HCO_3^-$

2. add H+ ions (for example $HCl \xrightarrow{H_2O} H^+ + Cl^-$)

3. $H^+ + HCO_3^- \rightarrow H_2CO_3$ (a weak acid, slow to ionize)

Examination of the stomach contents shows a low pH, a high degree of acidity caused by hydrochloric acid (HCl) secretion. HCl is a strong acid; that is, it ionizes almost 100%. If the intestinal contents beyond the stomach are examined, the pH is discovered to be about 7. What happened? An opening at the upper end of the intestine releases fluid from the pancreas. This fluid contains $NaHCO_3$, an efficient buffer.

B Early history and development of methods in biology

The refinement of methods of observing, collecting, and analyzing data led to the development of biology as a science. Beginning with the unaided eye, the science of biology has progressed to the modern laboratory where tools such as electron microscopes, chemical analyzers, complex recorders, and computers may be used.

EARLY RECORDS (before 200 A.D.)

The history and development of biology have passed through distinct periods. Since the beginning of history human beings have "studied" plants and animals and later tried to identify, name, classify, and use them. Early man associated closely with plants and animals in daily living. Because he depended upon them for food, shelter, clothing, and medicines, he had to know something about them.

Systems of medicine were used more than 5,000 years ago. Many names of plants were given to the early Greeks by the guild of root-cutters (*rhizotomoi*) who supplied ingredients for early medicines and many foods.

Many names of animals were supplied by early hunters, fishermen, and priests. Agriculture, hunting, and medicine of one type or another had origins with early man.

When Greek and Roman civilizations were at their height, the foundations of natural science were laid. The first scientific and systematic work was accomplished in Greece. The early Romans seemed to be interested more in the practical or applied side of science because they probably had to develop agriculture for their food supply.

Long before the Christian era various civilized people possessed considerable information about plants and animals that was of value for foods, medicines, and shelter. Some of this information, in the form of pictures and hieroglyphics painted on tombs or carved in stone, is available for study today. These records show that the Egyptians and Assyrians (4000 B.C.) were practical plant scientists of a rather high type, having cultivated medicinal, ornamental, and food plants. Among them were roses, apricots, figs, dates, grapes, olives, wheat, and barley. Early Egyptians had a wealth of knowledge about animals and possessed domesticated cattle, sheep, pigs, ducks, geese, and cats. Some of early man's impressions of animals have survived in the cave paintings of France and Spain. Man decorated pottery, cloth, and tools with animal figures.

The Chinese (2500 B.C.) had acquired a practical knowledge of plant uses, including such cultivated plants as rice, tea, oranges, and a plant that is the source of ephedrine (a medicine).

Ancient races in America also had considerable knowledge about plants and animals. The pre-Inca race of Peru (3000 B.C.) was apparently the first in American to cultivate maize (corn). From Peru its cultivation spread northward and southward, so that by 1492 it was a major crop from the St. Lawrence River to Argentina. The aboriginal Americans domesticated and cultivated such practical plants as potatoes, squashes, cacao trees, and avocados (alligator pear).

THE THIRD TO THE TWELFTH CENTURY (200-1200 A.D.)

After the Greek and Roman periods there was a decline in European science during the Middle Ages, or so-called Dark Ages, which extended from the third century to the twelfth century A.D. For more than 1,000 years after 200 A.D. much of the knowledge of the ancients was lost in Europe, and things were accepted without question, observation, or experimentation. During this time science was opposed by public opinion and authority. In general, science as known by the earlier Greeks did not exist.

The period of the Middle Ages was one of relative intellectual inactivity because of political, social, and psychological adjustments after the fall of the Roman Empire (476 A.D.) and the domination of much of Europe by barbarian tribes. During this unproductive period the emphasis was placed on "opinions" and "judgments" of a few so-called authorities. Their opinions were accepted, usually without question. Few if any investigators attempted to prove things for themselves. Much of the biologic work of this period involved mythology and superstition, with only a limited knowledge of biologic facts.

There was considerable scientific activity in Arabia (800-1300 A.D.), where much of the biologic interest centered around medicinal plants.

Botanical gardens were established, and the medicinal uses of plants and plant products flourished. The Arabian scientific activity was extensive, yet it was not original and progressive, since much of the work involved the translation of early Greek and Roman publications into the Arabic language in Baghdad.

THE THIRTEENTH TO THE SIXTEENTH CENTURY (1200-1600 A.D.)

After the many centuries of comparative inactivity during the Dark Ages there was a revival in science and other areas of man's activities. This period is referred to as the Renaissance (fr. *renaistre,* to revive or to be born again), which began in Europe about ten centuries after the fall of the Roman Empire.

There was a gradual revival of the scientific studies of plants and animals that had been started by the Greeks many years before. In Europe, universities were founded in which professors lectured from pulpit-like reading desks, and students in academic robes observed demonstrations by assistants. Students themselves did little or no dissecting or direct observing.

Botanical gardens were popular during the sixteenth century, and most European universities had gardens of medicinal and food plants by the mid-seventeenth century. During these two centuries many large books known as *herbals* were published, these containing descriptions of food and medicinal plants, illustrations of living plants, and descriptions of natural phenomena, some of which were purely superstitions and myths. Gaspard Bauhin published excellent descriptions of nearly 6,000 plants in his *Herbals* (1623).

Many fantastic stories and drawings appeared in the *Herbals,* and many descriptions were not very accurate. A so-called *Doctrine of Signatures,* advocated by Paracelsus (1493-1541), a Swiss physician and alchemist, stated that certain plant structures were modeled upon structural principles similar to those of human organs and that these plant structures supposedly constituted remedies for diseases of those human organs that they resembled most closely. For example, the sap of the blood-root plant was used as a blood tonic; walnuts with their numerous ridgelike convolutions were used in the treatment of brain diseases.

Other published volumes known as *Bestiaries* contained descriptions of animals and, although the earliest ones were somewhat allegorical and moralizing in their descriptions of beasts of the Middle Ages, they represented the best efforts of their authors to meet the needs of their times. Scientific societies and the publication of scientific journals were established in the seventeenth century.

DEVELOPMENT OF THE MICROSCOPE

It is not known who invented the first microscope, but in Nineveh, an ancient city in Assyria, a rock crystal was excavated that may have been used as a "lens" in the eighth century B.C. Euclid (of Megara) (440 B.C.-?), a Greek philosopher, investigated the properties of curved reflecting surfaces. His investigations were used later in connection with studies in magnification. Lucius Seneca (4 B.C.?-65 A.D.), a Roman philosopher, reported that water-filled glass globules would assist one in seeing small objects. Claudius Ptolemy (127-151 A.D.), a Greco-Egyptian astronomer, studied some problems of magnification by using curved surfaces. Even though burning glasses were used, magnifying glasses (lens thicker at the center than at the edge) were probably not used extensively until the invention of eyeglasses in the thirteenth century. Leonardo da Vinci (1452-1519), an Italian painter and architect, stressed the use of lenses in studying small objects. The early microscopes (magnifying glasses) were commonly called "flea microscopes," because a flea was a specimen commonly studied.

Because so much scientific progress of the past has been dependent upon the development and

Fig. B-1. Exact replica of the compound microscope made by Zaccharias Janssen and his son Hans of Middleburg, Holland, between 1590 and 1610.
(Courtesy The Armed Forces Institute of Pathology, Washington, D. C. No. 53-662.)

use of microscopes, a brief consideration of early and recent microscopes will be presented.

A compound microscope (Fig. B-1) is thought to have been invented by Zaccharias Janssen and his son Hans in Middleburg, Holland, between 1590 and 1610. These spectacle makers and lens grinders combined lenses when viewing objects and discovered that a second lens would magnify the enlarged image from a magnifying glass. Their microscope was made of tubes that slid together for focusing, had a size of 2 by 18 inches, and magnified about 9 times.

Fig. B-2. Hooke's microscope (1665). The body tube contained a series of lenses that magnified the image in the manner of the compound microscope. Illumination was provided by a lamp and bull's-eye condenser. The instrument was 16 inches high and had a maximum magnification of 42×.
(From The Evolution of the Microscope, American Optical Co., Instrument Division, Buffalo, New York.)

Robert Hooke (1635-1703), an English microscopist, constructed a microscope in 1665 that consisted of an objective lens, a field glass, and an eye lens (Fig. B-2). The latter two magnified the image of the former in the manner of a compound microscope. He provided a lamp for illumination and a bull's-eye condenser for intensifying the light. His microscope had magnifications of 14 to 42 diameters.

Antonj van Leeuwenhoek (1632-1723), a Dutch microscopist, developed a simple microscope (about 1673) by mounting a lens between two flat pieces of metal and adding a pivoted joint for holding the specimen (Fig. B-3). He ground lenses that had magnifications of up to 300 diameters. With his lenses he studied bacteria, molds, protozoans, red blood corpuscles, plants, animals, and the circulation of blood in the tadpole tail.

Bonannus improved the microscope in 1691 and developed a horizontal type that included a source of light, a condenser to concentrate light, and a rack-and-pinion mechanism for more efficient focusing (Fig. B-4).

Wilson, about 1710, developed a screwbarrel type of microscope (Fig. B-5). The body had threads on the observing end into which lenses of different magnifying powers might be screwed. The opposite end had a condensing lens to concentrate light. The specimen was pushed against a spring for focusing. A Wilson type of microscope was received at Harvard College in 1732 and may have been one of the first compound microscopes used in American colleges, although simple microscopes probably were used earlier.

There were only about a dozen microscopes in the United States in 1831. Instructors were using them by 1850, and students began using them as early as 1875, but they were not in general student use until about 1890.

Charles A. Spencer (1813-1881) built the first American microscope (1847) and several additional models of it. Robert B. Tolles (1824-1883) was another early American microscope builder who started as an apprentice of Spencer but established his own business (1858). He is famous for his improvements of objectives and for inventing the homogenous immersion objectives. In these a drop of the proper type of liquid is placed on the cover slip on the slide and the immersion objective is made to contact the liquid, which acts as a type of lens to assist in higher magnifications. Today, oil immersion objectives are used for high magnifications. Edward Bausch (1854-1944) made his first microscope in 1872. He was the son of J. J. Bausch (1830-1896), the founder of the Bausch & Lomb Optical Company.

Until the end of the nineteenth century, the making of complete microscopes was largely done by individuals who made one microscope at a time. The metal parts were made by hand and the lenses ground and polished with rather simple equipment. Increasing demands for more microscopes suggested to manufacturers that specialists (scientists, designers, engineers, specialized

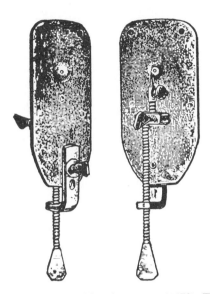

Fig. B-3. Leeuwenhoek's microscope (1673). This simple microscope consisted of a lens mounted between two flat pieces of metal, with an adjustable point for holding the specimen and for focusing purposes.
(From The Evolution of the Microscope, American Optical Co., Instrument Division, Buffalo, New York.)

workers) must be trained, that standards must be set, and that microscopes must be built on an assembly-line basis.

The twentieth century has seen many improvements in the manufacture and usefulness of the various types of microscopes (Figs. B-6 and B-7). Some of the more recent improvements include ultramicroscopes, ultraviolet microscopy, dark-field microscopy, phase microscopy, and electron microscopy.

When particles too small to be seen with a microscope under ordinary conditions are illuminated by a strong beam of light parallel to the surface of the stage (at right angles to the direction of vision through the microscope), they appear as bright specks because of their reflection of light, but do not show their outline or shape. The apparatus used for such study is called an ultramicroscope (L. *ultra*, beyond).

In an ultraviolet microscope, invisible ultraviolet rays (of shorter wavelengths and beyond the visible violet light waves) are used instead of ordinary light. Because of the invisibility of the ultraviolet rays, photographs must be made, since the image cannot be seen. Special quartz

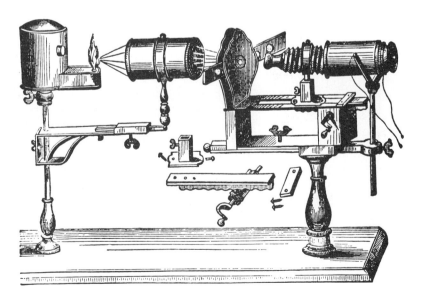

Fig. B-4. Bonannus' horizontal microscope (1691). This type included a light source, a condenser to concentrate light on the specimen, and rack-and-pinnion focusing mechanism on a horizontal stand.
(From The Evolution of the Microscope, American Optical Co., Instrument Division, Buffalo, New York.)

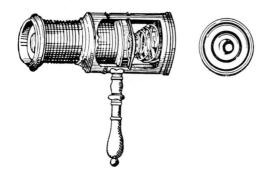

Fig. B-5. Wilson's microscope (about 1710). The model is made of ivory, and the body is cut open and the ends threaded for the attachment of lenses. The specimen is held by a spring for focusing. The handle is unscrewed when carried in the pocket.
(From The Evolution of the Microscope, American Optical Co., Instrument Division, Buffalo, New York.)

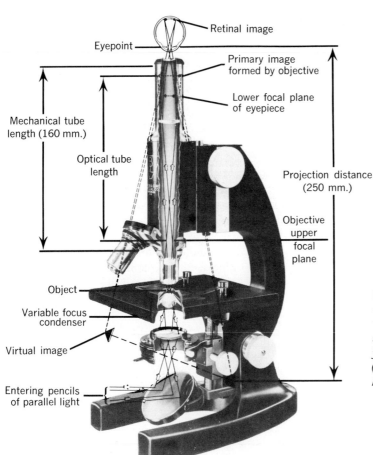

Retinal image

Eyepoint

Primary image formed by objective

Lower focal plane of eyepiece

Mechanical tube length (160 mm.)

Optical tube length

Projection distance (250 mm.)

Objective upper focal plane

Object

Variable focus condenser

Virtual image

Entering pencils of parallel light

Fig. B-6. Modern microscope showing mechanical and optical features. Retinal and virtual images are reversed (as shown by arrows). A substage condenser with adjustment knob is shown.
(Courtesy Bausch & Lomb Optical Co., Rochester, New York.)

lenses must be employed that permit the passage of the ultraviolet rays.

In dark-field microscopy the term "dark field" refers to a method of illuminating a specimen brightly while the surrounding background (field) remains dark. The most practical dark field is obtained by a special dark-field condenser (dark-field illuminator) whereby direct light rays do not enter the specimen. The oblique light rays are focused on the specimen, which thus appears as a luminous body against a dark field. A very small, bright object is more easily seen in a dark background (field) than is a very small, dark object in a bright field. This is similar to the phenomenon of seeing small dust particles in a beam of light when the region at the back of the light beam is dark.

Phase-contrast microscopy, proposed by Zernike of Holland in 1932, permits the study of living organisms and other transparent materials, some of which do not absorb visible light and hence are not visible under an ordinary lighter microscope.

Electron microscopy employs beams of electrons produced by special apparatus instead of light (Fig. B-7). Magnetic fields ("electron lenses") are used instead of glass lenses, and photographic films may be used to record the image, which may be magnified over 100,000 times. The specimens being photographed must be very thin and in a vacuum.

The fact that axially symmetrical magnetic and electric fields could be employed as lenses was discovered by H. Busch in 1926. hence, by the proper use of magnetic fields (acting as lenses), the charged particles (electrons) can be made to do what light waves accomplish in ordinary, optical microscopy. Electron microscopes were made by Knoll and Ruska (Germany) in 1932, by Marton (Belgium) in 1934, and by Prebus and Hillier (Canada) in 1938. The Radio Corporation of America in 1941 manufactured a commercial electron microscope of the magnetic type. Many kinds and models have been made and used in various parts of the world since that time.

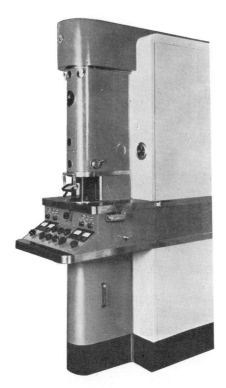

Fig. B-7. Electron microscope capable of magnifying thousands of times.
(Courtesy RCA Victor Corporation, Camden, New Jersey.)

Index